Phenolic Compounds - Chemistry, Synthesis, Diversity, Non-Conventional Industrial, Pharmaceutical and Therapeutic Applications

Edited by Farid A. Badria

Published in London, United Kingdom

IntechOpen

Supporting open minds since 2005

Phenolic Compounds - Chemistry, Synthesis, Diversity, Non-Conventional Industrial, Pharmaceutical and Therapeutic Applications
http://dx.doi.org/10.5772/intechopen.94825
Edited by Farid A. Badria

Contributors
Maria Adelina Jiménez-Arellanes, Mariana Z. Pérez-González, Lucienne Gatt, Pierre Schembri Wismayer, Sharad Srivastava, Bhanu Kumar, Ankita Misra, Elba Lucia Cavalcanti de Amorim, Patrícia Cruz, Jorge Veras-Filho, Jenifer Oliveira, Uyara Costa, Marcelino Diniz, Ítalo Caio Silva, Kivia Machado, Maria Santa Medeiros, Ana Caroline Xavier, Eman A Ibrahim, Doha Abou Baker, Zeinab A Salama, Farid A. Badria, Rizwan Rafique, Monis Shah, Tanzila Rafique, Mehwish Naseer, Uzman Khalil, Rehan Rafique, Jigna Tank, Rohan Pandya, Vibhakar Chowdhary, Sheena Alooparampil, Boudiba Sameh, Hanini Karima, Boudiba Louiza, Benahmed Merzoug, Saouane Izzeddine, Asma Nisar, Yu-Chiang Hung, Wen-Long Hu, Yu-Chen Cheng, Christian Drapeau, Veronique Traynard, Muk Wing Yuen, Hamad Al Mamari, Josephine Ampofo, Michael Ngadi, Jyoti V. Vastrad, Pratikhya Badanayak, Giridhar Goudar, Elham H. Fini, Farideh Pahlavan, Shakiba Ayat, Addepally Uma, Venkanna Banothu, Michael Jourdes, Zuriñe Rasines-Perea, Pierre-Louis Teissedre, Ruth Hornedo-Ortega, Ana B. Cerezo, Volkan Gelen, Abdulsamed Kükürt, Emin Şengül, Ömer Faruk Başer, Mahmut Karapehlivan, Ryan Lucas McKinley

First published in London, United Kingdom, 2021 by IntechOpen
IntechOpen is the global imprint of INTECHOPEN LIMITED, registered in England and Wales, registration number: 11086078, 5 Princes Gate Court, London, SW7 2QJ, United Kingdom
Printed in Croatia

British Library Cataloguing-in-Publication Data
A catalogue record for this book is available from the British Library

Additional hard and PDF copies can be obtained from orders@intechopen.com

Phenolic Compounds - Chemistry, Synthesis, Diversity, Non-Conventional Industrial, Pharmaceutical and Therapeutic Applications
Edited by Farid A. Badria
p. cm.

This title is part of the Biochemistry Book Series, Volume 26
Topic: Chemical Biology
Series Editor: Miroslav Blumenberg
Topic Editors: Şükrü Beydemir and Deniz Ekinci

Print ISBN 978-1-83969-346-5
Online ISBN 978-1-83969-347-2
eBook (PDF) ISBN 978-1-83969-348-9
ISSN 2632-0983

For EU product safety concerns:
IN TECH d.o.o., Prolaz Marije Krucifikse Kozulić 3, 51000 Rijeka, Croatia, info@intechopen.com or visit our website at intechopen.com.

IntechOpen Book Series

Biochemistry

Volume 26

Aims and Scope of the Series

Biochemistry, the study of chemical transformations occurring within living organisms, impacts all of the life sciences, from molecular crystallography and genetics, to ecology, medicine and population biology. Biochemistry studies macromolecules - proteins, nucleic acids, carbohydrates and lipids –their building blocks, structures, functions and interactions. Much of biochemistry is devoted to enzymes, proteins that catalyze chemical reactions, enzyme structures, mechanisms of action and their roles within cells. Biochemistry also studies small signaling molecules, coenzymes, inhibitors, vitamins and hormones, which play roles in the life process. Biochemical experimentation, besides coopting the methods of classical chemistry, e.g., chromatography, adopted new techniques, e.g., X-ray diffraction, electron microscopy, NMR, radioisotopes, and developed sophisticated microbial genetic tools, e.g., auxotroph mutants and their revertants, fermentation, etc. More recently, biochemistry embraced the 'big data' omics systems.
Initial biochemical studies have been exclusively analytic: dissecting, purifying and examining individual components of a biological system; in exemplary words of Efraim Racker, (1913 –1991) "Don't waste clean thinking on dirty enzymes." Today, however, biochemistry is becoming more agglomerative and comprehensive, setting out to integrate and describe fully a particular biological system. The 'big data' metabolomics can define the complement of small molecules, e.g., in a soil or biofilm sample; proteomics can distinguish all the proteins comprising e.g., serum; metagenomics can identify all the genes in a complex environment e.g., the bovine rumen. This Biochemistry Series will address both the current research on biomolecules, and the emerging trends with great promise.

Meet the Series Editor

Miroslav Blumenberg, Ph.D., was born in Subotica and received his BSc in Belgrade, Yugoslavia. He completed his Ph.D. at MIT in Organic Chemistry; he followed up his Ph.D. with two post-doctoral study periods at Stanford University. Since 1983, he has been a faculty member of the RO Perelman Department of Dermatology, NYU School of Medicine, where he is codirector of a training grant in cutaneous biology. Dr. Blumenberg's research is focused on the epidermis, expression of keratin genes, transcription profiling, keratinocyte differentiation, inflammatory diseases and cancers, and most recently the effects of the microbiome on the skin. He has published more than 100 peer-reviewed research articles and graduated numerous Ph.D. and postdoctoral students.

Meet the Volume Editor

Farid A. Badria, Ph.D., is a scholar of the Arab Development Fund, Kuwait; ICRO-UNESCO, Chile; and UNESCO Biotechnology France. He has submitted 47 patents, 20 of which have been granted final certificates. He has more than 250 publications, 12 books, and several marketed pharmaceutical products to his credit. He continues to lead research projects on developing new therapies for liver disease, skin disorders, and cancer. Dr. Badria was listed among the top 2% of most-cited scientists in medical and biomolecular chemistry by Stanford University in 2019 and 2020. He has received several awards including the TWAS Prize for "Public Understanding and Popularization of Science"; WIPO Gold Medal (Best Inventor); State Outstanding Award in Medicine; Outstanding Arab Scholar, Kuwait; and Khawrazmi International Award, Iran.

Contents

Preface

Phenolic compounds represent one of the major secondary metabolites of plants with pharmaceutical, industrial, and therapeutic applications. Flavonoids, quinones, lignans, bioflavonoids, neo-lignans, xanthones, curcuminoids, chalcones, stilbenes, phenylpropanoids, tannins, and coumarins are some examples of the major groups of commonly available phenolic compounds in our daily foods, beverages, and spices. A Scopus database search for phenolics for the period 2005–2015 showed more than 1800 review articles and approximately 50,000 articles having the word "phenolics." This portrays the importance of phenolic compounds in industry, health, and research. **Figure 1** shows the diversity of natural phenolic compounds used for novel drugs, environmental sustainability, and green industry.

From this standpoint, this book presents up-to-date research in natural products and utilization of environmental waste containing phenolic compounds in pharmaceutics for various health disorders (breast, colon, and liver cancer, cataract, degenerative diseases, hyperpigmentation, hyperglycemia, skin disorders, and others), synthesis of nano-silver, green chemistry, click chemistry, and chelating agents for iron overload.

Over four sections, this book addresses several new and controversial issues regarding these interesting and diverse natural molecules with impressive pharmacological and therapeutic effects. Each chapter has been reviewed and revised and the authors have brought current research to make the book more informative, illustrative, and easy to read.

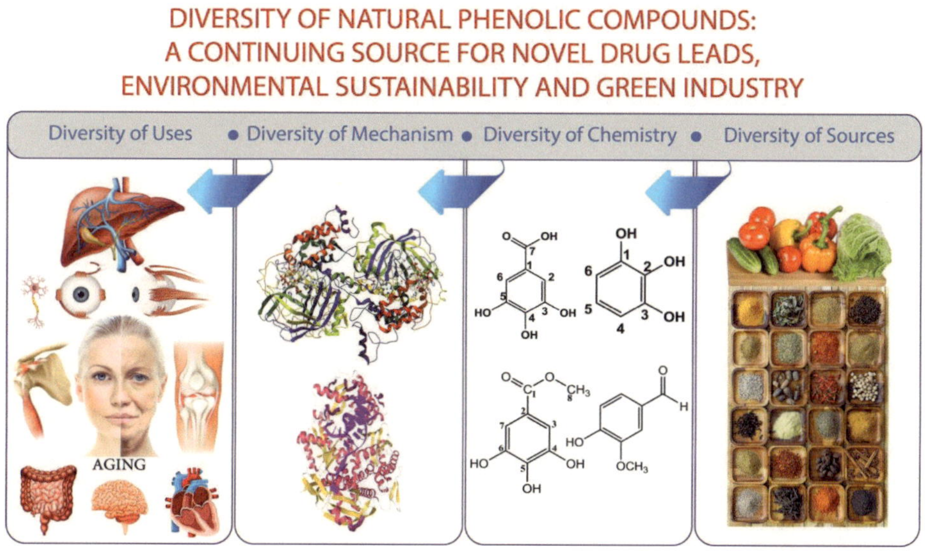

Figure 1.
The diversity of natural phenolic compounds used for novel drugs, environmental sustainability, and green industry.

Section 1 examines the nature and role of phenolic compounds, presenting ongoing and nonconventional research with a focus on marijuana. Section 2 discusses sources, biosynthesis, diversity, metabolic pathways, and the effect of biotic and abiotic stresses on the production of phenolic compounds. Section 3 tackles interesting issues of environmental sustainability, climate change, and green industrial applications of phenolic compounds. Section 4 presents *in vitro*, *in vivo*, preclinical, and clinical biological and therapeutic applications of phenolic compounds in many health disorders including leukemia, COVID-19, aging-associated cardiovascular diseases, and cancer.

This book strikes a balance between developments in scientific research and the premises that researchers must be able to absorb to link scientific advances with clinical practice so that the management of diseases can be based on sound physiological concepts. It is a useful resource for students, clinicians, nutrition specialists, and researchers.

Farid A. Badria Ph.D.
Liver Research Lab,
FAB-Lab,
Faculty of Pharmacy,
Pharmacognosy Department,
Mansoura University,
Mansoura, Egypt

Section 1

On-Going Research on Natural Phenolic Compounds: Past, Present and Future

Chapter 1

Research on Natural Phenolic Compounds in FAB-Lab: Nonconventional Industrial, Pharmaceutical, and Therapeutic Applications

Farid A. Badria

Abstract

Phenolic compounds represent one of the secondary metabolites of plants with pharmaceutical and therapeutic applications. Flavonoids, quinones, bioflavonoids, neo-lignans, xanthones, curcuminoids, tannins, and coumarins are some examples of the major groups of commonly available phenolic compounds in our daily foods, beverages, and spices. From this standpoint, the Liver Research Laboratory (FAB Lab) at Mansoura university, Egypt, established a multidisciplinary research (chemistry, molecular biology, bioinformatics, pharmacology, and pharmaceutics) based on utilization of commonly abundant natural products from plants and agricultural wastes, especially phenolic compounds to meet the goal of applied scientific research in pharmaceutical industry, environment, public health, and to furnish a sustainable well-developed globe. Examples of our concerted efforts, for over 30 years, are in the area of natural products and utilization of environmental waste containing phenolic compounds for various health disorders (cancer, cataract, degenerative diseases, hyperpigmentation, hyperglycemia, skin disorders), nano-, green and click chemistry. This chapter presents a practical model from FAB-Lab to maximize the benefits from phenolic natural products that have not been optimally exploited to establish meaningful scientific applied research. Patents, innovations, and significant publications indexed by the Web of Science and Scopus databases in the journals that occupy the 1st and the 2nd quartile will be presented.

Keywords: phenolic compounds, green chemistry, environment-friendly industry, therapeutic applications

1. Introduction

The Liver Research Laboratory (FAB-Lab) was established in 1990 in the Department of Pharmacognosy, Faculty of Pharmacy, Mansoura University, Egypt. FAB-Lab adopts various research projects aimed at innovation in health, environmental, and pharmaceutical industry issues.

To reach today's destination, FAB-Lab based the ongoing research on the following premises:

1. A sound healthy environment embraces a sound and healthy society.

2. A healthy environment provides us with a cure to any ailment.

3. An unhealthy environment is the source of all illnesses.

2. Vision and mission of FAB-Lab

The Liver Research Laboratory (FAB-Lab) was established in 1990 in the Department of Pharmacognosy, Faculty of Pharmacy, Mansoura University, Egypt. FAB-Lab adopts various research projects aimed at innovation in health, environmental, and pharmaceutical industry issues.

Vision: Conducting distinguished research in the field of drug discoveries from natural sources with tangible returns at the regional and international levels.

Message: FAB-Lab seeks to promote integration between different departments in the field of drug discovery to contribute in solving health problems.

The research strategy depends on shifting from conventional approaches in medicine (treating symptoms) to a more holistic approach that is patient/environment-centered.

The drug discovery team, Badria and coworkers, at Pharmacognosy department, Mansoura University presented many significant contributions in using simple, economic, and abundantly available phenolic compounds to produce a series of useful analogues that exhibited very promising applications industrial and therapeutic applications (as shown in **Figure 1**).

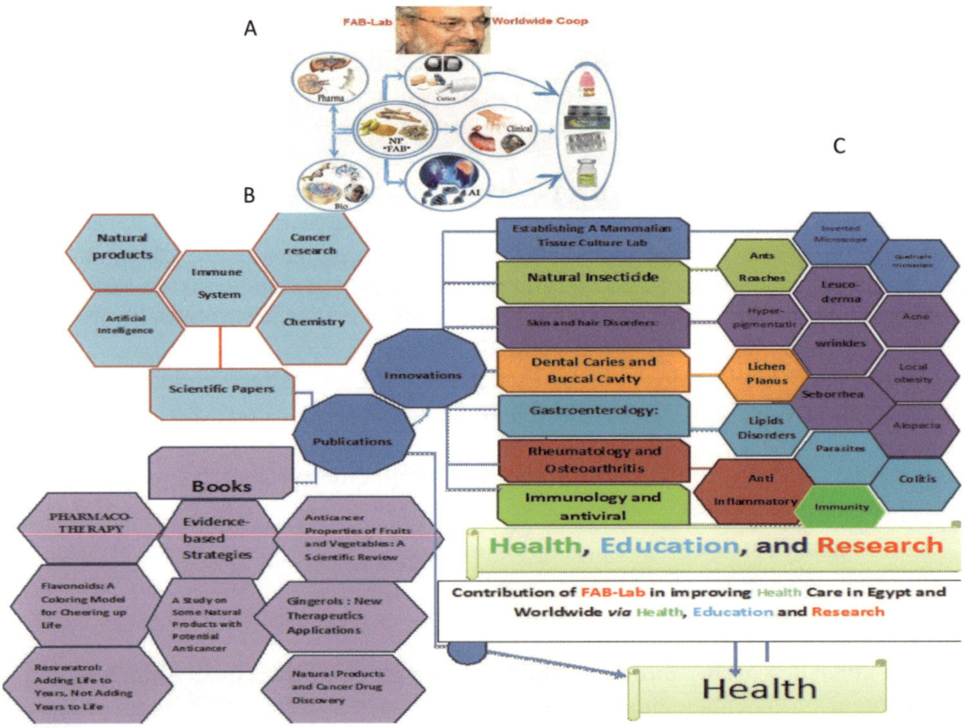

Figure 1.
FAB-lab multidisciplinary team. (A) Research was initiated in the Faculty of Pharmacy and then extended to worldwide cooperation. FAB-lab contributions in improving health care, education, and research collaboration were presented in research and books publications (B), and patents (C).

One of FAB-Lab's early papers entitled "Is Man Helpless Against Cancer: An Environmental Approach in "Cancer Letters" in 1994 [1] is a landmark in exploiting the role of foods, medicinal plants, and herbs rich in phenolic compounds in prophylaxis and treatment of cancer.

FAB-Lab stressed the role of green and conventional chemistry using natural, in-house resources and facilities as friendly environmental substances to construct new therapeutic agents for the well-being of society; e.g., for treatment of epidemics of neglected tropical diseases such as Schistosomiasis [2]. The extensive studies on environment-friendly chemistry revealed valuable information about the root cause of many complex health disorders. Therefore, a final highlight is considered a serious attempt toward solving local, regional, and global problem via designing and producing a drug from agricultural wastes, foods, and plants for treatment of cancer and hepatic disorders [3–9], degenerative diseases [10, 11], hair and skin [12], dentistry [13], ophthalmology [14], breast cancer [15–20], osteoarthritis [21], and bronchial asthma [22, 23].

Liver research laboratory (FAB-Lab, Faculty of Pharmacy, Mansoura University, Mansoura, Egypt) presented a multidisciplinary approach for nonconventional industrial, pharmaceutical, and therapeutic applications.

These efforts demonstrate that, despite the many limitations of NPs, reasonable modifications may lead to the discovery of a novel drug. From this standpoint, the Liver Research Laboratory (FAB Lab) was a pioneer in designing a system to meet the goal of the scientific research in order to serve society, the environment, public health, and to furnish a sustainable well-developed globe. Examples of our concerted efforts, for over 30 years, are in the area of natural products and utilization of environmental waste containing phenolic compounds as potential therapeutic agents in many health disorders; e.g., cancer, cataract, degenerative diseases, hyperpigmentation, hyperglycemia, skin disorders, and others, besides contribution in synthesis of nano-silver, green chemistry, click chemistry, and chelating agents for iron overload.

Structures and biological diversity of phenolic compounds isolated in Fab-Lab included but not limited to simple phenolics, e.g., gallic acid, methyl gallate, vaniline, and eugenol and polyphenolic compounds; e.g., flavonoids, curcuminoids, anthraquinones, gingerols, epigallicatectchins.

Therefore, this chapter will present some FAB-Lab's success and accomplishments in industrial and environment-friendly chemistry (e.g., green and click chemistry), modulation of enzymes in different biological systems (e.g., tyrosinase, alpha-amylase, hyaluroniase, aldose reductase, topoisomerase, and leukotriene hydrolase "LTH4ase," and therapeutic applications (e.g., hepatology, ophthalmology, nephrology, dermatology, dentistry, virology, and metabolic disorders).

3. Industrial and environment-friendly chemistry

3.1 Green chemistry and preparation of silver nanoparticles using natural phenolic compounds

Silver nanoparticles (AgNPs) exhibit unique chemical and biological properties and thus gained extensive interest in commercial applications including food, textiles, pharmaceuticals, and medical products. Green synthesis is a reliable and eco-friendly process for synthesis of AgNPs, which was reported by FAB-Lab team [24] based on the reducing power of different plant extracts. Forty-two aqueous plant extracts were investigated for their ability to produce AgNPs from aqueous solution of $AgNO_3$. Our study showed that the extracts of *Emblica officinalis* fruit, *Psidium guajava* leaves, and *Lawsonia inermis* leaves were able not only to produce AgNPs but also to stabilize the produced nanoparticle. Phytochemical study showed that these extracts contain tannins, polyphenolics, flavonoids, and naphthoquinones, which are responsible for the bioreduction of silver ions to AgNPs (As shown in **Figure 2**). The transmission electron microscopy (TEM) images revealed that the produced AgNPs are characterized by having a spherical and well-dispersed particles with size range from 5 to 30 nm. Interestingly, the AgNPs prepared by *E. officinalis* fruit, *P. guajava* leaves showed nearly the same cytotoxic activity effect as their plant extract. However, AgNPs capped with *L. inermis* exhibited cytotoxic effect against both colon and breast cancer cell lines. This study suggests that AgNPs synthesized and stabilized with *L. inermis* leaves extract could contribute to the development of an appropriate anticancer medication [24].

3.2 Click chemistry and preparation of potential therapeutic agents using natural phenolic compounds

Search for new compounds, e.g., commonly available phenolic compounds, which can be used by using small molecules (units) to join together as building block with heteroatom links.

A click reactions advantage may include high yield products, and less or no by-products, simple reaction conditions, simple available starting materials, simple reagents, and environment-friendly solvents (mainly water), which will help in easy and clean isolation, purification, and crystallization. Therefore, click reactions may offer a good alternative to conventional reactions. This prompted us to use vaniline, eugenol, and curcumin in azide-alkyne cycloaddition (AAC) as a major reaction of click chemistry based on the CuAAC [25].

Curcumin was used as a starting material to prepare three new triazole derivatives via 1,3-dipolar cycloaddition (CuAAC) click reaction to produce three triazole derivatives, which were tested against alpha-amylase, which is an essential metabolic enzyme for carbohydrate metabolism. Curcumin-benzyl triazole derivatives

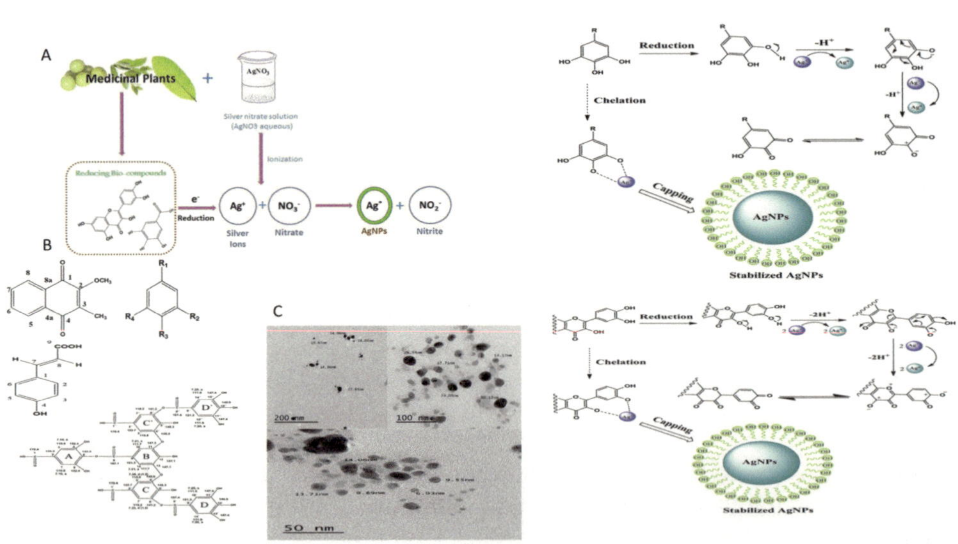

Figure 2.
Preparation of nano-silver particle using natural phenolic compounds; flavonoids (A), naphthoquinone, gallic acid, and hexagaloyl derivatives (B) via chelation, capping, and stabilized silver ion (C).

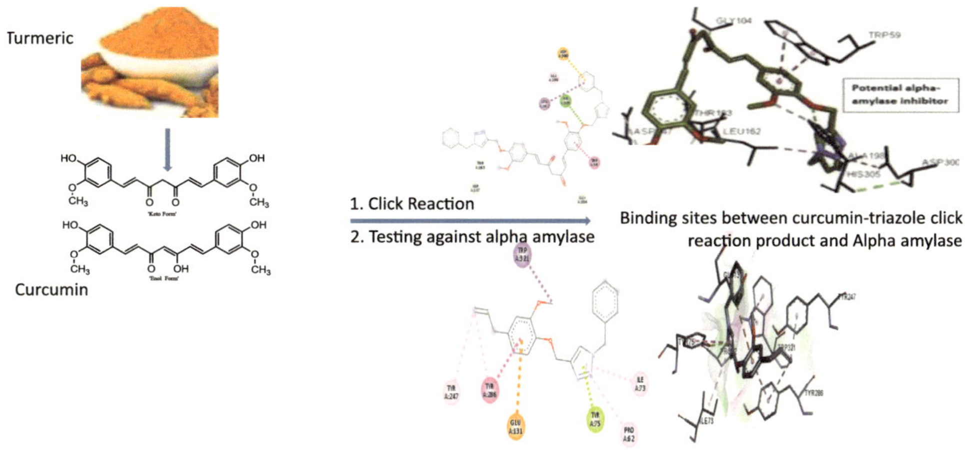

Figure 3.
Curcumin-triazole click reaction product and binding sites between curcumin-triazole click reaction product and alpha amylase.

showed effective inhibitory activity against alpha-amylase (83.9% inhibition at concentration 1 mg/ml). In silico studies were also performed to predict the binding affinity of the prepared triazoles toward human α-amylase (PDBID: 1u30) (as presented in **Figure 3**) [25].

3.3 Augmentation of phenolic compounds in edible seed sprouts

Elicitation and physical stimulation during germination are efficient tools to modulate both chemical and biological contents of many important functional foods and medicinal plants. Elicitors from different origins could be used either alone or with hydroponic sprays during germination and growth or right before harvest. A better knowledge on the effect of certain compounds on biosynthetic pathways in responding to specific treatments with elicitors would be a very useful way to augment the production of secondary metabolites or produce new metabolites. This will help in production of high-quality, healthy, and useful medicinal plants and foods. *Moringa oleifera* (MO) leaves extract contains a several active constituents; alkaloids, carotenoids, glucosinolates, polyphenols, tannins, and saponins and is considered as a good biotic elicitor. It was used in this study to enhance both phenolic and antioxidant contents in germinated alfalfa sprouts. Germination of alfalfa seeds in continuous light and soaking seeds in 0.0625, 0.03125 g/L MO extract before germination significantly increases the levels of total phenolics and their antioxidant activity. The maximum amount of flavonoids was exudated after 8 hours of germination. The optimal concentration to elicit maximum phenolic levels was further used to study the biological activities [26].

3.3.1 Histochemical localization of polyphenolic compounds

Localization of phenolic compounds in the different organs and tissues of cotton (*Gossypium barbadense* L. var. Giza 86) plant (seeds, stems, leaves, and roots) was conducted in FAB-Lab. The study revealed the presence of polyphenolic compounds as tiny particles in the cytoplasm of some parenchymatous cells surrounding the lysigenous glands. The obtained data shed more light on our previous results on the antimitotic activity of polyphenolic aldehyde gossypol and why does not affect the growing tip of the plant [27].

4. Developing new therapeutic agents via modulation of biological systems using phenolic compounds

Over 15 years, FAB-Lab team had developed several models for designing new therapeutic agents from phenolic compounds through inhibiting different target enzymes.

4.1 Tyrosinase

Flavonoids contain an alpha-keto group as new type of tyrosinase inhibitors from natural products as potential treatments for hyperpigmentation [28].

4.2 Hyalourindase

The extract of different organs of *Ravenala madagascariensis* (Sonn.) plant showed a good inhibitory activity against hyaluronidase enzyme. Both metabolic analysis and phytochemical studies disclosed the presence of 19 different phenolic compounds, which may present flavone, flavonol, and flavanol glycosides, and aglycone. Specifically, isorhamnetin-7-O-rutinoside, rutin, epiafzelechin, kaempferol, isorhamnetin 7-O-glucoside, and narcissin were isolated and characterized from the butanolic extract of leaves. In docking experiments, narcissin, quercetin 3-O-glucoside, and rutin may interact with enzymes via H-bonding with the Asp111, Gln271, and/or Glu113 residues. These interesting results could be used in pharmaceutical industry to develop new therapeutic agent(s) for skin wrinkles and other cosmetic purposes [29].

4.3 Aldose reductase

The olive and ginkgo leaves extracts with high phenolic contents was proved in an *in vitro* and *ex vivo* inhibitory activity against aldose reductase and could be used as promising therapy for cataract [30, 31].

4.4 Leukotriene hydrolyase

Leukotriene hydrolyase LT4: Synthesis, docking, cytotoxicity, and LTA4H inhibitory activity of phenolic gingerol derivatives as potential colorectal cancer therapy [32, 33].

4.5 Promiscuous multitarget inhibitors for treatment of chronic and complicated health disorders

Phenolic compounds proved to be safe and effective multitarget enzyme inhibitors. Screening of over 50 medicinal plant extracts revealed that the phenolic contents of *Punica granatum*, *Phyllanthus emblica*, and *P. guajava* showed good inhibitory activity on α-amylase tyrosinase and hyaluronidase enzymes (as shown in **Figure 4**). These raw data encouraged us in FAB-Lab to go further to develop and design a multitarget drug from phenolic compounds [34–36].

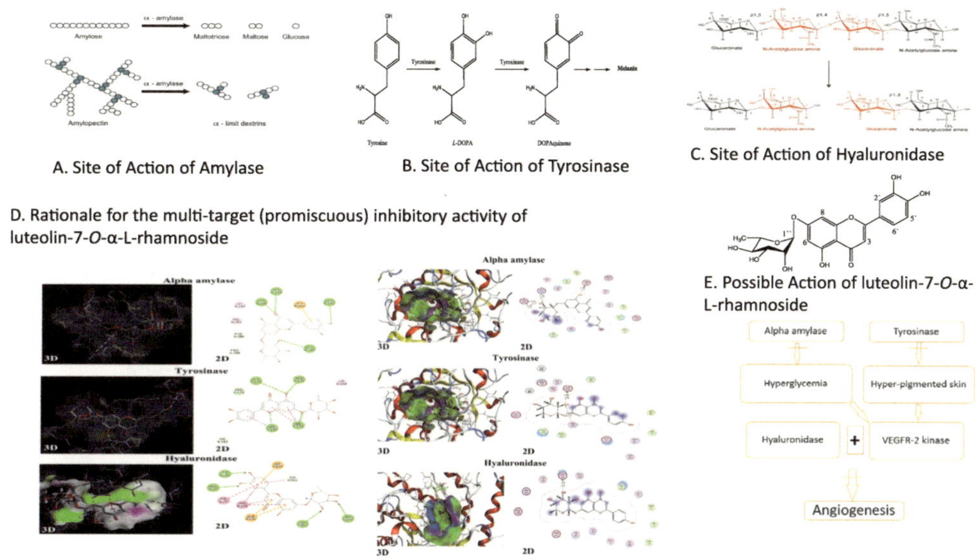

Figure 4.
Developing luteolin-7-O-α-L-rhamnoside as a promiscuous multitarget enzyme inhibitors at three different sites of amylase (A), tyrosinase (B), and hyaluronidase (C). Rationale for the multitarget (promiscuous) inhibitory activity of luteolin-7-O-α-L-rhamnoside (D) and possible action of luteolin-7-O-α-L-rhamnoside (E).

5. Designing a promiscuous multitarget inhibitor against three metabolic enzymes (alpha amylase, tyrosinase, and hyaluronidase)

In principle, we hypothesized the following:

a. Alpha amylase inhibitor could inhibit vascular endothelial growth factor receptor 2 (VEGFR-2) through inhibition of high glucose-induced [37]

b. Tyrosinase inhibitor could inhibit melanin synthesis and halt the expression of vascular endothelial growth factor (VEGF) [38]

c. Hyaluronidase inhibitor could reduce or stop hyaluronic acid fragmentation and subsequently halt the proliferation and endothelial cell migration and capillary formation [39].

Based upon this hypothesis, we have developed in FAB-Lab luteolin-7-O-α-L-rhamnoside, which could be used as a potential multitarget enzyme inhibitor (promiscuous inhibitor) for alpha amylase, tyrosinase, and hyaluronidase.

Luteolin-7-O-α-L-rhamnoside could be used as a potential multitarget enzyme inhibitor in another words, promiscuous enzyme inhibitor, for the possible treatment of various health disorders such as angiogenesis-related disorders.

Luteolin aglycone, when compared with its glycoside, can easily access the catalytic site through 3′ and 4′-hydroxy group in ring B (bonded to Cys83) and the 7-hydroxy in ring A (bonded to Gly245, Ala246, and Val248). These data are in agreement with other reports in which luteolin was proven to be a noncompetitive tyrosinase inhibitor.

While luteolin glycoside(5-O-β-D-glucopyranoside) could also be interacted close to Cu and HOO ions as kojic acid and L-tyrosine, luteolin aglycone and luteolin glycoside (7-O-β-D-glucopyranoside) could not. These findings support that sugar moiety at 7 position may have a role in the type of inhibition (i.e., noncompetitive).

Luteolin as a free aglycone has a very weak inhibitory activity toward hyaluronidase in comparison with luteolin-7-O-α-L-rhamnoside. This may refer to the importance of the hydroxyl groups in the rhamnose moiety at 7 position, and this was confirmed by the molecular docking simulation. Because there are two hydroxyl groups that bind to the amino acid residues Asp292 and Ser245 via hydrogen bond interactions.

There were not any reports about luteolin-7-O-rhamnoside inhibitory activity toward the three metabolic enzymes of interest.

In conclusion, more than 50 extracts of different medicinal plants were screened for their biological activities as inhibitors for some metabolism-related enzymes. Extracts with the highest activities were fractionated, and four compounds were isolated, which were found to be multitarget inhibitors for alpha amylase, tyrosinase, and hyaluronidase or at least two of them. Virtual screening and mechanism of action determination studies were performed also for these compounds.

6. Pharmaceutical and therapeutic applications

6.1 Liver research (liver fibrosis, liver cancer, interferon inducer)

Establishing *in vitro*, *in vivo*, and preclinical models for liver fibrosis, liver injury, schistosomiasis, iron-overloaded, fatty liver, and immunosuppression, FAB-Lab team showed that naringin (a flavanol isolated in FAB-Lab grape fruit) exhibited a potent hepatoprotective activity [40]. Recently, it was further prepared in a nanoscopic nanomicelle formula to improve its efficacy and bioavailability as antiulcer and anticancer [41]. The prepared formula showed a very good activity in protection of gastric mucosa and suppressed the release cytokines *in vivo* model using ethanol-induced ulcer in rats. Moreover, in *in vitro* cytotoxicity assay using cell lines and EAC-bearing mice, naringin nanomicelle showed an excellent result as cytotoxic and tumor agent. This may prompt us to propose naringin nanomicelles as a nanodrug with prolonged release and enhanced antiulcer as well as antitumor activities [41].

6.2 Colorectal cancer, human breast carcinoma, triple breast cancer

Even though many flavonoids proved their efficacy in different models/assays against colorectal cancer, no solid evidence was about the relation between SAR (Structure-Activity-Relationship) and colorectal cancer. This prompted the FAB-Lab team to examine the SAR of flavonoids and *in vitro* anticolon cancer using human colon cancer cell line (Caco-2). Surprisingly, the obtained results showed that the OH of C-5 and C-7 in A ring increasingly improved the anticancer effect of flavonoids when compared with 5-FU. In contrary, the presence of glucose moiety or OH— groups in B ring drastically reduced the anticancer activity. In conclusion, FAB-Lab team proposed a novel, hypothesis SAR of flavonoid-colorectal cancer therapy, which may provide a new horizon to better improve management of colorectal cancer [42].

6.3 Antimitotic activity

Several natural phenolic anthraquinone compounds were isolated from different plants and agriculture wastes in FAB-Lab. The salient features and the different biological activity of various anthraquinones compounds depend mainly on the distribution of OH groups in the basic skeleton. The study in FAB-Lab revealed that the presence of OH-group at O-position is a very important feature to portray a potent antimitotic activity (Badria assay) as seen in alizarin, which is the only OH-quinone having an OH— group in an ortho-position. The SAR study revealed that other compounds were active but in the following orders: emodin > aloe emodin > rhein > quinzarin. Interestingly, our results showed that 1-hydroxyanthraquinone could be a carcinogen [43].

Gingerol, a natural phenolic compound, and its derivatives exhibited a broad spectrum against different cancer cell lines and could be also used as a chemosensitizer with currently used anticancer drugs [44].

Ricinine, a simple alkaloid isolated from castor seeds in FAB-Lab, was used to prepare 16 derivatives and tested against SAS-oral cancer cell line in MTT assay versus 5-FU as possible agents for treatment of oral cancer. Sixteen new analogues were synthesized from ricinine. In contrary to 5-FU, the ricinine derivatives were able to suppress the growth of cancer cells at 25 mM [45].

6.4 Cataract

Phenolic compounds from both ginkgo (GK) and olive leaves (OL) extracts were examined both in *in vitro* and *ex vivo* assays against aldose reductase to find potential and selective inhibitors for the enzyme and possible use as a preventive or treatment for cataract. The promising results prompted FAB-Lab team to reveal the possible inhibitory mechanism of GK and OL by using computer-assisted programs. The results revealed that the phenolic compounds could inhibit the polyols pathways via halting the advanced glycation, which may have a crucial role in pathogenesis of cataract [46].

6.5 Idiopathic nephrotic syndrome

Phenolic contents of lyophilized *Citrus paradise* (grape fruit) were prepared in standardized soft gelatin capsules (GF) in FAB-Lab. The prepared capsules (GF) were coadministered with cyclosporine (CsA) for patients diagnosed with idiopathic nephrotic syndrome in Nephrology and Urology center at Mansoura University, Egypt. The coadministration of GF capsules resulted in increased CsA exposure in the range of 10–25%. Moreover, this clinical study proved that CsA-GF coadministration was found to be safe and well-tolerated as confirmed from laboratory and clinical studies [47].

6.6 Iron chelation (curcumin, mangiferin, catechins)

Catechins (from green tea leaves), curcumin (from turmeric rhizomes), and mangiferin (from mango leaves) were among many other phenolic compounds that had been isolated in FAB-Lab [48, 49]. Iron-overload disorder (hemochromatosis) is one of the major reasons of morbidity. Most of the commonly available iron-chelating agents suffer from many side effects, which pronounced the need for a safe and effective natural iron-chelating agent(s) alternative. Therefore, FAB-Lab team conducted several *in vitro* and *in vivo* studies on selected polyphenol compounds (catechin, curcumin, and mangiferin), which may serve excellent natural alternative for commonly synthetic iron-chelating agents. Different iron-overloaded experimental models were conducted and disclosed the high efficacy of catechin, mangiferin, and curcumin (as shown in **Figure 5**). Subsequently, three very promising alternatives could be used in both clinical and industrial iron-overload or oxidative stress health and/or environmental problems [49].

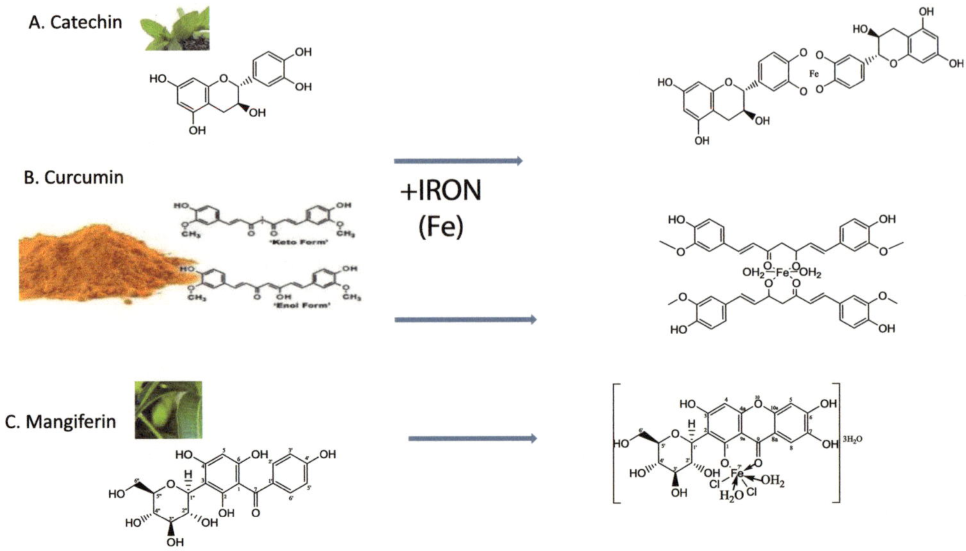

Figure 5.
Chelation of iron by catechin from green tea leaves (A), curcumin from turmeric rhizomes (B), and mangiferin from mango leaves (C).

6.7 Antiviral activity

Several formulae using proniosomes technique were proposed in FAB-Lab to prepare nanocurcumin (NC) as possible antiviral antiherpes (*Herpes simplex* type I) agent in comparison to a commonly used acyclovir (ACV). The results showed that NC exhibited a better and safe activity over ACV against Herpes-simplex Type I. Interestingly, NC-ACV combination reduced the toxicity and enhanced the efficacy that led to 100% plaque reduction over ACV alone [50].

7. Conclusions and future perspective

Phenolic compounds of natural sources, herbs, foods, marine organisms, insects, and other natural sources, still maintain a crucial role in our daily health life in prevention and/or treatment. Therefore, a better utilization of extracts rich in phenolic compounds and/or isolated pure phenolic compounds, e.g., alkaloids, flavonoids, stilbenes, tannins, curcuminoids, coumarins, lignans, quinones, may help in providing the community with chemopreventive properties (e.g., antioxidant, anticarcinogenic, or antimutagenic and antiinflammatory effects). Moreover, phenolic compounds may also contribute in inducing apoptosis by arresting cell cycle, regulating carcinogen metabolism and ontogenesis expression, inhibiting DNA binding and cell adhesion, migration, proliferation, or differentiation, and blocking signaling pathways.

This chapter covers the foremost recent preclinical and clinical research from FAB-Lab and summarizes structural categories and molecular mechanisms of phenolic compounds from medicinal herbs and dietary plants.

Accordingly, the founding factors of FAB-Lab vision were to:

- Protect the environment against all hazards.
- Treat the current diseases using the abundantly available phenolic compounds, whereas many people are suffering worldwide and inspiring from totally indigenous raw materials.

- Conduct basic and simple technology to produce a 100% natural medicine.

- Upon such vision, all research projects, scientific creations, trouble-shooting and problem-solving techniques were based so that the environment was the arena and the main assistant so long as we cared for it and lived in it and with it in harmony. In our turn, FAB-lab team proved the usefulness of phenolic compounds in different aspects by which we can fight contagious and cancerous diseases with hope to add life to our years rather than adding years to life and present a healthy model for cheering up life especially during the turmoil of COVID-19 pandemic [51–55].

Acknowledgements

The author acknowledges the support of Academy of Scientific Research and Technology ASRT, the Arab Development Fund, Kuwait, ICRO-UNESCO International, UNESCO Biotechnology, TWAS-ARO, CEC for both research and capacity building grants to FAB-Lab. Also, the author is deeply indebted to the unwavering support and sincere appreciation of Prof. Dr. W.T. Shier (University of Minnesota, USA, Prof. Dr. I. Khan, Prof. Dr. Charles Hufford, and Prof. Dr. M. El-Sohaly (University of Mississippi, USA), Prof. Dr. Francesca Pataculi (College of Medicine, University of Rome), Prof. A. Kumar (Assam University, India), Prof. Dr. Bharat B. Aggarwal (The University of Texas MD Anderson Cancer Center, Houston, Texas), Prof. Dr. Werner E. Müeller (Universität Mainz, Germany), Prof. Dirk Selmar (Institut für Pflanzenbiologie, Technische Universität Braunschweig, Germany), Prof. Dr. Amarullah H. Siregar ((Indonesian Community on Functional Advancement Medicine, Jakarta, Indonesia), Prof. Dr. Ibraheim Jantan (Kuala Lumpur, Malaysia), and Prof. Dr. Delgado (Havana, Cuba).

Conflict of interest

The author has declared that no competing or conflict of interests exists.

Appendix

Table 1. Examples of Different Classes of Phenolic Compounds isolated and prepared in FAB-Lab (Liver Research Lab, Mansoura University, Egypt).

Class	Structure and name		
Simple phenolic compounds	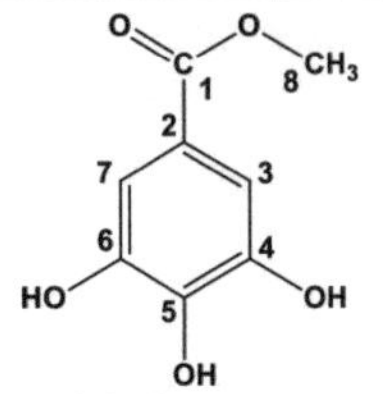 Methyl gallate	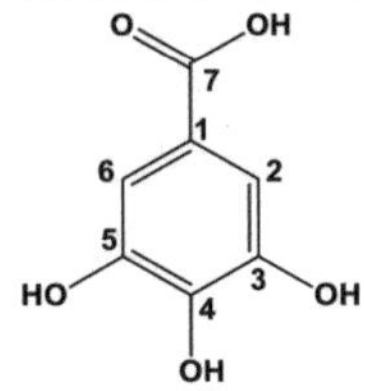 Gallic acid	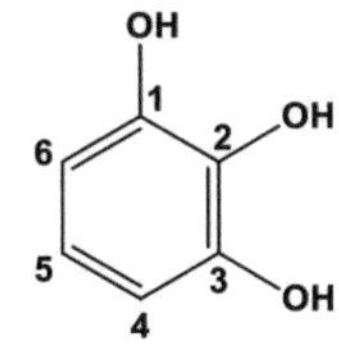 Pyrogallol
	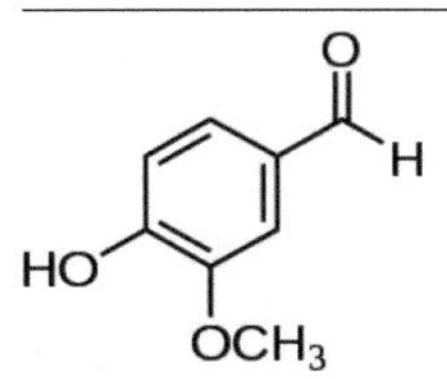 Vaniline	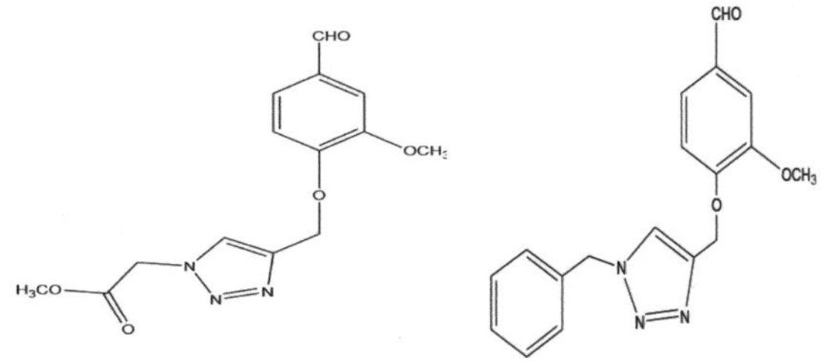 Vaniline triazoles (click reaction products)	
	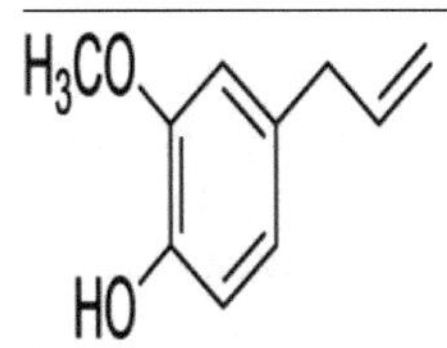 Eugenol	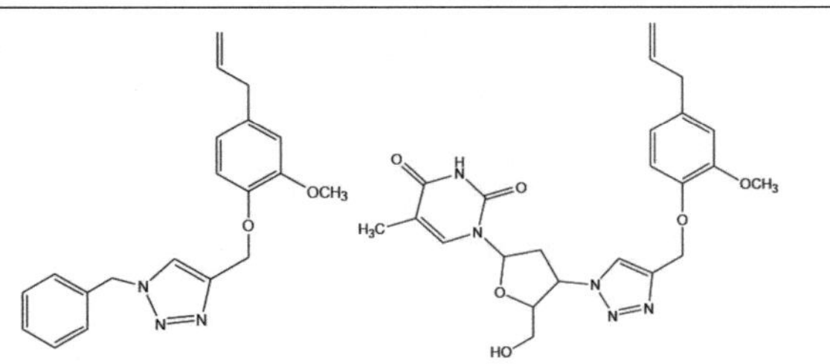 Eugenol triazoles (click reaction products)	
	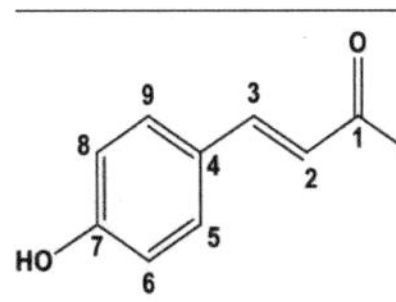 *p*-Coumaric acid	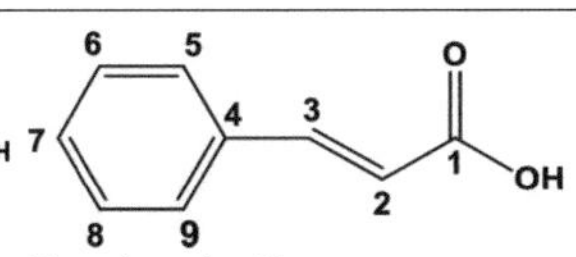 *Trans* cinnamic acid	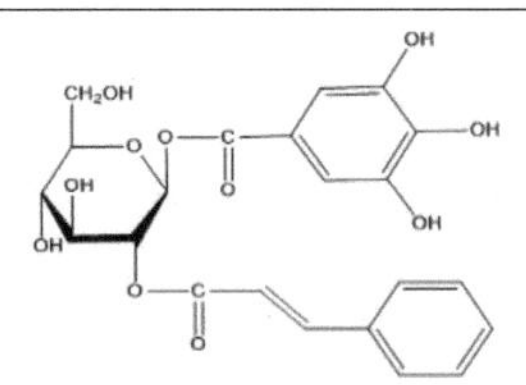 2-*O*-Cinnamoyl-1-*O*-galloyl-β-D-glucopyranoside
	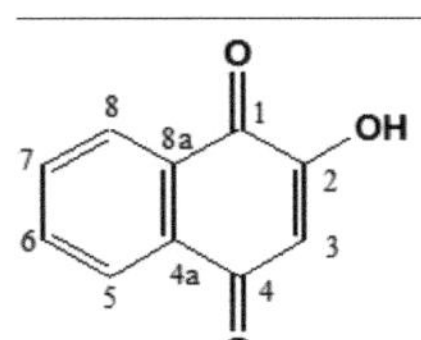 Lawsone		

Class	Structure and name	
Polyphenolic compounds		
1. Flavonoids	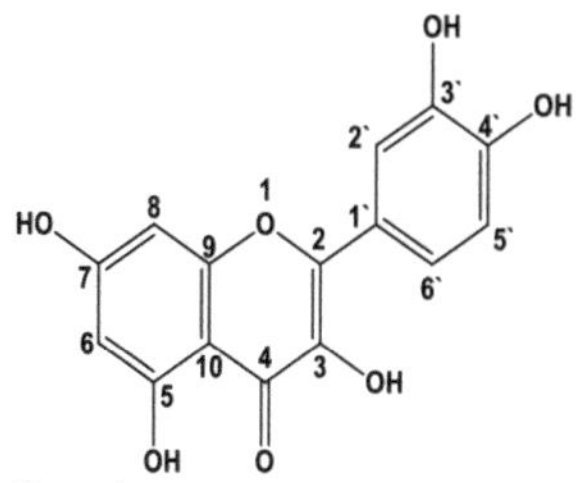 Quercetin	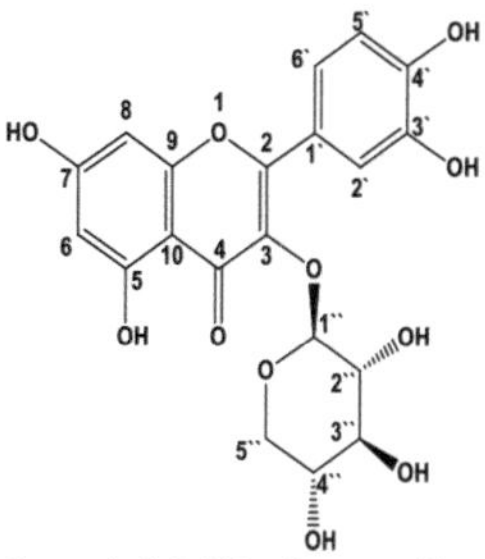 Quercetin-3-*O*-β-D-xylopyranoside (reynoutrin)
	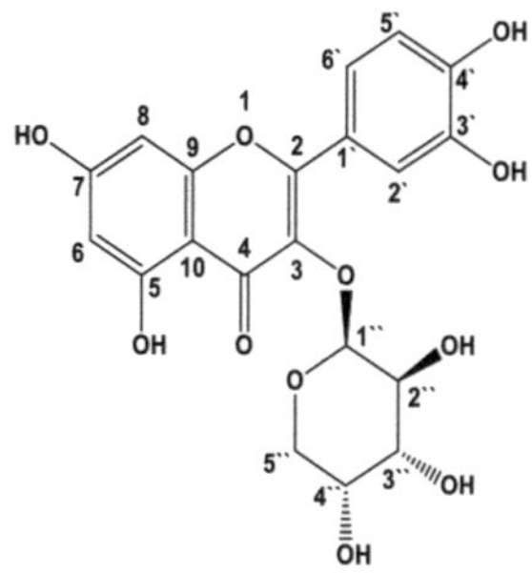 Quercetin-3-*O*-β-arabinopyranoside	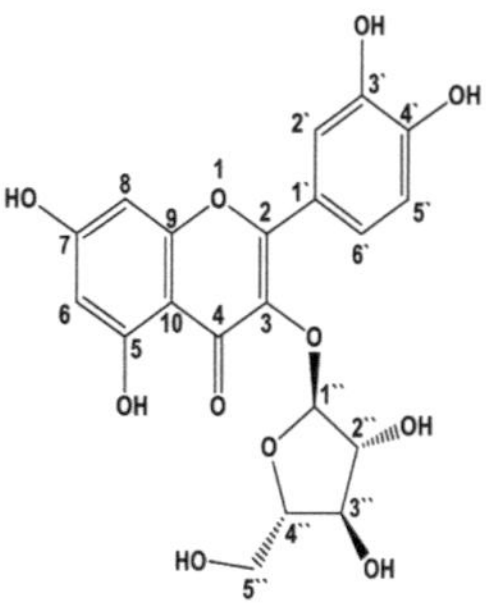 Quercetin 3-*O*-α-L- arabinofuranoside (avicularin)
	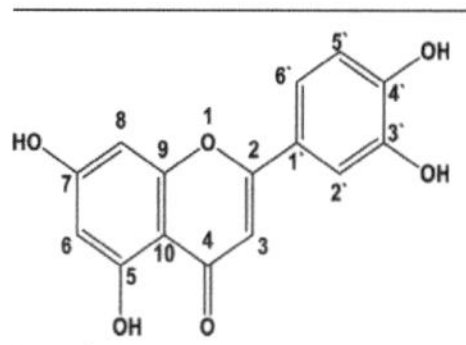 Luteolin	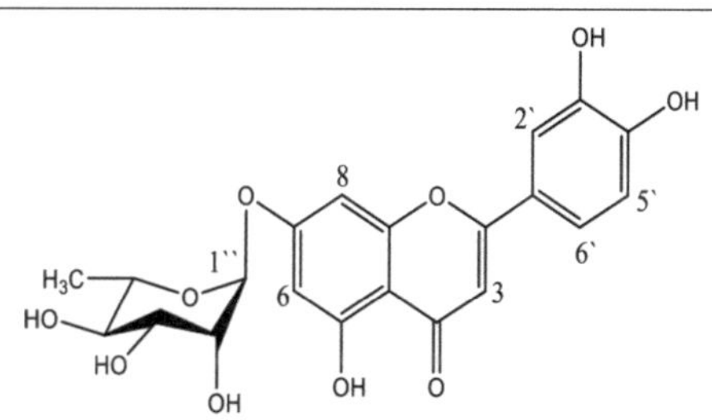 Luteolin-7-O-α-L-rhamnoside
	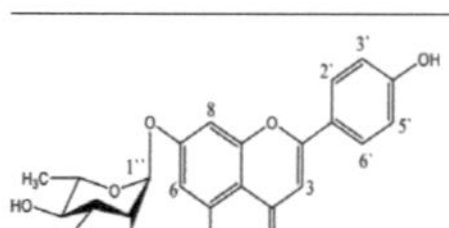 Apigenin-7-O-α-L-rhamnoside	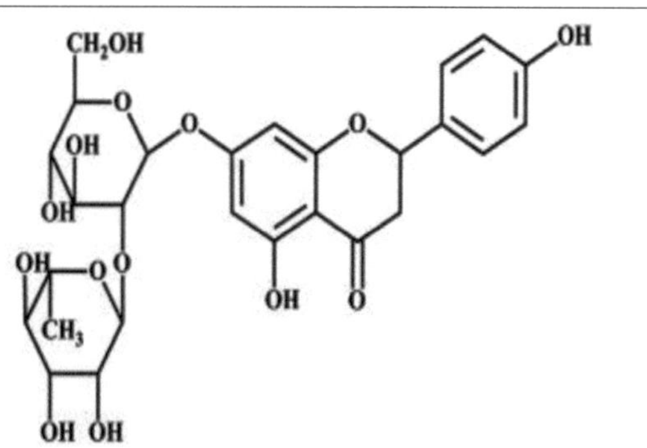 **Naringin (4', 5, 7-trihydroxy flavanone 7-rhamnoglucoside)** Naringin (4′,5,7-trihydroxy flavanone 7-rhamnoglucoside)

Class	Structure and name
2. Tannins	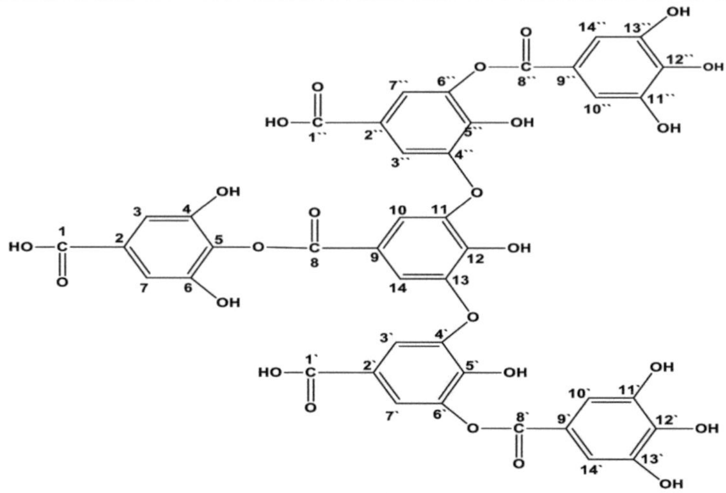 Emblifatmin
1,3,6,7-Tetrahydroxy-9H-xanthen-9-one	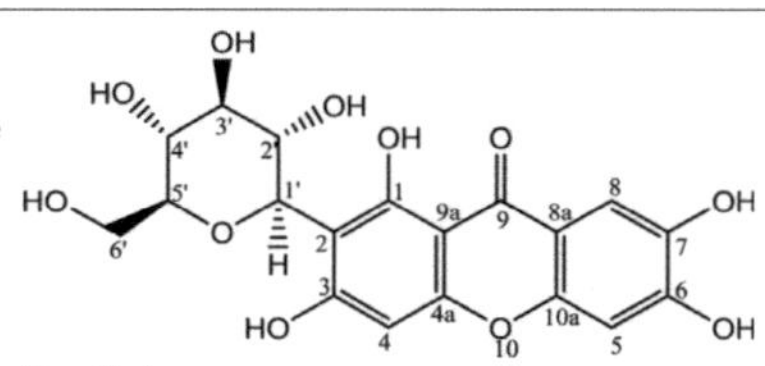 Mangiferin 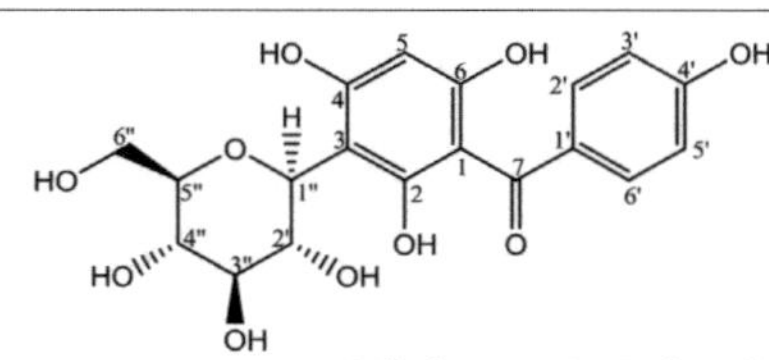 Iriflophenone-3-C-β-D-glucoside
3. Gingerol	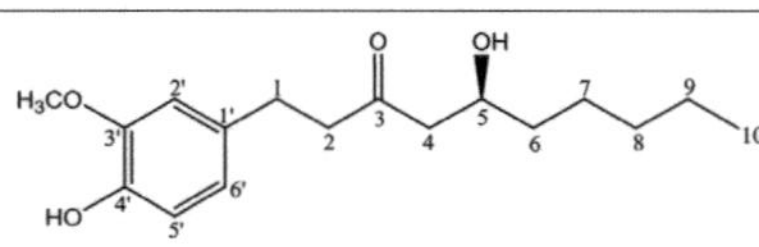
	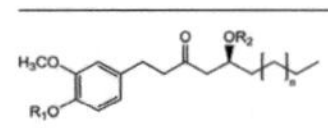 n, R_1, R_2 N1 2 H H N2 4 H H N3 6 H H N4 0 H H N5 2 CH_3 H D1 2 Allyl H D3 2 $COCH_3$ $COCH_3$ D5 2 H CH_3 D6 2 CH_3 CH_3 D10 2 Prenyl H 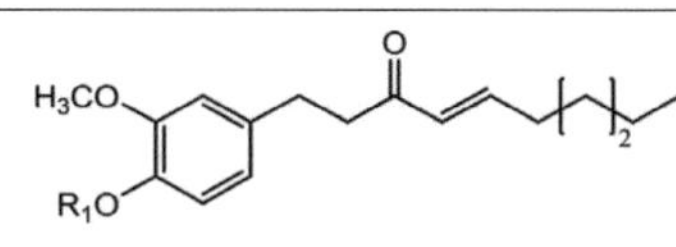 R_1 D2 $COCH_3$ D7 H D8 CH_3 D11 Prenyl
	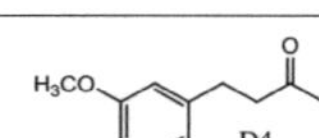 D4
	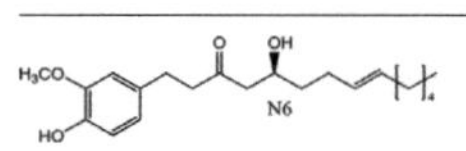 N6
	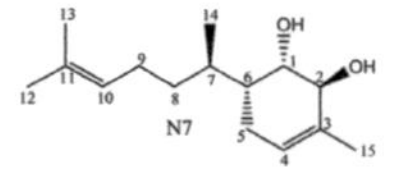 N7 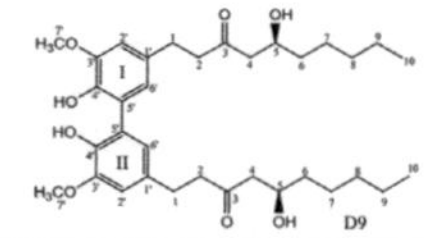 D9

Class	Structure and name
4. Ricinine	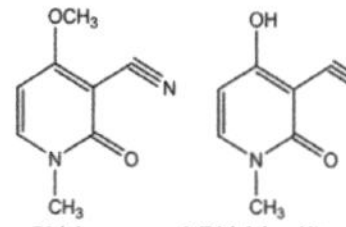 Ricinine 3 (Ricininic acid) 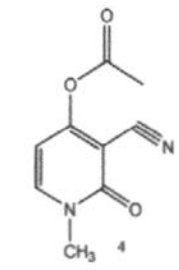 4 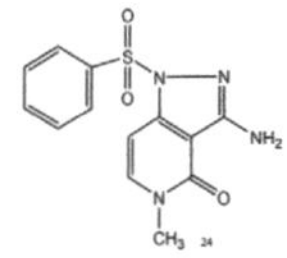 24
	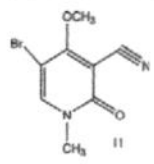 11 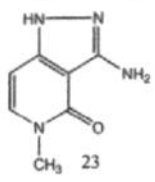 23 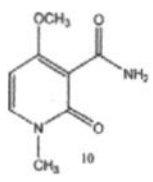 10
5. Curcumin	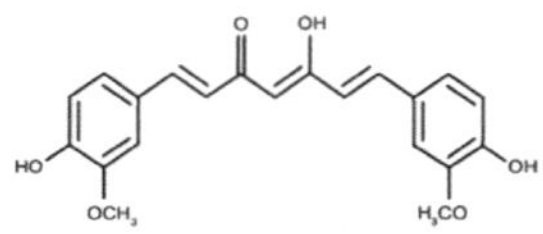
	a. Click Reaction Products (Curcumin-Triazole Deivatives) 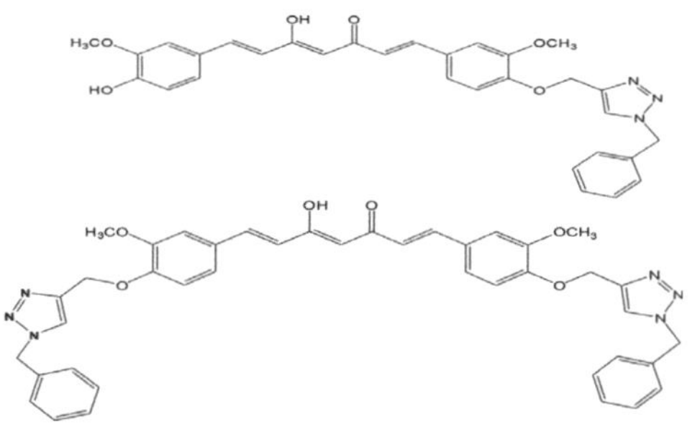
	b. Preparation of curcumin derivatives 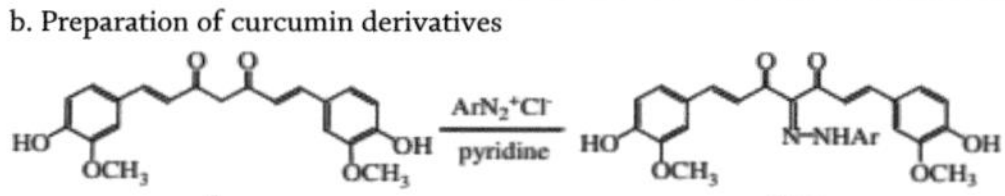 1 2a-e 2a, Ar = C_6H_4- OCH_3-*p* 2b, Ar = C_6H_4- CH_3-*p* 2c, Ar = C_6H_5 2d, Ar = C_6H_4- Cl-*p* 2e, Ar = C_6H_4- NO_2-*p*
6. Anthraquinones	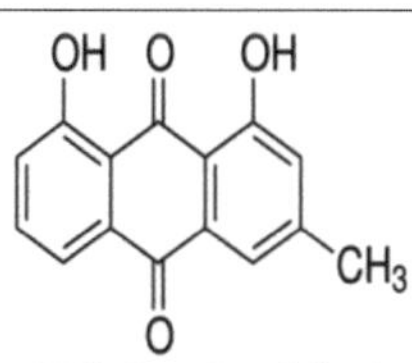 1,8-Dihydroxy-3-methyl anthraquinone (chrysophanol) 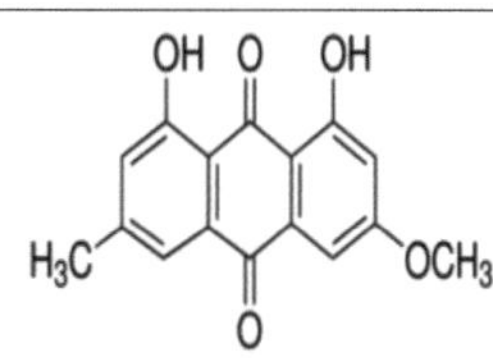 1,8-Dihydroxy-3-methyl-6-methoxy anthraquinone (physcion)
	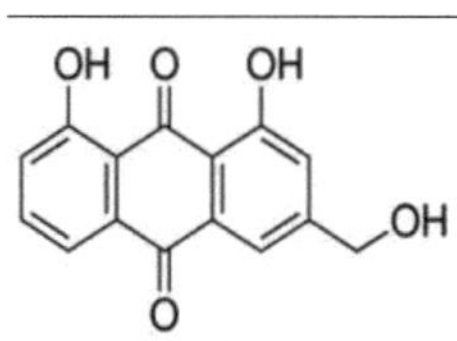 1,8-Dihydroxy-3-hydroxy methyl anthraquinone (aloe-emodin)
	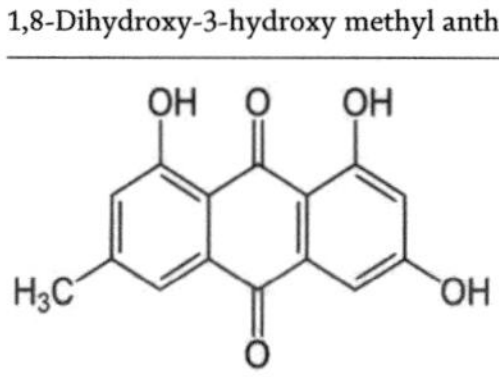 1,6,8-Trihydroxy-3-methyl anthraquinone (emodin)

Class	Structure and name
	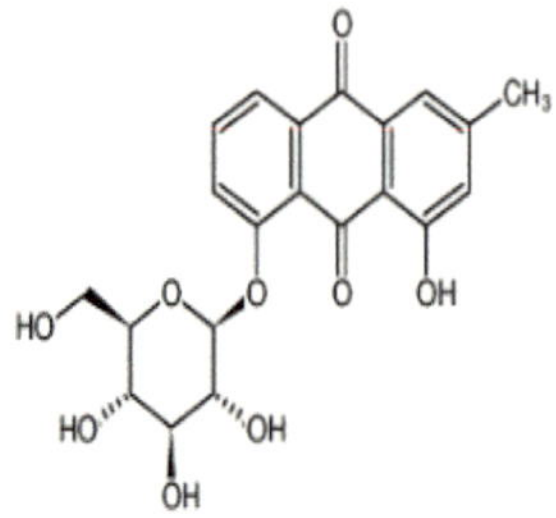 Chrysophanol-8-*O*-β-D-glucopyranoside (chrysophanein)
	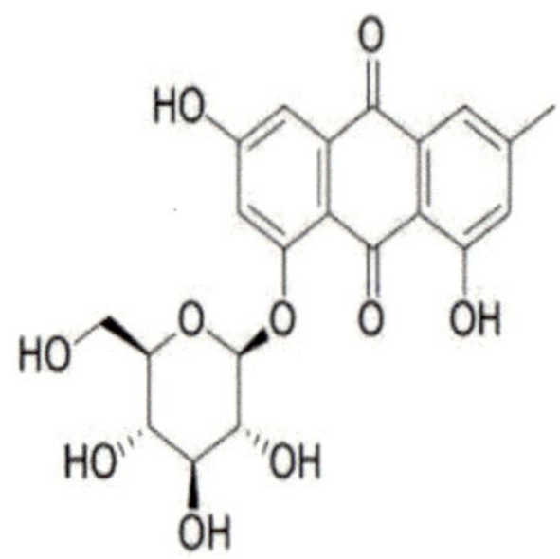 Emodin-8-O-β-D-glucopyranoside

Table 1.
Examples of different classes of phenolic compounds isolated and prepared in FAB-Lab (Liver Research Lab, Mansoura University, Egypt).

Author details

Farid A. Badria
Liver Research Lab, FAB-Lab, Pharmacognosy Department, Faculty of Pharmacy, Mansoura University, Mansoura, Egypt

*Address all correspondence to: badri002@mans.edu.eg

References

[1] Badria FA. Is man helpless against cancer? An environmental approach: Antimutagenic agents from Egyptian food and medicinal preparations. Cancer Letters. 1994;**84**:1-5

[2] Badria F, Abou-Mohamed G, El-Mowafy A, Masoud A, Salama O. Mirazid: A new schistosomicidal drug. Pharmaceutical Biology. 2001;**39**: 127-131

[3] Lao Y. Clinical study on effect of matrine injection to protect the liver function for patients with primary hepatic carcinoma after trans-artery chemo-embolization (TAE). Zhong Yao Cai. 2005;**28**(7):637-638

[4] Kwon KH, Barve A, Yu S, Huang MT, Kong AN. Cancer chemoprevention by phytochemicals: Potential molecular targets, biomarkers and animal models. Acta Pharmacologica Sinica. 2005;**28**(9): 1409-1421

[5] Rudolf E, Andelova H, Cervinka M. Activation of several concurrent proapoptic pathways by sulforaphane in human colon cancer cells SW620. Food and Chemical Toxicology. 2009;**47**(9): 2366-2373

[6] Scheckel KA, Degner SC, Romagnolo DF. Rosmarinic acid antagonizes activator protein-1-dependent activation of cyclooxygenase-2 expression in human cancer and nonmalignant cell lines. Journal of Nutrition. 2008;**138**(11): 2098-2105

[7] Sanchez-Fidalgo S, Cardeno A, Villegas I, Talero E, de la Lastra CA. Dietary supplementation of resveratrol attenuates chronic colonic inflammation in mice. European Journal of Pharmacology. 2010;**633**(1-3):78-84

[8] Narisawa T, Fukaura Y, Hasebe M, Ito M, Aizawa R, Murakoshi M, et al. Inhibitory effects of natural carotenoids, alpha-carotene, beta-carotene, lycopene and lutein, on colonic aberrant crypt foci formation in rats. Cancer Letters. 1996;**107**(1):137-142

[9] Sung B, Pandey MK, Ahn KS, et al. Anacardic acid (6-nonadecyl salicylic acid), an inhibitor of histone acetyltransferase, suppresses expression of nuclear factor-kappaB-regulated gene products involved in cell survival, proliferation, invasion, and inflammation through inhibition of the inhibitory subunit of nuclear factor-kappaB alpha kinase, leading to potentiation of apoptosis. Blood. 2008;**111**:4880-4891

[10] Vahid F, Zand H, Nosrat-Mirshekarlou E, et al. The role dietary of bioactive compounds on the regulation of histone acetylases and deacetylases: A review. Gene. 2015;**562**:8-15

[11] Di Paolo M, Papi L, Gori F, Turillazzi E. Natural products in neurodegenerative diseases: A great promise but an ethical challenge. International Journal of Molecular Sciences. 2019;**20**(20):5170

[12] Majtan J, Bucekova M, Jesenak M. Natural products and skin diseases. Molecules. 2021;**26**:4489

[13] Philip N, Leishman S, Walsh L. Potential role for natural products in dental caries control. Oral Health and Preventive Dentistry. 2019;**17**(5): 479-485

[14] Alex AF, Spitznas M, Tittel AP, Kurts C, Eter N. Inhibitory effect of epigallocatechin gallate (EGCG), resveratrol, and curcumin on proliferation of human retinal pigment epithelial cells in vitro. Current Eye Research. 2010;**35**(11):1021-1033

[15] Mitra S, Dash R. Natural products for the management and prevention of

breast cancer. Evidence-Based Complementary and Alternative Medicine. 2018;**2018**:8324696

[16] Nagle DG. A perspective on natural products in breast cancer carcinogenesis and chemoprevention. Annals of Carcinogenesis. 2016;**1**(1):1002

[17] Ko EY, Moon A. Natural products for chemoprevention of breast cancer. Journal of Cancer Prevention. 2015;**20**: 223-231

[18] Nagle DG, Ferreira D, Zhou YD. Molecules of interest: Epigallocatechin-3-gallate (EGCG): Chemical and biomedical perspectives. Phytochemistry. 2006;**67**:1849-1855

[19] Cragg GM, Pezzuto JM. Natural products as a vital source for the discovery of cancer chemotherapeutic and chemopreventive agents. Medical Principals and Practice. 2016;**25**:1-19

[20] Li Y, Li S, Meng X, Gan R-Y, Zhang J-J, Li H-B. Dietary natural products for prevention and treatment of breast cancer. Nutrients. 2017;**9**(7):728

[21] Elmali N, Esenkaya I, Harma A, Ertem K, Turkoz Y, Mizrak B. Effect of resveratrol in experimental osteoarthritis in rabbits. Inflammation Research. 2005;**54**:158-162

[22] Leung CY, Liu L, Wong RNS, Zeng YY, Li M, Zhou H. Saikosaponin-d inhibits T cell activation through the modulation of PKCθ, JNK, and NF-κB transcriptionfactor. Biochemical and Biophysical Research Communications. 2005;**338**(4):1920-1927

[23] Ammon HPT. Modulation of the immune system by *Boswellia serrata* extracts and boswellic acids. Phytomedicine. 2010;**17911**:862-867

[24] Badria F, Abouhabieb M, Elbar F. Synthesis of Nanoparticles Using Green Chemistry: Green Synthesis. Germany: Grin Verlag; 2019. ISSN 3346083462. Available from: https://www.amazon.com/Synthesis-Nanoparticles-Using-Green-Chemistry/dp/3346083462

[25] Badria F, Elattar E, Galala A, El-Rady H. Alpha-amylase inhibitory effect and docking studies of some new curcumin-triazole derivatives. Acta Scientific Nutritional Health. 2019; **3**(11):124-129

[26] Badria F, Abouzeid S. *Moringa oleifera* leaves extract: A potent elicitor for germination of alfalfa sprout with high antioxidant activity. Asian Journal of Research in Pharmaceutical Sciences and Biotechnology. 2019;**7**(4):74-83

[27] Badria F, Houssen W, Zaghlol M, Halim A. Histochemical localization of polyphenolic compounds in *Gossypium barbadense*. International Journal of Research in Pharmaceutical and Nano Sciences. 2018;**7**(5):200-204

[28] Badria F, ElGayyar M. A New type of tyrosinase inhibitors from natural products as potential treatments for hyperpigmentation. Bollettino Chimico Farmaceutico. 2001;**140**:267-271

[29] Mohamed E, Hetta M, Rateb M, Selim M, AboulMagd A, Badria F, et al. Bioassay-guided isolation, metabolic profiling, and docking studies of hyaluronidase inhibitors from *Ravenala madagascariensis*. Molecules. 2020;**25**: 1714

[30] Elimam IA, Liou G, Badria FA. Olive and ginkgo extracts as potential cataract therapy with differential inhibitory activity on aldose reductase. Drug Discoveries and Therapeutics. 2017;**11**(1):41-46

[31] Muhammad A, Muangchoo K, Muhammad IA, Ajingi Y'u S, Bello AM, Muhammad IY, et al. A molecular modeling study of novel aldose reductase (AR) inhibitors. AIMS Biophysics. 2020;**7**(4):380-392

[32] Badria F. New Gingerol Derivative against Human Colorectal Carcinoma Caco-2 Cells (German Edition). Germany: Grin Verlag; 2019. ISSN 3346078817. Available from: https://www.amazon.com/Gingerol-Derivative-against-Colorectal-Carcinoma/dp/3346078817

[33] Low C, Akthar S, Patel D, et al. The development of novel LTA4H modulators to selectively target LTB4 generation. Scientific Reports. 2017;7:44449

[34] Badria F, Ahmed M. Multi target drugs: A new therapeutic approach. International Journal of Research in Pharmaceutical and Nano Sciences. 2018;7(5):205-213

[35] Badria F, Omar R, Galala A, Badria F. Multi-target inhibitory activity of some medicinal plants on alpha-amylase, tyrosinase and hyaluronidase: Potential therapy for treatment diabetes and diabetic complications. Acta Scientific Nutritional Health. 2019; **3**(10):68-75

[36] Ramsay RR, Popovic-Nikolic MR, Nikolic K, et al. A perspective on multi-target drug discovery and design for complex diseases. Clinical and Translational Medicine. 2018;**7**:3

[37] Fernando K, Yang H-W, Jiang Y, Jeon Y-J, Ryu B. Diphlorethohydroxycarmalol isolated from *Ishige okamurae* represses high glucose-induced angiogenesis in vitro and in vivo. Marine Drugs. 2018; **16**:375

[38] Kim M, Shibata T, Kwon S, Park TJ, Kang HY. Ultraviolet-irradiated endothelial cells secrete stem cell factor and induce epidermal pigmentation. Scientific Reports. 2018; **8**:4235

[39] Park D, Kim Y, Kim H, Kim K, Lee Y-S, Choe J, et al. Hyaluronic acid promotes angiogenesis by inducing RHAMM-TGFβ receptor interaction via CD44-PKCδ. Molecules and Cells. 2012; **33**:563-574

[40] Badria F, ElKashef H. A potent hepatoprotective agent from group fruit. Alexandria Journal of Pharmaceutical Sciences. 1994;**8**(3): 165-169

[41] Mohamed EA, Abu Hashim II, Yusif RM, Shaaban AAA, El-Sheakh AR, Hamed MF, et al. Polymeric micelles for potentiated antiulcer and anticancer activities of naringin. International Journal of Nanomedicine. 2018;**13**: 1009-1027

[42] Ibrahim A, Sobeh M, Ismail A, Alaa A, Sheashaa H, Sobh M, et al. Free-B-ring flavonoids as potential lead compounds for colon cancer therapy. Molecular and Clinical Oncology. 2014 Jul;**2**(4):581-585

[43] Badria FA, Ibrahim AS. Evaluation of natural anthracene-derived compounds as antimitotic agents. Drug Discoveries and Therapeutics. 2013; **7**(2):84-89

[44] Ibrahim AS, Sobh MA, Eid HM, Salem A, Elbelasi HH, El-naggar MH, et al. Gingerol-derivatives: Emerging new therapy against human drug-resistant MCF-7. Tumor Biology. 2014; **35**(10):9941-9948

[45] El-Naggar M, Abdel Bar F, Choudhary H, Javadi M, Shimizu K, Kunnumakkara A, et al. Synthesis of new selective cytotoxic ricinine analogues against oral squamous cell carcinoma. Natural Product Research. 2021 Jul;**35**(13):2145-2156

[46] Badria F, Elgazar A. Chapter 37—Revealing the molecular mechanism of *Olea europaea* L. in treatment of cataract. In: Olives and Olive Oil in Health and Disease. UK, USA: Academic Press, Elsevier; 2021. pp. 445-456

[47] Badria FA, Elimam DM, Ibrahim AS, Hamdy AF, El-Husseini A, et al. Management of idiopathic nephrotic syndrome: Co-administration of cyclosporine and lyophilized grapefruit juice. Biomed Discoveries: BMDC-102. 2018;**1**(5):104-108. DOI: 10.29011/BMDC-102.100002

[48] Badria FA, Ibrahim AS, Badria AF, Elmarakby AA. Curcumin attenuates iron accumulation and oxidative stress in the liver and spleen of chronic iron-overloaded rats. PLoS One. 2015;**10**(17): e0134156

[49] Badria FA, Suliman SN, Elsbaey M, El-Naggar MH. Bio-guided fractionation and iron chelating activity of agricultural residues. Drug Discoveries and Therapeutics. 2018;**12**(5):299-303

[50] Badria FA, Abdelaziz AE, Hassan AH, Elgazar AA, Mazyed EA. Development of provesicular nanodelivery system of curcumin as a safe and effective antiviral agent: Statistical optimization, in vitro characterization, and antiviral effectiveness. Molecules. 2020;**25**:5668

[51] Paraiso IL, Revel JS, Stevens JF. Potential use of polyphenols in the battle against COVID-19. Current Opinion in Food Science. 2020;**32**:149-155

[52] Hendaus MA. Remdesivir in the treatment of coronavirus disease 2019 (COVID-19): A simplified summary. Journal of Biomolecular Structure and Dynamics. 2020:1-6

[53] Wu C, Liu Y, Yang Y, Zhang P, Zhong W, Wang Y, et al. Analysis of therapeutic targets for SARS-CoV-2 and discovery of potential drugs by computational methods. Acta Pharmaceutica Sinica B. 2020;**10**: 766-788

[54] Manoharan Y, Haridas V, Vasanthakumar KC, Muthu S, Thavoorullah FF, Shetty P. Curcumin: A wonder drug as a preventive measure for COVID19 management. Indian Journal of Clinical Biochemistry. 2020; **35**:373-375

[55] Williamson G, Kerimi A. Testing of natural products in clinical trials targeting the SARS-CoV-2 (Covid-19) viral spike protein-angiotensin converting enzyme-2 (ACE2) interaction. Biochemical Pharmacology. 2020;**178**:114123

Chapter 2

Marijuana, a Journey through the Endocannabinoid System: Unmasking the Paradoxical Effect - Part 1

Ryan Lucas McKinley

Abstract

This two-part section helps the reader to achieve a better understanding of how cannabis works as a viable medication for the endocannabinoid system (ECS) and central nervous system (CNS) in humans by identifying individual synergies between cannabinoids, or cannabinoids and terpenes in their journey through the ECS and CNS in various mammalian patient indicators to unmask this paradoxical effect. The specific biphasic/paradoxical manner in question was researched and inevitably identifies cannabis use that manipulates tryptophan uptake, serotonin release, and dopamine actuation. Therefore, a patient's diet may demand a higher tryptophan and dopa-L supplementation to avoid a paradoxical agitation on the receptor level. This chapter explains the pathology of how cannabis consistently reacts in the ECS for every individual, only separated by metabolism and disruption/trauma in the ECS and CNS, implying that there was no found paradoxical effect existing in cannabis, but in the patient, and thus is perceived the same in every individual, only mediated by metabolism, environment (surroundings), and the exception for individuals who process stimulants and tryptophan and/or serotonin in a disrupted manner causing a perceived paradoxical effect or the build-up to and/or what will be referred to as ASR/ATD. The cannabis industry, growers/breeders, interpeners/cannabis sommeliers/bud tenders, and dispensaries need to continue to constantly strive for more knowledge, just as the researchers and FDA need to continue their work to understand the benefits of cannabis, and most importantly, all must work together to remove cannabis from the Schedule I and Schedule 2 classification.

Keywords: advanced synergistic serotonin release (ASSR/ASR), advanced tryptophan depletion (ATD), Endo-cannabinoid system (ECS), central nervous system (CNS), psychoactive (PA), non psychoactive (N-PA), Cannabigerol (CBG), Tetrahydrocannabinol (THC), Tetrahydrocannabinolic acid (THCA), Tetrahydrocannabivarin (THC-V), Cannabinol (CBN), Cannabichromene (CBC), Cannabielsoin (CBE), Cannabicyclol (CBL), Cannabidiol (CBD), Cannabidiolic acid (CBDA), Cannabidivarin (CBDV), broad leaf marijuana (BLM), broad leaf marijuana dominant (BLMD), medium leaf marijuana (MLM), narrow leaf marijuana dominant (NLMD), narrow leaf marijuana (NLM), cannabinoid receptor type1 (CB1), cannabinoid receptor type 2 (CB2), Vanilloid receptor 1 (TRPV1), transient receptor potential Ankyrin 1(TRPA1), γ-Aminobutyric acid (GABBA-A),

Hydroxy-Tryptamine (5-HT), transient receptor potential cation channel, subfamily V, member 3 (TRPV3), attention deficit hyperactivity disorder (ADHD), γ-Aminobutyric acid (GABAA)

1. Introduction

When visiting a medical marijuana dispensary, it is common to hear, "try and see what works for you." Unfortunately, in today's cannabis industry, some physicians and most bud tenders or "patient care specialist" have to tell patients to go through a trial and error process until they figure out what works best for their indications. This can be very stressful and unfortunate since no single cannabis cultivar strain is the same; implying there is no consistent structure or knowledge or actual prescription while being treated with cannabis.

What numerous studies in this paper show are that specific indications (i.e., physical, mental, neurological disruptions that slowly degrade the quality of everyday life) seem to respond best to specific entourages of cannabinoids and terpenes within a sub-specie ballpark grouped cultivar similarity. Such Cannabis sub-specie groups will be described in the mannerism of an Interpener (Cannabis Sommelier) to guarantee fundamental accuracy of sub-specie variation, chemotype, phenotype, and genotype, which was modified by Trichome Institute based off the study of Clark and Merlin (Evolution and Ethnobotany).

There is a common misconception about what constitutes a Cannabis Indica strain and its sub-specie variations. To understand the mechanisms associated with cannabis, it is important to separate the whole to understand how to consume properly for any specific indication.

A substantial amount of named genetics from growers and their companies are unfortunately carried out through the whole seed-to-sale process claiming the term "Indica" when in actuality is most likely an Indica leaning Hybrid/BLMD. Reasons for Sativa not being a part of this paradox is that sativa is known to excite, and haze has been known as the couch lock of sativa (most likely due to specific terpene profiles). There has never been an identification for Indicas that cause stimulation (what is now known as the terpene profile and chemovar sub-specie to denote cannabis's therapeutic effects). This could be claimed as "stoner myth" since it may have been considered unfavorable cannabis that made people paranoid or anxious, hence another reason to look into the paradoxical effect.

This simple misconception causes improper strain speciation leaving a patient to improperly consume. Ultimately, cannabinoids, terpenes, and other minor phytochemicals are what dictate how cannabis will react in the human body. Ignoring that and only judging by genetic names or suggested sub-specie can result in unintentional wrong profile. This is obviously unacceptable for any terminal patient as much as it is unacceptable for patients with indications such as panic attacks, neurodegenerative disease, or those on the spectrum.

Individuals who have the propensity to experience the "paradoxical effect" where the patient experiences agitation from an implied "sedative and/or stimulant" may also need to consider how an individual metabolizes said entourage from any cultivar administered medically or recreationally; different cannabinoids and terpenes metabolically break down at varying rates within the body.

I hope this paper will provide the information that will pave a new road for patient care. Additional research is underway to identify those patients with the propensity toward a paradoxical effect or ASR/ATD from stimulants or sedatives depending on neurological and physiological disabilities that are tied to the brain and disrupt the regulatory process it takes for homeostasis in any human.

2. Defining a paradoxical effect

A paradoxical effect is an effect of a chemical substance, usually a medical drug or horticultural consumable that has the propensity to react opposite to the effect that would normally be expected. To understand why paradoxical effects happen for some and not for most calls for some examples to further understand this enigma. Specifically the paradox in question seems to act in a biphasic manner (having two phases), i.e., normal function to overabundance or a lack there of.

2.1 Benzodiazepine

A sparse example is benzodiazepine, intended to mildly sedate, wherein rare cases can cause excessive talkativeness, excitement, and increased movement. Benzodiazepine forms a pharmacological effect by actuating the γ-aminobutyric acid (GABA) receptor, this effect causes an elevated chloride channel opening with increased GABA-mediated inhibition giving the perception of sedation, anti-anxiety, and reports of amnesia [1]. In a 2004 study by Mancuso [2], it was reported that a very small percentage of patients experienced a paradoxical or biphasic reaction including acute excitement and hostility.

This could then imply any previous and/or present damage to reuptake pathology of serotonin and sedatives, systemic or invoked, could stop sedatives altogether from working via trauma, prolong depression, and/or abuse or natural tolerance of sedatives. Sedative tolerance may be due to poor uptake and reuptake including the nurtured abuse of dopamine and/or serotonin actuation; an abuse of the drug, exhausting the serotoninergic pool; i.e., a situation where there is not enough tryptophan in a diet to invoke the positive effects of medication.

2.2 Methylphenidate

A more common example would be the pronounced mediation between stimulating and sedating perceptions of psychostimulants in people who are prescribed Attention Deficit Hyperactivity Disorder (ADHD) medications. ADHD medications such as "methylphenidate" i.e., "Ritalin" are by nature stimulants and inhibit reuptake and stimulant release of dopamine in the Central Nervous System (CNS), thus giving increased temporal and spatial presence of dopamine at postsynaptic receptors [3–6]. The intent of Ritalin is to calm and focus patients in attempt to correct or alleviate cognitive dysfunction, whereas a non-ADHD person will simply experience Ritalin as a stimulant.

Use of the Spontaneously Hypertensive Rat (SHR) is widely accepted in the hypothesis of dysregulation and dopaminergic neurotransmission in line to the behavioral alterations in both ADHD patients and SHR [3]. Past reports have shown an imbalance in the pathophysiology of ADHD and SHR displaying altered functional adenosinergic neurotransmission and affinity of agonists to brain adenosine receptors [7–9]. Thus, adenosine, a neuromodulator in the CNS via cell surface receptors, may display a paradoxical effect at adenosine locations or disrupted locations. Adenosine was more recognized for the ability of caffeine as an A1 and A2 receptor antagonist [3, 9, 10]. Extensive evidence to date states that ADHD patients have formidable disadvantages with dopamine uptake, storage, and/or metabolism, [11–14]. In addition, that most, if not all, adenosine receptors are a prime target for treatment of diverse disorders in relation with a dysregulation and dopamine neural transmission that occurs in PD, schizophrenia, and ADHD [3, 15].

The paradoxical/biphasic effect displayed in the paragraph above shows a close relation between adenosinergic and dopaminergic transmission in which A1A, A2A are modulated by dopaminergic processes and down/upregulate glutamate. Where the medication is it meant to overwhelm and suppress; instead in a non-ADHD stays open and causes anxiety or stimulant experience.

2.3 Coffee

A closer look into coffee, a common household stimulant and an adenosine A2A receptor antagonist. Caffeine specifically is responsible for "antagonizing all types of adenosine receptors (ARs): A1, A2A, A3, A2B and, adenosine exerting effects on neurons and glial cells of all brain areas" [16]. In coffee, the natural psychoactive stimulant, caffeine, is well known for causing uplift and energy in the general population. This type of stimulant at first look would deter most in the effort for a calm and relaxed state due to being in an extended hyperactive state **Figure 1**.

However, past study results show that adenosine receptor antagonist, i.e., caffeine, might represent a very important therapeutic role for the treatment of ADHD [18–21] but would not be a substantial replacement for the current medications. This would then imply coffee moreover, caffeine, an A2A adenosine receptor antagonist has the propensity in humans and SHR to act in a biphasic manner much like Methylphenidate, in the efforts of either stimulating or sedating a specific patient indication.

2.4 Cocaine hydrochloride

One not so common household stimulant is cocaine hydrochloride. Technically, a cocaine alkaloid collection is also an A2A receptor antagonist. As an A2A antagonist, cocaine is known for its increased alertness, elevated body temperature, euphoria, excessive talkativeness, restlessness, irritability, pupil dilation, and decreased appetite [22].

A closer look in a more recent study will show that D2/A2A in its activation of cholinergic interneurons influences the excitatory synaptic transmission MSNs of direct and indirect pathways via a retrograde release of endocannabinoids, which in turn interacts with striatal glutamatergic (GABA) and dopaminergic transmission (Dopamine) [23, 24]. This implies that A2A antagonists affect retrograde cannabinoid release in the ECS allowing tryptophan use and serotonin release along with dopamine transmission, showing similar receptor affinity and excitatory properties much like methylphenidate and caffeine.

2.5 Cannabis

Another well-known plant across the globe is cannabis. The use of cannabis is known throughout history and in one study dates medicinal use back to 4000 BC via Carbon-14 dating [25]. More archeological research could be done to discover if cannabis or other mind-expanding drugs (with respective similarity) were implemented at even earlier dates. This could then give even stronger precedents for cannabis as evidence for a viable medication and/or diet additive toward a true non-synthetic homeostasis. Presently cannabis is understood as a medicine to be used in treatment for various indications ranging from cancer to neurodegenerative disorders. Vast studies show cannabis acting on the ECS, which is "comprised of cannabinoid receptors, endogenous cannabinoids (endocannabinoids), and the enzymes responsible for the synthesis and degradation of the endocannabinoids" [26].

Notorious and major parts that make up the female inflorescence of cannabis are cannabinoids, terpenoids, flavonoids, bracts (flower), pistils, styles and

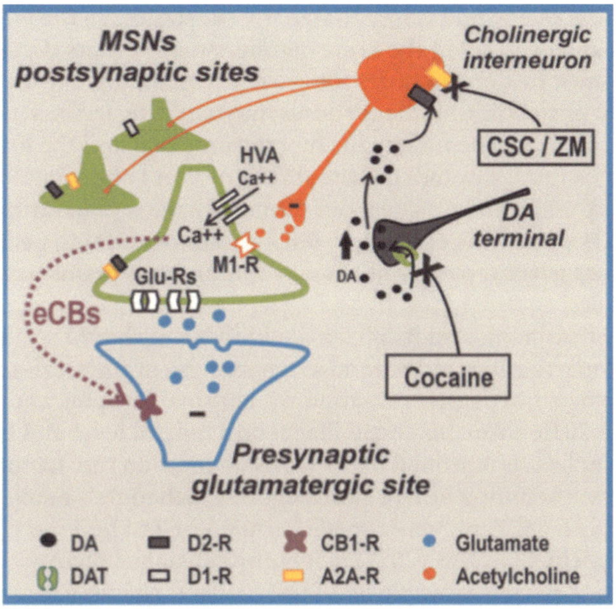

Figure 1.
Safer and Krager [6] Model of intrastriatal network during cocaine and A2A-R antagonists' exposure [17].

stigmas, trichomes, fan, and sugar leaves. Of these parts, their biochemistry is psychoactive, i.e., (the ability to pass through the blood-brain barrier and modulate brain chemistry) and non-psychoactive, synergistic and non-synergistic, and do so through the efficacy of specific synergies between bio-available phytochemicals such as cannabinoids and terpenes; thus, creating an entourage effect transmitting throughout the ECS and CNS that almost works in a harmonic and chaotic matrix of possible synapses.

New discoveries are constantly unfolding about this herbal Rubik's Cube as cannabis becomes accepted into society medically or recreationally. The paradoxical effect in question that cannabis is suspected to give has not gone through any pathological nor clinical study to present date, but has been said to have effects much like coffee commonly existing in the cultivar ranges of BLMD-NLM, where caffeine may stimulate or sedate some but still has been solely based out of hearsay and grapevine knowledge or "Stoner Mythology." Given the far-reaching medicinal properties of cannabis, one would be fair to assume that any medication acting on ECS and CNS would have the affinity to act like cannabis with the respective nature of cannabis and its ongoing discoveries. Therefore, receptors are going to respond no differently if cannabis, cocaine, or coffee triggers the receptor. Importantly, what separates cocaine and cannabis is what makes cannabis unique.

A published doctor of osteopathic medicine, Joseph Cohen of Holo Health explains that "what allows cannabis to be separated from most pharmaceuticals, especially opioids, is due to the natural scarcity of endocannabinoid in the brain stem, avoiding cardiac and respiratory centers entirely. Whereas narcotic analgesics (opioids), or any medicine affecting opioid receptors have a chance to manipulate dopamine and opioid receptors in the brain stem with a fair risk of overdosing." Thus, cannabis only affects a specific area of the body leaving alone parts that are crucial in the sustainability of continuing life and is probably one of the safest means of medication for the human mind and body than any pharmaceutical on the market when paired with a healthy diet and wholesome mindful coexistence or simply a perspective of livity.

Briefly, all the above would imply that many substances in the world can plausibly be medicine and even act in the same mannerisms as others. Undoubtedly more scientific study must be done in order for those things acting like medications, like pharmaceuticals, or possible medicine occurring naturally in the environment to be a safe viable and fundamental means for consumption and the longevity of the consumer. Furthermore, through the thousands of years of evolution engaging with intoxicants, many other herbs aside from cannabis have therapeutic involvement with the human body such as clary sage, mushrooms containing psilocybin, as to say that it may be a natural part of life and evolution for humans having the sentiments that they do.

In the study of cannabis and its pharmacokinetics, it should not be limited to just chemical properties but understood as a paradigm of physical-anthropology, nutrition, neurology, horticulture, taxonomy, chromatography, and herbalism. Unfortunately in 2019, cannabis is still illegal on a federal level and is described as a Schedule 1 drug, i.e., determined by the abuse/addiction rate factor of the drug, which results in a scheduling of five subcategories; Schedule 1 being the most restrictive; 21 U.S.C. §802, prevents any Schedule 1 or 2 to be a medication or used for clinical study. 21 USC § 813 (2011) states any substance pharmacologically substantially similar (a proper example would be Sativex, Nabilone, or any synthetic acting like cannabis in the United States) to a Schedule 1 or Schedule 2 substance will be carried out under the same extent of the law as said above in 21 U.S.C. §8029 (Pub. L. 91–513, title II, §203, as added Pub. L. 99–570, title I, §1202, Oct. 27, 1986, 100 Stat. 3207–13; amended Pub. L. 100–690, title VI, §6470), Nov. 18, 1988, 102 Stat. 4378).

3. Short history subsection

So starts Marijuana's long process down a road of a predetermined discrimination without proper fundamental scientific representation to mandate it as an illicit drug starting as early as the 1900s in America.

This repetitive historical adolescent or fearful behavior can date back to the early 1500s in Mexico during the Spanish occupation when Christianity was introduced, and hemp was promoted over the indigenous crop of cannabis. Together these factors inhibited native people from cultivating their own spiritual plants that were used for ceremonial purposes fighting their own prohibition centuries ago (Santiago Guerra). Mexico's reestablishment in 1810 “Rosa Maria” or “Mariguana” was added to the 1846 Mexican Pharmacopeia from the Mexican Medicinal Academy for medicinal purposes. For the next 172 years, Mexico will go through various political agendas, upper and lower-class segregation, and racism along with its own battle for the legalization of marijuana. By October 31, 2018, the Supreme Court of Mexico declared prohibiting its use was unconstitutional, therefore deeming cannabis as a recreational, legal medicine within the confines that constitute the law making it legal.

Circa 1910, the word marijuana begins to spread across the America via returning US soldiers and legal immigrant Mexicans fleeing the Mexican Revolution who also brought the term marijuana back with them. By 1930, prohibitionists and a handful of people in power such as FBN's (Federal Bureau of Narcotics which would eventually become the DEA) very own narcotics commissioner Harry Jacob Anslinger, who was unfortunately and constantly self-submerged into trying to put an end to the relentless violent and gruesome human behaviors of the international trafficking or smuggling of booze and later more known for taking on the narcotics circuit. Harry would later draft the 1937 Marijuana Tax Act. Harrys' mindset and perspective at the time can be understood from his article called “Marijuana,

Assassin of Youth" about a "marijuana addict" who was hung for a criminal assault of a 10-year-old girl, as Harry explains, "Those who first spread its use were musicians. They brought the habit northward with the surge of "hot" music demanding players of exceptional ability, especially in improvisation. Along the Mexican border and in southern seaport cities it had long been known that the drug has a strangely exhilarating effect upon the musical sensibilities. The musician who uses it finds that the musical beat seemingly comes to him quite slowly, thus allowing him to interpolate improvised notes with comparative ease. He does not realize that he is tapping the keys with a furious speed impossible for one in a normal state" [27].

With this manipulated perspective among the many others, this fear aimed to sway the masses, demonizing the term "marijuana" and anything or anyone that could be associated with cannabis (two prevalent examples would be Mexican workers or jazz musicians of the time like Billie Holiday, which Anslinger personally went after) as a pro-racism (i.e., separation of unity of the people) scare tactic to manipulate the masses for political purposes. Scare tactics such as movies like Reefer Madness, countless publications, and government reporters would continue to justify without representation and slander this medication for the next 110 years regardless of Marijuana's appearance in the 1851 US Pharmacopeia [28].

Contradictory to the laws and discrimination explained above, the US government started what was called the Investigational New Drug (IND) program. This program's itinerary seized marijuana all across America via the Drug Enforcement Agency (DEA) and housed marijuana cigarettes and plants at a highly secure facility called the Coy W. Waller Laboratory Complex to have approved researchers of the FDA conduct studies. The IND government-run program also consisted of subsidizing a large marijuana grow called the "Medicinal Plant Garden" located at the University of Mississippi, Oxford, since the late 1970s wherein was responsible for running test on genetics, bioavailability, and THC extraction from the harvested plants. The "Medicinal Plant Garden" in 2007 would produce 880 pounds worth of marijuana for the National Institute on Drug Abuse (NIDA). This facility would also send their research and the facilities grown marijuana to the Research Triangle Institute in North Carolina for the "Compassionate Investigational New Drug Program." A 1976 federal case involving a glaucoma patient by the name of Robert Randall, who was found not guilty on the charge of growing marijuana at home for the treatment of glaucoma. As a result, the federal government cooperatively allowed Mr. Robert Randall marijuana under FDA regulation creating the Compassionate Investigational New Drug Program.

By 1992, the IND program, the 35 patients and all its constituents were shut down due to the high frequency of new applicants and consequently by the W. H. Bush Administration. In 2018, under the Trump administration, attorney General Jeff Sessions issued a memo that effectively overturned the Cole memorandums guidance allowing prosecutors to include the law enforcement propriety set by the attorney general along with other relevant considerations when privatizing federal cannabis law enforcement. This allows federal law to out rank state policy and US federal government may now prosecute businesses and individuals for legal cannabis State-related activities under federal law at any time. The same Jeff Sessions was quoted in 2016 saying, "Good people don't smoke marijuana." Fear, ignorance, and the adolescent state of mind it creates can be guided in any direction and have been demonstrated at numerous points in this subsection to have a negative effect on mental health, society, and individual rights. The American governments' carelessness with the frailty of life can be referenced back to the CIA's MK-ULTRA program, which ran from 1953 to 1964 consisting of extremely unethical drug testing and LSD experiments; from mentally impaired boys at a state school, to American soldiers, to "sexual psychopaths" at a state hospital, MK-Ultra's programs often preyed on the most vulnerable members of society. The CIA considered prisoners especially good subjects, as they were willing to give consent in exchange for extra recreation time or commuted sentences [26].

To conclude this section, in the case of new developing medicines, the FDA should be the main deciding agency and the medicine be researched by all and not regulated by any law enforcement agency. Law enforcement should deal with illegal trafficking, as it does with opioids, methamphetamines, cocaine.

In the end, the study of cannabis and its medicinal use is critical. To limit this apple of Eden or any medicine in this garden is to hinder ones right for healing, knowledge, choice, and the choice of medical care on this "pale blue dot" [29].

4. Studied vertebrate

4.1 Inside the ECS

This section will go over pathologies in the ECS and CNS that cannabis, in study, has been proven to manipulate including the directives of cannabinoids, endogenous cannabinoids, and TRPV channels [1]. The ECS is a far-reaching neuromodulatory system having strong presence and significant roles in the CNS. The ECS consists of cannabinoid receptors, lipid-based retrograde neurotransmitters (endocannabinoids) heavily existing in the CNS including specific enzymes responsible for the synthesis and degradation of endocannabinoids.

4.2 2-AG and anandamide

2-Arachidonoyl glycerol (2-AG) and arachidonoyl ethanolamide (anandamide) are the best-studied endogenous cannabinoids and are synthesized and degraded by distinct pathways. 2-AG is an agonist for either CB1 or CB2 receptors. Interestingly, anandamide is a low-strength agonist at CB1 receptors and very low agonist at CB2 receptors [30–32]. "Implying systems with low receptor expression or when receptors couple weakly to signaling pathways anandamide can antagonize the effects of more efficacious agonists in efforts to maintain a directed homeostasis [33]." CB1 and CB2 receptors are primarily mediated by endocannabinoids, along with Transient Receptor Potential (TRP) channels. Primarily Anandamide degradation in the CNS is by the enzyme fatty acid amino hydrolase (FAAH) [34]. As its name suggests, FAAH degrades multiple fatty acid amides, including palmitoyl and ethanolamide. This has important experimental and therapeutic implications as inhibition of FAAH increases levels of these ethanolamides, which have widespread actions independent of cannabinoid receptors. It is important to note 2-AG and CB1 have interactions with serotonin via 5HT2C with a crucial participation from neuropeptide Y1 receptor (NPY1R) as explained in the article "Effect of cannabinoid-serotonin interactions in the regulation of neuropeptide Y1 receptors expression in rats: the role of CB1 and 5-HT2C receptor." Common precursors to the neurotransmitter serotonin, the hormone melatonin, and vitamin B3 are TRP channels specifically, TRPV1, that are activated by anandamide under certain conditions [35]. Anandamide also activates PPAR-alpha(responsible for cell division, cell growth, and cell death throughout life), a major overseer of lipid metabolism in the liver [32, 33]. PPAR-alpha goes active under energy deprivation and is necessary for the breakdown of fatty acids, which is a major adaptive response to prolonged fasting [36]. Moreover, increasing anandamide by decreasing its degradation by inhibition of FAAH also increases levels of other N-acylamides, in turn modulating

PPARα [37, 38]. To explain, anandamide has practical roles in modulating and regulating pain, depression, appetite, memory, and fertility.

Importantly 2-AG biology, as an endogenous ligand for cannabinoid receptors like CB1 and CB2 in the brain, liver, and lung, and a major source of arachidonic acid, is used for prostaglandin synthesis [39]. Since 2-AG is an intermediate metabolizer in fatty acid synthesis [39], any manipulation of 2-AG production and degradation will undoubtedly have vast reaching effects that can even be independent of the ECS but interestingly avoiding the gut, heart, kidney, and spleen. A sound representative case is that "the measurement of bulk tissue levels of 2-AG is an indirect measure of 'synaptically-active' or 'interstitial' 2-AG, which is most relevant for cannabinoid receptor signaling and might be more accurately measured by microdialysis [40]."

Furthermore, sourced on knockout mice data, DAGLα, a prime enzyme responsible for 2-AG synthesis in the postsynaptic neuron in response to increased synaptic activity [41–43], appears to be the isoform responsible for most 2-AG production that contributes to synaptic plasticity in the adult CNS [32]. Many studies show 2-AG, Anandamide, CB1, CB2, and TRP channels naturally affecting serotonin either directly or indirectly. Whether they have a part in the paradoxical effect has yet to be analyzed.

4.3 CB1 and CB2

CB receptors have existed long before cannabis evolved circa 25 million years ago, beginning in organisms such as sea squirts and fugu fish 600 million years ago [44–46], but have evaded non-chordate invertebrates, i.e., insects, hydra, nematodes, fungi, and plants. CB1 and CB2 receptors are G-protein-coupled receptors (GPCRs) and their activation obstructs the catalyzing chemical reaction cyclases, voltage-dependent calcium channels, activates several amino acids specific to the amino acids serine and threonine kinases, inwardly rectifying potassium channels, with some variation depending on cell type [45]. Thus, activation of CB1 or CB2 receptors exerts diverse consequences on cellular physiology, including synaptic function, gene transcription, cell motility, etc. [41]. CB1 receptors are exceptionally abundant in the cortex, basal ganglia, hippocampus, and cerebellum [47].

The majority of CB1 receptors are on nerve fibers, specifically axon terminals and pre-terminal axon segments, while avoiding the operational zones. Cortical and hippocampal CB1 receptor expression is particularly high on the direct pathway axons as they enter the globus pallidus heading toward the substantia nigra [48]. CB1 receptors are also expressed in glutamatergic neurons [49].

CB2 receptors in comparison with CB1 are expressed at much lower levels in the CNS. The CB2 receptor is primarily present in active immune defenses and vascular elements [50–52]. Interestingly, CB2 does appear to be expressed by nerve injury and has the potential to increase expression 100-fold post tissue injury or during inflammation [53]. It remains to be determined whether CB2 expression is increased in the CNS during brain injury. This is due to increased expression of CB2 on cells intrinsic to the CNS or is a result of the migration (e.g., CB2-expressing monocytes) of peripheral immune cells into the CNS.

Given that the paradoxical effect is found in the human and not the medicine, i.e., cannabis, it is important to shine light on the areas in which the ASR or ATD has functions. There are many studies that display the Endocannabinoid system manipulating serotonin/5HT. As understood in "Modulation of the Serotonin System by Endocannabinoid Signaling," serotonin can be actuated by the engagement of stress to constrain further activation of the HPA axis. The HPA axis is a group of closely knit influences and feedback interactions consisting of the Hypothalamic (CRH),

Anterior Pituitary (ACTH), and Adrenal Cortex (CORT) and controls reactions to stress and regulates many body processes, including digestion, the immune system, mood and emotions, sexuality, and energy storage and expenditure. In turn the HPA axis is also under the control of the serotonergic system. Studies have shown that 5-HT through the activation of 5-HT receptors located in the PVN regulates neuroendocrine responses to stress (for review see, [54]). For instance, activation of the 5-HT1A receptors has been shown to reduce the secretion of ACTH (often produced in response to biological stress) and corticosterone (affecting carbohydrate, potassium, and sodium metabolism, i.e., glucocorticoid) induced by an array of stressors. The general consensus is that the serotonergic system contributes in the ECS-induced modulation of the HPA axis and stress responses. Researched receptors of the CNS and ECS that take part in Serotonin Modulation relevant to cannabis psychopharmacological effects are 5-HT1A, reducing stress (passive coping), and 5-HT2A, attuning actively or pro-actively through Default-mode-network/stress (active coping) as illuminated by RL Carhart-Harris & DJ Nutt in "Serotonin and brain function: a tale of two receptors" [14, 17, 55].

5. Finding the paradox

To find, understand, or even combat the paradoxical effect, one must first determine why the incorporated cannabis profile produces the experienced reaction. Whether it is the profile itself or the person who is being treated, acknowledge preexisting psychological and/or physiological aspects systemic or invoked, and then figure out how to counteract the symptoms so the goal for homeostasis works in the way it is intended.

In this section, some of the most common indications that respond to cannabis will be discussed. In addition, the clinical research studies addressing the impact of various cannabinoids on pain have been conducted throughout the world including Bangladesh, Canada, Columbia, Finland, Germany, Italy, North Korea, Poland, Portugal, Spain, Sweden, the United Kingdom, Uruguay, and in much of the United States. The traditional approach to pain management has led to a significant increase in opioid abuse and addiction. More recent studies have focused on the use of marijuana and resulted in decreasing the use of opioids for pain, reducing the withdraw symptoms from opioid use, and increasing the quality of life for patients [56]. Many patient surveys have been conducted in the States that allow marijuana. These surveys clearly indicate that patients reduced their use of opioids; in a New England survey, the respondents reported using less opioids (a 75% reduction) as well as reducing other medications used to treat anxiety, migraines, and sleep disorders after starting medical cannabis [57].

6. Anxiety and post-traumatic stress disorder (PSTD)

Anxiety can be described as an inner emotion that can create a state of unease, usually correlated with future events. Biologically, anxiety is a response to a perceived danger or threat (i.e., hypervigilance) in the future using past key memories as validation (i.e., learned trauma) as opposed to an immediate threat (i.e., fear). Anxiety has many disorders that manifest in different forms such as Generalized Anxiety Disorder, Panic Attack Disorder, COPD, and asthma [58–61].

PTSD can be understood as a form of anxiety but its onset from specific traumatic events, which then the patient eventually experiences the constant state of the same symptoms. Like those mentioned above for anxiety with obvious commonalities such as panic attack and generalized anxiety disorder depending on the nature of the trauma and the psychological makeup of the patient. PTSD patients will display higher affinity of CB1 receptors but lower peripheral concentrations of anandamide or N -arachidonoylethanolamine (AEA), the endogenous ligand of CB1 [60–62].

In a study using THC and Cognitive Brain Therapy, it was demonstrated that THC prevented the recovery of learned fear. This was a randomized double-blind placebo-controlled study [63]. With the guidance of a psychiatrist/therapist, tetrahydrocannabinol moreover, cannabis could be used as a viable therapy additive for cognitive brain therapy, PTSD, and other psychological and sociological disadvantages. In an anxiety study using Nabilone, a synthetic THC, patients showed a dramatic improvement when compared with placebo. Side effects reported were dry mouth, dry eyes, and drowsiness. Patients did not report any psychotropic effects of Nabilone since it was synthesized to act like the non-intoxicating cannabinoid of cannabis [64]. Refer to Terpenes and Flavanoids chapter, Pinene section, and Unraveling Cannabis for potential paradox.

7. Multiple sclerosis/Parkinson's disease

MS is a disease where the myelin, a protectant surrounding nerves, is attacked by the immune system. This is a progressive disease and many patients have trouble walking, muscle weakness and spasms, pain, depression, problems focusing or remembering. Parkinson's disease (PD) is a neurodegenerative disease wherein the substantia nigra (i.e., a basal ganglia structure located in the midbrain) begins to deteriorate due to dopamine deficiency. Post synaptic results of this disease involve the extrapyramidal system (i.e., denoting parts of the nervous system dealing with motor function) wherein the central nervous system that mainly affects the motor system begins to cause stiffness, bradykinesia, resting tremor, speech in pediment, and postural instability. Symptoms will not appear until approximately 50% of the nigral dopamine (DA) neurons are lost in the substantia nigra and striatal dopamine deficiency.

Numerous studies have shown that Sativex, an oromucosal spray of cannabis-based medicinal extract (CBME), significantly reduced spasms and pain [65, 66] showing great promise for cannabis as a medication for calming and protecting the auto immune system even in damaged systems.

In a 2016 Survey conducted through the Michael J Fox Foundation and National Multiple Sclerosis Society, both PD and MS volunteers of both cannabis users and non-cannabis users participated. A total 85% reported cannabis effectiveness as moderate or above in relieving their symptoms. In this study MS participants found more relief than PD patients. Additional findings showed that people that suffer from MS reduced the use of prescription medications since beginning cannabis use. Both MS and PD participants that medicated with cannabis reported lower levels of their disability, mostly in regions of memory, mood, and fatigue [67].

Once again, the paradoxical effect is found in the human and not the medicine, i.e., cannabis, it is important to shine light on the areas in which the ASR or ATD has functions. The disruption from PD (dopamine deficiency) and MS (nerve protectants attacked by the immune system) in the CNS could be considered as focal points to the cascading effects where the paradoxical effect may have a hand in. As to say, it would be a progressive move to avoid extreme dopamine actuation in PD and suppress over responsive immune responses for MS that attack the nervous system through the blood-brain barrier (BBB). Though cannabis shows promising therapeutic responses via CBD < study>, those therapeutic responses depend on resources. This then opens the field to dietary supplementation for the symptom and the medicine. Once again, to shine a light on phytochemical entourages that could lead to a paradoxical effect, which are: Limonene being an antagonist via A2A actuating dopamine via D2 and serotonin agonist via 5HT1A as explained in "The Paradoxical Location." To counter the paradox in a lack there of, memory mood and fight fatigue, a high tryptophan diet, DOPA-L supplimentation, and proper cannabis dosing all play a part in supplementing homeostasis.

Tryptophan and DOPA-L supplementation work in replenishing and regulating serotonin and dopamine as an additive to any ongoing pharmaceutical regiment and both have a higher chance of efficacy with patients prescribed cannabis and in theory could combat the likelihood of a paradoxical effect happening at the Serotoninergic/HPA axis level and dopaminergic transmission. As dopamine is produced in the body, Tryptophan is a precursor to the neurotransmitter serotonin, a non-polar aromatic amino acid and is something humans cannot biologically make it and must get the essential tryptophan via diet.

8. Asperger's syndrome

Asperger's syndrome is classified as a subtype of the autism spectrum disorder that encompasses a spectrum of psychological conditions that are characterized by abnormalities in social interaction and communication that provide the individuals functioning and by restricted and repetitive interests and behavior as defined by World Health Organization (WHO). In 2015, an estimated 37.2 million people around the world suffer from this entourage of a disorder [68, 69].

The syndrome is lifelong and usually begins around the second year from birth and the effectiveness of interventions is supported by only limited data. Most treatments are geared toward improving communication skills, unhealthy and life hindering OCD or repetitive routines, and physical coordination. The methods that have proven they are worth include cognitive behavioral therapy (CBT), physical therapy, speech therapy, parental training, and medications for the associated problems, i.e., mood and/or anxiety. Medication for Asperger's includes but is not limited to Catapres Lamictal, Guanfacine, Oxcarbazepine, Zoloft, Buspar, CeleXA, Prozac, Lexapro, Klonopin, Strattera, Risperdal, Ritalin, Paxil from a 2019 national survey for psychiatric and seizure medications. Most, if not all, of these medications mentioned treat conditions ranging from anti-seizure, SSRI anti-psychotic, anti-seizure, and stimulants and are riddled with common and frequent side effects such as fatigue/drowsiness, depression, aggression, appetite loss, sleep problems, and general worsening [46, 70, 71].

When biochemically and neuropathically compared to cannabis (i.e., a specific cultivar that has been cultivated for specific cannabinoids, terpenoids, and flavonoids in the efforts for a higher chance at treating specific symptoms) for the purpose of alleviating Asperger's syndrome, the low side effects that can be avoided in cannabis and the vast medicinal properties of cannabis are unmatched and should be considered a dietary additive in some medical regimens. Since cannabis has infinite possible genetic outcomes, cloning and hybridizing and marijuana extraction methods would be the best means to find and maintain a specific cultivar/chemovar for any one person and their symptoms on the spectrum per harvest [70].

9. The entourage effect

The Entourage effect can be explained as a specific group of cannabinoids, terpenes, and flavonoids that have the ability to synergistically create specific effects on the endocannabinoid system. Some of these effects can magnify/desensitize the nervous system, force more CB1 and CB2 receptors on and/or off, reduce unwanted effects while amplifying wanted effects/vice versa. Through reviewing numerous studies, I have come to find that in every sub-specie of cannabis lies a unique terpene bouquet and a general entourage of cannabinoids specific to the cultivar and genetics).

Terpenoids and cannabinoids are present throughout the plant's flowering stage. Terpenes can also have the potential to determine what the most likely outcome of the plant's impact on the ECS and CNS will be. Another way to describe it is a gestalt, the whole plant being larger than the sum of its parts.

Not all terpenes contribute to the entourage affect [72]. As for as the ones that do, and that will be gone over, they do exist in cannabis. Phytocannabinoid-terpenoid synergy, if proven, increases the likelihood that an extensive pipeline of new therapeutic products is possible from this venerable plant [73]. In a more recent study, "Terpenoids From Cannabis Do Not Mediate an Entourage Effect by Acting at Cannabinoid Receptors," it is thoroughly explained that terpenes have modes of operation elsewhere outside of CB one and CB two signaling via 5HTs, A2A, TRP GPR, and many more [74].

The inter-entourage effect suggests that enhanced biological activity may be attributed to secondary metabolites—mainly terpenes—produced by cannabis strains. Terpenes are known for their medicinal properties including anti-inflammatory and anticancer activities [75–77], but here, in the general gist of the inter-entourage effect, they are considered as promoters and instigators of therapeutic phytocannabinoid activity.

Moreover, mixing co-related terpenes and phytocannabinoids (i.e., THCA. related terpenes with THC or CBDA related terpenes with CBD) at ratios close to the natural plants showed the strongest effect. This could then encourage research studies to look into multiple cultivars for treatment of an indicator.

10. The paradox

Once again to understand the paradoxical effect whether it is the cannabis itself or the person who is being treated, a closer look to what makes marijuana's entourage will display pathology and functionality in the ECS and CNS in the efforts of reaching homeostasis.

10.1 Unraveling cannibis (cannabinoids sub)

This section will educate the reader on the various bioavailable cannabinoids that reside in the five sub-species of cannabis, excluding Hemp/Sativa/NLH, i.e., industrial hemp. There are over 150 identified cannabinoids in in legal Medical Cannabis from past to recent study.

CBD, CBC, CBG, CBDA, CBD-V, CBN, THC-V, THC-A, THC, CBL, which are the most bioavailable cannabinoids that have beneficial health impacts ranging from anti-inflammatory, pain relief, anti-anxiety, neuroprotectant, anti-spasmodic, anti-cancer/tumor, analgesic that all have a place in the mammalian ECS. Henceforth each cannabinoid will be evaluated for its affinity to a possible biphasic ASR/ATD.

10.2 Cannabidiol

10.2.1 (CBD)N-PA

CBD, being one of the main cannabinoids in cannabis, possesses no intoxicating effects and works frequently with the CB-2 receptor, which interacts directly with the immune system via 5-HTA1 and combats inflammatory diseases [65, 73]. Other areas affected are not limited to Gastrointestinal via transient receptor potential (TRP) channels, specifically the TRP cation channel, subfamily V, member (TRPV3) treating IG inflammation. CBD can also act like an antagonist, blocking THC from binding to the CB2 receptor. This affect has the tendency to also reduce the anxiety associated with THC [45, 78]. This binding shows promise by lowering the rate of psychotic episodes of those individuals by using cannabis with higher levels of CBD [79]. A 1:1 ratio of CBD to THC and their respective constitutes would suffice depending on the metabolism of a patient.

Does CBD contribute to the paradoxical effect? Yes and no.

No. It helps alleviate side effects of the indication, therefore canceling out a possible ASR/ATD.

Yes. If serotonin levels are below a healthy level, it would be fair to assume nothing will most likely be felt since CBD is 5Htp-dependent.

Moreover, if THC is more abundant causing more serotonin depletion while tryptophan is already low in the body or below a healthy level, the use of CBD may be futile in the efforts of analgesia but may only have the ability in this state to counteract side effects of THC via CB1 and CB2 binding unless receptors have been exhausted.

Conclusion: More study must be done to understand tryptophan depletion in the body, metabolism of cannabinoids, and the medication needed to help it.

10.3 Cannabichromene

10.3.1 (CBC)N-PA

CBC is an abundant non-intoxicating cannabinoid due to a recessive gene [80] that modulates the vanilloid type-1 (TRPV1) and ankyrin type-1 (TRPA1) receptors and TRPV2,3,4 [45, 81]. Briefly, TRP channels and the ECS are involved in inflammation and have a role in pain [81, 82]. Modulation of these receptors can cause elevated endocannabinoid levels, thereby amplifying total cannabinoid availability via turning on more docile CB1 and CB2 receptors with more respectable affinity to the CB2 receptor. Health benefits range from anti-inflammatory and pharmacokinetics of other available cannabinoids [83]. CBC also has the ability to potentiate the analgesic effects of THC [45, 84, 85]. In one study CBC shows promise in positively affecting the viability of mammalian adult neural stem cell progenitor cells, i.e., an essential component of brain function in health and disease [82]. This particular cannabinoid could then be what allows a patient's "high" to then proverbially stack or amplify if given more of the same medication. This then opens the door for a possible addition to the paradox in question being an "agonist" via TRPV1 and possibly, marginally mediating CB1, thus amplifying GABA sensory inhibiting or prohibiting 5-HT.

One study indirectly shows CBCs' synergistic affinity with limonene [86], this could then mean if both are present in the ECS; limonene has a valid chance at having a synergistic paradoxical effect via "CBC modulating TRPV1" [81, 87, 88]. In another synergy with TRPV1, the synergistic manner of (the known receptor affinity to the monoterpene Limonene) adenosine A2A receptor modulating TRPV1 as documented by [89] raises a curiosity to the possible multitude of concurrently dependent inceptions of synergies and the affiliated synergies in between. Henceforth, these two synergies between CBC and TRPV1, and A2A and TRPV1 should definitely be researched to further understand the cannabis entourage and its effects in the human body.

Does CBC contribute to the paradoxical effect? May have amplifying properties when combined with THC and limonene and/or linalool.

10.4 Cannabidiolic acid

10.4.1 (CBDA)N-PA

CBDA (Cannabidiolic Acid) transforms into CBD through a process called decarboxylation. Baking, lighting, or heating cannabis removes the acid group from CBDA and transforms it into CBD [90]. Prolonged oxidation via sunlight (infrared/ultraviolet light) can also slowly change CBDA to CBD. The majority of cannabis research has focused on THC or CBD, not CBDA. Though one study shows that CBDA is a Cox-1 and Cox-2 inhibitor; an anti-inflammatory and analgesic, similar to ibuprofen [91]. The study shows CBD and its constituents to be a more effective analgesic than Ibuprofen with nonexistent to minimal side effects. CBDA, in a toxicology study, showed strong dependence on particular sesquiterpenoids, namely guiaolstol, γ-eudesmolstol, trans-α- bergamotenest, γ-elemenest, α-bisabololstol, and α-farnesenest.

Does CBDA contribute to the paradoxical effect? No, unlikely based on the current research, however, nothing is definitive at this time since minimal research is done.

10.5 Cannabinol (CBN)

CBN can be considered a time stamp and uses the same logic as carbon dating. This process is due to a degradation byproduct of various cannabinoids via oxidation. CBN can also cause drowsy-like effects like an analgesic, but at high doses [90], CBN is the cannabinoid that has been used to treat glaucoma; its anti-inflammatory properties reduce intraocular eye pressure (IOP). A reduction of 16–45% of IOP was initially documented in a 1971 study [92]. CBN is considered the natural decomposition byproduct of the three main phytocannabinoids (i.e., CBDA, THCA, and CBCA), the strongest correlation between two phytocannabinoids is between THC and CBN [53].

Does CBN contribute to the paradoxical effect? No evidence so far for ASR or ATD pathological indicator.

10.6 Tetrahydrocannabinol (Δ^9-THC)PA

THC is a Phytocannabinoid chiefly from the Cannabis Indica ssp. cultivar that actuates endogenous signaling in the ECS and CNS, predominantly known for its psychotomimetic effects. Receptors affected in the ECS and CNS are the CB1, CB2, GPR55, GPR18 receptors. From these cannabinoid receptors, extracellular signals trigger intracellular cascades. These cascades can represent behavior from cannabis with therapeutic effects. THC is an agonist of the CB1 receptors (Psychotomimetic effect) and CB2 receptors (Possible immunologic, anti-inflammatory effects) [93]. Numerous studies and current education show THC is also an agonist at GPR18 (most efficacious at), GPR55, and TRP ligands TRPA1,TRPV2, TRPV4 and TRPV3 while being an antagonist at TRPM8 and 5HT-3A known for treating long-term depression (LTD). The agonistic effects range from: GRP55 responsible for neuroimmunological regulation; GPR18 has been associated with numerous physiopathological processes, such as cellular migration, immunomodulation, sperm physiology, cardiac physiology, obesity, intraocular pressure, pain, and cancer, among others.

Does THC contribute to the paradoxical effect?

Yes, only via CB1 Signaling, especially when CBD is absent.

No, when a balanced amount/or more of CBD is present, but is a terpene carrier of limonene through the BBB when vaporized, smoked, taken as a tincture or orally ingested.

No. If THC is lower than CBD allowing the agonist ability of CBD to calm the usage with seritonin i.e., 5-HT, while only needing little for itself (CBD) to operate, then THC will most likely operate within a healthy ratio, aside from the metabolic breakdown within the endocannabinoid system. Also if limonene is not present.

10.7 Tetrahydrocannabivarin (THC-V)PA

THC-V Is an intoxicating cannabinoid mostly found in NLM (narrow leaf marijuana) (stimulating); small traces have been found in BLMD (sedating) strains as well. In a 2013 mouse pilot study, the purpose was to investigate the clinical effect and tolerability of THCV and CBD alone and in combination with patients with Type 2 diabetes [94]. THCV decreased plasma glucose and increased B-cell function (B-cells identify pathogens and produce antibodies). In the conclusion section of the study, it states "based on these data, it can be suggested that THCV may be useful for the treatment of the metabolic syndrome and/or type 2 diabetes, either alone or in combination with existing treatments. Given the reported benefits of another non-THC cannabinoid, CBD in type 1 diabetes, a CBD/THCV combination may be beneficial for different types of diabetes." Later in 2016, a human study was conducted, the same results were reported "compared with placebo, THCV significantly decreased fasting plasma glucose." The study concluded that THCV could be a new therapy for patients with type 2 diabetes [95, 96].

Does THC-V contribute to the paradoxical effect? Yes.

Conclusion: Plausible, if in combination with certain cannabinoids terpenes and depending on the indicator to the ECS.

10.8 Tetrahydrocannabinolic acid (THCA)*N-PA*

THC-A is a highly plentiful non-intoxicating cannabinoid lacking affinity to the CB1 receptor. Lacking affinity to the CB1receptor could imply that vaporization (heating and/or any processes of combustion) would be a non-effective delivery method for THC-A [44, 45]. Like CBD, THC-A can relieve inflammation being a viable neuroprotectant, therefore providing treatment for various neurological diseases such as MS, ALZ, Parkinson's, and has even shown to slow the expansion/multiplication/proliferation of cancer cells [87]. Given the cannabinoid life cycle, developing a stable version may be difficult because by its very nature THCA converts to THC easily. "Studies suggest that THCA may be more stable in herbal cannabis, where it is 'hermetically sealed' within glandular trichomes, along with terpenoids which serve as protective antioxidants. The same studies showed that THCA decarboxylated within minutes at temperatures above 80°C. At room temperature in glass bottles with limited exposure to light, THCA dropped to 80% of initial levels after 25 months. At refrigeration (4°C) temperatures, 94.7% of THCA was still present" [44].

The correlation plot in the study shows that while nerolidol has relative affinity to THCA, the other terpenoids described in their paper range from having no affinity to THCA to having minor affinity to CBDA, see also [7, 97]. This may explain the lack of activity observed when those specific terpenoids were added to THC. According to results, THC activity is enhanced only by its co-related terpenoids, while other terpenoids inhibit its biological activity [44, 45, 53].

Does THCA contribute to the paradoxical effect? No.

Conclusion: non-intoxicating and no synergistic entourage to Paradox.

10.9 Cannabivarin

10.10 (CBD-V)N-PA

CBDV has the affinity to inhibit the biosynthesis of the endocannabinoid 2-arachidonoylglycerol (diacylglycerol lipase/DAGL) [50% inhibitory concentration (IC50) 16.6 μM] and may decrease activity of its product, the endocannabinoid, 2-AG [81, 98]. An experiment on GABA receptors in the production of use-dependent GABA, a current after prolonged exposure to CBDV has shown great efficacy in the efforts as an anticonvulsant, especially for epilepsy, via GABAergic action. Therefore, a solid DAGL regulator to the endo cannabinoid system [99].

Does CBDV contribute to the paradoxical effect? No.

Conclusion: may show great promise to alleviate ATD.

10.10.1 Cannabicyclol (CBL)N-PA

Most cannabinoids are the chemical breakdown of CBG, whereas CBL starts its oxidizing life cycle from CBC. As observed by Shoyama, much larger amounts of CBLA can be harvested early in the vegetative phase and stored as opposed to harvesting in the reproductive phase. This prompted a quick conclusion that CBLA is a natural breakdown via ultraviolet light of CBCA [100–102]. Clearly more study must be done on this world-renowned medicinal herb to understand its true potential.

10.11 Cannabigerol (CBG) cannabigerolic acid (CBGA) N-PA

10.11.1 The mother cannabinoid

CBGA is of the first cannabinoids produced for the cannabis plant and births once geranyl pyrophosphate biosynthesizes with olivetolic acid through a prenyltransferase catalyst conversion, thus creating CBGA. From this momentary precursor begins several different syntheses, i.e., CBG, THC, CBD, and CBC wherein the number of biosynthesized and oxidized (aged) cannabinoids reaches the hundreds and from these will eventually create the cannabinoid side of the entourage effect [103].

Not much study has gone into CBGA but in numerous studies it has been noted as an analgesic and anti-inflammatory. Once CBGA loses its acid group via heat or oxidation, Cannabigerolic acid becomes Cannabigerolic (CBG).

CBG affects cannabinoid, serotoninergic, peroxisome proliferator-activated receptors ((PPARs), i.e., nuclear receptor proteins that act as a transcription factor of the expression of genes regulating cellular differentiation, development, and metabolism, and tumorigenesis; α2-adrenoceptors (norepinephrine (noradrenaline) and epinephrine (adrenaline) signaling); TRP, vanilloid, melastatin, and ankyrin channels. CBG inhibits dopamine norepinephrine, GABA, and serotonin reuptake [103]. Thus, utilizing 5HT as an anti-depressant [104]. So to say, CBG does take part in serotonin release and reuptake especially alongside THC more so in the instance where THC is higher creating a synergistic 5-hydroxytryptamine release/uptake/reuptake. For CBG, potential medicinal uses can range from analgesia/inflammation, feeding disorders, cancer, glaucoma, inflammatory bowel disease (IBD), psoriasis, Neuroinflammation (MS), bone healing, antibacterial; helping with testosterone balance and mood disorders [103]. In the case of mood disorders and the bidirectional influence at CB1, and an anxiogenic at TRPV1, other synergies along this pathology should be acknowledged.

In light of synergies, specifically terpenoid synergies, "CBGA was related to δ-selinenest, cis-α- bisabolenest, and α-famesenest, moreover, mixing co-related terpenoids and phytocannabinoids (i.e., THCA-related terpenoids with THC or CBDA-related terpenoids with CBD) at ratios close to the natural plants showed the strongest effect. This increased activity may be the result of some preferential pathway in which the given terpenoids enhance the absorbance or activity of phytocannabinoids in the cells" [53]. This proves that specific synergies from/between cannabinoids and terpenes can take place at different stages of biosynthesis and/or oxidation either via UV, heat, human influence, or natural degradation but still hold relative synergistic terpene relation. To further understand decarboxalation synergies, Cannabigerolic acid (CBGA) was related to δ-selinenest, cis-α- bisabolenest, and α-famesenest (subscript abbreviations: mt—monoterpene; st—sesquiterpenes; stol—sesquiterpenes; dt—diterpene), Ergo, guaiol and eudesmol derivatives showed strong positive correlation with CBD [53].

Though dependence on different terpenes and cannabinoids may vary from cultivar, it is only due to the bioavailability and genotype. Thus, more study must be done to fully understand synergistic properties between cannabinoids and terpenes at the various growth stages in cannabis.

In medicating with medical cannabis, specifically MLM sub-species, if multiple cannabinoid synergies are co-related with one terpene, i.e., B- Myrcene, a hypothesis could then be made; if multiple synergies to one terpene might cause a faster metabolization depleting that specific synergy resource for synthesis and if paired with D-Limonene, a known A2A antagonist, would then be the remainder, possibly giving a delayed agitation wherein the goal of this type of homeostasis (MLM) (a 1:1 of stimulating and sedating terpenes) balance/homeostasis would be lost toward the middle or end of the bell curve depending on the metabolization mediation of any one patient. So as to understand, CBG has a synergy with limonene, it is fair to assume between the oxidative life cycle and biosynthesis of CBG that is later counterparts may have synergies much like the one between CBG and limonene and that more study should go into this volatile mono terpenoid and its many possible synergistic effects throughout the human body.

Does CBG contribute to the paradoxical effect? Isolated, No.

Acknowledgements

Special thanks to Dr. Tamara A. Schiappa.
Special thanks to Dr. Joseph Hulihan.
special thanks to Brandon Allen.

Supporting data

Free access.

Author details

Ryan Lucas McKinley
Slippery Rock University, Slippery Rock, Pennsylvania, USA

*Address all correspondence to: dj8t6d@icloud.com

References

[1] Fluyau D, Revadigar N, Manobianco BE. Challenges of the pharmacological management of benzodiazepine withdrawal, dependence, and discontinuation. Therapeutic Advances in Psychopharmacology. 2018:147-168. DOI: 10.1177/2045125317753340

[2] Mancuso CE, Tanzi MG, Gabay M. Paradoxical reactions to benzodiazepines: Literature review and treatment options. Pharmacotherapy: The Journal of Human Pharmacology and Drug Therapy. 2004;**24**:1177-1185. DOI: 10.1592/phco.24.13.1177.38089

[3] Prediger RDS, Pamplona FA, Fernandes D, Reinaldo N. Caffeine improves spatial learning deficits in an animal model of attention deficit hyperactivity disorder (ADHD) – the spontaneously hypertensive rat (SHR). Takahashi International Journal of Neuropsychopharmacology. 2005;**8**:583-594. DOI: 10.1017/S1461145705005341

[4] Blaszczak-Boxe Agata. "Marijuana's History: How One Plant Spread Through the World. Available from: https://www.livescience.com/48337. 2014

[5] Krause KH, Dresel SH, Krause J, Kung HF, Tatsch K. Increased striatal dopamine transporter in adult patients with attention deficit hyperactivity disorder: Effects of methylphenidate as measured by single photon emission computed tomography. Neuroscience Letters. 2000;**285**(2):107-110

[6] Safer DJ, Krager JM. A survey of medication treatment for hyperactive/inattentive students. Journal of the American Medical Association. 1988;**260**(15):2256-2258. DOI: 10.1001/jama.1988.03410150104040

[7] Davies LP, Hambley JW, Johnston GAR. Reduced adenosine deaminase activity in the CNS of spontaneously hypertensive rats. Neurochemistry International. 1987;**10**:533-536

[8] Kamikawa Y, Cline WH Jr, Su C. Diminished purinergic modulation of the vascular adrenergic neurotransmission in spontaneously hypertensive rats. European Journal of Pharmacology. 1980;**66**:347-335

[9] Matias A, Zimmer FJ, Lorenzen A, Keil R, Schwabe U. Affinity of central adenosine A1 receptors is decreased in spontaneously hypertensive rats. European Journal of Pharmacology. 1993;**244**:223-230

[10] Snyder SH, Katims JJ, Annau Z, Bruns RF, Daly JW. Adenosine receptors and behavioral actions of methylxanthines. Proceedings of the National Academy of Sciences USA. 1981;**78**:3260-3264

[11] Castellanos FX, Tannock R. Neuroscience of attentiondeficit/hyperactivity disorder: The search for endophenotypes. Nature Reviews Neuroscience. 2002;**3**:617-628

[12] Ernst M, Zametkin AJ, Matochik JA, Jons PH, Cohen RM. DOPA decarboxylase activity in attention deficit hyperactivity disorder adults. A [fluorine-18] fluorodopa positron emission tomographic study. Journal of Neuroscience. 1998;**18**(15):5901-5907. DOI: 10.1523/JNEUROSCI.18-15-05901.1998

[13] Swanson JM, Flodman P, Kennedy J, Spence MA, Moyzis R, Schuck S, et al. Dopamine genes and ADHD. Neuroscience and Biobehavioral Reviews. 2000;**24**(1):21-25

[14] Tozzi A, de Iure A, Marsili V, Romano R, Tantucci M, Filippo M, et al. A2A adenosine receptor antagonism enhances synaptic and motor effects of

cocaine via CB1 cannabinoid receptor activation. PLoS One. 2012;7:e38312. DOI: 10.1371/journal.pone.0038312

[15] Ribeiro JA, OAM S, de Mendonc AA. Adenosine receptors in the nervous system: Pathophysiological implications. Progress in Neurobiology. 2003;**68**:377-392

[16] Cornelis MC, El-Sohemy A, Campos H. Genetic polymorphism of the adenosine A2A receptor is associated with habitual caffeine consumption 13. The American Journal of Clinical Nutrition. 2007;**86**:240-244

[17] Carhart-Harris RL, Nutt DJ. Serotonin and brain function: A tale of two receptors. Journal of Psychopharmacology. 2017;**31**(9):1091-1120. DOI: 10.1177/0269881117725915

[18] Dalby JT. Will population decreases in caffeine consumption unveil attention deficit disorders in adults? Medical Hypotheses. 1985;**18**:163-167. DOI: 10.1016/0306-9877(85)90049-0

[19] Garfinkel BD, Webster CD, Sloman L. Responses to methylphenidate and varied doses of caffeine in children with attention deficit disorder. Canadian Journal of Psychiatry. 1981;**26**(6):395-401

[20] Reichard CC, Elder T. The effects of caffeine on reaction time in hyperkinetic and normal children. American Journal of Psychiatry. 1977;**134**(2):144-148

[21] Schecter MD, Timmons GD. Objectively measured hyperactivity-II Caffeine and amphetamine effects. Journal of Clinical Pharmacology. 1985;**25**:276-280

[22] National Institute on Drug Abuse. 2015 European Drug Report 2015: Trends and Developments, Lisbon June 2915 European Monitoring Centre for Drugs and Drug Addiction. European Monitoring Center for Drugs and Drug Addiction; 2016

[23] Tozzi A, de Iure A, Marsili V, Romano R, Tantucci. The distinct role of medium spiny neurons and cholinergic interneurons in the D2/A2A receptor interaction in the striatum: Implications for Parkinson's disease. The Journal of Neuroscience. 2011;**31**: 1850-1862

[24] Ingla S, Kreitzer AC, Malenka RC. Mechanisms for synapse specificity during striatal long-term depression. The Journal of Neuroscience. 2007;**27**:5260-5264

[25] Li H-L. An Archaeological and Historical Account of Cannabis in China. Economic Botany. 1974;**28**(4) New York Botanical Garden Press:437-448 http://www.jstor.org/stable/4253540

[26] Linville TM. Project MKULTRA and the Search for Mind Control: Clandestine Use of LSD Within the CIA. History Capstone Research Papers 6; 2016 Available from: http://digitalcommons.cedarville.edu/history_capstones/6

[27] Available from: "https://timeline.com/harry-anslinger-racist-war-ondrugs-prison-industrial-complexfb5cbc281189; and https://deamuseum.org/anslinger/beginnings-moving-tothe-treasury-department/

[28] United States Pharmacopeial Convention. The Pharmacopeia of the United States of America. Easton, PA: Hathi Trust Digital Library; 1851. p. 50

[29] Sagan C. Speech. Cornell University; 1994

[30] Lu H-C, Mackie K. An introduction to the endogenous cannabinoid system. Biological Psychiatry. 2016;**79**(7):516-525

[31] Gonsiorek W, Lunn C, Fan X, Narula S, Lundell D, Hipkin RW. Endocannabinoid 2-arachidonyl glycerol is a full agonist through human type 2 cannabinoid receptor: Antagonism by anandamide. Molecular Pharmacology. 2000;**57**:1045-1050

[32] Luk T, Jin W, Zvonok A, Lu D, Lin XZ, Chavkin C, et al. Identification of a potent and highly efficacious, yet slowly desensitizing CB1 cannabinoid receptor agonist. British Journal of Pharmacology. 2004;**142**:495-500

[33] Mackie K, Devane WA, Hille B. Anandamide, an endogenous cannabinoid, inhibits calcium currents as a partial agonist in N18 neuroblastoma cells. Molecular Pharmacology. 1993;**44**:498-503

[34] Cravatt BF, Giang DK, Mayfield SP, Boger DL, Lerner RA, Gilula NB. Molecular characterization of an enzyme that degrades neuromodulatory fatty-acid amides. Nature. 1996;**384**:83-87

[35] Slominski A, Semak I, Pisarchik A, Sweatman T, Szczesniewski A, Wortsman J. Conversion of L-tryptophan to serotonin and melatonin in human melanoma cells. FEBS Letters. 2002;**511**(1-3):102-106. DOI: 10.1016/s0014-5793(01)03319-1)

[36] Kersten S, Seydoux J, Peters JM, Gonzalez FJ, Desvergne B, Wahli W. Peroxisome proliferator-activated receptor alpha mediates the adaptive response to fasting. The Journal of Clinical Investigation. 1999;**103**(11):1489-1498. DOI: 10.1172/JCI6223

[37] Poirier H, Braissant O, Niot I, Wahli W, Besnard P. 9-cis-retinoic acid enhances fatty acid-induced expression of the liver fatty acid-binding protein gene. FEBS Letters. 1997;**412**:480-484

[38] Tontonoz P, Hu E, Devine J, Beale EG, Spiegelman BM. PPAR gamma 2 regulates adipose expression of the phosphoenolpyruvate carboxykinase gene. Molecular and Cellular Biology. 1995;**15**:351-357

[39] Nomura DK, Morrison BE, Blankman JL, Long JZ, Kinsey SG, Marcondes MC, et al. Endocannabinoid hydrolysis generates brain prostaglandins that promote neuroinflammation. Science. 2011;**334**:809-813

[40] Caille S, Alvarez-Jaimes L, Polis I, Stouffer DG, Parsons LH. Specific alterations of extracellular endocannabinoid levels in the nucleus accumbens by ethanol, heroin, and cocaine self-administration. The Journal of Neuroscience. 2007;**27**:3695-3702

[41] Bisogno T, Howell F, Williams G, Minassi A, Cascio MG, Ligresti A, et al. Cloning of the first sn1-DAG lipases points to the spatial and temporal regulation of endocannabinoid signaling in the brain. J Cell Biol. 2003;**163**(3):463-468. DOI: 10.1083/jcb.200305129

[42] Tanimura A, Yamazaki M, Hashimotodani Y, Uchigashima M, Kawata S, Abe M, et al. The endocannabinoid 2-arachidonoylglycerol produced by diacylglycerol lipase alpha mediates retrograde suppression of synaptic transmission. Neuron. 2010;**65**(3):320-327. DOI: 10.1016/j.neuron.2010.01.021

[43] Shonesy BC, Bluett RJ, Ramikie TS, Báldi R, Hermanson DJ, Kingsley PJ, et al. Genetic disruption of 2-arachidonoylglycerol synthesis reveals a key role for endocannabinoid signaling in anxiety modulation. Cell Rep. 2014;**9**(5):1644-1653. DOI: 10.1016/j.celrep.2014.11.001

[44] McPartland JM, MacDonald C, Young M, Grant PS, Furkert DP, Glass M. Affinity and efficacy studies of tetrahydrocannabinolic acid A at

cannabinoid receptor types one and two. Cannabis and Cannabinoid Research. 2017;**2**(1):87-95. DOI: 10.1089/can.2016.0032

[45] McPartland JM, Glass M, Pertwee RG. Meta-analysis of cannabinoid ligand binding affinity and receptor distribution: Interspecies differences. British Journal of Pharmacology. 2007;**152**:583-593. DOI: 10.1038/sj.bjp.0707399

[46] McPartland J, Klin A. Asperger's syndrome. Adolescent Medicine Clinics. 2006;**17**(3):771-788, abstract xiii. DOI: 10.1016/j.admecli.2006.06.010

[47] Munro S, Thomas KL, AbuShaar M. Molecular characterization of a peripheral receptor for cannabinoids. Nature. 1993;**365**:61-65

[48] Tsou K, Brown S, SanudoPena MC, Mackie K, Walker JM. Immunohistochemical distribution of cannabinoid CB1 receptors in the rat central nervous system. Neuroscience. 1998;**83**:393-411

[49] Marsicano G, Lutz B. Expression of the cannabinoid receptor CB1 in distinct neuronal subpopulations in the adult mouse forebrain. The European Journal of Neuroscience. 1999;**11**:4213-4225

[50] Ramirez SH, Hasko J, Skuba A, Fan S, Dykstra H, McCormick R, et al. Activation of cannabinoid receptor 2 attenuates leukocyte-endothelial cell interactions and blood-brain barrier dysfunction under inflammatory conditions. The Journal of Neuroscience. 2012;**32**:4004-4016

[51] Walter L, Franklin A, Witting A, Wade C, Xie Y, Kunos G, et al. Nonpsychotropic cannabinoid receptors regulate microglial cell migration. The Journal of Neuroscience. 2003;**23**: 1398-1405

[52] Kanoh H, Yamada K, Sakane F. Diacylglycerol kinases: Emerging downstream regulators in cell signaling systems. Journal of Biochemistry. 2002;**131**:629-633

[53] Namdar D, Voet H, Ajjampura V, Mazuz M, Shalev N, Koltai H. Terpenoids and phytocannabinoids co-produced in cannabis sativa strains show specific interaction for cell cytotoxic activity. Molecules. 2019;**24**(17):3031

[54] Lanfumey L, Mongeau R, Cohen-Salmon C, Hamon M. Corticosteroid-serotonin interactions in the neurobiological mechanisms of stress-related disorders. Neurosci Biobehav Rev. 2008 Aug;**32**(6): 1174-1184. DOI: 10.1016/j.neubiorev.2008.04.006

[55] Haj-Dahmane S, Shen RY. Modulation of the serotonin system by endocannabinoid signaling. Neuropharmacology. 2011;**61**(3): 414-420. DOI: 10.1016/j.neuropharm.2011.02.016

[56] Boehnke KF, Litinas E, Clauw DJ. Medical cannabis use is associated with decreased opiate medication use in a retrospective cross-sectional survey of patients with chronic pain. The Journal of Pain. 2016;**17**(6):739-744

[57] Wiese B, Wilson-Poe AR. Emerging evidence for cannabis' role in opioid use disorder. Cannabis and Cannabinoid. Research. 2018;**3**(1). DOI: 10.1089/can.2018.0022

[58] Barlow DH, Gorman JM, Shear MK, Woods SW. Cognitive-behavioral therapy, imipramine, or their combination for panic disorder a randomized controlled trial. JAMA. 2000;**283**(19):2529-2536. DOI: 10.1001/jama.283.19.2529

[59] Barlow DH. Unraveling the mysteries of anxiety and its disorders from the perspective of emotion theory. American Psychologist.

2000;**55**(11):1247-1263. DOI: 10.1037/0003-066X.55.11.1247

[60] Kamal F, Lantela DE. Cannabis and the anxiety of fragmentation-a systems approach for finding an anxiolytic cannabis chemotype. Frontiers in Neuroscience. 2018;**12**:730. DOI: 10.3389/fnins.2018.00730

[61] Kamal BS, Kamal F, Lantela DE, et al. Whistler Therapeutics. Whistler, BC, Canada: Whistler Medical Marijuana; 2018

[62] Lisboa SF, Gomes FV, Terzian AL, Aguiar DC, Moreira FA, Resstel LBM, et al. The endocannabinoid system and anxiety. Vitamins and Hormones. 2017;**103**:193-279. DOI: 10.1016/bs.vh.2016.09.006

[63] Rabinak CA, Angstadt M, Sripada CS, Abelson JL, Liberzon I, Milad MR, et al. Cannabinoid facilitation of fear extinction memory recall in humans. Neuropharmacology. 2012;**64**(1):396-402

[64] Fabre LF, McLendon D. The efficacy and safety of nabilone (a synthetic cannabinoid) in the treatment of anxiety. Journal of Clinical Pharmacology. 1981;**21**(8-9 Suppl):377S-382S Available from: https://www.cannabis-med.org/studies/ww_en_db_study_show.php?s_id=229&&search_pattern=Fabre

[65] Russo M, Naro A, Leo A, Sessa E, D'Aleo G, Calabrò BP. Evaluating Sativex® in neuropathic pain management: A clinical and neurophysiological assessment in multiple sclerosis. RS2. Pain Medicine. 2016;**17**(6):1145-1154. DOI: 10.1093/pm/pnv080

[66] Wade DT, Makela P, Robson P, House H, Bateman C. Do cannabis-based medicinal extracts have general or specific effects on symptoms in multiple sclerosis? A double-blind, randomized, placebo-controlled study on 160 patients. Multiple Sclerosis. 2004;**10**(4):434-441

[67] John H, Kindreda KL, Ketelhuta NB, Proessla F, Flinga BW, Honceb JM, et al. Cannabis use in people with Parkinson's disease and multiple sclerosis: A webbased investigation. Complementary Therapies in Medicine. 2017;**33**:99-104. DOI: 10.1016/j.ctim.2017.07.002

[68] Vos T, Allen C, Arora M, Barber RM, Bhutta ZA, Brown A, et al. GBD 2015 Disease and Injury Incidence and Prevalence Collaborators. 2016

[69] Vos T, Allen C, Arora M. R. Global, regional, and national incidence, prevalence, and years lived with disability for 310 diseases and injuries, 1990-2015: A systematic analysis for the Global Burden of Disease Study 2015". Lancet. 2016;**388**(10053):1545-1602. DOI: 10.1016/S0140-6736(16)31678-6

[70] National Institute of Neurological Disorders and Stroke (NINDS). Asperger syndrome fact sheet". Archived from the original on 21 August 2007. Retrieved 24 August 2007. NIH Publication No. 05-5624.(8). 2007

[71] Coleman DM, Adams JB, Anderson AL, Frye RE. Rating of the of the effectiveness of 26 psychiatric and seizure medications for autism spectrum disorder: Results of a national survey results of a national survey. Journal of Child and Adolescent Psychopharmacology. 2019;**29**(2):107-123. DOI: 10.1089/cap.2018.0121

[72] Fischedick JT, Hazekamp A, Erkelens T, Choi YH, Verpoorte R. Metabolic fingerprinting of Cannabis sativa L., cannabinoids and terpenoids for chemotaxonomic and drug standardization purposes. Phytochemistry. 2010 Dec;**71**(17-18):2058-2073. DOI: 10.1016/j.phytochem.2010.10.001

[73] Russo EB. Taming THC: Potential cannabis synergy and phytocannabinoid-terpenoid entourage effects. British Journal of Pharmacology. 2011;**163**(7):1344-1364. DOI: 10.1111/j.1476-5381.2011.01238.x

[74] Finlay David B, Sircombe Kathleen J, Mhairi N, Callum J, Michelle G. Terpenoids From Cannabis Do Not Mediate an Entourage Effect by Acting at Cannabinoid Receptors. Frontiers in Pharmacology. 2020;**11**:359. https://www.frontiersin.org/article/10.3389/fphar.2020.00359. DOI: 10.3389/fphar.2020.00359

[75] RJS V, Xolalpa NC, AJA C, Pérez González C, Pérez Ramos J, Pérez Gutiérrez S. Terpenes from natural products with potential anti-inflammatory activity. In: Perveen S, Al-Taweel A, editors. Terpenes and Terpenoids. London, UK: IntechOpen; 2018. pp. 59-85

[76] Guesmi F, Tyagi AK, Prasad S, Landoulsi A. Terpenes from essential oils and hydrolate of Teucrium alopecurus triggered apoptotic events dependent on caspases activation and PARP cleavage in human colon cancer cells through decreased protein expressions. Oncotarget. 2018;**9**:32305-32320

[77] de las Heras B, Hotelano S. In: Rahman A, Choudhary MI, editors. Terpenoids and Cancer: Molecular Targets and Clincal Perspectives. In Frontiers in Anti-Cancer Drug Discovery. Sharjah, UAE: Bentham Science Publisher; 2015. pp. 137-195

[78] Kogan NM, Mechoulam R. Cannabinoids in Health and Disease. LLS SAS; 2007 Available from: www.dialogues-cns.org

[79] Schubart DC, Sommer IE, Yan Gastel WA, Goetgebuer RL, Boks MP. Cannabis with high cannabidiol content is associated with fewer psychotic experiences. Schizophrrenia Research. 2011;**130**(1-3):216-221

[80] Holley JH, Hadley KW, Turner CE. Constituents of Cannabis sativa L. XI: Cannabidiol and cannabichromene in samples of known geographical origin. J Pharm Sci. 1975;**64**(5):892-894. DOI: 10.1002/jps.2600640546

[81] De Petrocellis L, Ligresti A, Moriello AS, Allarà M, Bisogno T, Petrosino S, et al. Effects of cannabinoids and cannabinoid-enriched Cannabis extracts on TRP channels and endocannabinoid metabolic enzymes. British Journal of Pharmacology. 2011;**163**(7):1479-1494. DOI: 10.1111/j.1476-5381.2010.01166.x

[82] Shinjyo N, Di Marzo V. The effect of cannabichromene on adult neural stem/progenitor cells. Neurochemistry International. 2013;**63**(5):432-437. DOI: 10.1016/j.neuint.2013.08.002

[83] Maione S, Piscitelli F, Gatta L, Vita D, De Petrocellis L, Palazzo E, et al. Non-psychoactive cannabinoids modulate the descending pathway of antinociception in anaesthetized rats through several mechanisms of action. Br J Pharmacol. 2011 Feb;**162**(3):584-596. DOI: 10.1111/j.1476-5381.2010.01063.x

[84] Russo EB, Marcu J. Cannabis pharmacology: The usual suspects and a few promising leads. Advances in Pharmacology. 2017;**80**:67-134. DOI: 10.1016/bs.apha.2017.03.004

[85] Davis WM, Hatoum NS. Neurobehavioral actions of cannabichromene and interactions with delta 9-tetrahydrocannabinol. Gen Pharmacol. 1983;**14**(2):247-252. DOI: 10.1016/0306-3623(83)90004-6

[86] Lee GY, Lee C, Park GH, Jang J. Amelioration of scopolamine-induced learning and memory impairment by α-pinene in C57BL/6 mice. Evidencebased Complementary and

Alternative Medicine. 2017;**2017**: 4926815. DOI: 10.1155/2017/4926815

[87] De Petrocellis L, Ligresti A, Moriello AS, Iappelli M, Verde R, Stott CG, et al. Non-THC cannabinoids inhibit prostate carcinoma growth in vitro and in vivo: Pro-apoptotic effects and underlying mechanisms. British Journal of Pharmacology. 2013;**168**(1):79-102. DOI: 10.1111/j.1476-5381.2012.02027.x

[88] De Petrocellis L, Orlando P, Moriello AS, Aviello G, Stott C, Izzo AA, et al. Cannabinoid actions at TRPV channels: Effects on TRPV3 and TRPV4 and their potential relevance to gastrointestinal inflammation. Acta Physiologica (Oxford, England). 2012;**204**(2):255-266. DOI: 10.1111/j.1748-1716.2011.02338.x

[89] Xing J, Li J. The role played by adenosine in modulating reflex sympathetic and pressor responses evoked by stimulation of TRPV1 in muscle afferents. Cellular Physiology and Biochemistry. 2016;**40**(1-2):39-48

[90] Wang M, Wang Y-H, Avula B, Radwan MM, Wanas AS, van Antwerp J, et al. Decarboxylation study of acidic cannabinoids: A novel approach using ultra-high-performance supercritical fluid chromatography/photodiode array-mass spectrometry. Cannabis and Cannabinoid Research. 2016;**1**(1):262-271. DOI: 10.1089/can.2016.0020

[91] Takeda S, Misawa K, Yamamoto I, Watanabe K. Cannabidiolic acid as a selective cyclooxygenase-2 inhibitory component in cannabis. Drug Metabolism and Disposition September. 2008;**36**(9):1917-1921. DOI: 10.1124/dmd.108.020909

[92] Hepler RS, Frank IR. Marihuana smoking and intraocular pressure. Journal of the American Medical Association. 1971;**217**(10):1392

[93] Boggs DL, Peckham A, Boggs AA, Ranganathan M. Delta-9-tetrahydrocannabinol and cannabidiol: Separating the chemicals from the "weed," a pharmacodynamic discussion. Ment Health Clin. 2016;**6**(6):277-284. DOI: 10.9740/mhc.2016.11.277

[94] Wargent ET, Zaibi MS, Silvestri C, Hislop DC, Stocker CJ, Stott CG, et al. The cannabinoid Δ(9)-tetrahydrocannabivarin (THCV) ameliorates insulin sensitivity in two mouse models of obesity. Nutr Diabetes. 2013;**3**(5):e68. DOI: 10.1038/nutd.2013.9

[95] Jadoon KA, Ratcliffe SH, Barrett DA, Thomas EL, Stott C, Bell JD, et al. Efficacy and safety of cannabidiol and Tetrahydrocannabivarin on glycemic and lipid parameters in patients with type 2 diabetes: A randomized, double-blind, Placebo-Controlled, Parallel Group Pilot study. Diabetes Care. 2016:dc160650. DOI: 10.N37/dc16-0650

[96] Avraham Y, Latzer Y, Hasid D, Berry EM. The impact of THC on the psychological symptoms of ANOREXIA nervosa: A pilot study. The Israel Journal of Psychiatry and Related Sciences. 2017;**54**(3):44-51

[97] França AP, Takahashi RN, Cunha RA, Prediger RD. Promises of caffeine in attention-deficit/hyperactivity disorder: From animal models to clinical practice. Journal of Caffeine and Adenosine Research. 2018;**8**(4)

[98] Russo EB, Burnett A, Hall B, et al. Agonistic properties of cannabidiol at 5-HT1a receptors. Neurochemical Research. 2005;**30**:1037-1043. DOI: 10.1007/s11064-005-6978-1

[99] Hill AJ, Mercier MS, Hill TD, Glyn SE, Jones NA, Yamasaki Y, et al. Cannabidivarin is anticonvulsant in mouse and rat. British Journal of

Pharmacology. 2012;**167**(8):1629-1642. DOI: 10.1111/j.1476-5381.2012.02207

[100] Shoyama Y, Fujita T, Yamauchi T, Nishioka I. Cannabichromenic acid, a genuine substance of cannabichromene. Chemical and Pharmaceutical Bulletin. 1968;**16**:1157-1158. DOI: 10.1248/cpb.16.1157

[101] Shoyama Y, Oku R, Yamauchi T, Nishioka I. Cannabis. VI. Cannabicyclolic Acid. Chemical and Pharmaceutical Bulletin. 1972;**20**:1927-1930. DOI: 10.1248/cpb.20.1927

[102] Elsohly MA, Slade D. Chemical constituents of marijuana: The complex mixture of natural cannabinoids. Life Sci. 2005;**78**(5):539-548. DOI: 10.1016/j.lfs.2005.09.011

[103] Deiana S. CNS Diseases Research Department, Boehringer Ingelheim Pharma GmbH & Co. KG, Birkendorfer straße, Biberach an der Riss, Germany

[104] Cascio MG, Gauson LA, Stevenson LA, Ross RA, Pertwee RG. Evidence that the plant cannabinoid cannabigerol is a highly potent alpha2-adrenoceptor agonist and moderately potent 5HT1A receptor antagonist. British Journal of Pharmacology. 2010;**159**(1):129-141. DOI: 10.1111/j.1476-5381.2009.00515.x

Chapter 3

Marijuana, a Journey through the Endocannabinoid System: Unmasking the Paradoxical Effect - Part 2

Ryan Lucas McKinley

Abstract

Here in part two, a brief explanation in essential oil/ terpene administration as well as cover the medicinal effects of terpenes focusing on biphasic pharmacokinetics and possible paradoxical reactions and molecular sites of interest, including the medicinal properties of a specific flavonoid; an explanation into the paradoxical entourage and identifying common misconceptions from cannabis use and education; we finalize our look into the paradoxical location learning biphasic and paradoxical reactions from cannabis with an in-depth look into the cause of ASR/ATD following with a fundamental explanation how stress with the wrong medication can instigate the situation. The **Multi Cultivar Entourage Effect Chart** (**MCEEC**) directed goal was to unravel multiple cultivars bioavailability to then combine and create a more robust and stronger entourage being pulled from multiple cultivars with specific bioavailability of cannabinoids, terpenoids, and flavonoids necessary to treat any specific indication. Indirectly the chart also identified inter-entourages, more importantly, “antagonistic” inter-entourages. By helping a patient describe their reactions, understand, identify and track terpenes and cannabinoids that cause specific reactions, the patient will be able to identify a profile that works for them, which gives an explanation and solution to identifying how to manage cannabis medication for the patient along with conclusion and thoughts.

Keywords: Advanced Synergistic Serotonin Release (ASSR/ASR), Advanced Tryptophan Depletion(ATD), Endo-Cannabinoid System (ECS), Central Nervous System (CNS), Psychoactive (PA), Non Psychoactive (N-PA), Cannabigerol (CBG), Tetrahydrocannabinol (THC), Tetrahydrocannabinolic Acid (THCA), Tetrahydrocannabivarin (THC-V), Cannabinol (CBN), Cannabichromene (CBC), Cannabielsoin (CBE), Cannabicyclol (CBL), Cannabidiol (CBD), Cannabidiolic Acid (CBDA), Cannabidivarin (CBDV), Broad Leaf Marijuana (BLM), Broad Leaf Marijuana Dominant (BLMD), Medium Leaf Marijuana (MLM), Narrow Leaf Marijuana Dominant (NLMD), Narrow Leaf Marijuana (NLM), Cannabinoid Receptor type1(CB1), Cannabinoid Receptor type 2 (CB2), Vanilloid Receptor 1 (TRPV1), Transient Receptor Potential Ankyrin 1(TRPA1), γ-aminobutyric acid (GABBA-A), Hydroxy-Tryptamine (5-HT), Transient Receptor Potential cation channel, subfamily V, member 3 (TRPV3), Attention Deficit Hyperactivity Disorder (ADHD), γ-aminobutyric acid (GABAA)

IntechOpen

1. Introduction

1.1 Terpenes and flavonoids

This section's objective is to weed out possible terpene synergies which may actuate a biphasic experience of either ASR/ATD. Over 200 terpenoids are primarily responsible for the many fragrances of cannabis and may represent 10% of trichome content [1–7]. Monoterpenoids, containing oxygen functionality or missing a methyl group, are commonly composed of limonene, myrcene, pinene, linalool [8–11]. Terpenes are widely known to cross the blood–brain barrier due to their chemical makeup to be lipophilic like cannabinoids [9]. Terpenoids will begin to break down before the processing stage at a rate of about 5%. After curing processes and in time (1–6 months) terpenoids will have diminished significantly [12, 13]. With this in mind, any further actions taken to separate the whole cultivars' phytochemistry undoubtedly weakens the quantum function of the entire medicinal chain, ie extraction and or isolation processes. Terpenes in their natural state, are incorporated in the trichomes of cannabis with a fundamental directive whether it's to keep predators at bay, maintain general cultivar homeostasis, or the more commonly known use, as a medication for homeostasis in most vertebrate species.

Any medicinal flower sold past 6 months from the end of the curing stage, will most likely be under the promised genomes bioavailability or true to the cultivar medicinal properties. The cannabis flower should be tested at the 6 month and interpened to either be converted to extracts or discarded unless properly stored (i.e., time would vary depending on consumer storing methods). In this time (1–6 > months), the main subsidies of most terpenes begin to chemically change and fall under generally categorized oxidized terpenoids or hydro-carbon terpenes. Cannabinoids in contrast are more resilient in this oxidative manner but should be understood that the intentional entourage from any cannabis cultivar should be consumed as a whole, not a hand-me-down to what was. Other constituents of terpenoids such as caryophyllene, geraniol, humulene, limonene, linalool, myrcene, ocimene, pinene, terpineol, and terpinolene have beneficial health properties that help to treat indications ranging from neurodegenerative disorders to cancer. Though terpenes never directly affect CB1 or CB2, their presence in an entourage may mediate other possible synergistic effects, along with serotonin release.

Flavonoids, a secondary metabolite and interestingly anthocyanin (i.e., blue, violet, and red plant pigments that exist in fruit, vegetables, tea, wine, and "more recently researched," cannabis) intake have had extensive research ranging from anti-inflammatory to pro-cognitive explained characteristics which also include passing the blood–brain barrier [9]. Thus, it has a viable and crucial part in medical cannabis as a whole product. Specific fruits with high levels of anthocyanins will display dark purple, blue, and perceptively black pigments through inflorescence; like in cherry juice or from Japanese plums, similar phenolic compound levels can be comparable to hemp seed extract. Cannabis displaying plentiful purple or deep red inflorescence could then be suggested for a different or new method of extracting with possible uses such as tinctures, teas, or juices as to not cause excessive degradation to the bioavailability of said cultivar medical administration.

2. α-Pinene

α-Pinene is a bicyclic unsaturated hydrocarbon with two isomers being α- & β-pinene that makes up the whole [14, 15]. This particular terpene is vastly known throughout nature but in cannabis acts as an/a anti-inflammatory, a bronchodilator, MRSA treatment, antibiotic [16], and even improve cognitive ability and memory retention in lieu of THC's supposed side effect of short-term memory loss [17, 18].

In a recent study, α-pinene's memory retention ability may add the concern to PTSD memory triggers, causing a tougher time disassociating traumatic memories with the trigger [19].

One trait of pinene interestingly stops excitation of a nerve after transmission of an impulse, in short, acetylcholinesterase [20]. In "Cannabis Pharmacology: The Usual Suspects and a Few Promising Leads," Russo and Marcu state, α-Pinene "... serves to reduce or eliminate one of the primary adverse events associated with THC, that of short-term memory impairment. This ability may also serve admirably in the treatment of dementia, a syndrome in which THC has already produced benefits in counteracting agitation".

Henceforth, Pinene, acetylcholinesterase [21], I believe α-pinene to contrary belief may be pertinent to patients with indicators such as PTSD, ADD/ADHD, OCD, panic disorders, spectrum disorders, and epilepsy, when paired with cognitive brain therapy (CBT) breaking an "adolescent fear loop" as thoroughly explained in the study, dynamic changes in neural circuitry during adolescence are associated with persistent attenuation of fear memories [22].

So in a more readable way, pinene has more efficacy during times of positive mental healing than as a "take as needed/ smoke-em-if-ya-got-'em; pill-popping mentality frequently associated with addictive/non-addictive pharmaceuticals" and the patients who take them (i.e., stress). Cannabis is psychoactive and intoxicating and thus has the potential to be mind-expanding ergo the reason for set and setting including CBT. Together, could prove more appropriate in guided treatment, furthermore, that THC lower than CBD and paired with proper synergistic terpenes, would be a safer means of medicating to avoid any excessive serotonin use, aside from what is needed from the HPA axis for the general operation of cannabinoids via pre and post-receptor Synoptics in regions of the brain and body. It is also important to understand pinene is a known characteristic of NLM varieties which are actually the ones provoking most episodes of PTSD and anxiety among many other side effects that stimulants may provide, i.e., pinene exhibits no such stimulation pharmacokinetically.

Hypothesis: So, when paired with a medical chemovar possessing a specific entourage, including α- or β-pinene, a patient can then efficiently break the fear loop cycle and the memory trigger associated with the traumatic cycle creating a new positive loop to trigger [22].

Does α-pinene contribute to the paradoxical effect? No perspective paradox or biphasic ASR manner but an understanding of "when and how to use, for specific neurological conditions."

3. β-Myrcene

β-Myrcene, a monoterpenoid with analgesic, muscle relaxant, and sedative-like properties with many cannabinoid synergies i.e., CBD, THC, CBG. Myrcene can display analgesia in mice, but synthetic drugs that block opioid receptors in the nervous system can be blocked, perhaps via the α-2 adreno-receptor [15, 23, 24], which is responsible for inhibiting the release of norepinephrine (noradrenaline) in a form of negative feedback i.e., "sedation" [25]. Agonists of these receptors have been used to treat mainstream medical conditions such as hypertension, ADHD, various pain and panic disorders; symptoms of opioid, benzodiazepine, and alcohol withdrawal; and surprisingly nicotine cravings, which is one of the most addictive chemicals known to humans [26].

Does β-myrcene contribute to the paradoxical effect? Yes, if there are sub-par levels of serotonin in the body.

Conclusion: A given sedative, myrcene does depend on serotonin for endogenous opioid production, hence, if there is a depletion in tryptophan, then there will most likely be a paradoxical reaction.

4. D-linalool

A similar fragrance is found in lavender, but in a cannabis cultivar and when phytochemically available, the synergy between cannabinoids and this monoterpenoid reveals treatments such as sedative-like effects. Linalool is used as a local anesthetic; an anti-convulsant, a powerful antileishmanial agent. Linalool is an antinociceptive, reversing defects and spatial memory and learning at high doses with a respectable contradiction in short- and long-term recognition memory. This implies detrimental to cognitively impaired sentient beings, though studies were done on healthy and cognitive impaired rats [9, 27, 28].

The NMDA receptor is very important for controlling synaptic plasticity and memory function. Specifically, linalool showed strong efficacy in inhibiting glutamate uptake in cortical synaptosomes and decreased extracellular glutamate availability via inhibiting the release or adding to the uptake [29, 30]. NMDA affinity means GABA will be either used or suppressed; and in the study, reduced morphine opioid dependency [29, 31].

Does D-Linalool contribute to the paradoxical effect? Linalool acts as a competitive antagonist of [^{3}H] glutamate binding and as a noncompetitive agonist of [^{3}H] dizocilpine (NMDA antagonist) [32, 33].

Conclusion (plausible): More study must be done to further identify linalool pathologies and how they may have representation in a paradox in brain plasticity.

5. Beta-Caryophyllene

Beta-caryophyllene sesquiterpenoid, in studies, has shown to operate in a "Phytochemical Polymorphism" manner [34]. With this in mind, Ethan Russo cites, "Terpenoids are pharmacologically versatile: they are lipophilic, interact with cell membranes, neuronal and muscle ion channels, neurotransmitter receptors, G-protein coupled (odorant) receptors, second messenger systems, and enzymes" [35, 36]. To be understood as a helper to bioavailable cannabinoids and could be thought of as an oil change for a car.

Does caryophyllene contribute to the paradoxical effect?

Conclusion: No pathology to denote a paradoxical behavior unless serotonin levels are sub-par.

6. D-limonene (the energetic uplifting agitator)

The volatile monoterpene, limonene, one of the most abundant terpenes in cannabis, and its perceived effects can be summed up as uplifting (as to correct a depressed mood) and energetic (as to cure slothfulness); having antioxidant, anti-inflammatory, and neuroprotective properties [37]. These hyperactive characteristics of D-limonene indicate it to be a prime candidate for A2a receptor affinity, and thus, "It plays an important role in many biological functions, such as cardiac rhythm and circulation, cerebral and renal blood flow, immune function, pain regulation, and sleep. It has been implicated in pathophysiological conditions such as inflammatory diseases and neurodegenerative disorders" [38].

In an *in vitro* dose-dependent study of D- and L-limonene, effects on the pregnant rat myometrium (mid-layer of the uterine wall and the smooth muscle tissue), D- and L-limonene caused myometrial contractility (i.e., increases the contractions of a pregnant uterus); interestingly, L-limonene caused myometrial smooth muscle contraction independent of A_{2A} receptors. Due to the subsequent findings of D- and

L-limonene causing myometrial contractility via activation of the A_{2A} receptor and opening of the voltage-gated Ca^{2+} channel, D- and L-limonene should be avoided during any pregnancy [39].

To indulge in the topic of the A_{2A} receptor and its synergistic affinity with limonene, the above-mentioned study was focused on "in vitro" muscle contractility, thus would be fair to assume that any humans suffering from constant or frequent agitation from symptoms/indicators like MS, PMDD, PD, spectrum disorders, or neurodegenerative diseases may also want to avoid this terpene. In cases where this is difficult to avoid, it would then be crucial to understand how a person metabolizes both sedatives and stimulants. Curiosity in this area may open more doors into the point of agitation i.e., invoked and/or systemic and could then be regulated with the correct level of sedating to stimulating terpenes and aim for a genetic strain with a similar entourage. In the effort to stimulate without agitation, cannabis entourages with sedative terpenes could be added into the stimulating entourage via cultivation/hybridizing and/or multi-cultivar entourage dosing.

A good rule of thumb about the terpene limonene is to understand the biphasic modes in which this part of the entourage manipulates; a handful of studies dissect the physiology of citrus fruit-bearing plants and all come to a consensus that limonene affects serotonin via 5HT1A and dopamine via D2, thus giving the cascade of both stimulating and suppressant like effects.

However, the patient should be made aware of any synergies that may use 5HT or suppress and should supplement the 5HT usage via diet to avoid accelerated tryptophan depletion (ATD).

Additionally, humans suffering from indications such as PTSD, spectrum disorders, and general anxiety should also be wary of this terpene due to its excitatory tendencies via the A_{2A} receptor having excitatory biphasic responses. The goal for the vast genetic variety of "balanced" cannabis cultivars, i.e., hybrids (BLMD, MLM, NLMD), has been utilized from BLM's(medicinal cannabis Indica subspecies Afghanica) genetics and crossed with NLM (*Cannabis indica* sub-species Indica) or any other sub-species to then aim to produce (in this case) a vast range of sedative leaning MLM's (marijuana hybrids) i.e., BLMD. Henceforth, giving agricultural/horticultural cannabis growers the ability to hybridize and clone for a more viable chance at an endless possibility of medicinal cannabis strains. This being said, a BLMD with deviating genetics from the true "Indica," i.e., the Afghanica sub-species, has the chance to contain the volatile monoterpene limonene. In spite of the excitatory ability of limonene, this could then be a perfect supplement or addition to pharmaceutical ADHD medication.

Furthermore, "All the terpenoids discussed herein are Generally Recognized as Safe, as attested by the US Food and Drug Administration as food additives or by the Food and Extract Manufacturers Association and other world regulatory bodies." [40].

Does limonene contribute to the paradoxical effect? Highly plausible, but clinical study must still be done to truly understand the all working mechanisms of its cascade effects within the entourage effect.

Plausible location (A2A, D2 dopaminergic receptors), (5HT1A, serotonergic receptors).

7. The paradoxical entourage

To find, understand, or even combat the paradoxical effect, one must figure out how to counteract the symptoms so the medication or goal for homeostasis works in the way it is intended.

7.1 Misconceptions and experiences: A paradoxical effect from Cannabis

There is a common misconception about what constitutes *C. indica* and its sub-species variation. Specifically, medicinal cannabis commonly described as "calming, couch-locked, sedating," and/or claiming the original term, "Indica" (i.e., BLM) may actually fall under BLMD where the distant genetics of an NLM is still relevant among the BLM genetics when hybridized. Thus, having a chance at agitation depends on the NLM genetics. A proper example of this miscommunication would be a BLM crossed with "Green Crack," a known Sativa cultivar creating a hybrid of stimulating and sedating effects and then sold as an Indica. To further explain the *C. indica* ssp. Afghanica/BLM is of the genotype sub-specie Afghanica i.e., Indica ssp. Afghanica; the plant structure is of the shortest growing species revealing the broadest leaves accompanied with the most round and dense flower structure containing very petite pistils permeating a terpene bouquet from trichomes of deep sugary warmth, earthy spiced leather, chocolate, tobacco, and mushroom perceptive smells. *Indica* ssp. *Indica* (NLM), i.e., original term "Sativa" a misclassification by Jean Baptiste Lamarck, (1802); permeating more volatile aromas like grapefruit, tangerine, diesel, solvents, lemon, and pine equally showing polar opposite inflorescence.

What may be perceived as a paradoxical effect, is in fact a misconception of what the medicine actually contains past its genetic name and suggested effects from cannabis. Also, a common misconception when dealing with MLM "Hybrids," is the extreme ebb and flow from BLMD and NLMD sub-species variation alone, making up a galaxy of possibilities. Within the infinite genetic possibilities of hybrids, the common misconception validifies a relative vice versa, where an NLMD i.e., "sativa dominant hybrid strain," for instance, "Purple Haze" genetics from Prime Wellness of PA will pleasantly contain predominantly sedating terpenes resulting in a metabolic paradox of a stimulating cultivar and predominantly sedating terpenes. Since cannabinoids are less volatile, terpenes and their chemical makeup break down faster. Thus, resulting in a premature depletion of part of the whole medicine possibly resulting in stimulation or agitation toward the end of the medicated bell curve; to counteract this paradox the simple solution would be to add more bioavailable sedative terpenoids along with proper levels of CBD to combat any excessive psychoactive imbalance such as the 2019 "Freedom blend distillate" produced by ILERA to continue the cultivars intended medicinal entourage.

The vice versa misconception plays out similarly wherein an Indica leaning hybrid (BLMD) claiming the label indica i.e., BLM, may carry enough traces of Sativa genetics possibly causing agitation to hypersensitive patients in efforts of sedation. When dealing with a Ruderalis (AFM) specific plant speciation can be guaranteed and determined in a chromatography test to identify where the phytonutrients land on the spectrum of sedating, null, and stimulating effects. Thus, the infinite possibilities of hybridization and the cannabinoid and terpene profiles that can be created, have viable means of documentation. So as to understand when medicating with cannabis, separate the whole to understand how to medicate properly for any specific indicator.

To further understand, cannabis speciation (BLM-NLM) controls sedating/stimulating properties; terpenes and cannabinoids, and their synergies, carry out special tasks manipulating the physical and psychological state prolonging or exhausting a patient's balance of homeostasis depending on the accuracy of correct strain name choice to the indicator. Meaning the "engine size" of psychoactivity; the "transportation" of cannabinoids and the ebb, and flow of stimulating and sedating terpenes are what "drive" any entourage to actuate specific tasks throughout the ECS and CNS.

Thus, looking past strain names, subspecies claim, suggested effects, and acknowledging the cannabinoid and terpene profile of any medicinal cannabis would be a safer guarantee of medicating properly.

8. The paradoxical location

Through the research conducted in the ECS and CNS receptors, I have found a specific entourage of cannabinoids and specific terpenes that are the likely cause of a constant fundamental reaction resulting in this easily avoidable paradoxical effect.

In this paradoxical entourage, I believe the terpenoid D-limonene to possess a prime fundamental pathology to modulate D2 respectively via adenosine A2A receptor dealing with motor behavior, emotional reward, and behavior motivation mechanisms as one of its synergistic post reactions; and an agonist at $5HT_{1A}$ with effects to counteract excitation; to explain further, limonene could be stimulating and consecutively sedating but that if certain biosynthesis pools are depleted or below average to achieve homeostasis then D-limonene will most likely cause excitation, thus a paradox. With synergistic cannabinoids, not limited to, <CBC, THC, THC-V, CBG, CBD>, and possibly other terpenes to help trigger and prolong the biphasic cycle/paradoxical effect more specifically ATD/ASR or avoid the paradox entirely. This biphasic actuation of D-limonene will have either sedating presents or stimulating presents depending on the entourage it is coupled with and the human physiology it is metabolized by. Patients with PTSD or patients under systemic or triggered anxiety/depression-like symptoms, ongoing physical trauma, will most likely have an affinity to the paradoxical effect or the buildup to an advanced serotonin release (ASR) and should consult their doctor before consuming any unintentional stimulants [41].

In a study where patients had their tryptophan artificially suppressed, it was reported that the depletion/reduction of tryptophan caused a severe decrease in mood [42]. There are significant ethical considerations as that tryptophan depletion can have a profound negative impact on the patient. " ... a recovered depressed patient from the acute tryptophan depletion study by Delgado" reported,"... she began to cry inconsolably and described her emotions as being out of control", continued explanations of feeling "as if all the gains she had made over the past few weeks had evaporated." After the tryptophan levels were restored, the patient reported feeling "back to herself."

A disrupted balance of serotonin is an important risk factor for depressive mood, also a common symptom in the later course of treatment of chronic disorders such as cancer, infections, and autoimmune syndromes [15, 37, 43] with indications like autoimmune disease depression autism, epilepsy, HD, or any indicators residing in the ECS and some parts of the CNS will have a higher risk of this happening.

I refer to this function as an ATD or ASR, which happens when specific synergistic cannabinoids and terpenes already using serotonin, uses reserves consecutively at a more advanced rate of depletion due to the chemical nature of another synergistic action of the same source of depletion or; whilst a consecutive respectively similar (i.e., dopamine and serotonin) depletion ensues.

It is interesting to note that acute tryptophan depletion techniques (referred to by Simon N. Young, PhD (2013) as ATD studies) were first applied by Concu in 1977 and have been used for over 25 years. This technique requires that the patient have their tryptophan levels artificially and intentionally suppressed in an attempt to document the cause and effect. ATD and ASR mentioned in this chapter, describe the postreactions and neurological processes that are documented through studies discovered and cited in this chapter. The information provided is a compilation of information gathered from clinical studies, scientific papers, and the study of cannabis.

In this paradoxical entourage, I believe D-limonene to be one of the prime terpenoids with the ability to excite and release GABA and dopamine as its post-reaction with synergistic cannabinoids, but not limited to CBC, THC, THC-V, and possibly other terpenes to help trigger and prolong the biphasic cycle/paradoxical effect. More specifically aiding in an ATD/ASR. D-limonene being an A2A antagonist in the presence of THC with excitatory properties would then have a much similar reaction like caffeine, methylphenidate, and certain modes of activation of a cocaine alkaloid collection. This allows an accelerated rate of D2/5-HT_{A1} and serotonin release via GABA-A, which can then cause patients with PTSD or patients under systemic or triggered anxiety/depression-like symptoms, ongoing physical trauma, an affinity to the paradoxical effect, or the buildup to an ASR. [43] cb1 PTSD). Other more serious symptoms if left untreated can result in tumors, cancer, autoimmune deficiencies, often antagonizing the main purpose for medicating.

In this hypothesis, the entourage in question plays out similarly from a study by J Marcel, et al. 2010 [44], "...On the one hand, both THC and CBD were shown to decrease TNF-α production in human NK cells and peripheral blood mononuclear cells (PBMC), whereas THC was demonstrated to increase TNF-α production in human monocytes [15, 19]. Treatment of human PBMC with low doses of THC or CBD, comparable to plasma levels detectable after smoking marijuana (10–100 ng/mL), was demonstrated to stimulate interferon (IFN)-γ production, while higher concentrations of these cannabinoids (5–20 μg/mL) efficiently suppressed formation of this cytokine [19]. These contradictory findings are suggested to be based on a biphasic response relative to the cannabinoid ligand concentration applied, since most of reports showing stimulatory capacities were reported at lower doses, in the nanomolar concentration range, whereas inhibitory activities of cannabinoids

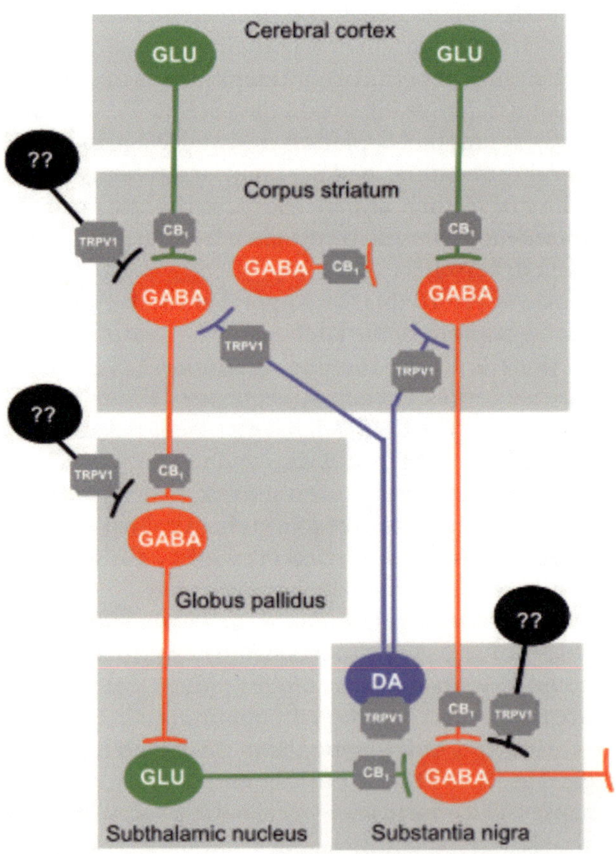

Figure 1.
Location of CB_1 and TRPV1 receptors in specific neuronal subpopulations within basal ganglia circuits. Regulatory pathways are indicated in blue, whereas inhibitory and excitatory inputs are indicated in red and green respectively. Unknown neurons are shown in black. CB_1, cannabinoid receptor type 1; GABA, γ-aminobutiric acid; GLU, glutamate; TRPV1, transient receptor potential vanilloid type 1.

were found in the micromolar concentration range [22, 25]. These concentration dependent effects of cannabinoids could be demonstrated for Th1- as well as Th2-type cytokines [26]." Marcel continues, "The suppressive effect of THC and CBD on cytokine-induced tryptophan degradation may constitute an additional mechanism by which anti-depressant effects of cannabinoids might be linked to the serotonergic system."

Disturbed balance of serotonin levels is an important risk factor for depressive mood, which is also a common symptom in the later course of chronic disorders such as cancer, infections, and autoimmune syndromes [6, 18, 45, 46]. Many patients with chronic inflammatory diseases show accelerated depressive mood, implicating a role of cytokine-induced IDO enzyme activity in psychiatric diseases [3, 19]. Additionally, several studies showed that mood is negatively influenced by the depletion of tryptophan [39, 47]. Since tryptophan is essential for the biosynthesis of serotonin, the decreased availability of tryptophan during inflammatory conditions as a result of degradation by IDO may negatively affect the biosynthesis of this neurotransmitter [37].

The charts below are taken from "Endocannabinoids and Motor Disorders" by J. Fenardez-Ruiz, British Journal of Pharmacology (2009) 1561029–1040 se.mcu.dem@rfjj or se.denrebic@ziur-zednanref.J (**Figures 1** and **2**).

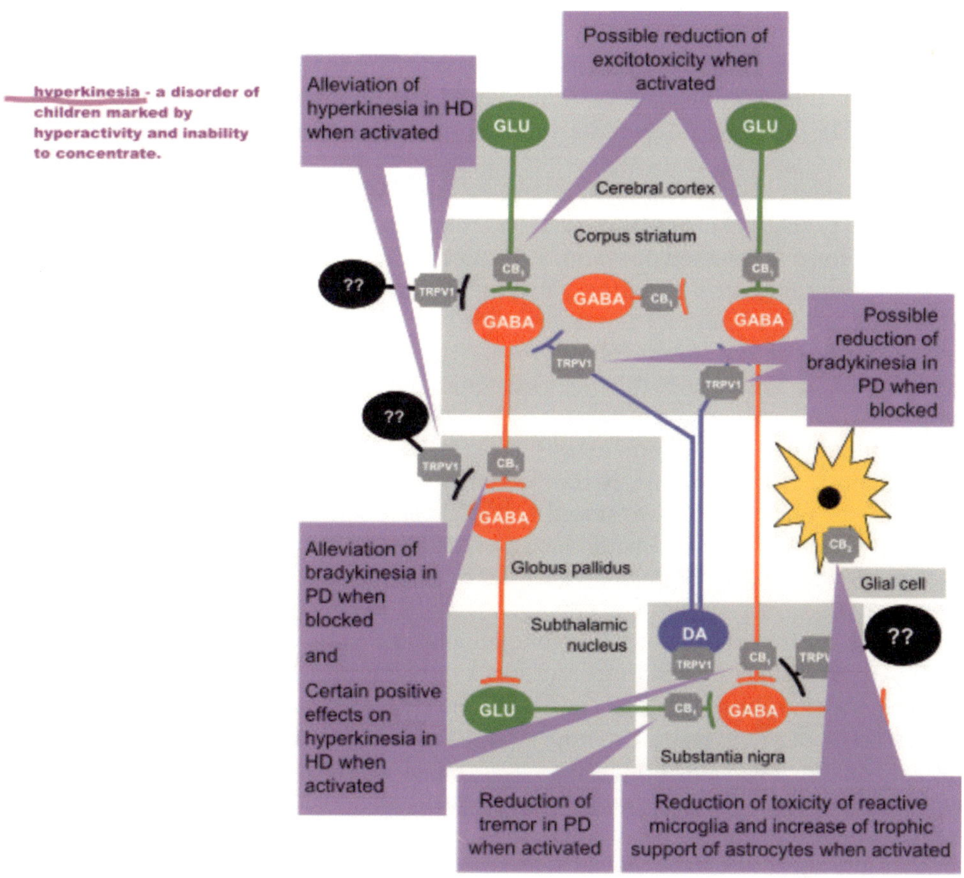

Figure 2.
*On the scheme shown in **Figure 1**, a diagram has been superimposed to show the different targets (CB_1, CB_2 and TRPV1 receptors) that might mediate the ability of cannabinoid-based medicines to alleviate specific symptoms, or to delay/arrest the progression of the disease in basal ganglia disorders. CB_1, cannabinoid receptor type 1; DA, dopamine; GABA, γ-aminobutiric acid; GLU, glutamate; HD, Huntington's disease; PD, Parkinson's disease; TRPV1, transient receptor potential vanilloid type 1.*

The connection between cannabis and serotonin release is that cannabinoids affect GABA which releases serotonin via CB1 and CB2. To further explain this catch 22 situation with cannabinoids and synergistic terpenes as cleanly said by an article in Proof of Pot Writer [48], "Low dose THC and FAAH inhibitors can have anti-anxiety effects. A 2007 study showed that the anti-anxiety effects of THC depended on the 5-HT1A receptor [49], although a 2015 study [50] demonstrated a dependence on the 5-HT2A receptor."

"Both THC and FAAH inhibitors, which raise levels of anandamide, can improve animal models of depression. The antidepressant effects of these molecules went away when animals were depleted of serotonin (2016 study, 2018 study), indicating that they are working at least partially through increasing serotonin release" [51, 52]. In addition, the antidepressant effect of CBD in animal models depended on activation of the 5-HT1A receptor (2016 study) [48, 51].

9. Fundamental and Hypothetical Solution

A fundamental fact to keep in mind is many foods like turkey, cheese, eggs, salmon, broccoli, or over the counter 5HT/ 5–hydroxytryptophan pills, mainly things that either contain or biosynthesize tryptophan for consumption is the only way to get it as humans do not produce tryptophan. Tryptophan is converted to 5-hydroxytryptophan by the hydroxyls enzyme (i.e., the rate-limiting step of serotonin synthesis) (l-tryptophan: basic metabolic functions, behavioral research, and therapeutic indications; 2009 Dawn Richard, Michael Dawes PMID20651948). Therefore, without tryptophan, there will be no serotonin production.

In the efforts to understand, supplement, and avoid an ASR or ATD; I would like to give a hypothetical example or scenario that briefly describes the ways tryptophan is utilized in invoked or systemic trauma.

For this hypothetical scenario, your directed goal is to always keep your tryptophan tank full or commonly at a respectable balance. If you take damage via physical, mental trauma and/or develop systemic trauma, this will then create an ongoing tank depletion. Thus, more damage will equal more tryptophan depletion, plus tryptophan dependency (due to prolonged systemic or invoked damage) in efforts to reach a targeted 100% homeostasis or maintain minimal levels. Cannabis with high amounts of D-limonene, in this scenario, would be the medication that gives you one step forward and two steps back given no tryptophan has been consumed and depending on how well your character utilizes tryptophan. In this reaction to stimuli via A2A, the body may react out of survival, fear, or a mindful manner response [22].

The longer the tank is empty or struggling to maintain a minimal equilibrium or homeostasis during excitation, the faster the body and mind will suffer due to the absence of tryptophan. Due to this absence of serotonin production, the mind will go through a depressive prolonged shutdown, almost forced into a default mode. The inevitable side effects open doors such as susceptibility to autoimmune deficiencies, cancer, more depression, and eventually your life would be over before your actual intended life expectancy.

Possibly supporting the case of some medical marijuana patients/MCP's peeking, wherein the normal amounts consumed no longer have the intended medicinal effect and may be due to an ATD/ASR without recovery or healthy diet to supplement cannabis amount into the diet.

The combined knowledge indicates that the paradoxical effect does not exist in cannabis but in an individual, and how they metabolize a certain entourage with any disruption to the CNS and ECS that has the potential to exist across the BLM-NLM spectrum of cannabis. The entourage in question has an affinity to use up tryptophan/release serotonin in the efforts to achieve homeostasis via GABA-A modulation through multiple networks in a biphasic/paradoxical manner. More so in those that constantly suffer from hyperkinesia, prone to anxiety and depression, ADD/ADHD, and or in constant extreme pain or from neurological indicators that depend on the presence of tryptophan and the release of serotonin and other dopaminergic reactions.

10. An entourage for everyone

In most cases, the dependability on strain availability and ebb and flow of cannabis morphology throughout each year will be tough to maintain for any cannabis industry grower or dispensary to keep consistency stocked or any cannabis patient to even have. Everyone has their own entourage that creates homeostasis in one's body whether it mediates heavily from one to another or hardly at all. A unique solution would be to compare and contrast multiple cultivars' bioavailability, synergies. In end creating a whole new cascade of newly perceived entourages that have a higher chance at meeting homeostasis directed goals for a patient and their ECS directed by their psychiatrist, cannabis physician, or therapist regardless of genetic availability.

I have created The **Multi Cultivar Entourage Effect Chart** (MCE^{2}C) that unravels multiple cultivars' bioavailability to then combine and create a more robust and stronger entourage being pulled from multiple cultivars with specific bioavailability of cannabinoids, terpenoids, and flavonoids necessary to treat any specific indication.

11. Multi-cultivar entourage effect chart

This chart is able to track a patient's intended intake of cannabinoids and terpenes; displaying what total entourage effect the patient is actually administering when using multiple cultivars to treat a whole indicator/or many; track over long periods of time showing conscious, subconscious, and habitual tendencies, and may even aid in the help of showing if a certain terpene or cannabinoid synergy is a causation to an adverse effect/ susceptibility through post and present

	Total	Morning			Total	Noon			Total	Night	
		CHD	RF			PHZ	FB			HBO	AFG
THC	73.800%	25.589%	73.800%	THC	41.144%	32.304%	41.144%	THC	68.692%	64.109%	68.692%
THCA	0.856%	0.000%	0.856%	THCA	42.400%	42.400%	0.000%	THCA	4.587%	3.596%	0.991%
THCV	3.562%	0.632%	2.930%	THCV	0.563%	0.563%	0.000%	THCV	0.390%	0.000%	0.390%
CBD	43.973%	42.035%	1.938%	CBD	47.346%	0.000%	47.346%	CBD	0.000%	0.000%	0.000%
CBDA	1.601%	1.601%	0.000%	CBDA	0.000%	0.000%	0.000%	CBDA	0.000%	0.000%	0.000%
CDBV	0.406%	0.406%	0.000%	CDBV	0.000%	0.000%	0.000%	CDBV	0.000%	0.000%	0.000%
CBN	1.860%	0.358%	1.502%	CBN	0.372%	0.372%	0.000%	CBN	0.896%	0.000%	0.896%
CBG	6.411%	1.621%	4.790%	CBG	0.815%	0.815%	0.000%	CBG	2.608%	2.208%	0.400%
CBC	6.598%	3.536%	3.062%	CBC	0.387%	0.387%	0.000%	CBC	1.990%	0.675%	1.315%

	Total	Morning			Total	Noon			Total	Night	
		CHD	RF			PHZ	FB			HBO	AFG
B-Carophyllene	1.405%	0.909%	0.496%	B-Carophyllene	1.276%	0.525%	0.751%	rophyllene	3.259%	1.503%	1.756%
B-Myrcene	3.129%	2.538%	0.591%	B-Myrcene	12.026%	8.279%	3.747%	B-Myrcene	3.991%	1.346%	2.645%
B-Pinene	0.362%	0.294%	0.068%	B-Pinene	0.316%	0.204%	0.112%	B-Pinene	0.923%	0.267%	0.656%
a-Pinene	0.734%	0.693%	0.041%	a-Pinene	1.327%	1.260%	0.067%	a-Pinene	2.264%	0.729%	1.535%
Limonene	0.746%	0.746%	0.000%	Limonene	1.678%	0.998%	0.680%	Limonene	1.161%	0.819%	0.342%
Linalool	0.000%	0.000%	0.000%	Linalool	0.734%	0.575%	0.159%	Linalool	0.562%	0.228%	0.334%
Bisabolo	0.385%	0.252%	0.133%	Bisabolo	0.295%	0.203%	0.092%	Bisabolo	0.463%	0.239%	0.224%
Humulene	0.732%	0.408%	0.324%	Humulene	0.819%	0.154%	0.665%	Humulene	0.896%	0.464%	0.432%
Terpinolene	0.842%	0.041%	0.801%	Terpinolene	0.070%	0.057%	0.013%	erpinolene	0.169%	0.054%	0.115%
Valancene	0.000%	0.000%	0.000%	Valancene	0.000%	0.000%	0.000%	Valancene	0.214%	0.000%	0.214%

perceptions with DATA to track primary, ancillary, and supplementary levels of an encourage (**Figure 3**).

Here in (**Figure 4**) every cultivar is broken down into the accompanying cannabinoids to be mixed and matched among other cultivars and their accompanying cannabinoids.

In (**Figure 5**) all terpenoids are separated to then be compared and contrasted to either amplify or avoid a specific terpene encourage.

Cultivator	Strain	Acronym	Plant speciation (BLM-NLM)
Prime	Purple Haze	PHZ	NLMD
Cresco	Rocket Fuel	RF	MLM
Ilera	Freedom blend	FB	1:1 BLMD
Standard Farms	Afgooey	AFG	BLM
Cresco	Honey Boo	HBO	BLM
Prime	Cherry CBD	CHD	1:1 BLMD

Figure 3.
Genetic cultivar key Ryan McKinley 2020.

	Strain					
Cannabinoids	PHZ	RF	FB	AFG	HBO	CHD
THC	32.304%	73.800%	41.144%	68.692%	71.652%	25.589%
THCA	42.400%	0.856%	0.000%	0.991%	0.398%	0.000%
THCV	0.563%	2.930%	0.000%	0.390%	0.640%	0.632%
CBD	0.000%	1.938%	47.346%	0.000%	0.372%	42.035%
CBDA	0.000%	0.000%	0.000%	0.000%	0.000%	1.601%
CDBV	0.000%	0.000%	0.000%	0.000%	0.000%	0.406%
CBN	0.372%	1.502%	0.000%	0.896%	0.364%	0.358%
CBG	0.815%	4.790%	0.000%	0.400%	3.088%	1.621%
CBC	0.387%	3.062%	0.000%	1.315%	1.004%	3.536%

Figure 4.
Unraveled cannabinoids from six different chemovars to be mixed and matched for a new entourage or combined synergistic experience (Ryan McKinley 2020).

	Strain					
Terpenoids	PHZ	RF	FB	AFG	HBO	CHD
B-Carophyllene	0.525%	0.496%	0.751%	1.756%	1.503%	0.909%
B-Myrcene	8.279%	0.591%	3.747%	2.645%	1.346%	2.538%
B-Pinene	0.204%	0.068%	0.112%	0.656%	0.267%	0.294%
a-Pinene	1.260%	0.041%	0.067%	1.535%	0.729%	0.693%
Limonene	0.998%	0.000%	0.680%	0.342%	0.819%	0.746%
Linalool	0.575%	0.000%	0.159%	0.334%	0.228%	0.000%
Bisabolo	0.203%	0.133%	0.092%	0.224%	0.239%	0.252%
Humulene	0.154%	0.324%	0.665%	0.432%	0.464%	0.408%
Terpinolene	0.057%	0.801%	0.013%	0.115%	0.054%	0.041%
Valancene	0.000%	0.000%	0.000%	0.214%	0.000%	0.000%

Figure 5.
Unraveled terpenoids from six different chemovars to be mixed and matched (Ryan McKinley, 2020).

In a MCE2C schedule, a patient can visually see the expected entourage from multiple cultivars, patients can add or remove as many strains to achieve a desired balance/homeostasis.

This schedule lasted until certain strains were no longer readily available (i.e. PHZ & AFG) but was later documented after the original multi cultivar entourage was concluded. that patient one replaced PHZ with RF and FB replaced AFG.

12. Conclusion and future research

In bringing this infinitely hybridized herbal Rubik's cube full "Square," the cascade of entourages displayed from cannabis and the cannabinoids, terpenes, and flavonoids therein have potential and viable medicinal purposes. The research studies used in this chapter cover centuries of cannabis usage since 4000 BC through modern pharmaceutical research and its history in North and South American society circa 2019.

This research identified people under systemic, invoked, or prolonged mental or physical trauma will be susceptible to agitation from certain cannabis inter-entourages consisting of stimulating terpenes (specifically D-limonene and its co-related cannabinoid synergies) more so than healthy individuals, thusly experiencing a paradoxical effect.

Individuals who have a propensity to experience stimulants in a biphasic manner have formidable disadvantages with dopamine uptake, storage, and/or metabolism. Limonene affects A_{2A} allowing a catch 22 of symptoms ranging from a general excitability/alertness via D-limonene. D-Limonene is also an anti-depressant via CBD from 5-HT_{1A} (serotonin, for reduction in neuronal excitability and firing), unless serotonin levels are low or depleted then perceptions may shift to the more agitative especially with an A_{2A} agitator present in the entourage.

ATD or ASR, happens when specific synergistic cannabinoids and terpenes already using serotonin and use reserves consecutively, at a more advanced rate of depletion due to the chemical nature of another synergistic action of the same source of depletion or; whilst a consecutive respectively similar (i.e., dopamine and serotonin) depletion ensues. It is fundamentally understood, that without tryptophan, most, if not all serotonin production would cease in the human body. Henceforth, studies state respectable levels of serotonin in the body inhibit dopamine production, calming down impulsive behavior and side effects of dopamine abuse either natural (i.e., habit) or foreign (i.e., drug).

The 5-HT_{2A} dependent cannabinoid THC uniquely being a synergistic vehicle for a few terpenes and a prime dictator of the psychoactive engine size guarantees the highly lipophilic D-limonene transportation through the blood–brain barrier. These two powerhouses in their own synergistic pathological right actuate dopamine and use of the serotonergic pool. Therefore paramount to uphold a strict healthy diet to supplement basic natural dopamine use and tryptophan depletion that may happen in patients with said indications is pertinent while medicating with cannabis to ensure true beneficial homeostasis from any entourage.

Acknowledgements

Special thanks to Dr. Tamara A. Schiappa.
Special thanks to Dr. Joseph Hulihan.
Special thanks to Brandon Allen.
Declaration Section

Authors' contributions

Not Applicable.

Funding

Not Applicable.

Competing interests

Not Applicable.

Consent for publication

Not Applicable.

Ethical approval and consent to participate

Not Applicable.

Availability of supporting data

Free Access.

Abbreviation

ASSR/ASR	Advanced Synergistic Serotonin Release
ATD	Advanced Tryptophan Depletion
ECS	Endo-Cannabinoid System
CNS	Central Nervous System
PA	Psychoactive
N-PA	Non-Psychoactive
CBG	Cannabigerol
THC	Tetrahydrocannabinol
THCA	Tetrahydrocannabinolic Acid
THC-V	Tetrahydrocannabivarin
CBN	Cannabinol
CBC	Cannabichromene
CBE	Cannabielsoin
CBL	Cannabicyclol
CBD	Cannabidiol
CBDA	Cannabidiolic Acid
CBDV	Cannabidivarin
BLM	Broad Leaf Marijuana
BLMD	Broad Leaf Marijuana Dominant
MLM	Medium Leaf Marijuana
NLMD	Narrow Leaf Marijuana Dominant

NLM	Narrow Leaf Marijuana
CB1	Cannabinoid Receptor type1
CB2	Cannabinoid Receptor type 2
TRPV1	Vanilloid Receptor 1
TRPA1	Transient Receptor Potential Ankyrin 1
GABBA-A	γ-aminobutyric acid
5-HT	Hydroxy-Tryptamine
TRPV3	Transient Receptor Potential cation channel, subfamily V, member 3
ADHD	Attention Deficit Hyperactivity Disorder
GABAA	γ-aminobutyric acid

Author details

Ryan Lucas McKinley
Slippery Rock University, Slippery Rock, Pennsylvania, USA

*Address all correspondence to: dj8t6d@icloud.com

References

[1] Langenheim JH. Higher plant terpenoids: A phytocentric overview of their ecological roles. J Chem Ecol. 1994;**20**:1223-1280. DOI: 10.1007/BF02059809

[2] Hendriks H, Malingre TM, Batterman S, Bos R. Mono-terpene and sesqui-terpene hydrocarbons of essential oil of cannabis-SATIVA. Phytochemistry. 1975;**14**(3):814-815

[3] Hendriks HFJ et al. Alkanes of the essential oil of Cannabis sativa. Phytochemistry. 1977;**16**:719-721

[4] Malingré T, Hendriks H, Batterman S, Bos R, Visser J. The essential oil of Cannabis sativa. Planta Med. 1975 Aug;**28**(1):56-61. DOI: 10.1055/s-0028-1097829

[5] Guerrero Davalos S, Boucher F, Fournier G, Paris M. Analyse d'une population de Cannabis sativa L. originaire du Mexique et cultivé en France [Analysis of a population of Cannabis sativa L. originating from Mexico and cultivated in France (author's transl)]. Experientia. 1977;**33**(12):1562-1563. French. DOI: 10.1007/BF01933997

[6] Ross SA, ElSohly MA. The volatile oil composition of fresh and air-dried buds of Cannabis sativa. J Nat Prod. 1996; **59**(1):49-51. DOI: 10.1021/np960004a

[7] Mediavilla V, Steinemann S. Essential oil of Cannabis sativa L. strains. J. Ind. Hemp. 1997;**4**:80-82

[8] Cravatt BF, Giang DK, Mayfield SP, Boger DL, Lerner RA, Gilula NB. Molecular characterization of an enzyme that degrades neuromodulatory fatty-acid amides. Nature. 1996;**384**: 83-87. [PubMed: 8900284]

[9] Katrina W.-G. The united chemicals of cannabis: Beneficial effects of cannabis phytochemicals on the brain and cognition. 2018

[10] Poirier H, Braissant O, Niot I, Wahli W, Besnard P. 9-cis-retinoic acid enhances fatty acid-induced expression of the liver fatty acid-binding protein gene. FEBS Letters. 1997;**412**:480-484

[11] Hood LV, Dames ME, Barry GT. Headspace volatiles of marijuana. Nature. 1973;**242**:402-403

[12] Turner CE, Elsohly MA, Boeren EG. Constituents of Cannabis sativa L. XVII. A review of the natural constituents. J Nat Prod. 1980;**43**:169-234

[13] Ross SA, ElSohly MA. The volatile oil composition of fresh and air-dried buds of Cannabis sativa. J Nat Prod. 1996;**59**(1):49-51. DOI: 10.1021/np960004a

[14] ElSohly MA, Slade D. Chemical constituents of marijuana: The complex mixture of natural cannabinoids. Life Sciences. 2005;**78**:539-548

[15] Russo EB, Marcu J. Cannabis Pharmacology: The Usual Suspects and a Few Promising Leads. Los Angeles, CA, United States; Noma & Asakawa: PHYTECS;

[16] Köse EO, Deniz IG, Sarıkürkçü C, Aktaş O, Yavuz M. Chemical composition, antimicrobial and antioxidant activities of the essential oils of Sideritis erythrantha Boiss. and Heldr. (var. erythrantha and var. cedretorum P.H. Davis) endemic in Turkey. Food Chem Toxicol. 2010;**48**(10):2960-2965. DOI: 10.1016/j.fct.2010.07.033. Epub 2010 Jul 27

[17] Russo EB, Guy GW, Robson PJ. Cannabis, pain, and sleep: lessons from therapeutic clinical trials of Sativex, a cannabis-based medicine. Chem Biodivers. 2007;**4**(8):1729-1743. DOI: 10.1002/cbdv.200790150

[18] Volicer L, Stelly M, Morris J, McLaughlin J, Volicer BJ. Effects of

dronabinol on anorexia and disturbed behavior in patients with Alzheimer's disease. Int J Geriatr Psychiatry. 1997;**12**(9):913-919

[19] Lee GY, Lee C, Park GH, Jang J. Amelioration of scopolamine-induced learning and memory impairment by α-pinene in C57BL/6 mice. Evidencebased Complementary and Alternative Medicine. 2017;**2017**:4926815. DOI: 10.1155/2017/4926815. Epub 2017 Nov 1

[20] Perry NS, Houghton PJ, Theobald A, Jenner P, Perry EK. In-vitro inhibition of human erythrocyte acetylcholinesterase by salvia lavandulaefolia essential oil and constituent terpenes. J Pharm Pharmacol. 2000;**52**(7):895-902. DOI: 10.1211/0022357001774598. Erratum in: J Pharm Pharmacol 2000 Dec;52(12):203

[21] Miyazawa M, Yamafuji C. Inhibition of acetylcholinesterase activity by bicyclic monoterpenoids. J Agric Food Chem. 2005;**53**(5):1765-1768. DOI: 10.1021/jf040019b

[22] Pattwell SS, Liston C, Jing D, Ninan I, Yang RR, Witztum J, et al. Dynamic changes in neural circuitry during adolescence are associated with persistent attenuation of fear memories. Nature Communications. 2016;**7**:11475

[23] Russo EB, Burnett A, Hall B, et al. Agonistic properties of cannabidiol at 5-HT1a receptors. Neurochemical Research. 2005;**30**:1037-1043

[24] Hill AJ, Mercier MS, Hill TD, Glyn SE, Jones NA, Yamasaki Y, et al. Cannabidivarin is anticonvulsant in mouse and rat. Br J Pharmacol. 2012;**167**(8):1629-1642. DOI: 10.1111/j.1476-5381.2012.02207. https://www.ncbi.nlm.nih.gov/pubmed/22970845)

[25] Levick JR. Chapter 14.1, Sympathetic vasoconstrictor nerves. In: Levick JR, editor. Cardiovascular Physiology. 3rd ed. Arnold Publishers; 2000

[26] Giovannitti JA, Thoms SM, Crawford JJ. Alpha-2 adrenergic receptor agonists: A review of current clinical applications. Anesthesia Progress. 2015;**62**(1):31-38. DOI: 10.2344/0003-3006-62.1.31

[27] Coelho VR, Gianesini J, Von Borowski R, Mazzardo-Martins L, Martins DF, Picada JN, et al. (−)-linalool, a naturally occurring monoterpene compound, impairs memory acquisition in the object recognition task, inhibitory avoidance test and habituation to a novel environment in rats. Phytomedicine. 2011;**18**(10):896-901. DOI: 10.1016/j.phymed.2011.02.010

[28] Costain WJ, Laprairie RB, editors. Recent Advances in Cannabinoid Research. London, United Kingdom: IntechOpen; 2018. pp. 83-100 (Neurochemical Research, Vol. 26, No. 3, 2001, pp. 191-194

[29] Silva Brum LF, Emanuelli T, Souza DO, Elisabetsky E. Effects of linalool on glutamate release and uptake in mouse cortical synaptosomes. Neurochemical Research. 2001;**26**(3):191-194. DOI: 10.1023/a:1010904214482

[30] Li F, Tsien JZ. Memory and the NMDA receptors. The New England Journal of Medicine. 2009;**361**(3):302-303. DOI: 10.1056/NEJMcibr0902052

[31] Wong C-S, Cherng C-H, Luk H-N, Ho S-T, Tung C-S. Effects of NMDA receptor antagonists on inhibition of morphine tolerance in rats: Binding at μ-opioid receptors. European Journal of Pharmacology. 1996

[32] Brum LF, Emanuelli T, Souza DO, Elisabetsky E. Effects of linalool on glutamate release and uptake in mouse cortical synaptosomes. Neurochemical Research. 2001;**26**:191-194

[33] Meldrum BS. Update on the mechanisms of action of antiepileptic drugs. Epilepsia. 1996;**37**:4

[34] Franz C, Novak J. Sources of essential oils. In: KHC B, Buchbauer G, editors. Handbook of Essential Oils: Science, Technology, and Applications. Boca Raton, FL: CRC Press; 2010. pp. 39-82

[35] Bowles EJ. The Chemistry of Aromatherapeutic Oils. Routledge; 2020

[36] Adorjan B, Buchbauer G. Biological properties of essential oils: An updated review. Flavour and Fragrance Journal. 2010;**25**(6):407-426

[37] Bigdeli Y, Asle-Rousta M, Rahnema M, et al. Effects of limonene on chronic restraint stressinduced memory impairment and anxiety in male rats. Neurophysiology. 2019;**51**:107-113

[38] ADORA2A adenosine A2a receptor [omo sapiens (human); Gene ID: 135, updated on 13 January 2020

[39] Hajagos-Tóth J, Hódi Á, Seres AB, Gáspár R et al. Department of Pharmacodynamics and Biopharmacy, Faculty of Pharmacy, University of Szeged, Szeged, Hungary _"Effects of d- and l-limonene on the pregnant rat myometrium in vitro". 2015

[40] Russo EB. Taming THC: Potential cannabis synergy and phytocannabinoid-terpenoid entourage effects. British Journal of Pharmacology. 2011;**163**(7): 1344-1364. DOI: 10.1111/j.1476-5381. 2011.01238.x

[41] Davies LP, Hambley JW, Johnston GAR. Reduced adenosine deaminase activity in the CNS of spontaneously-hypertensive rats. Neurochemistry International. 1987;**10**(4):533-536. ISSN: 0197-0186. (https://www.sciencedirect.com/science/article/pii/0197018687900817). DOI: 10.1016/0197-0186(87)90081-7

[42] Young SN. Acute tryptophan depletion in humans: A review of theoretical, practical and ethical aspects. Journal of Psychiatry & Neuroscience: JPN. 2013;**38**(5):294

[43] Namdar D, Voet H, Ajjampura V, Nadarajan S, Mayzlish-Gati E, Mazuz M, et al. Terpenoids and phytocannabinoids co-produced in Cannabissativa strains show specific interaction for cell cytotoxic activity. Molecules. 2019;**24**:3031

[44] Jenny M, Schroecksnadel S, Florian Ü, Fuchs D. The potential role of cannabinoids in modulating serotonergic signaling by their influence on tryptophan metabolism. Pharmaceuticals. 2010;**3**. DOI: 10.3390/ph3082647

[45] Rothschild JM, Keohane CA, Cook EF, Orav EJ, Burdick E, Thompson S, et al. A controlled trial of smart infusion pumps to improve medication safety in critically ill patients. Crit Care Med. 2005;**33**(3):533-540. DOI: 10.1097/01.ccm.0000155912.73313.cd

[46] Potter DJ. The propagation, characterisation and optimisation of Cannabis sativa L. as a phyto-pharmaceutical [PhD], London: King's College; 2009

[47] Fernández-Ruiz J. The endocannabinoid system as a target for the treatment of motor dysfunction. British Journal of Pharmacology. 2009;**156**(7):1029-1040. DOI: 10.1111/j.1476-5381.2008.00088.x

[48] 6 Effects of Cannabinoids That Involve Serotonin Receptors, Proof of Pot. Available from: https://profofpot.com/serotonin-cannabinoids/. 2019

[49] Velenovská M, Fišar Z. Preclinical study: Effect of cannabinoids on platelet serotonin uptake. Addiction Biology. 2007;**12**:158-166. DOI: 10.1111/j.1369-1600.2007.00065.x

[50] Viñals X, Moreno E, Lanfumey L, et al. Cognitive impairment induced by delta9-tetrahydrocannabinol occurs through heteromers between cannabinoid CB1 and serotonin 5-HT2A receptors. PLoS Biology.

2015;**13**(7):e1002194. Published 2015 July 9. DOI: 10.1371/journal.pbio.1002194

[51] De Gregorio D et al. Cannabidiol modulates serotonergic transmission and reverses both allodynia and anxietylike behavior in a model of neuropathic pain. Pain. 2019;**160**(1):136-150. DOI: 10.1097/j.pain.0000000000001386

[52] Sartima AG, Guimar FS, Joca RL. Antidepressant-like effect of cannabidiol injection into the ventral medial prefrontal cortex—Possible involvement of 5-HT1A and CB1 receptors. Behavioural Brain Research. 2016;**303**(15):218-227

Section 2

Sources, Biosynthesis, Diversity and Chemistry of Phenolic Compounds

Chapter 4

Phenolic Compounds: Classification, Chemistry, and Updated Techniques of Analysis and Synthesis

Hamad H. Al Mamari

Abstract

Phenolic compounds are vast, diverse, ubiquitous and widespread in nature. The biological significance of bioactive phenolic natural secondary metabolites is immense and of high and significant importance. Phenolic compounds are known to exhibit various biological activities such as antimicrobial, antioxidant and anti-inflammatory properties. This book chapter begins with classification of phenolic compounds in concise manner followed by going through their chemical properties that are essential for their biological activities. Some chemical properties such as acidity and formation of radicals are directly linked with their important and key biological activities such as antioxidant properties. The chapter covers methods and updated techniques of analysis of phenolic compounds. Finally, biosynthesis of such important organic molecules is covered going through some of their current synthesis methods in the laboratory, methods of their synthetic elaboration. Due to the high potential of phenolic compounds for applications in various industries such as pharmaceutical and food industries, the search for the development of efficient methods for their synthesis as well as modern and accurate methods for their detection and analysis will continue.

Keywords: Phenolic compounds, chemistry of phenolic compounds, biosynthesis of phenolic compounds, phenolics, classification of phenolic compounds

1. Introduction

Phenolic compounds are a diverse class of bioactive secondary metabolites and are of high and significant importance [1–7]. They can be described as compounds that contain a phenol moiety. Phenol itself is a benzene ring that is substituted with a hydroxyl group (**Figure 1**). Thus, its systematic name is hydroxybenzene.

Phenolic compounds display a wide range of biological activities. For instance, they are known to exhibit antioxidants, antimicrobial, and anti-inflammatory properties. They are ubiquitous in nature. For instance, they are present in various types of fruits such as apple, banana, orange, mango, peach, papaya, strawberry, pomegranate, watermelon, and pineapple. For example, myricetin (a flavonol) is found in apple, gallic acid (a hydroxybenzoic acid) is found in banana, quercetin (a flavonol) and cyanidin (an anthocyanin) are found in pomegranate, *p*-coumaric

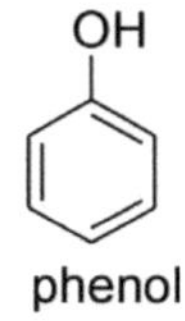

Figure 1.
The structure of phenol.

acid (a hydroxycinnamic acid) and naringenin (a flavanone) are found in orange, vanilic acid (a hydroxybenzoic acid) and resveratrol (a stilbene) are found in strawberry, ferulic acid (a hydroxycinnamic acid) and apigenin (a flavone) are found in mango, and luteolin (a flavone) is found in watermelon and pineapple [8]. The health benefits of phenolic compounds are immense. Since original research articles and reviews on phenolic compounds are numerous in the literature, this chapter is not intended to be a comprehensive review of phenolic compounds. It is rather an attempt to complement existing article and reviews and to serve as a brief reference on their classification, analysis, chemistry, and synthesis.

2. Classification

Phenolic compounds can generally be classified into simple and polyphenolic compounds [9–12].

2.1 Simple phenolic compounds

Phenolic compounds that contain one phenol unit (or a derivative of it) are considered "simple". Fundamentally, they are substituted phenol compounds. Simple phenolic compounds have C_6 general skeleton representation. The general structure is shown below (**Figure 2**). The group denoted by "R" (an organic group which could be alkyl, alkenyl, aryl ...etc. or hydroxy, alkoxy, amino ...etc) which can be in the *ortho* (*o*), *meta* (*m*), or *para* (*p*) positions of the aromatic ring. These descriptors refer, with respect to the position of the hydroxyl group constituting phenol which is given position 1, to 1,2-, 1,3, and 1,4-carbon relationship respectively.

Below are some simple phenolic compounds.

2.1.1 Simple phenolics

Simple substituted phenol compounds can be hydroxyphenols or dihydroxybenzenes. Examples are catechol (1,2-dihydroxybenzene), resorcinol (1,3-dihydroxybenzene), and hydroquinone (1,4-dihydroxybenzene) (**Figure 3**).

Other simple substituted phenol compounds can also be dihydroxyphenols or trihydroxybenzenes. Examples are pyrogallol (1,2,3-trihydroxybenzene), hydroxyquinol (1,2,4-trihydroxybenzene), and phloroglucinol (1,3,5-trihydroxybenzene) (**Figure 4**).

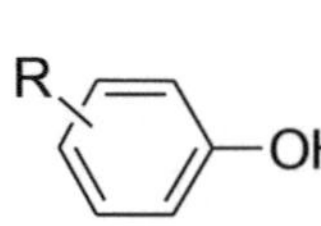

Figure 2.
General structure of simple substituted phenols.

Figure 3.
The structure of hydroxyl-substituted phenols.

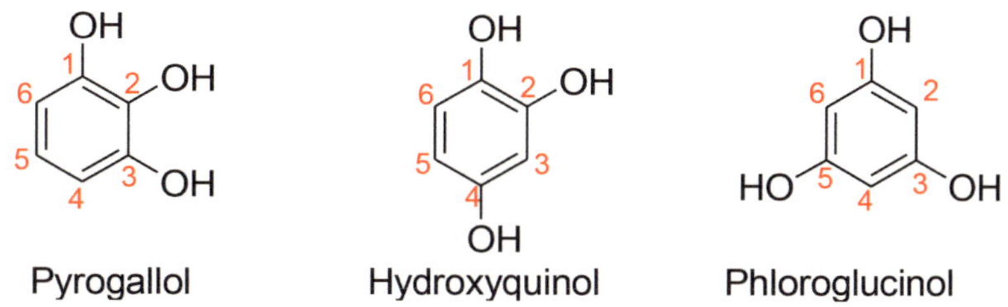

Figure 4.
The structure of dihydroxyl-substituted phenols.

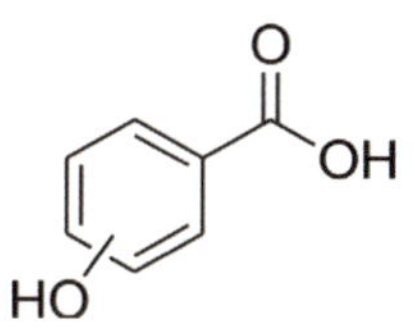

Figure 5.
General structures of hydroxyl-substituted benzoic acids.

2.1.2 Phenolic acids

Phenols that contain a carboxylic acid are termed as phenolic acids. If the carboxylic acid functional group is directly bonded to the phenol ring, the phenolic compound is termed as hydroxybenzoic acid. When carboxylic acid functional group and the phenol ring are separated by two doubly bonded carbons (a C=C bond), phenolic compounds are termed as hydroxycinnamic acids.

2.1.2.1 Hydroxybenzoic acids

Hydroxybenzoic acids are benzoic acids substituted with a hydroxyl group. Alternatively, they can be viewed as phenols that are substituted with a carboxylic acid functional group that is directly bonded to the phenol ring (**Figure 5**).

The hydroxyl group in hydroxybenzoic acids can be *ortho* (*o*) (salicylic acid), *meta* (*m*), or *para* (*p*). The structures are shown below (**Figure 6**).

Dihdyroxybenzoic acids are benzoic acids that are substituted with two hydroxyl groups. The two hydroxyl groups can mainly be in 2,3-, 2,4-, 2,5-, 2,6-, 3,4-, and 3,5-relative positions (**Figure 7**).

Trihdyroxybenzoic acids are benzoic acids that are substituted with three hydroxyl groups. Examples include 2,4,6-trihydroxybenzoic acid and 3,4,5- trihydroxybenzoic acid (gallic acid) (**Figure 8**).

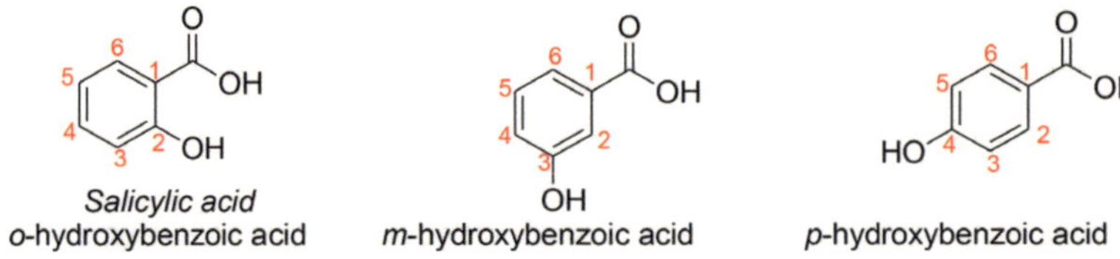

Figure 6.
Structures of hydroxybenzoic acids.

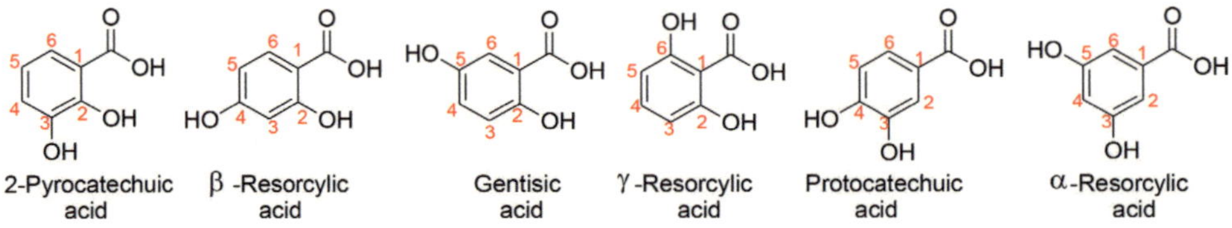

Figure 7.
Structures of main dihydroxybenzoic acids.

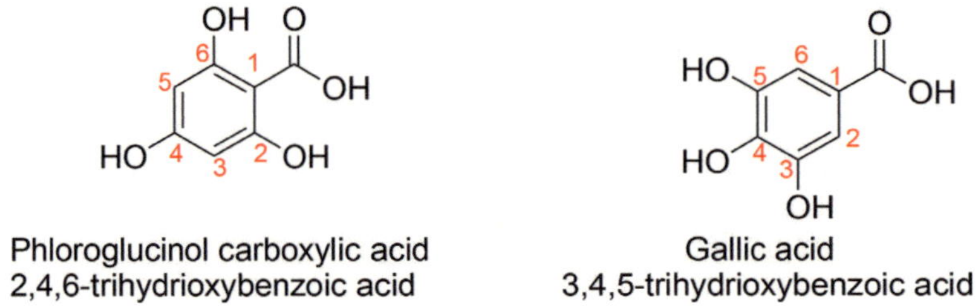

Figure 8.
Examples of triihydroxybenzoic acids.

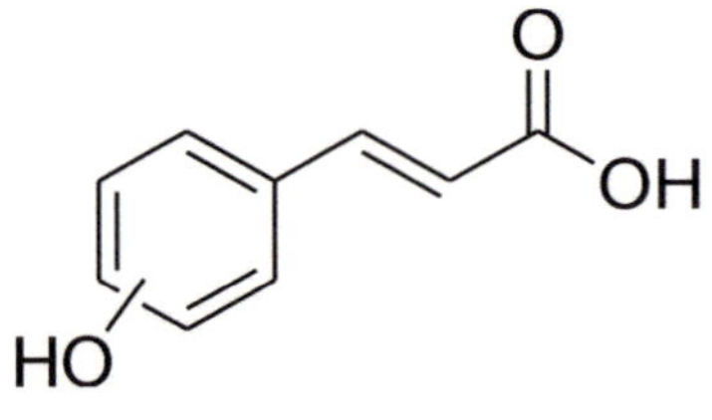

Figure 9.
General structures of hydroxyl-substituted cinnamic acids.

2.1.2.2 Hydroxycinnamic acids

When the carboxylic acid functional group is separated from the phenol ring by a C=C bond, phenolic acids are described as hydroxycinnamic acids (**Figure 9**).

Examples of hydroxycinnamic acids are 2-, 3-, and 4-hydroxycinnmaic acid shown below (**Figure 10**).

Other common examples of cinnmaic acids are caffeic acid, ferulic acid, and sinapic acids shown below (**Figure 11**).

2.1.2.3 Coumarins

Hydroxycoumarins are hydroxyl-substituted coumarins (**Figure 12**). They are examples of phenolic compounds.

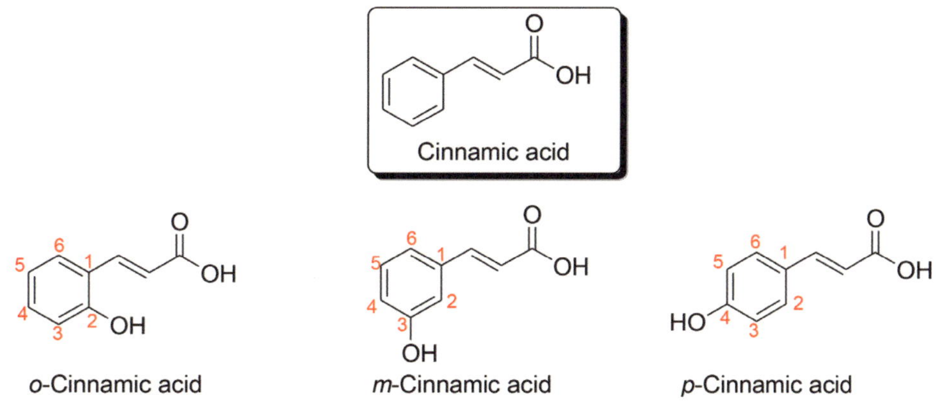

Figure 10.
Examples of hydroxycinnamic acids.

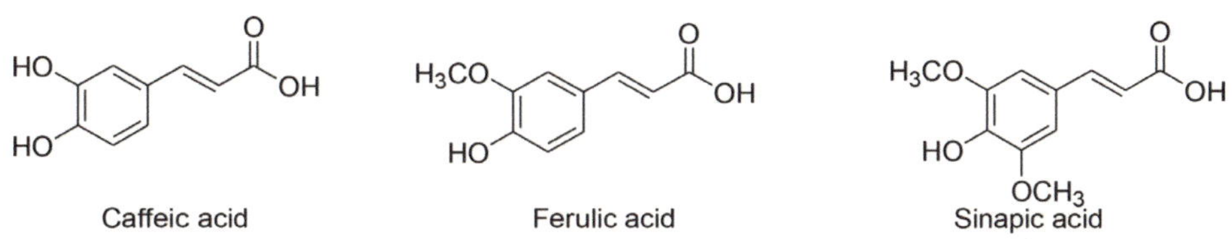

Figure 11.
Common examples of hydroxycinnamic acids.

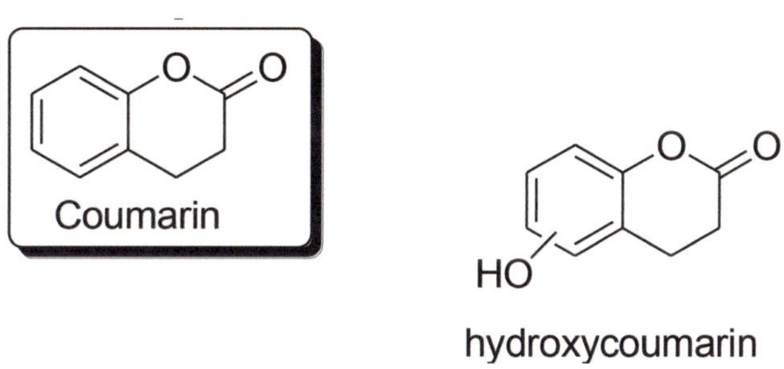

Figure 12.
General structure of hydroxycoumarins.

Examples of hydroxycoumarins are scopoletin and auraptene (**Figure 13**).

2.2 Polyphenols

Phenolic compounds that contain more than one phenol unit are considered "polyphenol". Polyphenolic compounds have C_{15} general skeleton representation.

2.2.1 Flavonoids

Flavonoids are polyphenolic compounds with the general structure shown below (**Figure 14**).

Generally, rings A and C are either mono, di, or trihydroxylated. The *O*-heterocycle B is usually a pyrone ring as in Luteolin but could also be a pyrlium ring as in delphinidin (**Figure 15**). If ring C is attached to C2 of ring B, the flavonoid is a flavone (as Luteolin), flavonol (as kaempferol), an anthocyanin (as delphindin) or a flavanone (as naringenin). If the ring C is attached to C3 of ring B, then the

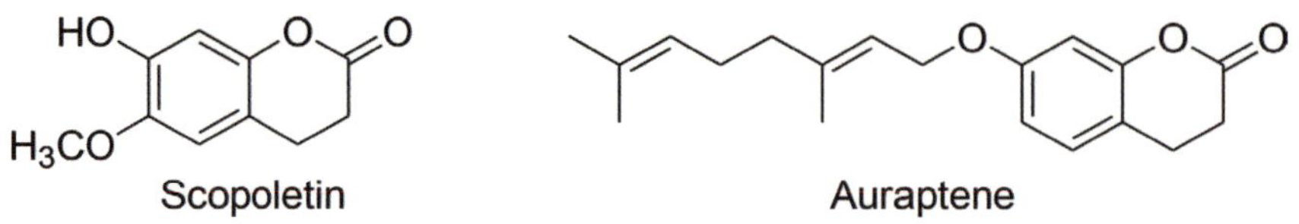

Figure 13.
Examples of hydroxycoumarins.

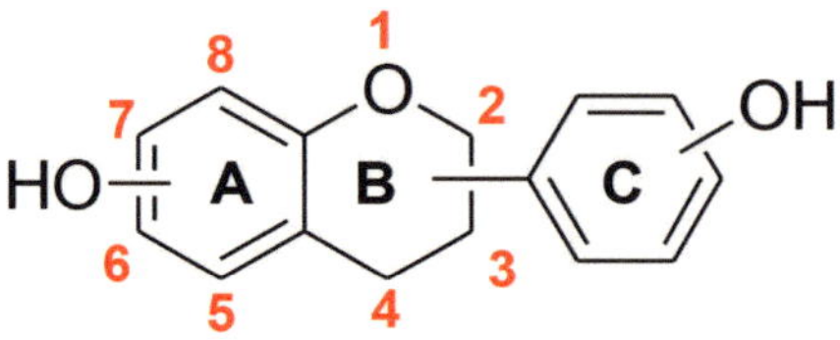

Figure 14.
General structure of flavonoids.

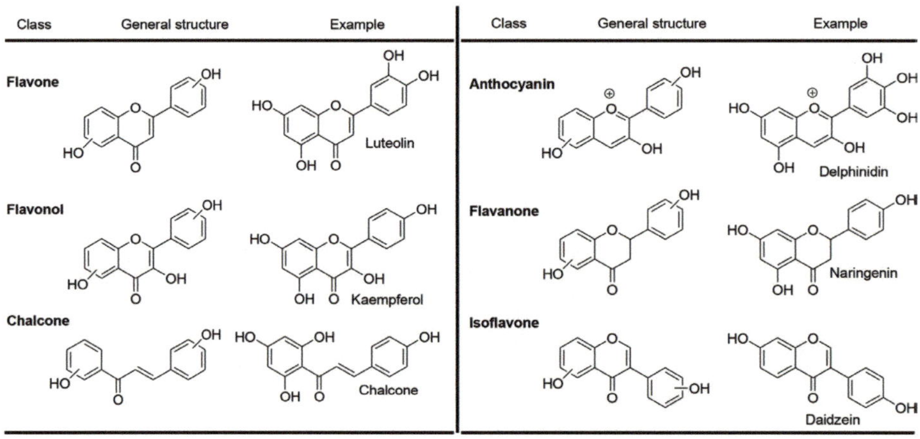

Figure 15.
Classification of falvonoids.

flavonoid is an isoflavone such as daidzein. Chalcones such as chalcone, are a class of flavonoids in which rings A and C are separated by 3-carbon linear chain rather than a ring. The bond between C2 and C3 of ring B is commonly double as in flavones, flavonols, chalcones and isoflavones. However, the C2-C3 bond could be single as in flavanones.

2.2.2 Tannins

Tannins are known to bind to and precipitate proteins and amino acids. They are subdivided into three types; hydrolyzable, condensed and complex. Hydrolyzable tannins can be gallotannins or ellagitannins. Gallotannins are polyols that are substituted with gallic acid units. The galloyl units in gallotannins are linked by depside (ester) linkages. Commonly the polyol core is a D-glucose that is substituted with gallic acid units. Tannic acid is an example of gallotannins (**Figure 16**).

Similar to gallotannins, ellagitannins are hydrolysable 1,2,3,4,6-pentagalloyl-glucose. However, unlike gallotannins characterized by depside linkages, adjacent galloyl groups in ellagitannins are linked by C-C bonds (**Figure 17**).

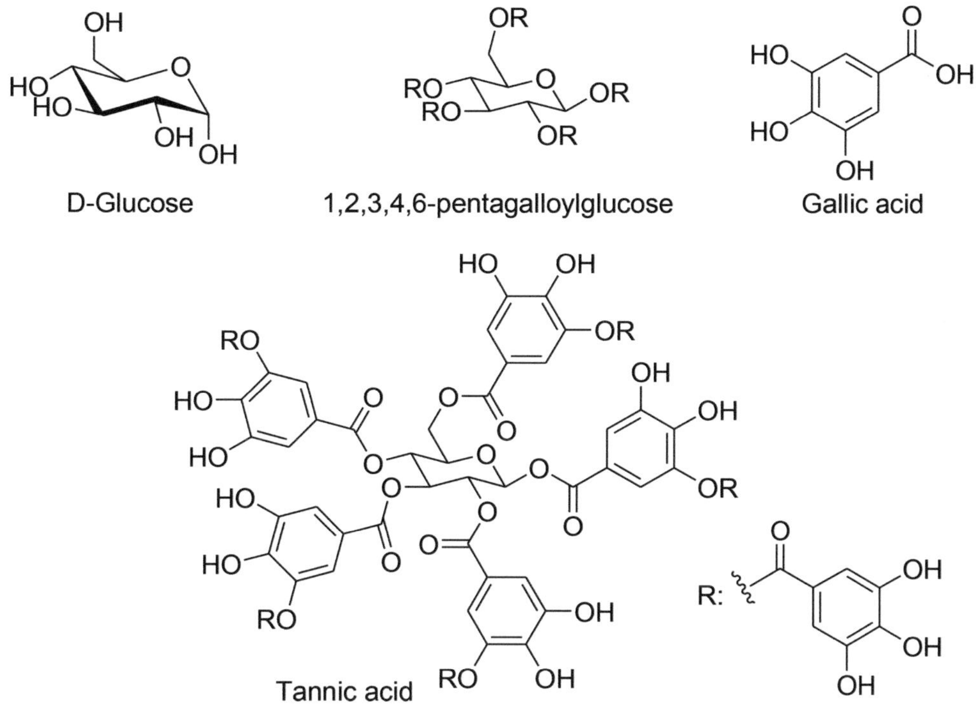

Figure 16.
Gallotannins.

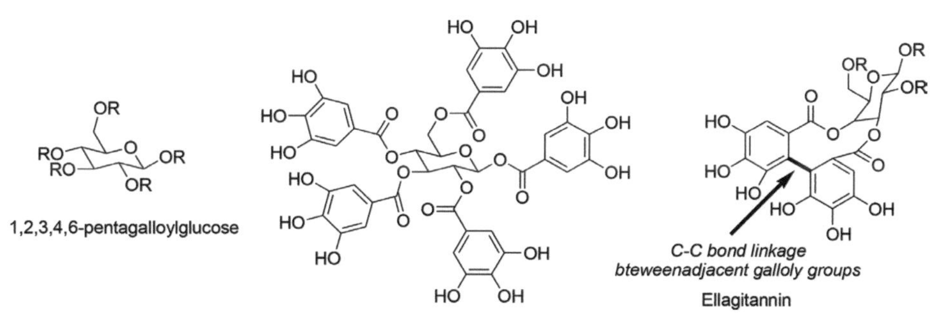

Figure 17.
Structure of ellagitannin.

Condensed tannins (**Figure 18**) are polymeric phenolic compounds that consist of catechin units. When depolymerized, they give anthocyanidin. Thus condensed tannins are called proanthocyanidins.

Complex tannins are gallotannins or ellagitannins bonded to a catechin unit (**Figure 19**).

2.3 Others phenolic compounds

2.3.1 Stilbenes

Stilbenes are phenolic compounds in which two phenol units are linked by two-doubly bonded carbons (**Figure 18**). Examples of stilbenes include resveratrol, pterostilbene and piceatannol shown (**Figure 20**).

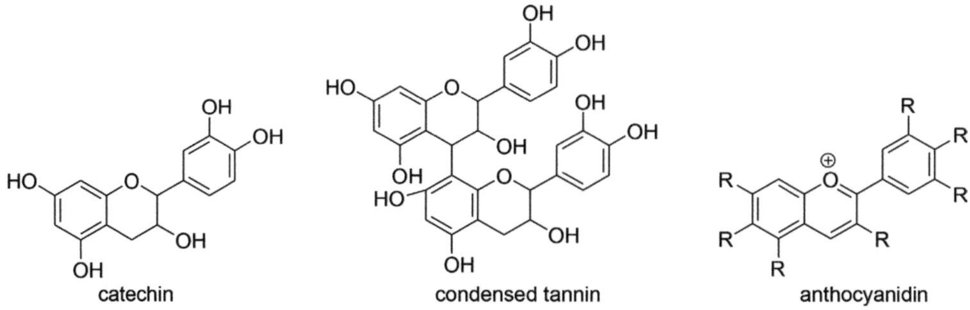

Figure 18.
General structure of condensed tannins.

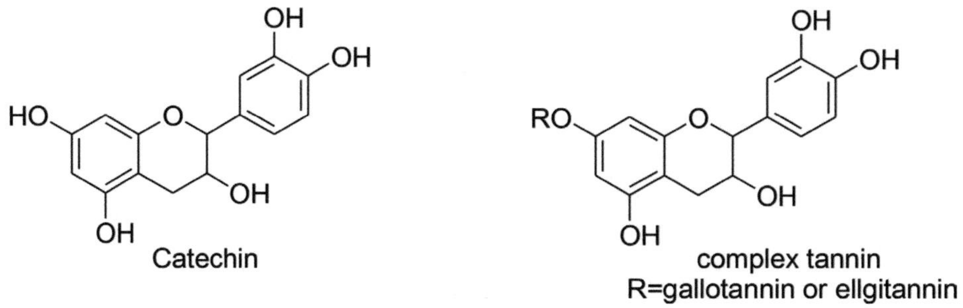

Figure 19.
General structure of complex tannins.

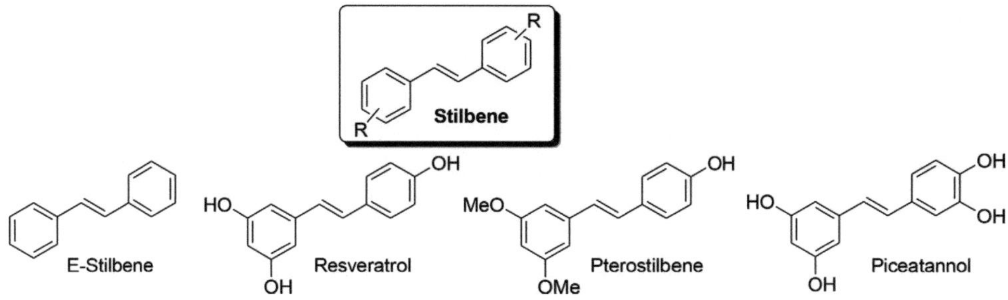

Figure 20.
Stilbenes.

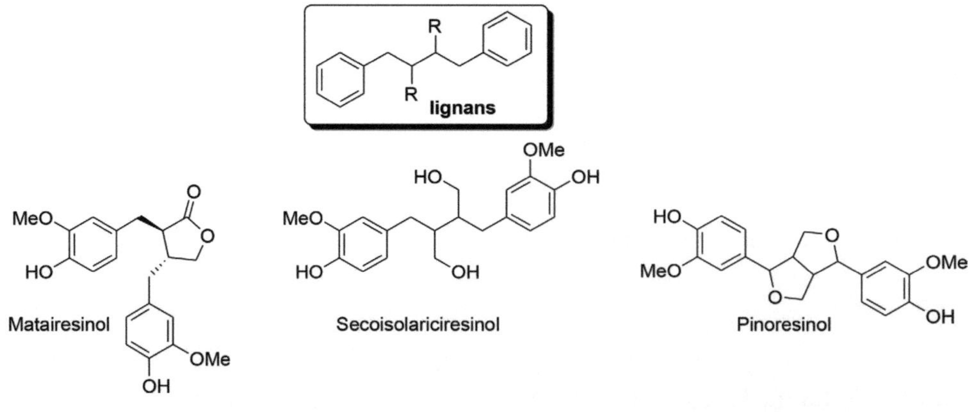

Figure 21.
General structure of lignans and examples.

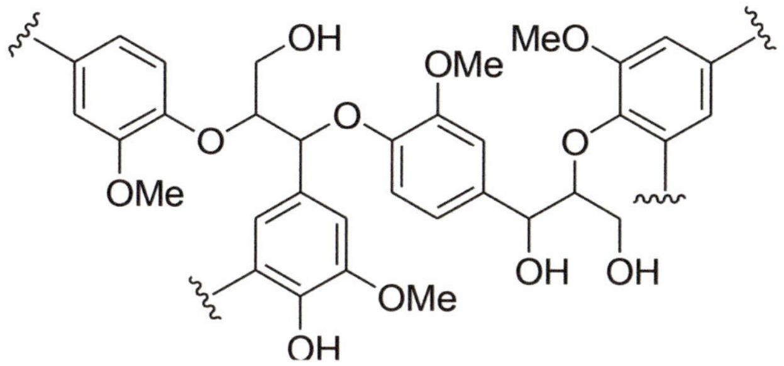

Figure 22.
A segment of lignins.

2.3.2 Lignans

Lignans consist of two phenol units linked by four carbons. Examples include matairesinol, secoisolariciresinol and pinoresinol (**Figure 21**).

2.3.3 Lignins

Lignins consist of phenol units or phenolic compounds that are linked with each other by carbon chains (**Figure 22**). Lignins are high molecular weight polymers.

3. Chemistry

3.1 Acidity

An important chemical feature of phenolic compounds is the acidity of the phenol moiety. The unequal shift of electrons in the O-H bond in phenol is caused by the difference in electronegativity between H and O. The arbitrary electronegativity values according to Pauling scale are 2.1 and 3.5 respectively. Thus the formed inductive effect imparts a positive partial charge on the H atom (**Figure 23**). Thus the H atom is removable in the form of a proton by a suitable base. The pKa of phenol is 9.9, relatively stronger as an acid than aliphatic alcohols (pKa ca. 16) [13].

The resultant conjugate base, the phenoxide ion, is further stabilized by resonance (**Figure 24**). The lone pair placed as a result of proton abstraction is delocalized over the phenyl ring. Electron delocalization by resonance results in stabilization of the phenoxide ion.

Substituents on the phenol ring can have a significant effect on the acidity of phenol (**Figure 25**). For instance, electron withdrawing groups (EWG) increase

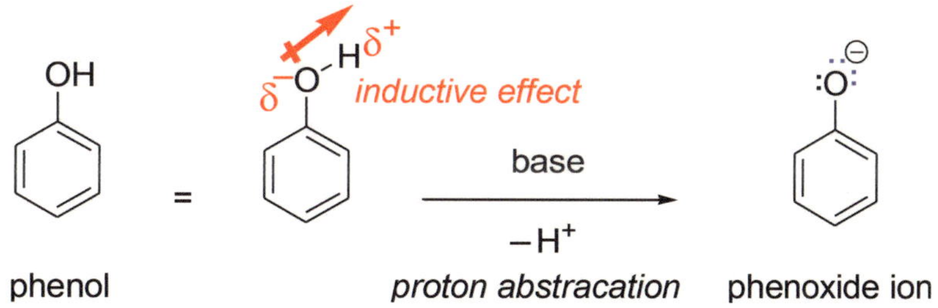

Figure 23.
Acidity of phenol caused by inductive effect.

the acidity of phenol. EWG stabilize the phenoxide ion further by inductive and resonance effects. On the other hand, electron donating groups (EDG) decrease the acidity of phenol. EDG lower the stability of the phenoxide ion by donating of electrons by inductive or resonance effects.

For example, 4-nitrophenol with a pKa of around 7 is more acidic than phenol itself. The nitro group withdraws electrons by resonance and thus imparts an additional resonance stabilization of the phenoxide ion (**Figure 26**). Thus, 4-nitrophenoxide ion is more stable than the simple unsubstituted phenoxide ion.

For example, 4-aminophenol with a pKa of around 10.3, is less acidic than phenol itself. The nitro group imparts an additional resonance stabilization of the phenoxide ion which is then more stable than the simple unsubstituted phenoxide ion (**Figure 27**).

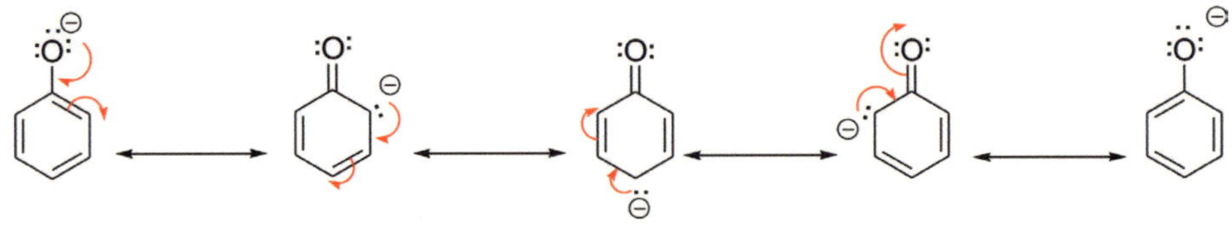

Figure 24.
Resonance stabilization of the conjugate base of phenol, the phenoxide ion.

Figure 25.
Effects of substituents on the acidity of phenol.

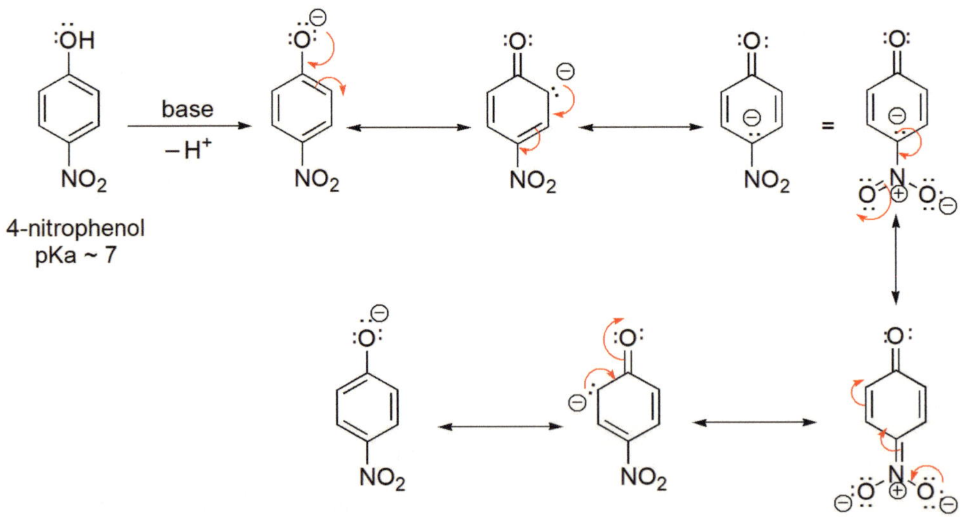

Figure 26.
Resonance stabilization of the conjugate base of 4-nitrophenol.

3.2 Hydrogen bonding

The inductive effect of the O-H bond in phenol induces a negative partial charge on O and a positive partial charge on H. Therefore, the hydrogen (H) atom can interact with heteroatoms possessing nonbonding electrons, such as O, N, F. This type of interaction is noncovalent and rather electrostatic and constitutes hydrogen-bonding (H-bonding) [14]. The H atom of the O-H bond in phenol can form a H-bond with the O atom in another phenol molecule, constituting intramolecular H-boning (**Figure 28**). In addition, the H atom is also capable of interacting with heteroatoms in other molecules to form intermolecular H-bonding.

Phenolic compound with adjacent hydroxyl groups such as protocathechuic acid, can exhibit intramolecular H-bonding (**Figure 29**). Phenolic compounds with adjacent hydroxyl and alkoxy groups are also capable of intramolecular H-bonding.

Another structural possibility for intramolecular H-bonding is the presence of a hydroxyl group *ortho* to a carbonyl group as in butein (a chalcone type) (**Figure 30**).

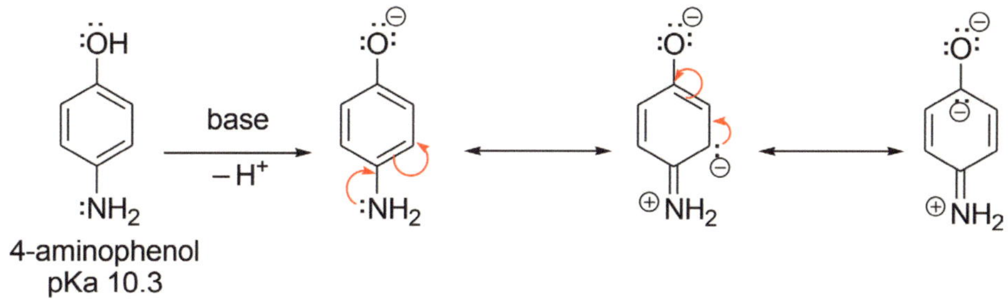

Figure 27.
Resonance stabilization of the conjugate base of 4-aminophenol.

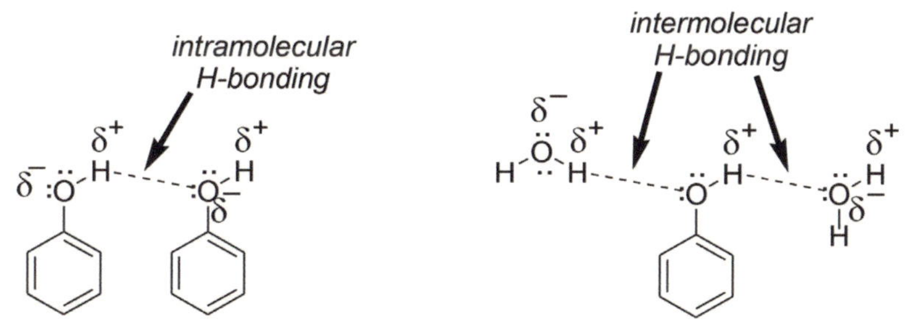

Figure 28.
Intramolecular and intermolecular H-bonding of phenol.

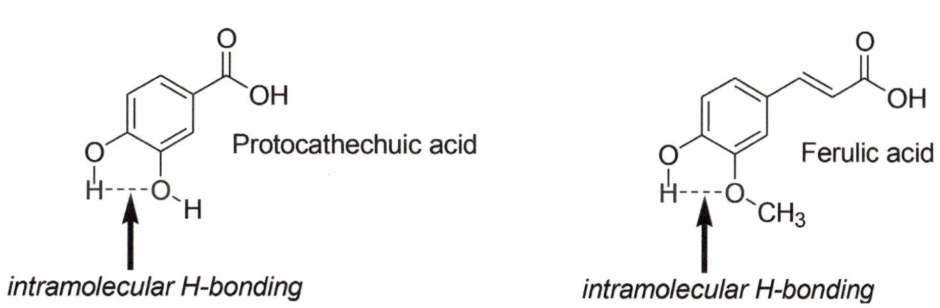

Figure 29.
Intramolecular H-bonding of phenolic compounds.

Another structural possibility is the presence of a hydroxyl group and a carbonyl group separated by a ring junction as in flavanone.

Intramolecular H-bonding can result in formation of five-membered rings as in Ferulic acid (**Figure 29**) and six-membered rings as in Flavanone (**Figure 30**). Such rings are inherently stable which would consequently influence the chemistry of phenolic compounds. For instance, intramolecular H-bonding can lower the solubility and reactivity of phenolic compounds in esterification and etherification reactions.

3.3 Esterification reactions

Phenolic compounds can take part in esterification reactions. They can contribute the phenolic hydroxyl group upon reactions with a carboxylic acid or a carboxylic acid derivative such as acid anhydrides or acid halides typified by acid chlorides (**Figure 31**), forming phenolic esters.

The other esterification possibility of phenolic compounds is for them to contribute their carboxyl group upon reactions with alcohols to produce the corresponding phenolic ester (**Figure 32**).

3.4 Etherification reactions

Phenolic compounds can undergo etherification reactions. Thus they can react with alcohols to produce the corresponding phenolic ether (**Figure 33**).

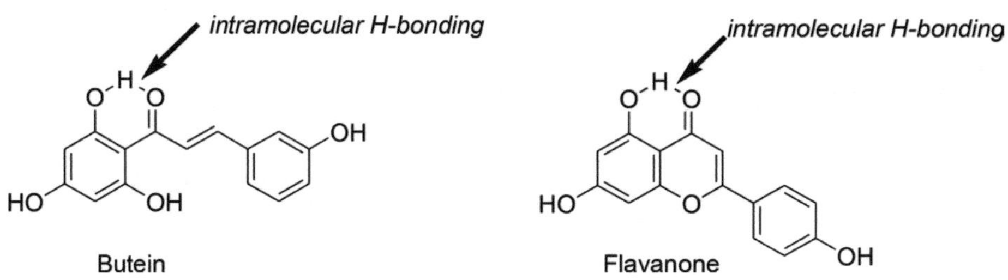

Figure 30.
Intramolecular H-bonding of phenolic compounds.

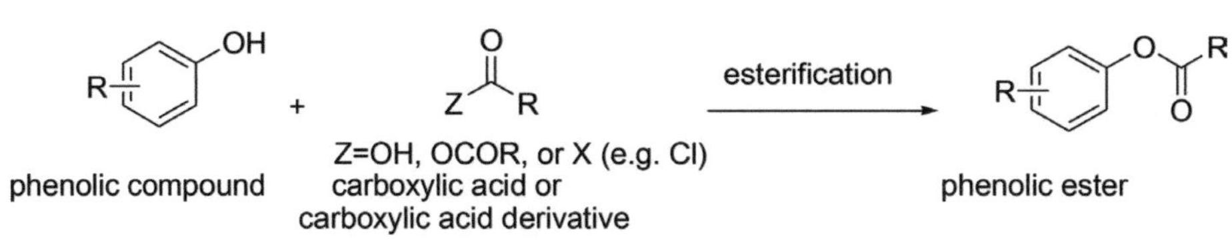

Figure 31.
Esterification of phenolic compounds.

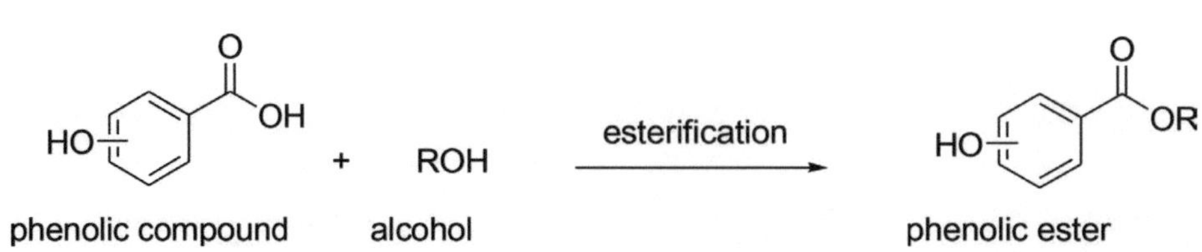

Figure 32.
Esterification of phenolic compounds.

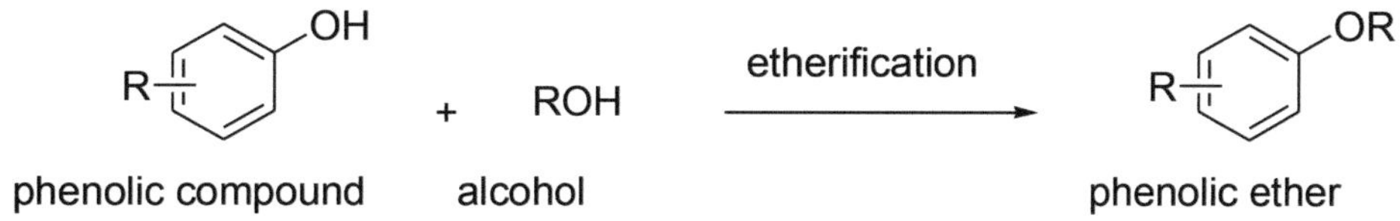

Figure 33.
Etherification of phenolic compounds.

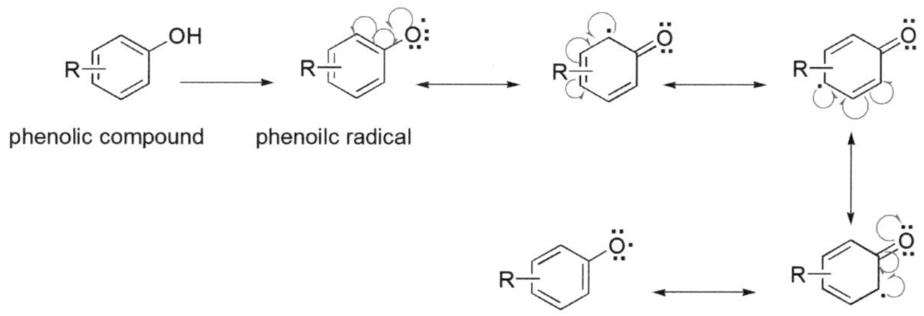

Figure 34.
Oxidation of phenolic compounds to form phenolic radicals.

3.5 Oxidation

Phenolic compounds can undergo oxidation reactions. Homolytic (symmetrical) oxidative O-H bond cleavage gives rise to a phenolic radical (**Figure 34**). Such radicals are stabilized by resonance by delocalization of the resultant single electron over the ring.

4. Analysis

Phenolic compounds can be analyzed using various techniques. Mass spectrometry (MS), high performance liquid chromatography (HPLC), gas chromatography (GC), GC–MS, calorimetry, ultraviolet (UV), ultraviolet–visible (UV/VIS) spectrophotometry, and other spectrophotometric techniques represent examples of such techniques [15–18]. Total phenolic content (TPC) of phenolic compounds in plants is commonly measured using spectrophotometry techniques such as Folin-Denis and Folin–Ciocalteu methods [19]. The latter method which is based on electron-transfer, was found to be more preferable and thus more common [15, 16]. GC has been used to analyze phenolic acids, condensed tannins, flavones, and falvonoids. HPLC has been used to analyze anthocyanins, hydrolysable tannins, phenolic acids, cinnamic acids and favonoids [18]. Anthocyanins have been analyzed using UV absorption at a wavelength range of 489–550 nm [18]. Hydrolysable tannins have been also analyzed at a wavelength of 500–550 nm [18]. Based on their ability to bind proteins, tannins have been analyzed using protein-binding methods [20, 21]. Calrimetric methods have been used to determine the TPC in flavonoids and tannins. Other techniques for analysis of phenolic compounds include capillary electrophoresis (CE) and micellar electro-kinetic chromatography [22].

Recent methods of analysis include ultra-high performance liquid chromatography (UHPLC) [23], ultra-high performance liquid chromatography-quadrupole-orbitrap (UHPLC-Q-Orbitrap) [24], high performance liquid chromatography coupled to tandem mass spectrometry (HPLC-MS/MS) [25], liquid chromatography coupled with electrospray- ionization triple quadrupole time- of- flight mass spectrometry (LC- ESI- QTOF- MS) and high- performance liquid chromatography- photo diode array (HPLC- PDA) [26], ESI-Ms/MS [27].

5. Synthesis

5.1 Biosynthesis

There are two general routes for the biosynthesis of phenolic compounds; shikimic acid pathway and the acetic acid pathway [12, 28]. In the shikimic acid pathway (**Figure 35**), hosphoenolpyruvate and erthrose-4-phosphate react in few steps to provide 3-dehydroquinate. Dehydration with shikimate dehydrogenase gives 3-dehydroshikimic acid. Reduction with NADPH gives shikimic acid. 3-Dehydroshikimic acid could lead to gallic acid in several steps. Shikimic acid is then converted into chorismic acid which undergoes Claisen rearrangement to afford prephenic acid. The product is then converted in several steps into tyrosine. The amino acid serves as a central point and a crucial precursor for the biosynthesis of various phenolic compounds (**Figure 35**).

Another route toward phenolic compounds, is the phenylpropanoid pathway (**Figure 36**). This route is essentially similar to the shikimic acid pathway until L-phenylalanine stage where the phenylpropanoid pathway takes form. L-Phenylalanine undergoes deamination catalyzed by phenylalanine ammonia lyase (PAL) enzyme to give cinnamic acid. Hydroxylation followed by conversion to the Coenzyme A provides *p*-coumaroyl Coenzyme A. This molecule serves as a central point toward various phenolic compounds.

5.2 Synthesis

There have been many methods that are used to synthesize phenolic compounds in the laboratory. For instance, phenolic compounds were obtained using Copper-catalyzed synthesis from 1,3-dicarbonyl compounds employing dimethylsulfoxide (DMSO) as a methylene source (**Figure 37**) [29].

Phenolic compounds have also been obtained using biocatalysis. Thus phenolic compounds were synthesized by lipase-catalyzed synthesis (**Figure 38**) [30].

Various imine phenolic compounds were synthesized starting from 3-aminobenzoic acids as schematically represented below (**Figure 39**) [31].

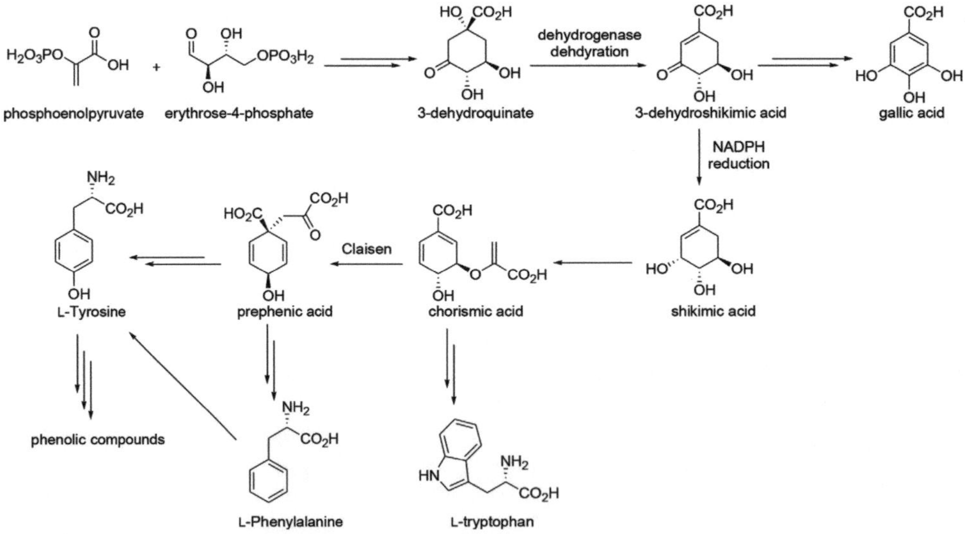

Figure 35.
Shikimic acid pathway toward phenolic compounds.

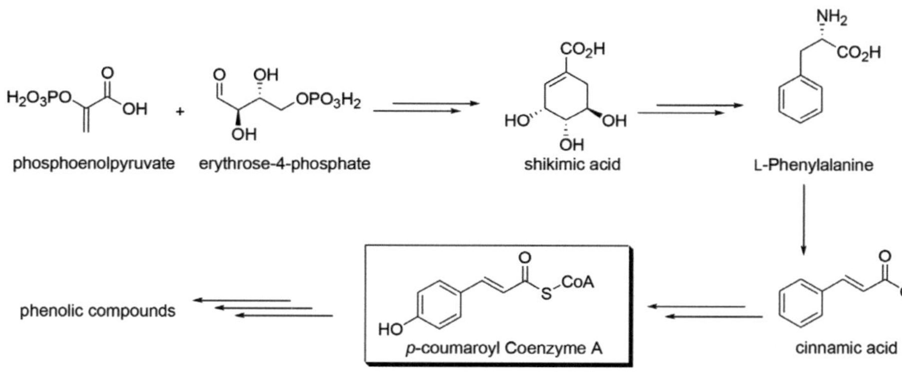

Figure 36.
The phenylpropanoid pathway toward phenolic compounds.

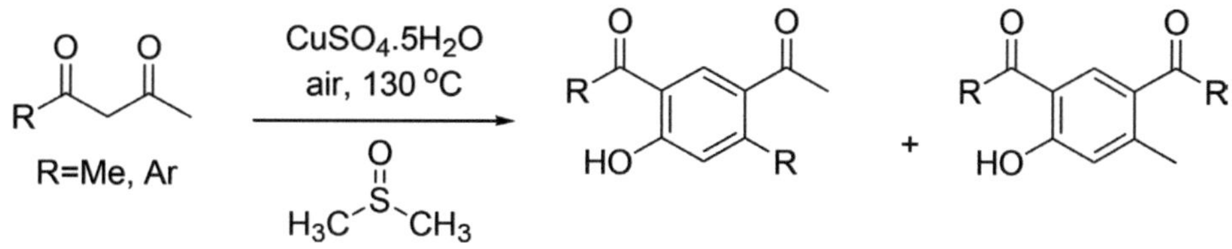

Figure 37.
Copper-catalyzed synthesis of phenolic compounds from 1,3-dicarbonyl compounds.

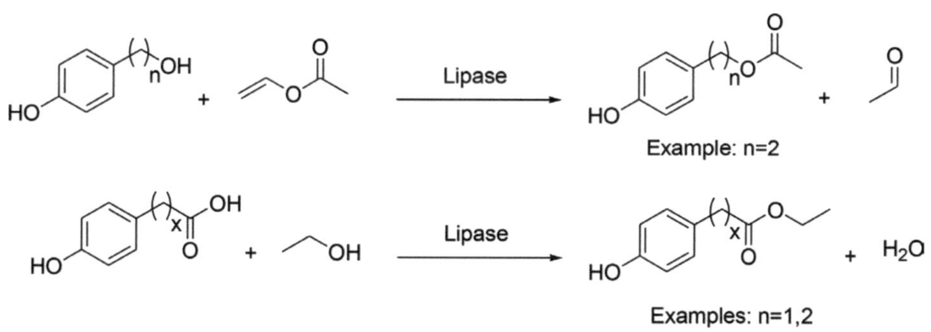

Figure 38.
Lipase-catalyzed synthesis of phenolic compounds.

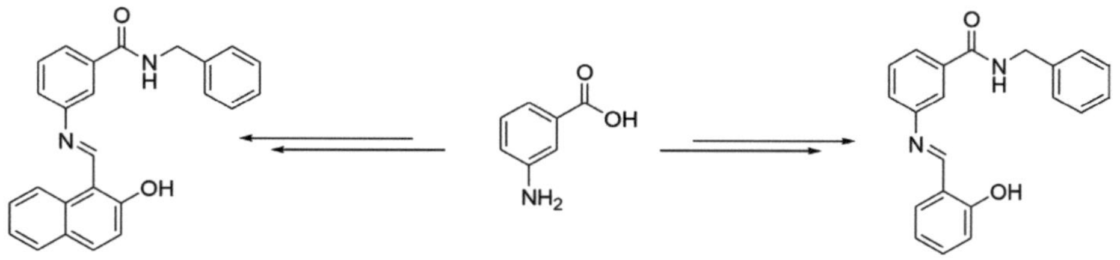

Figure 39.
Synthesis of imine phenolic compounds.

Various Schiff bases were also accessed from 3-nitroaniline (**Figure 40**) [31].

Other azomethine-based phenolic compounds were prepared from 3-nitroacetophenone (**Figure 41** i) [31].

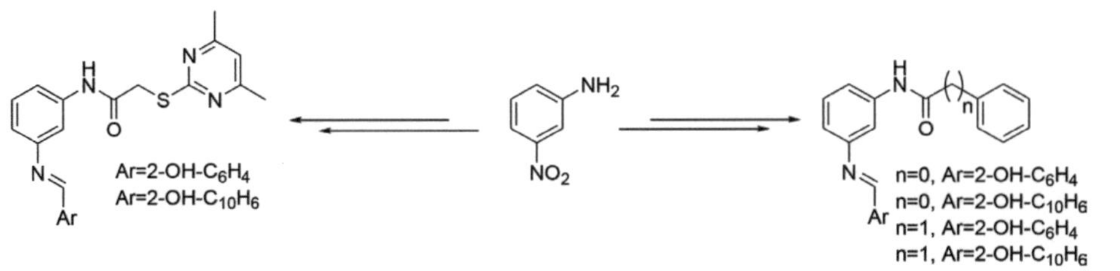

Figure 40.
Synthesis of imine phenolic compounds.

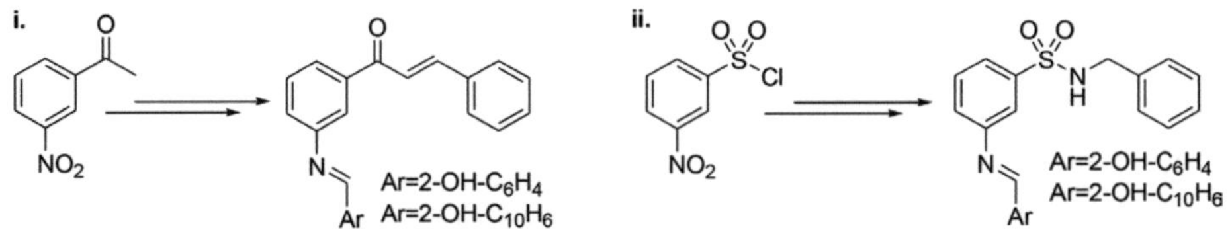

Figure 41.
Synthesis of imine phenolic compounds.

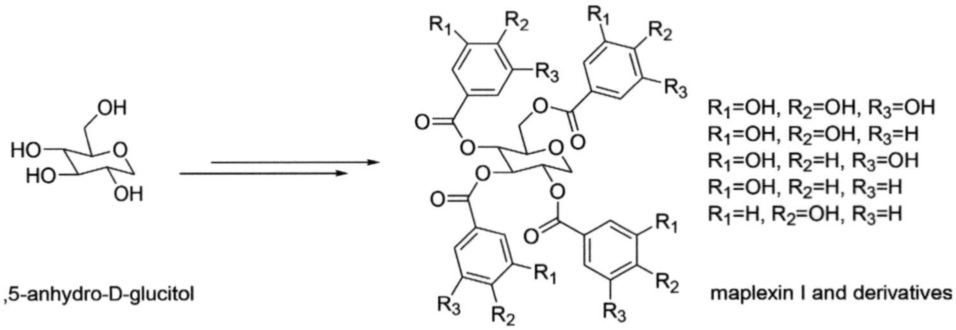

Figure 42.
Synthesis of maplexin J and its derivatives from 1,5-anhydro-D-glucitol.

Sulfonyl amide phenolic compounds were prepared from 3-nitrobenzenesulfonyl chloride (**Figure 41** ii) [31].

Carbohydrate-based polyphenolic compounds were synthesized from 1,5-anhydro-D-glucitol as schematically shown below (**Figure 42**) [32]. Thus maplexin J (R1 = R2 = R3 = OH) and its derivatives were synthesized using this route.

Another example is the synthesis of an analog of tellimagrandin I from benzyl glucoside (**Figure 43**) [32].

Functionalization and expeditious transformation of phenol derivatives into new functional molecules have been made possible with metal-catalyzed C-H bond functionalization [28, 29]. The C-H activation science has allowed accessing new and further functionalized phenol derivatives in an expedient and efficient manner (**Figure 44**). Thus catalysts based on various transition metals such as Pd, Rh, Ru, Ir, Au and Fe have allowed functionalization of inert C-H bonds in simple phenolic compounds and subsequently their transformation into new functionalized molecules [33, 34].

Recently, 1,3-dipolar cycloaddition (Click chemistry) of cellulose-based azides with alkynes derived from phenolic compounds were transformed into new

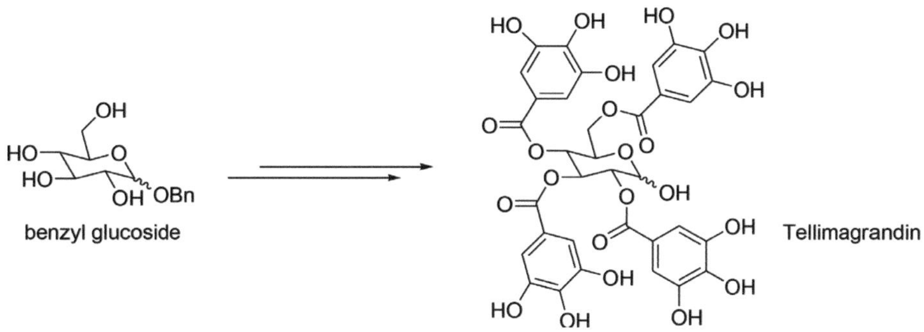

Figure 43.
Synthesis of tellimagrandin I analog.

Figure 44.
C-H bond functionalization of phenol derivatives.

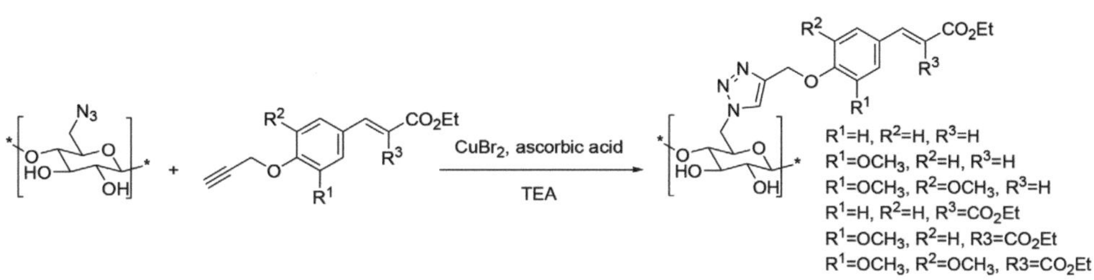

Figure 45.
1,3-Dipolr cycloaddition of cellulose-based azides with alkynes obtained from phenolic compounds.

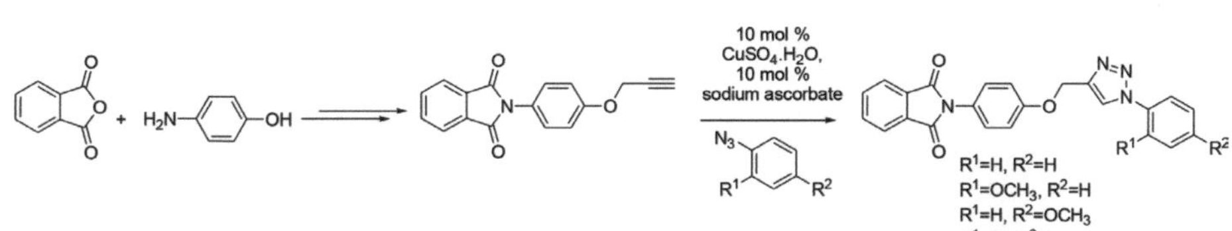

Figure 46.
1,3-Diploar cycloaddition of alkynes obtained from 4-aminphenol with aryl azides.

phenolic compounds-based adducts (**Figure 45**). The new triazole products display some applicable anti UV properties [35].

Simple phenolic compounds such as 4-aminophenol were transformed, *via* Click chemistry between their alkynes and aryl azides, into new triazole-containing isoindoline derivatives (**Figure 46**). The products obtained were of potential biological activities [36].

Natural phenolic compounds were transformed via ther alkyne-derivatives into the corresponding triazole adducts by the click reaction with aryl azides (**Figure 47**). The method demonstrates an example of synthetic elaboration of phenolic compounds into new ones of potential biological functions [37].

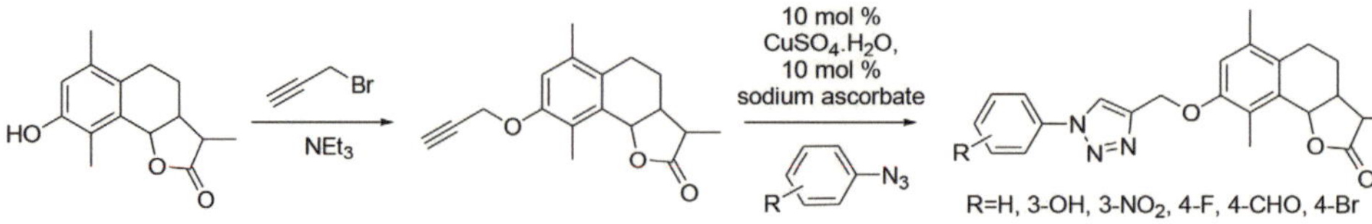

Figure 47.
1,3-Diploar cycloaddtion of alkynes derived from natural phenolic compounds and aryl azides.

6. Future perspective and industrial applications

Phenolic compounds possess a wide range of biological activities such as antioxidant, anti-inflammatory, and antimicrobial properties. Such properties allow phenolic compounds be able to reduce various illnesses and diseases such as cardiovascular diseases, diabetes, cancer, and hypertension. Therefore, they can be used in pharmaceutical industry as therapeutic agents. The antioxidant and antimicrobial properties enable phenolic compounds to function as food preservatives and additives. Thus they have also applications in food industry. In addition, phenolic compounds have applications in cosmetic and packaging industries. Exploitation of the full potential of phenolic compounds lies in the development of prudent and efficient methods for their detection, isolation, and analysis. Also, a key direction is their synthetic transformation and elaboration into new and potentially more biologically active molecules. Moreover, another prospect is the developments of new methods for the expeditious synthesis of phenolic compounds.

7. Conclusions

Phenolic compounds are ubiquitous in nature. Phenolic compounds display a wide range of biological activities such as antioxidant, antimicrobial and anti-inflammatory properties. Therefore, they have versatile applications in various industries such as pharmaceutical and food industries. Developments of efficient methods and protocols for the identification of phenolic compounds, their detection and analysis should continue. Due to their high potential and applications in various industries, development of efficient methods for their synthesis and synthetic elaborations into new phenolic compounds is sought. This chapter has been an attempt to provide the reader with a quick guide and reference for the classification, chemistry, analysis and synthesis of phenolic compounds.

Author details

Hamad H. Al Mamari
Department of Chemistry, College of Science, Sultan Qaboos University, Muscat, Sultanate of Oman

*Address all correspondence to: halmamari@squ.edu.om

References

[1] Albuquerque B R, Heleno S A, Oliveira M B P P, Barros L, Ferreira I C F R: Phenolic compounds: current industrial applications, limitations and future challenges. Food Funct. 2021;12:14-29. https://doi.org/10.1039/D0FO02324H.

[2] Cheynier V: Phenolic compounds: from plants to foods. Phytochem Rev. 2012;11:153-177. https://doi.org/10.1007/s11101-012-9242-8.

[3] Servili M, Sordini B, Esposto S, Urbani S, Veneziani G, Di Maio H, Selvaggini R, Taticchi A: Biological activities of phenolic compounds of extra virgin olive oil. Antioxidants. 2014;3:1-23. https://doi.org/10.3390/antiox3010001.

[4] Lin D, Xiao M, Zhao J, Li Z, Xing B, Li X, Kong M, Li L, Zhang Q, Liu Y, Chen H, Qin W, Wu H, Chen S: An overview of plant phenolic compounds and their importance in human nutrition and management of type 2 diabetes. Molecules. 2016;21:1374-1393. https://doi.org/10.3390/molecules21101374.

[5] Durrazo A, Caiazzo E, Lucarini M, Cicala C, Izzo A: Polyphenols: a concise overview on the chemistry, occurrence, and human health. Phytother Res. 2019;33:2221-2243. https://doi.org/10.1002/ptr.6419.

[6] Mark R, Lyu X, Lee J J L, Parra-Sladivar R, Chen W N: Sustainable production of natural phenolics for functional food applications. J. Funct. Foods. 2019;57:233-254. https://doi.org/10.1016/j.jff.2019.04.008.

[7] Santos S C, Fortes G A C, Camargo L T F M, Camargo A J, Ferri P H: Antioxidant effects of polyphenolic compounds and structure-activity relationship predicted by multivariate regression tree. LWT Food Science and Technology. 2021;137:110366-110381. https://doi.org/10.1016/j.lwt.2020.110366.

[8] Haminiuk C W, Maciel G M, Plata-Oviedo M S V, Peralta R M: Phenolic compounds in fruits-an overview. Food Sci. Technol. Int. 2012;47:2023-2044. https://doi.org/10.1111/j.1365-2621.2012.03067.x.

[9] Harborne J B, Simmonds N W. Biochemistry of Phenolic Compounds. Academic Press; London; 1964, P 101.

[10] Vuolo M M, Lima V S, Junior M R M. Bioactive Compounds: Health Benefits and Potential Applications. 2019; p 33-50.

[11] Tsimogiannis D, Oreopoulou V. Classification of Phenolic Compounds in Plants in Polyphenols in Plants. 2nd ed; 2019; p 263-284.

[12] Vermerris, W, Nicholson R. Phenolic Compounds Biochemistry. Springer; 2009; p 2-10.

[13] Vollhardt K P C, Schore N E Organic Chemistry: Structure and Function. 3rd ed; W H Freeman and Company, New York; 1999.

[14] Jeffrey G A. An Introduction to Hydrogen Bonding. Oxford University Press; Oxford; 1997; p 11-54.

[15] Khoddami A, Wilkes M A, Roberts T H: Techniques for Analysis of Plant Phenolic Compounds. Molecules. 2013;18:2328-2375. https://doi.org/10.3390/molecules18022328.

[16] Lapornik B, Prošek M, Wondra A G: Comparison of extracts prepared from plant by-products using different solvents and extraction time. J. Food Eng. 2005;71:214-222. https://doi.org/10.1016/j.jfoodeng.2004.10.036.

[17] Naczk M, Shahidi F: Phenolics in cereals, fruits and vegetables: Occurrence, extraction and analysis. J. Pharmaceut. Biomed. Anal. 2006;41:1523-1542. https://doi.org/10.1016/j.jpba.2006.04.002.

[18] Jin D, Mumper R J: Plant phenolics: Extraction, analysis and their antioxidant and anticancer properties. Molecules. 2010; 15:7313-7352. https://doi.org/10.3390/molecules15107313.

[19] Singleton V L, Orthofer R, Lamuela-Raventos R M: Analysis of total phenols and other oxidation substrates and antioxidants by means of Folin-Ciocalteu reagent. Methods Enzymol. 1999; 299:152-178. https://doi.org/10.1016/S0076-6879(99)99017-1.

[20] Hagerman A E, Butler L G: Choosing appropriate methods and standards for assaying tannin. J. Chem. Ecol. 1989;15:1795-1810. https://doi.org/10.1007/BF01012267.

[21] Hagerman A E, Butler L G: Assay of condensed tannins or flavonoid oligomers and related flavonoids in plants. Meth. Enzymol. 1994;234:429-437. https://doi.org/10.1016/0076-6879(94)34113-3.

[22] Caridi D, Trenerry V C, Rochfort S, Duong S, Laugher D, Jones R: Profiling and quantifying quercetin glucosides in onion (*Allium cepa* L.) varieties using capillary zone electrophoresis and high performance liquid chromatography. Food Chem. 2007;105:691-699. https://doi.org/10.1016/j.foodchem.2006.12.063.

[23] Gośliński M, Nowak, D, Szwengiel, A. Multidimensional Comparative Analysis of Bioactive Phenolic Compounds of Honeys of Various Origin. Antioxidants. 2021;10:530-544. https://doi.org/10.3390/antiox10040530.

[24] Hu B, Jin Q, Wang J, Wu C, Luo Z: The analysis of phenolic compounds in Walnut Husk and Pellicle by UPLC-Q-Orbitrap HRMS and HPLC. Molecules. 2021;26:3013-3031. https://doi.org/10.3390/molecules26103013.

[25] Cozzolino R, Malorni L, Martignetti A, Picariello G, Siano F, Forte G, De Giulio B: Comparative analysis of volatile profiles and phenolic compounds of four southern Italian onion (*Allium cepa*, L.). J. Food. Comp. Anal. 2021;101:103990-1031000. https://doi.org/10.1016/j.jfca.2021.103990.

[26] Lee F Y, Vo G T, Barrow C J, Dunshea F R, Suleria H A R: Mango rejects and mango waste: characterization and quantification of phenolic compounds and their antioxidant potential. J. Food Process. Preserv. 2021:ahead of print. https://doi.org/10.1111/jfpp.15618.

[27] Leah P, Oh W Y, Y, J, S F: Determination of soluble and insoluble-bound phenolic compounds in dehulled whole and hulls of green and black lentils using electrospray ionization (ESI)-MS/MS and their inhibition in DNA strand scission. Food Chem. 2021;361:130083-130091. https://doi.org/10.1016/j.foodchem.2021.130083.

[28] Marchica A, Cotrozzi L, Detti R, Lorenzini G, Pellegrini E, Petersen M, Nali C: The biosynthesis of phenolic compounds is an integrated defence mechanism to prevent ozone injury in *Salvia officinalis*. Antioxidants. 2020;9:1274-1290. https://doi.org/10.3390/antiox9121274.

[29] Luo R, Guo L, Liu W, Wang S: Copper-catalyzed synthesis of phenolic compounds with DMSO as the methylene source. Synth. Commun. 2021;1712-1722. https://doi.org/10.1080/00397911.2021.1902536.

[30] Zieniuk B, Wolozynowska M, Bialecka-Floria E, Fabiszewska A: Synthesis of industrially useful phenolic compounds esters bymeans of

biocatalysts obtained with waste fish oil utilization. Sustainability, 2020;12:5804-5832. https://doi.org/10.3390/su12145804.

[31] Ibrahim M A, George R F, Abou-Sri S M, El-Moghazy S M: Synthesis of new phenolic compounds and biological evaluation as antiproliferative agents. J. Chem. Res. 2020;44:181-192. https://doi.org/10.1177%2F1747519819895238.

[32] Machida S, Mukai S, Kono R, Funato M, Saito H, Uchiyama T: Synthesis and comparative structure activity study of carbohydrate-based phenolic compounds as α–glucosidase inhibitors and antioxidants. Molecules. 2019;24:4340-43-67. https://doi.org/10.3390/molecules24234340.

[33] Huang Z, Lumb J P: Phenol-directed C-H functionalization. ACS Catal. 2019;9:521-555. https://doi.org/10.1021/acscatal.8b04098.

[34] Al Mamari H H, Štefane B, Žugelj H B: Metal-catalyzed C-H bond functionalization of phenol derivatives. Tetrahedron. 2020;76:130925-130946. https://doi.org/10.1016/j.tet.2020.130925.

[35] Mendoza D J, Mouterde L M M, Borwne C, Raghuwanshi V S, Simon G P, Garnier G, Allais F: Grafting nature-inspired and bio-based phenolic esters onto cellulose nanocrystals gives biomaterials with photostable anti-UV properties. ChemSusChem. 2020;13:6552-6561. https://doi.org/10.1002/cssc.202002017.

[36] Boddu L, Potlapati V, Subhashini N J P: Synthesis of novel isoindolone-based medium-sized macromolecules and triazole containing heterocyclic compounds. J. Heterocycl. Chem. 2019;56:3197-3205. https://doi.org/10.1002/jhet.3704.

[37] Dangroo N A, Dar A A, Dar B A: An efficient protocol for domino one pot synthesis of 1,2,3-triazoles from natural organic acids and phenols. Tetrahedron Lett. 2014;55:6729-6733. https://doi.org/10.1016/j.tetlet.2014.09.123.

Chapter 5

Anthocyanins: Dietary Sources, Bioavailability, Human Metabolic Pathways, and Potential Anti-Neuroinflammatory Activity

Ruth Hornedo-Ortega, Zuriñe Rasines-Perea, Ana B. Cerezo, Pierre-Louis Teissedre and Michael Jourdes

Abstract

The objectives of this chapter are to summarize and discuss (i) the anthocyanins structure and content in foodstuffs and their dietary intake (ii) the anthocyanins bioavailability and human metabolic pathways and (iii) the *in vitro* and *in vivo* potent anti-neuroinflammatory effects of anthocyanins and their metabolites. Indeed, anthocyanins are polyphenolic compounds belonging to the group of flavonoids, and are one of the most commonly consumed polyphenols in a normal diet. They are responsible of red, blue and purple color of several fruits and vegetables and their intake has been related with several human health benefits. The anthocyanins structures diversities as well as their content in various fruits, vegetables and cereals is addressed. Moreover, despite the growing evidence for the protective effects of anthocyanins, it is important to highlight that the *in vivo* bioavailability of these compounds is relatively low in comparison to their more stable metabolites. Indeed, after consumption, these bioactives are subjected to substantial transformations in human body. Phase I and II metabolites generated by intestinal and hepatic enzymatic reactions, and phenolic acids produced by gut microbiota and their metabolized forms, are the most important metabolic anthocyanins forms. For this reason, the study of the biological properties of these circulating metabolites represents a more *in vivo* realistic situation. Although the anthocyanin bioavailability researches in humans are limited, they will be discussed together with a global metabolic pathway for the main anthocyanins. Moreover, several works have demonstrated that anthocyanins can cross the blood brain barrier, and accumulate in brain endothelial cells, brain parenchymal tissue, striatum, hippocampus, cerebellum and cortex. Consequently, the study of anthocyanins as potent therapeutic agents in neurodegenerative diseases has gained relevance and the principal and the most recent studies are also discussed in the book chapter.

Keywords: anthocyanin, metabolites, neuroinflammation, phenolic acids, bioactives

1. Introduction

Anthocyanins (deriving from the Greek *anthos* means flower, and *kyanos* means blue) are one of the most important pigments in the plant kingdom after chlorophyll. Anthocyanins belong to the widespread family of flavonoid polyphenolic compounds and are responsible of red, purple and blue colors of a great numbers of vegetables and fruits [1]. Although several hundred of natural anthocyanins has been identified (more than 600), they all derived from 31 naturally known anthocyanidins (anthocyanins aglycone) [2, 3]. When looking at the human diet (fruits, vegetables and cereals), the number of anthocyanidins can be reduce to only six different anthocyanidins which are: pelargonidin, cyanidin, peonidin, delphinidin, petunidin, and malvidin. Among these, cyanidin represents the most widespread anthocyanidin in plants (50%). Cyanidin, delphinidin and pelargonidin are the non-methylated anthocyanidins whereas peonidin, malvidin and peonidin possess *O*-methylation. However, as free aglycones are considerable unstable, anthocyanins (the glycosylated forms) are more usually present in natural sources [1, 4].

Apart of being responsible for the color of many foods and beverages, anthocyanins also have numerous health benefits resulting of their antioxidant and anti-inflammatory activities, among others. Although the dietary intake of anthocyanins depends on the nutritional habits [5], they have received less attention than other flavonoids compounds. This may be due to the fact that anthocyanins are poorly absorbed, highly metabolized, and rapidly excreted in the urine [6]. In addition, their bioavailability and the metabolites formed by intestinal, hepatic enzymatic reactions, and gut microbiota depend on the chemical structure of anthocyanin.

This book chapter will summarize and discuss (i) the anthocyanins structure and content found in fruits, vegetables and cereals as well as the global dietary intake (ii) the anthocyanin bioavailability and human metabolic pathways and (iii) the *in vitro* and *in vivo* anti-neuroinflammatory effects of anthocyanins and their metabolites.

2. Anthocyanins: chemistry, intake and dietary sources

From a structural point of view, anthocyanins are glycosylated, polyhydroxy or polymethoxy derivatives of 2-phenylbenzopyrylium (or flavylium cation) containing two benzoyl rings (A and B) separated by a heterocyclic (C). The number of hydroxyl groups and their degree of methylation, the nature and number of the sugar and the position of the attachment, as well as the nature and number of aliphatic or aromatic acids attached to the sugars, determine their different structural variations [3, 7].

Regarding sugars, they can be attached at different positions: 3-monoglycosides, 3-diglycosides, or 3-triglycosides, 3,5-diglycosides and to a lesser extent 3,7-diglycosides. Glucose is the most common sugar moiety but other monosaccharides as rutinoside, rhamnose, galactose, arabinose, xylose are found. Furthermore, the disaccharides as sambubioside or sophoroside and as well as trisaccharides like as xylosilrutinoside or glucosylrutinoside can also be present [8, 9]. The linking of acyl substituents to sugars make possible a further degree of complexity of anthocyanins. Among them, aliphatic (acetic, malonic, succinic, malic) and cinnamic acids (*p*-coumaric, ferulic, sinapic) are the most predominant [10]. Both glycosylation and acylation affects the chemical and physical properties of anthocyanins. Thus, glycosylation improves water solubility whereas acylation have the contrary effect [11].

Anthocyanins are sensible to different factors such as temperature, light, oxygen or enzymes but pH represent one of the most important factors affecting them. Four different equilibrium species can co-exist, the flavylium cation (red; pH 1), the quinonoidal base (bleu, pH 4), the carbinol pseudobase (colorless or pale yellow, pH 5) and the chalcone (*-cis* and *-trans*) (colorless; pH 6). At pH values higher than 7, anthocyanis are degraded. Generally, anthocyanins are more stable and more soluble at low pH [12]. Among anthocyanidins, pelargonidin is the most stable compounds because of its B ring substituents and the presence of hydroxyl or methoxyl groups decrease the stability. However, the glycosylation confers to the molecules a higher stability at neutral pH, since the presence of sugar avoid the degradation into phenolic acid and aldehyde compounds [13].

The determination of the dietary intake of flavonoids, and among them, the mean consumption of anthocyanins has been the subject of several studies over the last two decades. In United States the daily consumption of these compounds in adults has been estimated in 12.5 mg/day, representing cyanidin anthocyanins the 44.7% of the total intake followed by delphinidin, malvidin, petunidin, peonidin and pelargonidin anthocyanidins [14]. Another study in adults (17900 individuals) showed a lower anthocyanidin intake, 9.20 ± 0.79 mg/day. In addition, they stated differences among anthocyanin consumption according to gender (women's consume higher anthocyanins than men's) and sociodemographic and lifestyle factors such as education, alcohol consumption and activity levels [15]. Concerning European data, the European Prospective Investigation into Cancer and Nutrition (EPIC) study estimated a mean of anthocyanin intake of 31 mg/day. At the same time, they also observed that these values vary according to the country, age, sex, body mass index (BMI), level of education, smoking status and physical activity level [5]. Between European countries, significant differences were reported. Indeed, Italy, France and Germany displays the greater mean values, from 35.1 to 42.3 mg/day, whereas Netherlands and Sweeden are the countries with a lower anthocyanin consumption (22.6 and 20.9 mg/day, respectively). More recently, after the study of the dietary habits of 30000 subjects in 14 European countries the mean intake of anthocyanins was estimated to be 19 mg/day [16]. In other continents and countries such as Australia (12153 subjects) or China (1393 subjects) the estimated anthocyanin mean intake was calculated at 24.2 mg/day and 27.6 mg/day, respectively, which are very closed values to European levels [17, 18].

Apart of being present in many colored fruits and vegetables they appeared also in beverages as red wine or juices and in processed foods as jams. Both, the type and the concentrations of anthocyanins are influenced by genetics (cultivar, species), cultivation, climate, soil, and processing [19]. However, one of the best sources of anthocyanins are berries. Among them, bilberries, blueberries and blackcurrants can be reach values greater than 1000 mg/100 g of fresh weight (FW) (**Table 1**). Among vegetables and cereals, red cabbage, cauliflower and colored corn and rice represent good sources of anthocyanins. The most common anthocyanins are cyanidin glucosides, but some fruits contain other predominant anthocyanin (**Table 1**). For example, pelargonidin-3-*O*-glucoside is the principal anthocyanin of strawberries, whereas malvidin-3-*O*-glucoside predominates in grapes and cyanidin-3,5-diglucoside is the major one in pomegranates [29–31]. Generally, the main anthocyanins in vegetables and cereals are chemically more complex in comparison with fruits. In fact, acylated and diglucosylated anthocyanins such as cyanidin-3-(*p*-coumaroyl)-diglucoside-5-glucoside can be found [33]. In addition, others less conventional sugars like sophoroside (cauliflower and radish) or laminiaribioside (red onions) can be present [35, 36, 39]. Interestingly, the most common anthocyanin type in sorghum, are the 3-deoxyanthocyanidins luteolinidin and apigeninidin (characterized by the lack of hydroxyl group at C3 position) and their derivatives, which are not commonly found in higher plants [44].

Fruit	Content	Main anthocyanin	Ref
Apples (Red)	0.1–315 mg/Kg peel	Cy-3-gal	[20]
Apricot	1.9–230.4 mg/100 g FW	Cy-3-rut	[21]
Bilberry	933–1017 mg/100 g FW	Delp-3-gluc/ Delp-3-ara/ Delp-3-gal	[22, 23]
Blackberry	84–201 mg/100 g FW	Cy-3-gluc/ Cy-3-rut	[24]
Blueberry	232–438 mg/100 g FW	Mv-3-gluc/ Mv-3-gal/ Delp-3-gal/ Delp-3-ara	[22]
Cherry	6.3–60 mg/100 g FW	Cy-3-gluc/ Cy-3-rut	[25, 26]
Cranberry	12.4–207.3 mg/100 g FW	Cy-3-gal / Cy-3-ara/ Peo-3-gal / Peo-3-ara	[27]
Blackcurrant	146.15–403.66 mg/100 g FW	Cy-3-rut/Cy-3-gluc/Delp-3-rut/ Delp-3-gluc	[28]
Red grapes	11.5–29.8 g/Kg DM	Mv-3-gluc/ Mv-3-acetylgluc (*V. vinifera*)	[29]
		Mv-3,5-digluc (other than *V. vinifera*)	
Pomegranate juice	8.9–346.6 mg/L	Cy-3,5-digluc/ Cy-3-gluc/ Delp-3,5-digluc/ Delp-3-gluc/ Pel-3-gluc	[30]
Strawberry	8.5–66 mg/100 g FW	Pel-3-gluc/ Cy-3-gluc/ Pel-3-rut	[31]
Vegetables	**Content**	**Main anthocyanin**	**Ref**
Black beans	32 mg/g DW	Delp-3-gluc/ Pet-3-gluc/ Mv-3-gluc	[32]
Red cabbage	2.32 mg/g DW	Cy-3-digluc-5-gluc/ Cy-3-coumaroyldigluc-5-gluc/ Cy-3-sinapoyldigluc-5-gluc	[33]
Purple carrot	168.7 mg/100 g FW	Cy-3-xylosyl-coumaroylglucosyl-gal/ Cy-3-xylosyl-feruloylglucosyl-gal/ Cy-3-xylosyl-gal	[34]
Purple cauliflower	7.18–201 mg/100 g FW	Cy-3-coumarylsoph-5-gluc/ Cy-3-coumarylsoph-5-sinapylgluc	[35, 36]
Eggplant (skin)	12.1 mg/ 100 g DW	Delp-3-rut	[37]
		Delp-3-coumaroylrut-5-gluc	
Colored potatoes	14.42–25.79 mg/g DW	Pel-3-coumaroylrut-5-gluc/ Pel-3-feruloylrut-5-gluc (red)	[38]
		Pet-Pe and Mv-3-coumaroylrut-5-gluc (blue–purple)	
Red onions	48.5 mg/ 100 g FW	Cy-3-gluc/ Cy-3-laminaribioside/ Cy-3-malonylgluc/ Cy-3-malonyllaminaribioside	[14]
Radish	32 mg/100 g FW	Pel-3-coumaroylsoph-5-gluc/ Pel-3-feruloylsoph-5-gluc/Pel-3-feruloylsoph-5-malonylgluc/ Pel-3-coumaroylsoph-5-malonylgluc	[39]
Cereals	**Content**	**Main anthocyanin**	**Ref**
Colored Barley	8–679 mg/Kg DW	Cy-3-gluc/ Peo-3-gluc (purple and blue)	[40, 41]
		Dep-3-gluc/Peo-3-gluc/ Mv-3-gluc (purple)	
Purple, blue, Red, black corn	27–1439 mg/Kg DW	Cy-3-gluc/Cy-3-malonylgluc/ Cy-3-dimalonylgluc	[41, 42]

Fruit	Content	Main anthocyanin	Ref
Purple, red, black rice	68–5101 mg/Kg	Cy-3-gluc/Peo-3-gluc (black)/Mv (red)	[43]
		Cy-3-gluc/Peo-3-gluc/Cy-3-gal/Cy-3-rut (purple)	
Black and red sorghum	32–680 μg/g DW	3-deoxyanthocyanins (Luteolinidin and apigeninidin)	[44]
Purple, blue, black wheat	10–212 mg/Kg DW	Cy-3-gluc/Peo-3-gluc/ Cy-malonylgluc/ Cy-succinylgluc	[45]

Cy: cyanidin; Delp: delphinidin; Mv: malvidin; Peo: peonidin; Pel: pelargonidin; Pet: petunidin; gluc: glucoside; digluc: diglucoside; sam: sambubioside; gal: galactoside; ara: arabinoside; rut: rutinoside; soph: sophoroside; DW: dry weight; FW: fresh weight.

Table 1.
Content and main anthocyanins in foodstuffs.

3. Anthocyanins bioavailability and human metabolic pathways

To validate the prominent health-promoting effects revealed in many *in vitro* and *in vivo* models, it is necessary to consider the anthocyanin bioavailability. Anthocyanin bioavailability has been reported to be very low, with recoveries of less than 1% of the ingested anthocyanin dose. However, higher values have been reported reaching recoveries values of 12.4% [46, 47]. As will be described later, anthocyanin can be absorbed from the stomach and small intestine, but a non-negligible part of them can reach the large intestine where they undergo also an extensive catabolism resulting in several metabolites (phenolic acids, propionic acids). For this reason, anthocyanin bioavailability is estimated much greater taking into account not only the phase I and phase II metabolites but also the microbiota catabolites [6]. Although the currently anthocyanin bioavailability researches in humans are limited, they will be discussed below.

3.1 Anthocyanins absorption

Despite having different molecular sizes and types of sugars or acetylated groups attached, anthocyanins can be absorbed intact [48, 49]. Moreover, anthocyanins were found in the blood stream within minutes of consumption in humans [6] suggesting that they can be quickly absorbed from the stomach. This fact is supported by the fact that anthocyanin urine concentrations were fivefold higher when introduced through nasal tubes into the stomach as opposed to the jejunum in patients with colorectal liver metastases after administration of a bilberry extract [50].
In fact, thanks to the low stomach pH (1.5–4) the anthocyanin stability increase permitting their absorption under their glycoside forms. Because anthocyanins are hydrophilic molecules, an organic anion membrane carrier named bilitranslocase, which is expressed in the gastric mucosa has been proposed to mediate anthocyanin transport [51]. Another hypothesis is the involvement of glucose transporter 1 in the transport of anthocyanin glucosides [52]. However, the main site of anthocyanin absorption is the small intestine. They undergo deglycosylation mediated by β-glucosidase in the intestinal lumen and lactasephloridzin hydrolase in the brush border of the intestinal epithelial cells. Alternatively, anthocyanins can enter the enterocyte without deglycosylation via the sodium-coupled glucose transporter after which deglycosylation can occur by cytosol β-glucosidase [51]. These proposed mechanisms are based, in contrast, on *in vitro* studies. Thus, more studies are required in order to gain insight in human anthocyanin absorption.

3.2 Anthocyanins metabolism

Anthocyanin aglycones that enter the intestinal epithelial are metabolized before reaching portal circulation. This metabolism includes oxidation, reduction, and hydrolysis reactions (phase I metabolism) and conjugation reactions (phase II metabolism). In the intestine, anthocyanins can undergo methylation, sulfation, and glucuronidation by catechol-*O*-methyltransferase, sulfotransferase, and uridine-5′-diphospho-glucuronosyltransferase enzymes [53]. These reactions can also take place in the liver and the kidneys.

Anthocyanin aglycones can alternatively undergo degradation rendering different phenolic compounds within the intestinal lumen or epithelial cells. Anthocyanin fragmentation can also be a result of the colonic microbiota activity. The microbiota gut can release many deglycosylation enzymes giving rise to aglycones that further undergo ring-opening to produce different benzoic acids or aldehydes such as gallic, vanillic, protocatechuic and syringic acids or aldehydes [46, 54]. Consequently, the phenolics acids portion increases whereas ingested anthocyanin forms portion decreases along the gastrointestinal tract. These products of anthocyanin degradation may be absorbed from the intestine and be transported and further metabolized in the liver and kidneys [55]. The specific anthocyanins metabolism will be described below.

3.3 Anthocyanin's distribution

The protective effects of flavonoids have been associated with diseases occurring in various tissues, but such claims are mainly based on *in vitro* evidence using different types of cell lines.

Anthocyanin distribution in tissues has been evaluated in rodents and pig models but never in humans [56–59]. In a study in which Wistar rats were fed during 15 days with blackberry extract (370 nmol anthocyanin/day), total averaged anthocyanins concentrations were found in jejunum (605 nmol/g), in stomach (68.6 nmol/g), in kidney (3.27 nmol/L), in liver (0.38 nmol/g) and in brain (0.25 nmol/g) [60]. In pigs, anthocyanins were identified in the liver (1.30 pmol/g), in eyes (1.58 pmol/g), in cortex (0.878 pmol/g) and in cerebellum (0.664 pmol/g) after being supplemented with 0, 1, 2, or 4% w/w blueberries for 4 weeks [61]. In anesthetized rats received cyanidin-3-*O*-glucoside by intravenous injection, this compound has been detected within 15 seconds in the brain tissue and a concentration comparable to that in serum [62]. The results suggested that anthocyanins may provide protection for brain and eye tissues after crossing the blood–brain and blood-retinal barriers.

3.4 Anthocyanin excretion

Anthocyanins can be excreted in urine, bile and even though in air. Around 5% of ^{13}C-label was recovered from urine after the [^{13}C]-cyanidin-3-*O*-glucoside administration in humans [46]. The urinary excretion of pelargonodin-3-*O*-glucoside seems to be higher than that of cyanidin-3-*O*-glucoside [63, 64]. This may be related to the stability of pelargonidin-3-*O*-glucoside than its real higher absorption. Furthermore, anthocyanins can undergo extensive bile secretion in their original forms or as their phase II metabolites. In human studies enterohepatic recycling of a several xenobiotic could be revealed by a second peak on the plasma concentration *versus* time curve; This phenomenon can be observed in the literature for several anthocyanins (cyanidin-3-*O*-glucoside, peonidin-3-*O*-glucoside, delphinidin-3-*O*-glucoside) [49, 65].

Finally, volatile metabolites produced from [^{13}C]-cyanidin-3-*O*-glucoside have also been found in large quantities in breath (6.9% of the administrated dose) following oral administration of [^{13}C]-cyanidin-3-*O*-glucoside [46].

3.5 Anthocyanin's behavior *in vivo*

Researching the xenobiotic methylation and hydroxylation of anthocyanins is challenging based on MS/MS because anthocyanidins are themselves differentiated by hydroxyl and methyl groups on the B-ring. For example, 3'-*O*-methylation can convert cyanidin to peonidin, and delphinidin to petunidin and 5'-*O*-methylation converts petunidin to malvidin [66]. Moreover, the removal of functional groups will interconvert anthocyanidins. For example, if cyanidin loses the hydroxyl group in position 2″ from the B-ring, it gives rise to pelargonidin (**Figure 1**) [67]. As methylation and glucuronidation occurs on hydroxyl groups, abundant in anthocyanins, positional isomers of anthocyanin and anthocyanidin conjugates can be predicted and are indeed detected [64–68]. As a consequence, data on anthocyanins bioavailabitily in humans after ingestion is potentially more straight forward to interpret.

3.5.1 Cyanidin metabolism

Cyanidin is the best-studied anthocynidin as it is the most widely distributed. Isotopically-labeled cyanidin-3-*O*-glucoside (C3g) was used to examine the absorption and metabolism of ^{13}C cyanidin-3-*O*-glucoside in humans [46]. In this study, 44% of the ^{13}C label has been excreted in urine (5.4%), breath (6.9%) and feces (32.1%) at 48 hours after intake. That implies also that more than 50% of the ^{13}C label was still inside the body at that moment. The absorption, digestion, metabolism and excretion of cyanidin-3-*O*-glucoside concur that methylation and glucuronidation are major routes of cyanidin-3-*O*-glucoside conjugation *in vivo* [46, 67]. The metabolites detected in these studies included methyl and glucuronide conjugates of cyanidin-3-*O*-glucoside, methyl cyanidin-3-*O*-glucoside (peonidin-3-*O*-glucoside), and their aglycones cyanidin and peonidin.

Recently, a human study has been carried on to investigate the metabolic pathways and human bioavailability of anthocyanins of red-fleshed apple in which 22% of phenolic compounds are anthocyanins and the main is cyanidin-3-*O*-galactoside. As a result, cyanidin glycosides (galactoside and arabinose) have been detected in plasma and urine samples. Moreover, peonidin-3-*O*-galactoside as phase II metabolite of cyanidin-3-*O*-galactoside methylation by the action of catechol-*O*-methyltransferase enzyme has been also detected [69]. Methylation, as one of the first

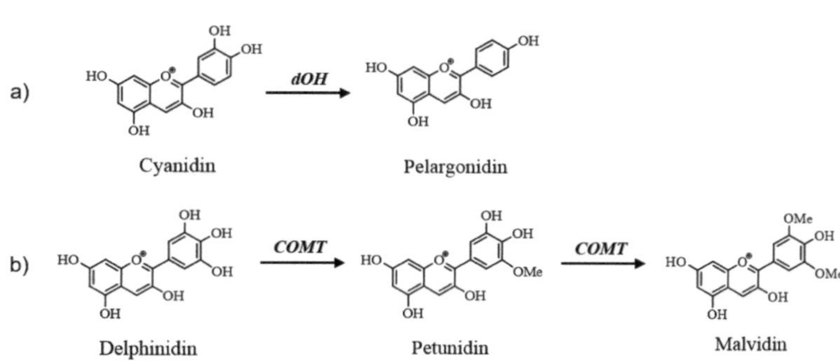

Figure 1.
Interconversion reactions between anthocyanins: (a) dehydroxylation reaction to arise pelargonidin from cyanidin; (b) methylation pathway that could be carried on by the action of catechol-O-methyltransferase enzyme. Reactions: dOH, *dihydroxylation;* COMT, *catechol-O-methyltransferase.*

metabolic reaction of cyanidin glycosides was also reported after the oral ingestion of 500 mg of ^{13}C-labeled cyanidin-3-*O*-glucoside [55].

Protocatechuic acid (PCA) and dihydroxyphenylpropionic acid (dihydrocaffeic acid) were respectively detected in these studies [55, 69]. PCA has been observed at maximum concentrations of 147 nM, thus suggesting that it is not a major metabolite of anthocyanins. The A-ring-derived degradation product, phloroglucinolaldehyde, was present at concentrations greater than either cyanidin-3-*O*-glucoside or PCA in the serum [55].

Hippuric acid has been identified as the major metabolite of anthocyanins, reaching a maximum concentration of 1962 nM in serum [55]. The detection of ^{13}C2-labeled hippuric acid in this study indicates that PCA and its conjugates are likely further metabolized to form benzoic acid, which is conjugated with glycine to form hippuric acid, or alternatively, formed from the α-oxidation and dihydroxylation of hydroxyphenylacetic acids [64]. PCA might have been formed by β-oxidation of dihydroxyphenylpropionic acid. Then, this phenolic acid could either be further degraded by the action of the gut microbiota to catechol metabolites (α-oxidation), pyrogallol metabolites (hydroxylation) and hydroxybenzoic acid (dehydroxylation), or methylated to vanillic acid [55, 69].

Colonic metabolism has long been speculated to be a major contributor to the overall metabolism of anthocyanins [70]. It has been proposed that phenylpropenoic acids arise from cyanidin-3-*O*-glucoside as a result of bacterial cleavage of the C-ring in the colon [71], which is supported by the detection of caffeic acid and its methyl metabolite, ferulic acid [55].

On the basis of the findings of these studies, the metabolic pathway of cyanidin-3-*O*-glucoside and peonidin-3-*O*-glucoside can be summarized as undergoing multiple biotransformation (**Figure 2**).

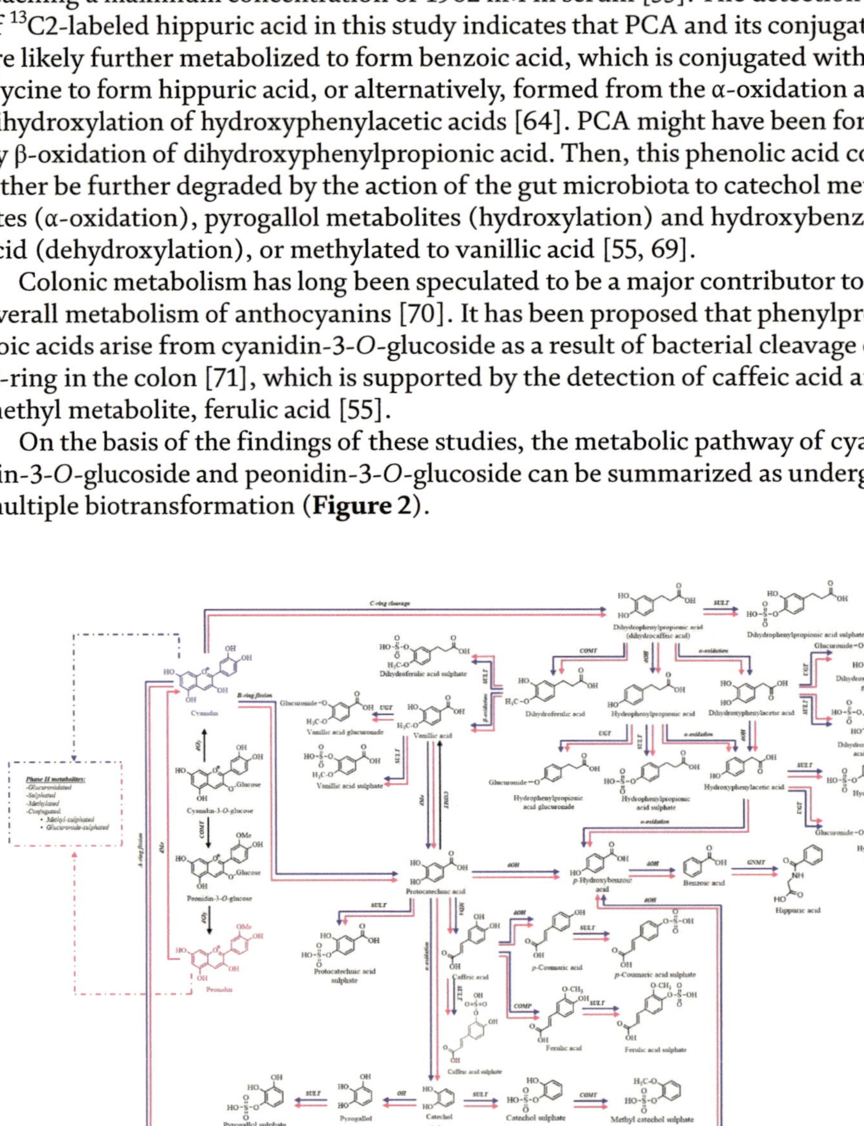

Figure 2.
Proposed metabolic pathway for cyanidin and peonidin glucosides. Reactions: dH, *dehydrogenation;* SULT, *sulphotransferase;* UGT, *glucuronosyl-transferase;* COMT, *catechol-O-methyltransferase;* dOH, *dehydroxylation;* dMe, *demethylation;* α-oxidation, *one decarboxylation;* β-oxidation, *two decarboxylation.*

3.5.2 Pelargonidin metabolism

As it was shown before, demethylation and dihydroxylation of highly substituted anthocyanins gives rise to pelargonidin, that helps to explain the high apparent recovery of pelargonidin-based metabolites [63]. Indeed, pelargonidin glucuronide has been detected in urine after the ingestion of boysenberry (rich in four cyanidin glycosides and without pelargonidin) in humans [67]. Furthermore, strawberry pelargonidin was found to be metabolized to 4-hydroxybenzoic acid in humans when 13 healthy volunteers consumed 300 g of fresh or stored strawberries [72]. In which 4-hydroxybenzoic acid plasma recovery was 23 and 17 mmol, corresponding to the percentages of 54 and 56% of pelargonidin-3-*O*-glucoside.

3.5.3 Delphinidin, petunidin and malvidin metabolism

After administration of Concord grape juice in humans, delphinidin-3-*O*-glucoside, petunidin-3-*O*-glucoside and malvidin-3-*O*-glucoside were found in blood or urine. Glucuronidated metabolites of aglycones have been identified as their major metabolites in urine [49]. In the urine of volunteers administered bilberry-lingonberry puree, a small amount of syringic acid, a potential metabolite of malvidin glycosides, was detected [73]. Recently, in a long-term study with humans consuming blueberry juice, 55 anthocyanin metabolites have identified. Among them, malvidin-3-*O*-glucoside, malvidin-3-*O*-galactoside and malvidin-3-*O*-arabinoside have been described representing around 5% of the total excretion [68]. *In vitro* experiments state that gallic acid is the major degradation product of delphinidin-3-*O*-glucoside. Moreover, syringic acid was described as the mean metabolite for malvidin-3-*O*-glucoside [13].

4. Anti-neuroinflammatory effects on anthocyanins and their metabolites

As it was discussed above, several works have demonstrated that anthocyanins can cross the blood brain barrier, and accumulate in brain endothelial cells, brain parenchymal tissue, striatum, hippocampus, cerebellum and cortex [74–76]. Consequently, the study of anthocyanins as therapeutic agents in neurodegenerative diseases has gained relevance.

Neuroinflammation is a common physiopathological hallmark in neurodegenerative diseases as Alzheimer, Parkinson or amyotrophic lateral sclerosis, among others. This process is mediated by microglial cells, the immune cells of central nervous system. Their functions are related with the host defense by destroying pathogens, promoting tissue repair and facilitating tissue homoeostasis [77]. Nowadays it is well establish that these cells can adopt different phenotypes depending on the brain environment to shift into pro-inflammatory/neurotoxic or anti-inflammatory/neuroprotective phenotypes. The stimulation agent will be the responsible of trigger one or another phenotype. Thus, when microglial cells are stimulated with lipopolysaccharide (LPS) and interferon gamma (IFN-γ), microglia develop a classically phenotype or M1, while when it is activated with IL-4 microglia show an alternative activated phenotype or M2 [78]. On the one hand, M1 microglia type is characterized by the production of nitric oxide (NO) by the inducible nitric oxide synthase (iNOS) [79, 80] and by the expression of inflammatory chemokines and cytokines, such as interleukin (IL)-6, IL-12, IL-1β, IL-23, and tumor necrosis factor (TNF)-α. All this culminates in the influx of new immune system cells to combat the infection. When neuroinflammation becomes chronic, it can ultimately lead to neuronal cell death. On the other hand, M2 microglia is characterized by a suppression of IL-12 secretion and an induction of the release of IL-10, transforming growth factor beta (TGB-β), IL-1R [81]. Furthermore, the expression of arginase-1 instead of iNOS, switching arginine metabolism from production of NO to ornithine, and also the increase of polyamines production for extracellular matrix and collagen synthesis, promotes the neuroregeneration and tissue repair [82].

Several *in vitro* and *in vivo* studies have shown that anthocyanins, overall rich anthocyanins extracts, are able to be neuroprotective and counteract neuroinflammation [83, 84]. Regarding *in vitro* studies, a blueberry extract (25–50 μg/mL) have demonstrated to be able to diminish the release of NO, TNF-α, iNOS and cyclooxygenase-2 (COX-2) protein expression in LPS-stimulated BV2 cells [85, 86] and in LPS or IFNγ-stimulated N9 cells [87]. In addition, they proved that this effect is mediated by NF-κB signaling pathway, via the inhibition of Nuclear Factor Kappa B (NF-κB) nuclear translocation [88]. NF-κB is a key inflammation regulator located in the cell cytoplasm and their nuclear translocation trigger the expression of inflammation-related genes. Likewise, the anti-inflammatory effect of blueberry extract has been related with the activation of janus kinase/signal transducer and activator of transcription (JAK/STAT) signaling (pathway activated after IFN-γ stimulation) [87]. Other study that evaluated the potential anti-neuroinflammatory effect of a large variety berries extracts (blackberry, black raspberry, blueberry, cranberry, red raspberry, and strawberry), showed that the cranberry extract (20 μg/mL) was the most active diminishing the NO production and inhibiting the fibrillation of amyloid-β peptide (peptide responsible of the formation of senile plaques in brain Alzheimer's patients) [89]. Moreover, elderberry extracts (400 μg/mL, ethanol or ethyl acetate extracts) has also been proposed as a potent suppressor of NO release [90]. Mitogen-activated protein kinases (MAPKs) are a family of serine/threonine protein kinases that mediate fundamental cellular responses to external stress signals. In particular, p38 MAPK, is involve in the regulation of the synthesis of inflammation mediators being for that a potential target for anti-inflammatory therapeutics. In this context, an anthocyanin-enriched extract of acai berry and a mixture of anthocyanins isolated from black soybean seed coats (cyanidin-3-*O*-glucoside (72%), delphinidin-3-*O*-glucoside (20%) and petunidin-3-*O*-glucoside (6%)) have demonstrated that MAPK pathways can be also implicated in the decrease in inflammatory mediators and cytokines [91, 92]. Finally, this year, an article have been published showing that a black raspberry extract reduced the production of IL-18, IL-1β and reactive oxygen species (ROS) in LPS-induced BV2 microglia by down-regulating the level of NADPH oxidase 2 (NOX2) and its downstream factors, including thioredoxin-interacting protein (TXNIP) and NOD-like receptor protein 3 (NLRP3) inflammasome [93]. The complexity of these extracts containing several structurally diverse anthocyanins makes difficult the interpretation of results. For this, some papers have been published concerning the evaluation of the activity of pure anthocyanins. This type of studies provide insight into the plausible mechanism of single compounds facilitating the understanding. Some (although very few) studies, have been performed with pure anthocyanins. An interesting work published by Miraeles and collaborators demonstrated that cyanidin-3-*O*-glucoside, (1 μM) and also cyanidin-3-*O*-glucoside and a mixture of 3′-methyl-cyanidin-3-*O*-glucoside and 4′-methyl-cyanidin-3-*O*-glucoside, were able to decrease a great number of pro-inflammatory mediators. Indeed, TNF-α and IL-6 mRNA expression was decrease by and methyl-cyanidin-3-*O*-glucoside. Moreover, cyanidin reverted the IL-1β expression. This paper also shows that even though cyanidin and theirs different chemical forms, are not able to shift microglia to an M2, they can interact with microglia biology increasing CX3C Motif Chemokine Ligand 1 (CX3CL1) expression [94]. Neurons can express this chemokine, which mediates microglial activation via interacting with its sole receptor CX3CR1 in microglia (axis CX3CL1/CX3CR1). Comparable results have been recently published, showing that the underlying responsible anti-neuroinflammatory mechanism of cynidin-3-*O*-glucoside is related with suppression of NF-κB and p38 MAPK signaling pathways [95]. Other pure anthocyanins as delphinidin-3-*O*-glucoside, malvidin-3-*O*-glucoside (20 μM) [86] and pelargonidin-3-*O*-glucoside (100 μM) [96] are also shown to be able to suppress the LPS/IFN-γ -induced phosphorylation of p38, p42/44 and MAPKs in BV2 cells and mouse C8-4B microglial cells.

Concerning *in vivo* studies, only around ten papers have been published about the effect of anthocyanins extracts/pure compounds in microglia-related diseases. The first paper published in 2015, evaluate the effect of a blackberry extract consumption at a dose of 25 mg/Kg in an standard or in a high fat diet, during 17 weeks in Wistar rats. The results showed that the intake of this fruit, in both dietary conditions, modulates CX3CL1 expression and the thymus chemokine TCK-1. In addition, they also found that blueberry can ameliorate synapse connectivity by regulating platelet-derived growth factor (PDGF)-AA, activin, vascular endothelial growth factor (VEGF) and agrin [97]. Another three works proved that the consumption of anthocyanins extracted of Korean black soybean (24–100 mg/Kg) inhibited the activation of astrocytes and neuroinflammation via suppression NF-κB, iNOS and TNF-α in the hippocampus and cortex regions of D-galactose and LPS treated rats brain [98–100].

Not only the reduction of IL-1β and TNF-α but also the reduction of IL-10 induced by LPS was observed after the treatment with 100 mg/Kg of anthocyanin obtained from *V. vinifera* grapes in mice [101]. Moreover, the addition of an enriched anthocyanin extract from purple corn in water (mean of 53 mg/Kg body weight) has proved to be able to reduce microglia size and Iba1 staining (marker of microglia activation) and IL-6, TNF-α, IL-1β, MCP-1 and iNOS. Interestingly, this papers showed that purple corn anthocyanins not only inhibit microglia activation but also promote their shift towards the production of anti-inflammatory mediators, such as arginase-1, IL-10, Fizz1, IL-13 and YM-1 (a marker of M2 microglia phenotype) [102]. In agreement, a diet based on anthocyanin-rich wheat during 6 months on Alzheimer and Parkinson disease mouse models, reduced the α-synuclein accumulation (protein responsible of the formation of Lewy bodies in Parkinson patients) [103].

Other rich anthocyanins fruits as bilberry has exhibited promising results. In fact, the administration in food or in water of an bilberry extract (20 mg/Kg day) on APP/PSEN1 mice and their littermates downregulates the expression of several inflammatory factors (TNF-α, NF-κβ, IL-1β, IL-6, COX-2, iNOS and cluster of differentiation 33 (CD33), the chemokine receptor CX3CR1, but also and for the first time, the microglia homeostatic factors (TREM2 and TYROBP) and the Toll-like receptors (TLR2 and TLR4) [104].

As was explained above, circulating concentrations of phenolic acid metabolites derived from anthocyanin degradation such as protocatechuic, gallic, syringic and ferulic acids have been observed at up to eight times to that of the parent anthocyanins [72]. Two papers have been very recently published showing that a mixture of anthocyanin metabolites can have anti-neuroainflammatory activities. Indeed, an *in vitro* digested blueberry and raspberry extracts (1.25–10 μg/mL) proved be able to reduce some key inflammatory markers (TNF-α and NO) and ROS in N9 cell line exposure to LPS and IFN-γ. This bioactivity has been related with the NF-κB and STAT1 molecular pathways [87, 105]. By using pure compounds, ferulic, caffeic and protocatechuic acids have been the most studied metabolites on neurodegenerative diseases with an inflammatory component. The pre-treatment of BV2 microglial cells (1 and 4 hours) with PCA (2.5–10 μM) attenuated microglial activation by suppressing TLR4-mediated NF-κB and MAPKs (JNK, p38, ERK) activation and SIRT1 pathway [106, 107]. Other interesting paper displayed that PCA (3,4-dihydroxybenzoic acid), ant not 4-hydroxybenzoic acid can reduced NO production of BV2 cells, however, in this case, PCA concentrations are ten times higher (100 μM) [108]. Furthermore, Koga and their co-workers demonstrated that caffeic acid-treated mice exhibited significantly lower levels of 4-hydroxynonenal (oxidative stress marker) and fewer activated microglia [109]. A long-term treatment (4-weeks) with ferulic acid (in drinking water (0.006%)) for male mice prevented the $A\beta_{1-42}$-induced activation of microglia [110]. Ferulic acid has also demonstrated interfered with TLR4 interaction sites in mouse hippocampus and in BV2 cells by down streaming iNOS, COX-2, TNF-α, and IL-1β via JNK and NF-κB phosphorylation [111]. Furthermore, the intra-peritoneal injection of 30 mg/Kg of vanillic acid reversed LPS-induced glial cells activation, neuroinflammation (TNF-α, IL1-β, and COX-2) and amyloidogenic markers (β-site amyloid precursor protein (APP)–cleaving enzyme 1 (BACE1) and amyloid-β [112]. Finally, concerning gallic acid, two articles can be highlighted. This compound (at 5–50 μM concentration) in a co-culture system consisted on BV2 and Neuro-2A cells and in primary microglia resulted on the diminution of cytokine production induced by the Aβ peptide [113]. After the orally administration of gallic acid (100 mg/Kg) 1 hour prior to the LPS infusion and daily afterwards for 7 days, an attenuation of LPS-induced elevation in heme oxygenase-1 level and α-synuclein aggregation was observed. Moreover, this same work revealed that gallic acid diminished the iNOS gene expression and the NO production *in vitro* [114].

However, any anti-neuroinflammatory activity has been reported for glucuronidated, sulfated and *O*-methylated anthocyanins and their corresponding metabolites. This lack of studies can be explained due to the lack of commercial compounds which makes the chemical synthesis or hemy-synthesis as the only alternative available. Even if it is a challenge to obtain glucuronidated, sulfated and *O*-methylated anthocyanins some methodologies can be used and are reported in the literature. For example methylation of cyandin-3-*O*-glucoside can be carried out by the reaction with dimethylcarbonate [115]. Regarding the hemisynthesis of sulfated derivatives several approaches are possible by bringing the anthocyanins into contact with chlorosulfonic acid [116] or even with sulfur trioxide-N-triethylamine [117]. Finally, glucuronidated anthocyanin can be obtain by acidic aldo-condensation between trihydroybenzalhaldehyde and acetophenone which have been previously functionalized with the expected OH and OMe group as well as the glucuronic acid at the proper position [118].

5. Conclusion

Anthocyanins represents one the most consumed polyphenols in human diet. However, their type, complexity and quantities depends on the foodstuff. For example, anthocyanins in vegetables and cereals are chemically more complex in comparison with fruits, but berries are the major source of these compounds. Anthocyanin bioavailability has been reported to be very low, with recovery of less than 1% of the ingested anthocyanin dose. However, nowadays much greater bioavailability values have reported taking into account not only the phase I and phase II metabolites but also the microbiota catabolites. One of the peculiarities of anthocyanin metabolism is their capacity of interconversion between them. For example, dehydroxylation reaction can arise pelargonidin from cyanidin and methylation reactions can convert delphinidin into petunidin and malvidin. For this reason, metabolism data after anthocyanin ingestion is more straight forward to interpret. Regarding metabolism, cyanidin is the most studied anthocyanin due to ubiquitous character in the nature. However, more studies are necessary to better understand the similarities and differences with the other less studied anthocyanins. Even though several papers have reported the potential anti-neuroinflammatory effect of rich anthocyanin extracts, anthocyanins or their metabolites, the number of papers are very scarce. The most important limitation to study the activity of anthocyanin metabolites is the lack of commercial phase II and microbiota catabolites compounds. Thus, the chemical synthesis is the most employed technique to obtain standards although more developments are requires in order to obtain greater quantities. Moreover, little is known about the molecular mechanisms implicated in the observed effects. Furthermore, the majority of works are based on the study of the microglia M1 phenotype, so more studies are necessary to know if anthocyanins and their metabolites are able to induce an anti-inflammatory phenotype. To sum up, more research is necessary to stablish if anthocyanins and their metabolites are efficacious in slowing the progression of brain aging or of neurodegenerative diseases with an inflammatory component.

Author details

Ruth Hornedo-Ortega[1], Zuriñe Rasines-Perea[2], Ana B. Cerezo[1], Pierre-Louis Teissedre[2] and Michael Jourdes[2*]

1 Departamento de Nutrición y Bromatología, Toxicología y Medicina Legal, Facultad de Farmacia, Universidad de Sevilla, Sevilla, Spain

2 Axe Qualité et Identité du Vin, Unité de Recherche Œnologie, Institut des Sciences de la Vigne et du Vin (ISVV), Université de Bordeaux, Villenave d'Ornon, France

*Address all correspondence to: michael-jourdes@u-bordeaux.fr

References

[1] Kong JM, Chia LS, Goh NK, Chia TF, Brouillard R. Analysis and biological activities of anthocyanins. Phyto chemistry. 2003;64:923-933. DOI: 10.1016/s0031-9422(03)00438-2. Erratum in: Phytochemistry. 2008;69: 1939-1940.

[2] Anderson, OM & Jordheim, M. The anthocyanins. In Flavonoids: Chemistry, Biochemistry and Applications. Boca Raton, FL: CRC Press/Taylor & Francis Group. 2006.

[3] Mazza G, Miniati E. Anthocyanins in fruits, vegetables, and grains. 1st ed. Boca Raton: CRC Press; 1993. DOI: 10.1201/97813510697000

[4] Smeriglio A, Barreca D, Bellocco E, Trombetta D. Chemistry, pharmacology and health benefits of anthocyanins. Phytotherapy Research. 2016;30:1265-1286. DOI: 10.1002/ptr.5642.

[5] Zamora-Ros R, Knaze V, Luján-Barroso L, Slimani N, Romieu I, Touillaud M, Kaaks R, Teucher B, Mattiello A, Grioni S, Crowe F, Boeing H, Förster J, Quirós JR, Molina E, Huerta JM, Engeset D, Skeie G, Trichopoulou A, Dilis V, Tsiotas K, Peeters PHM, Khaw K-T, Wareham N, Bueno-de-Mesquita B, Ocké MC, Olsen A, Tjønneland A, Tumino R, Johansson G, Johansson I, Ardanaz E, Sacerdote C, Sonestedt E, Ericson U, Clavel-Chapelon F, Boutron-Ruault M-C, Fagherazzi G, Salvini S, Amiano P, Riboli E, González CA. Estimation of the intake of anthocyanidins and their food sources in the European Prospective Investigation into Cancer and Nutrition (EPIC) study. The British Journal of Nutrition. 2011:1-10. DOI: 10.1017/S0007114511001437

[6] Fang, J. Bioavailability of anthocyanins. Drug Metabolism Reviews. 2014;46:508-520. DOI: 10.3109/036002532.2014.978080

[7] Konczak I, Zhang W. Anthocyanins—more than nature's colours. Journal of Biomedicine and Biotechnology. 2004;5:239-240. DOI: 10.1155/s1110724304407013

[8] He J, Giusti MM. Anthocyanins: natural colorants with health-promoting properties. Annual Review of Food Science and Technology. 2010;1:163-187. DOI: 10.1146/annurev.food.080708.100754

[9] Tulio AZ Jr, Reese RN, Wyzgoski FJ, Rinaldi PL, Fu R, Scheerens JC, Miller AR. Cyanidin 3-rutinoside and cyanidin 3-xylosylrutinoside as primary phenolic antioxidants in black raspberry. Journal of Agricultural and Food Chemistry. 2008;56:1880-1888. DOI: 10.1021/jf072313k

[10] Clifford MN. Anthocyanins–nature, occurrence and dietary burden. Journal of Science of Food and Agriculture. 2000;80:1063-1072. DOI: 10.1002/(SICI)1097-0010(20000515)80:7<1063::AID-JSFA605>3.0.CO;2-Q

[11] Borkowski T, Szymusiak H, Gliszczynska-Swiglo A, Tyrakowska B. The effect of 3-O-beta-glycosylation on structural transformations of anthocyanins. Food Research International. 2005; 38:1031-1037. DOI: 10.1016/j.foodres.2005.02.020

[12] Fossen T, Cabrita L, Andersen OM. Colour and stability of pure anthocyanins influenced by pH including the alkaline region. Food Chemistry. 1998;63:435-440. DOI: 10.1016/s0308-8146(98)00065-x

[13] Fleschut I, Kratzer F, Rechkemmer G, Kulling SE. Stability and biotransformation of various dietary anthocyanins in vitro. European

Journal of Nutrition. 2006;45:7-18. DOI: 10.1007/s00394-005-0557-8

[14] Wu X, Beecher GR, Holden JM, Haytowitz DB, Gebhardt SE, Prior RL. Concentrations of anthocyanins in common foods in the United States and estimation of normal consumption. Journal of Agricultural and Food Chemistry. 2006;54:4069-4075. DOI: 10.1021/jf060300l

[15] Bai W, Wang C, Ren C. Intakes of total and individual flavonoids by US adults. International Journal of Food Science and Nutrition. 2014;65:9-20. DOI: 10.3109/09637486.2013.832170

[16] Vogiatzoglou A, Mulligan AA, Lentjes MA, Luben RN, Spencer JP, Schroeter H, Khaw KT, Kuhnle GG. Flavonoid intake in European adults (18 to 64 years). PLoS One. 2015;10:e0128132-e0128154. DOI: 10.1371/journal.pone.0128132

[17] Igwe EO, Charlton KE, Probst YC. Usual dietary anthocyanin intake, sources and their association with blood pressure in a representative sample of Australian adults. Journal of Human Nutrition Dietetics. 2019;32:578-590. DOI: 10.1111/jhn.12647

[18] Li G, Zhu Y, Zhang Y, Lang J, Chen Y, Ling W. Estimated daily flavonoid and stilbene intake from fruits, vegetables, and nuts and associations with lipid profiles in Chinese adults. Journal of the Academy of Nutrition and Dietetics. 2013;113:786-794. DOI: 10.1016/j.jand.2013.01.018

[19] Patras A, Brunton NP, Colm O'Donnell, B.K. Tiwari. Effect of thermal processing on anthocyanin stability in foods; mechanisms and kinetics of degradation. Trends in Food Science & Technology. 2010;21:3-11. DOI: 10.1016/j.tifs.2009.07.004

[20] Bars-Cortina D, Macià A, Iglesias I, Romero MP, Motilva MJ. Phytochemical profiles of new red-fleshed apple varieties compared with traditional and new white-fleshed varieties. Journal of Agricultural and Food Chemistry. 2017;65:1684-1696. DOi: 10.1021/acs.jafc.6b02931

[21] Bureau S, Renard CMGC, Reich M, Ginies C, Audergon JM. Change in anthocyanin concentrations in red apricot fruits during ripening. LWT - Food Science and Technology. 2009; 42:372-377. DOI: 10.1016/j.lwt.2008.03.010

[22] Müller D, Schantz M, Richling E. High performance liquid chromatography analysis of anthocyanins in bilberries (Vaccinium myrtillus L.), blueberries (Vaccinium corymbosum L.), and corresponding juices. Journal of Food Science. 2012 Apr;77(4):C340-C345. DOI: 10.1111/j.1750-3841.2011.02605.x.

[23] Pires TCSP, Caleja C, Santos-Buelga C, Barros L, Ferreira ICFR. Vaccinium myrtillus L. Fruits as a novel source of phenolic compounds with health benefits and industrial applications - a Review. Current Pharmaceutical Design. 2020;26:1917-1928. DOI: 10.2174/1381612826666200317132507

[24] Fan-Chiang H, Wrolstad RE. Anthocyanin pigment composition of blackberries. Journal of Food Science. 2005;70:C198-C202. DOI: 10.1111/j.1365-2621.2005

[25] Gao L, Mazza G, Characterization, quantitation, and distribution of anthocyanins and colorless phenolics in sweet cherries. Journal of Agricultural and Food Chemistry.1995;43:343-346. DOI: 10.1021/jf00050a015

[26] Grigoras C, Destandau E, Zubrzycki S, Elfakir C. Sweet cherries anthocyanins: An environmental friendly extraction and purification method. Separation and Purification

Technology. 2012;100:51-58. DOI: 10.1016/j.seppur.2012.08.032

[27] Cesoniene L, Jasutiene I, Sarkinas A. Phenolics and anthocyanins in berries of European cranberry and their antimicrobial activity. Medicina (Kaunas). 2009;45(12):992-999. PMID: 20173403.

[28] Chen X, Parker J, Krueger C, Shanmuganayagam D, Reed J. Validation of HPLC assay for the identification and quantification of anthocyanins in black currants. Analytical Methods. 2014;6:8141-8147. DOI: 10.1039/C4AY01500B

[29] Ky I, Lorrain B, Kolbas N, Crozier A, Teissedre PL. Wine by-products: Phenolic characterization and antioxidant activity evaluation of grapes and grape pomaces from six different French grape varieties. Molecules. 2014;19:482-506. DOI: 10.3390/molecules19010482

[30] Vardin H, Fenercioğlu H. Study on the development of pomegranate juice processing technology: clarification of pomegranate juice. Nahrung. 2003;47:300-303. DOI: 10.1002/food.200390070

[31] Aaby K, Mazur S, Nes A, Skrede G. Phenolic compounds in strawberry (Fragaria x ananassa Duch.) fruits: Composition in 27 cultivars and changes during ripening. Food Chemistry. 2012;132:86-97. DOI: 10.1016/j.foodchem.2011.10.037

[32] Mojica L, Berhow M, Gonzalez de Mejia E. Black bean anthocyanin-rich extracts as food colorants: Physicochemical stability and antidiabetes potential. Food Chemistry. 2017;229:628-639. DOI: 10.1016/j.foodchem.2017.02.124

[33] Wiczkowski W, Dorota Szawara-Nowak, Joanna Topolska. Red cabbage anthocyanins: Profile, isolation, identification, and antioxidant activity. Food Research International. 2013;51:303-309. DOI: 10.1016/j.foodres.2012.12.015

[34] Assous M, Abdel-Hady M, Medany GM. Evaluation of red pigment extracted from purple carrots and its utilization as antioxidant and natural food colorants. Annals of Agricultural Science. 2014;59:1-7. DOI: 10.1016/j.aoas.2014.06.001

[35] Kapusta-Duch J, Szeląg-Sikora A, Sikora J, Niemiec M, Gródek-Szostak Z, Kuboń M, Leszczyńska T, Borczak B. Health-promoting properties of fresh and processed purple cauliflower. Sustainability. 2019;11:4008-4013. DOI: 10.3390/su11154008

[36] Lo Scalzo R, Genna A, Branca F, Chedin M, Chassaigne H. Anthocyanin composition of cauliflower (Brassica oleracea L. var. botrytis) and cabbage (B. oleracea L. var. capitata) and its stability in relation to thermal treatments. Food Chemistry. 2008;107:136-144. DOI: 10.1016/j.foodchem.2007.07.072

[37] Lo Scalzo R, Fibiani M, Francese G, D'Alessandro A, Rotino GL, Conte P, Mennella G. Cooking influence on physico-chemical fruit characteristics of eggplant (Solanum melongena L.). Food Chemistry. 2016;194:835-842. DOI: 10.1016/j.foodchem.2015.08.063

[38] Ji X, Rivers L, Zielinski Z, Xu M, MacDougall E, Stephen J, Zhang S, Wang Y, Chapman RG, Keddy P, Robertson GS, Kirby CW, Embleton J, Worrall K, Murphy A, De Koeyer D, Tai H, Yu L, Charter E, Zhang J. Quantitative analysis of phenolic components and glycoalkaloids from 20 potato clones and in vitro evaluation of antioxidant, cholesterol uptake, and neuroprotective activities. Food Chemistry. 2012;133:1177-1187. DOI: 10.1016/j.foodchem.2011.08.065

[39] Koponen JM, Happonen AM, Mattila PH, Törrönen AR. Contents of anthocyanins and ellagitannins in selected foods consumed in Finland. Journal of Agricultural and Food Chemistry. 2007;55:1612-1619. DOI: 10.1021/jf062897a

[40] Bellido GG, Beta T. Anthocyanin composition and oxygen radical scavenging capacity (ORAC) of milled and pearled purple, black, and common barley. Journal of Agricultural and Food Chemistry. 2009;57:1022-1028. DOI: 10.1021/jf802846x

[41] Francavilla A, Joye IJ. Anthocyanins in whole grain cereals and their potential effect on health. Nutrients. 2020;12:2922-2942. DOI: 10.3390/nu12102922

[42] Harakotr, B., Suriharn, B., Tangwongchai, R., Scott, M. Paul, & Lertrat, K. (2014). Anthocyanins and antioxidant activity in coloured waxy corn at different maturation stages. Journal of Functional Foods, 9, 109-118. DOI: 10.1016/j.jff.2014.04.012

[43] Chen XQ, Nagao N, Itani T, Irifune K. Anti-oxidative analysis, and identification and quantification of anthocyanin pigments in different coloured rice. Food Chemistry. 2012;135:2783-2788. DOI: 10.1016/j.foodchem.2012.06.098

[44] Awika JM, Yang L, Browning JD, Faraj A. Comparative antioxidant, antiproliferative and phase II enzyme inducing potential of sorghum (Sorghum bicolor) varieties. LWT - Food Science and Technology. 2009;42:1041-1046. DOI: 10.1016/j.lwt.2009.02.003

[45] Abdel-Aal el-SM, Young JC, Rabalski I. Anthocyanin composition in black, blue, pink, purple, and red cereal grains. Journal of Agricultural and Food Chemistry. 2006;54:4696-4704. DOI: 10.1021/jf0606609

[46] Czank C, Cassidy A, Zhang Q, Morrison DJ, Preston T, Kroon PA, Botting NP, Kay CD. Human metabolism and elimination of the anthocyanin, cyanidin-3-glucoside: a (13)C-tracer study. The American Journal of Clinical Nutrition. 2013;97:995-1003. DOI: 10.3945/ajcn.112.049247

[47] Lila MA, Burton-Freeman B, Grace M, Kalt W. Unraveling anthocyanin bioavailability for human health. Annual Review of Food Science and Technology. 2016;7:375-394. DOI: 10.1146/annurev-foof-041715-033346

[48] Kurilich AC, Clevidence BA, Britz SJ, Simon PW, Novotny JA. Plasma and urine responses are lower for acylated versus nonacylated anthocyanins from raw and cooked purple carrots. Journal of Agricultural and Food Chemistry. 2005;53:6537-6542. DOI: 10.1021/jf050570o

[49] Stalmach A, Edwards CA, Wightman JD, Crozier, A. Gastrointestinal stability and bioavailability of (poly)phenolic compounds following ingestion of Concord grape juice by humans. Molecular Nutrition & Food Research. 2012;56,497-509. DOI: 10.1002/mnfr.201100566

[50] Cai H, Thomasset SC, Berry DP, Garcea G, Brown K, Steward WP, Gescher AJ. Determination of anthocyanins in the urine of patients with colorectal liver metastases after administration of bilberry extract. Biomedical Chromatography. 2011;25:660-663. DOI: 10.1002/bmc.1499

[51] Hribar U, Ulrih NP. The Metabolism of Anthocyanins. Current Drug Metabolism. 2014;15:3–13. DOI: 10.2174/1389200214666131211160308

[52] Oliveira H, Fernandes I, Brás NF, Faria A, De Freitas V, Calhau C,

Mateus N. Experimental and theoretical data on the mechanism by which red wine anthocyanins are transported through a human MKN-28 gastric cell model. Journal of Agricultural and Food Chemistry. 2015;63:7685–7692. DOI: 10.1021/acs.jafc.5b00412

[53] Zhong S, Sandhu A, Edirisinghe I, Burton-Freeman, B. Mint: Characterization of wild blueberry polyphenols bioavailability and kinetic profile in plasma over 24-h period in human subjects. Molecular Nutrition & Food Research. 2017;61:1700405-1700418. DOI: 10.1002/mnfr.201700405

[54] Vitaglione P, Donnarumma G, Napolitano A, Galvano F, Gallo A, Scalfi L, Fogliano V. Protocatechuic acid is the major human metabolite of cyanidin-glucosides. The Journal of Nutrition. 2007;137:2043-2048. DOI: 10.1093/jn/137.9.2043

[55] de Ferrars R, Czank C, Zhang Q, Botting N, Kroon P,Cassidy A, Kay C. The Pharmacokinetics of anthocyanins and their metabolites in humans. British Journal of Pharmacology. 2014;171:3268–3282. DOI: 10.1111/bph.12676

[56] Sandoval-Ramírez BA, Catalán Ú, Fernández-Castillejo S, Rubió L, Macià A, Solà R. Anthocyanin tissue bioavailability in animals: Possible implications for human health. A systematic review. Journal Agricultural and Food Chemistry. 2018;66:11531-11543. DOI: 10.1021/acs.jafc.8b04014

[57] Kirakosyan A, Seymour EM, Wolforth J, McNish R, Kaufman PB, Bolling SF. Tissue bioavailability of anthocyanins from whole tart cherry in healthy rats. Food Chemistry. 2015;171:26-31. DOI: 10.1016/j.foodchem.2014.08.114

[58] Chen TY, Kritchevsky J, Hargett K, Feller K, Klobusnik R, Song BJ, Cooper B, Jouni Z, Ferruzzi MG, Janle EM. Plasma bioavailability and regional brain distribution of polyphenols from apple/grape seed and bilberry extracts in a young swine model. Molecular Nutrition & Food Research. 2015;59:2432-2447. DOI: 10.1002/mnfr.201500224

[59] Aqil F, Vadhanam MV, Jeyabalan J, Cai J, Singh IP, Gupta RC. Detection of anthocyanins/anthocyanidins in animal tissues. Journal of Agricultural and Food Chemistry. 2014;62:3912-3918. DOI: 10.1021/jf500467b

[60] Talavéra S, Felgines C, Texier O, Besson C, Gil-Izquierdo A, Lamaison J-L, Rémésy C. Anthocyanin metabolism in rats and their distribution to digestive area, kidney, and brain. Journal of Agricultural and Food Chemistry. 2005;53:3902-3908. DOI: 10.1021/jf050145v

[61] Kalt W, Blumberg JB, McDonald JE, Vinqvist-Tymchuk MR, Fillmore SAE, Graf BA, O'Leary JM, Milbury PE. Identification of anthocyanins in the liver, eye, and brain of blueberry-fed pigs. Journal of Agricultural Food Chemitry. 2008 ;56:705-12. DOI : 10.1021/jf071998l

[62] Fornasaro S, Ziberna L, Gasperotti M, Tramer F, Vrhovšek U, Mattivi F, Passamonti S. Determination of cyanidin 3-glucoside in rat brain, liver and kidneys by UPLC/MS-MS and its application to a short-term pharmacokinetic study. Scientific Reports. 2016;6:1-11. DOI: 10.1038/srep22815

[63] Carkeet C, Clevidence BA, Novotny JA. Anthocyanin excretion by humans increases linearly with increasing strawberry dose. Journal ol Nutrition. 2008;138:897-902. DOI: 10.1093/jn/138.5.897

[64] Mullen W, Edwards CA, Serafini M, Crozier A. Bioavailability of pelargonidin-3-O-glucoside and its

metabolites in humans following the ingestion of strawberries with and without cream. Journal of Agricultural and Food Chemistry. 2008;56:713-719. DOI: 10.1021/jf072000p

[65] Charron CS, Kurilich AC, Clevidence BA, Simon PW, Harrison DJ, Britz SJ, Baer DJ, Novotny JA. Bioavailability of anthocyanins from purple carrot juice: effects of acylation and plant matrix. Journal of Agricultural and Food Chemistry. 2009;57:1226-1230. DOI: 10.1021/jf802899s

[66] Crozier A, Del Rio D, Clifford MN. Bioavailability of dietary flavonoids and phenolic compounds. Molecular Aspects of Medicine. 2010;31:446-467. DOI: 10.1016/j.mam.2010.09.007

[67] Cooney JM, Jensen DJ, McGhie TK. LC-MS identification of anthocyanins in boysenberry extract and anthocyanin metabolites in human urine following dosing. Journal of the Science of Food and Agriculture. 2004;84:237-245. DOI: 10.1002/jsfa.1645

[68] Kalt W, McDonald JE, Liu Y, Fillmore SAE. Flavonoid metabolites in human urine during blueberry anthocyanin intake. Journal of Agricultural and Food Chemistry. 2017;65:1582-1591. DOI: 10.1021/acs.jafc.6b05455

[69] Yustea S, Ludwiga IA, Rubió L, Romero M-P, Pedret A, Valls R-M, Solà R, Motilva M-J, Macia A. In vivo biotransformation of (poly)phenols and anthocyanins of red-fleshed apple and identification of intake biomarkers. Journal of Functional Foods. 2019;55:146-155. DOI: 10.1016/j.jff.2019.02.013

[70] Cardona F, Andrés-Lacueva C, Tulipani S, Tinahones FJ, Queipo-Ortuño MI. Benefits of polyphenols on gut microbiota and implications in human health. J Nutrition Biochemistry. 2013;24:1415-1422. DOI: 10.1016/j.jnutbio.2013.05.001

[71] Gonzalez-Barrio R, Edwards C, Crozier A. Colonic catabolism of ellagitannins, ellagic acid, and raspberry anthocyanins: in vivo and in vitro studies. Drug Metabolism & Disposition. 2011;39:1680-1688. DOI: 10.1124/dmd.111.039651

[72] Azzini E,Vitaglione P, Intorre F, Napolitano A, Durazzo A, Foddai MS, Fumagalli A, Catasta G, Rossi L, Venneria E, Raguzzini A, Palomba L, Fogliano V, Maiani G. Bioavailability of strawberry antioxidants in human subjects. British Journal of Nutrition. 2010;104:1165-1173. DOI: 10.1017/S000711451000187X

[73] Nurmi T, Mursu J, Heinonen M, Nurmi A, Hiltunen R, Voutilainen S. Metabolism of berry anthocyanins to phenolic acids in humans. Journal of Agricultural and Food Chemistry. 2009;57:2274-2281. DOI: 10.1021/jf8035116

[74] Andres-Lacueva C, Shukitt-Hale B, Galli RL, Jauregui O, Lamuela-Raventos RM, Joseph JA. Anthocyanins in aged blueberry-fed rats are found centrally and may enhance memory. Nutrition Neuroscience. 2005;8:111-120. DOI: 10.1080/10284150500078117

[75] El Mohsen MA, Marks J, Kuhnle G, Moore K, Debnam E, Kaila Srai S, Rice-Evans C, Spencer JP. Absorption, tissue distribution and excretion of pelargonidin and its metabolites following oral administration to rats. British Journal of Nutrition. 2006;95:51-58. DOI: 10.1079/BJN20051596

[76] Mullen W, Larcombe S, Arnold K, Welchman H, Crozier A. Use of accurate mass full scan mass spectrometry for the analysis of anthocyanins in berries and berry-fed tissues. Journal of Agricultural and Food Chemistry.

2010;58:3910-3915. DOI: 10.1021/jf902267v

[77] Glass CK, Saijo K, Winner B, Marchetto MC, Gage FH. Mechanisms underlying inflammation in neurodegeneration. Cell. 2010;19;140(6):918-34. DOI: 10.1016/j.cell.2010.02.016.

[78] Martinez FO, and Gordon S. The M1 and M2 paradigm of macrophage activation: time for reassessment. F1000Prime Rep. 2014; 6:13. doi: 10.12703/P6-13

[79] MacMicking J, Xie QW, Nathan C. Nitric oxide and macrophage function. Annual Review of Immunology. 1997;15:323-350. DOI: 10.1146/annurev.immunol.15. 1.323

[80] Arnold CE, Whyte CS, Gordon P, Barker RN, Rees AJ, Wilson HM. A critical role for suppressor of cytokine signalling 3 in promoting M1 macrophage activation and function in vitro and in vivo. Immunology. 2014;141:96-110. DOI: 10.1111/imm.12173

[81] Brancato S K, Albina JE. Wound macrophages as key regulators of repair: origin, phenotype, and function. The American Journal of Pathology. 2011;178:19-25. DOI: 10.1016/j.ajpath.2010.08.003

[82] Gordon S, Martinez FO. Alternative activation of macrophages: mechanism and functions. Immunity. 2010;32:593-604. DOI: 10.1016/j.immuni.2010.05.007

[83] Salehi B, Sharifi-Rad J, Cappellini F, Reiner Ž, Zorzan D, Imran M, Sener B, Kilic M, El-Shazly M, Fahmy NM, Al-Sayed E, Martorell M, Tonelli C, Petroni K, Docea AO, Calina D, Maroyi A. The Therapeutic Potential of Anthocyanins: Current Approaches Based on Their Molecular Mechanism of Action. Frontiers in Pharmacology. 2020;11:1300. DOI: 10.3389/fphar.2020.01300.

[84] Henriques JF, Serra D, Dinis TCP, Almeida LM. The Anti-Neuroinflammatory Role of Anthocyanins and Their Metabolites for the Prevention and Treatment of Brain Disorders. International Journal of Molecular Sciences. 2020.21(22):8653. DOI: 10.3390/ijms21228653.

[85] Lau FC, Bielinski DF, Joseph JA. Inhibitory effects of blueberry extract on the production of inflammatory mediators in lipopolysaccharide-activated BV2 microglia. Journal of Neuroscience Research. 2007;85:1010-1017. DOI: 10.1002/jnr.21205

[86] Carey AN, Fisher DR, Rimando AM, Gomes SM, Bielinski DF, Shukitt-Hale B. Stilbenes and anthocyanins reduce stress signaling in BV-2 mouse microglia. Journal of Agricultural and Food Chemistry. 2013;61:5979-5986. DOI: 10.1021/jf400342g

[87] Serra D, Henriques JF, Serra T, Bento Silva A, Bronze MR, Dinis TCP, Almeida LM. An anthocyanin-rich extract obtained from portuguese blueberries maintains its efficacy in reducing microglia-driven neuroinflammation after simulated digestion. Nutrients. 2020;12:3670-3692. DOI: 10.3390/nu12123670

[88] Lau FC, Joseph JA, McDonald JE, Kalt W. Attenuation of iNOS and COX2 by blueberry polyphenols is mediated through the suppression of NF-B activation. Journal of Functional Foods. 2009;1:274-283. DOi: 10.1016/j.jff.2009.05.001

[89] Ma H, Johnson SL, Liu W, DaSilva NA, Meschwitz S, Dain JA, Seeram NP. Evaluation of polyphenol anthocyanin-enriched extracts of blackberry, black raspberry, blueberry, cranberry, red raspberry, and

strawberry for free radical scavenging, reactive carbonyl species trapping, anti-glycation, anti-β-amyloid aggregation, and microglial neuroprotective effects. International Journal of Molecular Sciences. 2018;19:461-780. DOI: 10.3390/ijms19020461

[90] Simonyi A, Chen Z, Jiang J, Zong Y, Chuang DY, Gu Z, Lu CH, Fritsche KL, Greenlief CM, Rottinghaus GE, Thomas AL, Lubahn DB, Sun GY. Inhibition of microglial activation by elderberry extracts and its phenolic components. Life Sciences. 2015;128:30-38. DOI: 10.1016/j.lfs.2015.01.037

[91] Poulose SM, Fisher DR, Larson J, Bielinski DF, Rimando AM, Carey AN, Schauss AG, Shukitt-Hale B. Anthocyanin-rich acai (Euterpe oleracea Mart.) fruit pulp fractions attenuate inflammatory stress signaling in mouse brain BV-2 microglial cells. Journal of Agricultural and Food Chemistry. 2012;60:1084-1093. DOI: 10.1021/jf203989k

[92] Jeong JW, Lee WS, Shin SC, Kim GY, Choi BT, Choi YH. Anthocyanins downregulate lipopolysaccharide-induced inflammatory responses in BV2 microglial cells by suppressing the NF-kappaB and Akt/MAPKs signaling pathways. International Journal of Molecular Sciences. 2013;14:1502-1515. DOI: 10.3390/ijms14011502

[93] Mu T, Guan Y, Chen T, Wang S, Li M, Chang AK, Yang Z, Bi X. Black raspberry anthocyanins protect BV2 microglia from LPS-induced inflammation through down-regulating NOX2/TXNIP/NLRP3 signaling. Journal of Berry Research. 2021;2:333-347. DOI: 10.3233/jbr-200692

[94] Meireles M, Marques C, Norberto S, Santos P, Fernandes I, Mateus N, Faria A, Calhau C. Anthocyanin effects on microglia M1/M2 phenotype: Consequence on neuronal fractalkine expression. Behavioural Brain Research. 2016;305:223-228. DOI: 10.1016/j.bbr.2016.03.010

[95] Kaewmool C, Udomruk S, Phitak T, Pothacharoen P, Kongtawelert P. Cyanidin-3-O-glucoside protects PC12 cells against neuronal apoptosis mediated by LPS-stimulated BV2 microglial activation. Neurotoxicity Research. 2020;37:111-125. DOI: 10.1007/s12640-019-00102-1

[96] Zhao L, Chen S, Liu T, Wang X, Huang H, Liu W. Callistephin enhances the protective effects of isoflurane on microglial injury through downregulation of inflammation and apoptosis. Molecular Medicine Reports. 2019;20:802-812. DOI: 10.3892/mmr.2019.10282

[97] Meireles M, Marques C, Norberto S, Fernandes I, Mateus N, Rendeiro C, Spencer JP, Faria A, Calhau C. The impact of chronic blackberry intake on the neuroinflammatory status of rats fed a standard or high-fat diet. Journal of Nutritional Biochemistry. 2015;26:1166-1173. DOI: 10.1016/j.jnutbio.2015.05.008

[98] Rehman SU, Shah SA, Ali T, Chung JI, Kim MO. Anthocyanins reversed D-galactose-induced oxidative stress and neuroinflammation mediated cognitive impairment in adult rats. Molecular Neurobiology. 2017;54:255-271. DOI: 10.1007/s12035-015-9604-5

[99] Khan MS, Ali T, Kim MW, Jo MH, Jo MG, Badshah H, Kim MO. Anthocyanins protect against LPS-induced oxidative stress-mediated neuroinflammation and neuro degeneration in the adult mouse cortex. Neurochemistry International. 2016;100:1-10. DOI: 10.1016/j.neuint.2016.08.005

[100] Khan MS, Ali T, Kim MW, Jo MH, Chung JI, Kim MO. Anthocyanins improve hippocampus-dependent memory function and prevent

neurodegeneration via JNK/Akt/GSK3β signaling in LPS-Treated adult mice. Molecular Neurobiology. 2019;56:671-687. DOI: 10.1007/s12035-018-1101-1

[101] Carvalho FB, Gutierres JM, Bueno A, Agostinho P, Zago AM, Vieira J, Frühauf P, Cechella JL, Nogueira CW, Oliveira SM, Rizzi C, Spanevello RM, Duarte MMF, Duarte T, Dellagostin OA, Andrade CM. Anthocyanins control neuro inflammation and consequent memory dysfunction in mice exposed to lipopolysaccharide. Molecular Neurobiology. 2017;54:3350-3367. DOI: 10.1007/s12035-016-9900-8

[102] Magni G, Marinelli A, Riccio D, Lecca D, Tonelli C, Abbracchio MP, Petroni K, Ceruti S. Purple corn extract as anti-allodynic treatment for trigeminal pain: Role of microglia. Frontiers in Cellular Neuroscience. 2018;12:378-390. DOI: 10.3389/fncel.2018.00378

[103] Tikhonova MA, Shoeva OY, Tenditnik MV, Ovsyukova MV, Akopyan AA, Dubrovina NI, Amstislavskaya TG, Khlestkina EK. Evaluating the effects of grain of isogenic wheat lines differing in the content of anthocyanins in mouse models of neurodegenerative disorders. Nutrients. 2020;12:3877-3898. DOI: 10.3390/nu12123877

[104] Li J , Zhao R , Jiang Y , Xu Y , Zhao H , Lyu X , Wu T. Bilberry anthocyanins improve neuroinflammation and cognitive dysfunction in APP/PSEN1 mice via the CD33/TREM2/TYROBP signaling pathway in microglia. Food & Function. 2020;11:1572-1584. DOI: 10.1039/c9fo02103e

[105] Garcia G, Nanni S, Figueira I, Ivanov I, McDougall GJ, Stewart D, Ferreira RB, Pinto P, Silva RF, Brites D, Santos CN. Bioaccessible (poly)phenol metabolites from raspberry protect neural cells from oxidative stress and attenuate microglia activation. Food Chemistry. 2017;215:274-283. DOI: 10.1016/j.foodchem.2016.07.128

[106] Wang HY, Wang H, Wang JH, Wang Q, Ma QF, Chen YY. Protocatechuic acid inhibits inflammatory responses in LPS-Stimulated BV2 microglia via NF-κB and MAPKs signaling pathways. Neurochemical Research. 2015;40:1655-1660. DOI: 10.1007/s11064-015-1646-6

[107] Kaewmool C, Kongtawelert P, Phitak T, Pothacharoen P, Udomruk S. Protocatechuic acid inhibits inflammatory responses in LPS-activated BV2 microglia via regulating SIRT1/NF-κB pathway contributed to the suppression of microglial activation-induced PC12 cell apoptosis. Journal of Neuroimmunology. 2020;341:577164-577175. DOI: 10.1016/j.jneuroim.2020.577164

[108] Winter AN, Brenner MC, Punessen N, Snodgrass M, Byars C, Arora Y, Linseman DA. Comparison of the neuroprotective and anti-inflammatory effects of the anthocyanin metabolites, protocatechuic acid and 4-hydroxybenzoic acid. Oxidative Medicine and Cellular Longevity. 2017;2017:6297080-6297094. DOI: 10.1155/2017/6297080

[109] Koga M, Nakagawa S, Kato A, Kusumi I. Caffeic acid reduces oxidative stress and microglial activation in the mouse hippocampus. Tissue and Cell. 2019;60:14-20. DOI: 10.1016/j.tice.2019.07.006

[110] Cho JY, Kim HS, Kim DH, Yan JJ, Suh HW, Song DK. Inhibitory effects of long-term administration of ferulic acid on astrocyte activation induced by intracerebroventricular injection of beta-amyloid peptide (1-42) in mice. Prog Neuropsychopharmacol Biological Psychiatry. 2005;29:901-907. DOI: 10.1016/j.pnpbp.2005.04.022

[111] Rehman SU, Ali T, Alam SI, Ullah R, Zeb A, Lee KW, Rutten BPF, Kim MO. Ferulic acid rescues LPS-induced neurotoxicity via modulation of the TLR4 receptor in the mouse hippocampus. Molecular Neurobiology. 2019;56:2774-2790. DOI: 10.1007/s12035-018-1280-9

[112] Ullah R, Ikram M, Park TJ, Ahmad R, Saeed K, Alam SI, Rehman IU, Khan A, Khan I, Jo MG, Kim MO. Vanillic acid, a bioactive phenolic compound, counteracts LPS-induced neurotoxicity by regulating c-Jun N-terminal kinase in mouse brain. International Journal of Molecular Sciences. 2020;22:361-382. DOI: 10.3390/ijms22010361

[113] Kim MJ, Seong AR, Yoo JY, Jin CH, Lee YH, Kim YJ, Lee J, Jun WJ, Yoon HG. Gallic acid, a histone acetyltransferase inhibitor, suppresses β-amyloid neurotoxicity by inhibiting microglial-mediated neuroinflammation. Molecular Nutrition & Food Research. 2011;55:1798-1808. DOI: 10.1002/mnfr.201100262

[114] Liu YL, Hsu CC, Huang HJ, Chang CJ, Sun SH, Lin AM. Gallic acid attenuated LPS-induced neuroinflammation: Protein aggregation and necroptosis. Molecular Neurobiology. 2020;57:96-104. DOI: 10.1007/s12035-019-01759-7

[115] Straßmann S, Brehmer T, Passon M, Schieber A. Methylation of cyanidin-3-*O*-glucoside with dimethyl carbonate. Molecules. 2021;26:1342. DOI: 10.3390/molecules26051342

[116] Soidinsalo O, Wähälä K. Synthesis of phytoestrogenic isoflavonoid disulfates. Steroids. 2004;69:613-616. DOI: 10.1016/j.steroids.2004.03.01

[117] Straßmann S, Passon M, Schieber A. Chemical hemisynthesis of sulfated cyanidin-3-*O*-glucoside and cyanidin metabolites. Molecules, 2021;26:2146. DOI: 10.3390/molecules26082146

[118] Cruz L, Basílio N, Mateus N, Pina F, de Freitas V. Characterization of kinetic and thermodynamic parameters of cyanidin-3-glucoside methyl and glucuronyl metabolite conjugates. The Journal of Physical Chemistry B. 2015; 119:2010-2018 DOI:10.1021/jp511537e

Chapter 6

Effect of Biotic and Abiotic Stresses on Plant Metabolic Pathways

Venkanna Banothu and Addepally Uma

Abstract

Plants are prone to encounter some environmental stresses that include both biotic and abiotic. Plants in response to these stress conditions alter their metabolism at the genetic level with consequential effects at the metabolite production. Phenolic compounds, which are secondary metabolites are one such chemical entity which plays a significant role in various physiological processes of the plant. They are mainly formed by three different types of metabolic pathways that produce phenyl propanoid derivatives, flavonoids, terpenoids based on the needs of the plant and the rate of their production is solely dictated by the type of stress condition. A number of phenolic compounds like phytoalexins, phytoanticipins and nematicides exhibit negative response to biotic stress against several soil borne pathogens and nematodes. But some of the phenolic compounds like acetosyringone, umbelliferone, vanillyl alcohol, *p*-hydroxybenzoic acid, 3,4-dihydroxybenzoic acid, apigenin and luteolin are found to exhibit beneficial effects to plants by encouraging rhizosphere formation particularly in Leguminosae family. Some of the ROS produced in various stress conditions are effectively dealt by various phenolics with antioxidant activity like hydroxyl benzoic acids and hydroxyl cinnamic acids. As the in vivo production of phenolics in plants is influenced by external factors it can certainly provide information for the adoption of agronomic practices to yield the full befits of commercial exploitation. As the in vivo production of phenolics in plants is influenced by external factors it can certainly provide information for the adoption of agronomic practices to yield the full befits of commercial exploitation.

Keywords: Phenolics, Plant Secondary metabolites, Plant physiology, Natural bioactive molecules, Antioxidant activities

1. Introduction

Every living organism shares some basic features like order, sensitivity or response to the environment, reproduction, growth and development, regulation, energy processing, and evolution with adaptation [1]. A basic concept in classical genetics emphasizes that the phenotype of the organism is based on the interaction of genotype with the environment. The emergence of specific natural products is dependent on the highly ordered interaction between plants with the biotic and abiotic environments around them [2]. Plants are sessile organisms and respond to the stress conditions by changing the expression levels of certain genes involved in the production of metabolites which are secreted in response to its interaction with its environment [3]. The various metabolic pathways produce different types of metabolites and based on the pathway they are classified as primary and secondary, although a strict demarcation is difficult to draw between them. The primary metabolites are essential to the cellular growth and reproduction whereas the secondary metabolites although not required directly for the same, these are the compounds that are synthesized in response to any biotic or abiotic stress which may be exogenous or endogenous the cell [3].

The production of secondary metabolites is infact influenced by primary metabolites. Some of the C and N fluxes can be diverted for the production of secondary metabolites during the stressful conditions and there is always a dynamic balance maintained between the two based on the cellular needs. In comparison with the primary metabolites the concentration of secondary metabolites is low and the type of secreted plant secondary metabolite (PSM) is based on the type of stressful physiology induced by biotic and abiotic stress condition. Some of the secondary metabolites are acts as regulators of development, growth and defense. Some of these compounds can be reintegrated into plant primary metabolism [4] *senu lato*.

The regulation of production of PSM involves extensive cross talk and signaling pathways with the key roles played by molecules like salicylic and jasmonic acids, calcium, abscisic acid, spolyamines and nitric oxides [5–8].

Over 2,14,000 types of secondary metabolites are known and are commonly classified according to their structure, function and biosynthetic pathway. Plant secondary metabolites can be classified into four major classes: i) Terpenoids, ii) Phenolic compounds, iii) Alkaloids and iv) Sulfur-containing compounds [9].

2. Types of stresses

The environment around the plant can influence the physiological condition of the plant and any disturbance in the external environment including physical factors and biological factors can influence the metabolic pathways in the cell. Accordingly, the types of stress conditions are defined as biotic and abiotic stresses, such as pathogen infection, water deprivation, salinization, high/low temperature stress, heavy metal toxicity, nutrient deficiency, atmospheric pollution and UV irradiation [10]. In addition to this endogenously generated stress also can influence the production of PSM which are consequential effect of external factors or molecules generated due to various physiological activities within the plant like reactive oxygen and nitrogen species (RONS) burden [10].

3. Phenolic compounds as secondary metabolites

The adoption of plants from aquatic habitat to terrestrial occurred at about 480–300 million years ago and to cope up with this change they adopted protective UV screens called phenolic compounds [11]. They constitute an important set of secondary metabolites which are ubiquitously spread in plant kingdom and the type of phenolic compounds differs in different genre of plant kingdom. Polyphenols are one of the important classes of specialized metabolites that play crucial physiological roles throughout the plant life cycle including responses to stress [12]. Therefore, as an adaptive response to adverse environmental conditions, phenolics are accumulated in various plant tissues which confers evolutionary fitness to the plant. Plants are continuously exposed to various biotic and abiotic stresses like intense

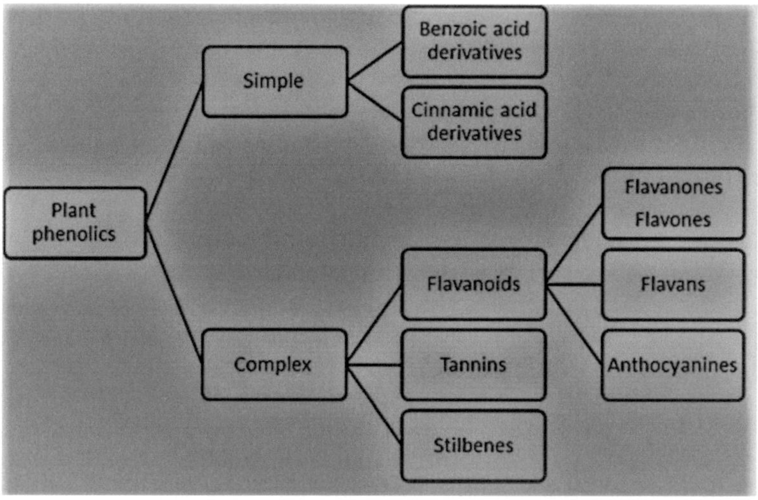

Figure 1.
Classifications of the plant phenols based on their structure.

light, low temperature nutrient deficiency, microbial infections with increased free radical and other oxidative species and plant phenolics are compounds which play a defense role by scavenging these high reactive species [13]. Plants adapt themselves their phenolic patterns to a changing environment through the emergence of new genes brought about by gene duplication and mutation and subsequent recruitment for adaptation to specific functions [14]. Many of the genes related to secondary metabolism are duplicated in plant genome and many of these secondary metabolites production demands change in the amounts of precursors supplied by primary metabolism to balance the perturbations in chemical ecology [15, 16]. The induction of secondary metabolism gene expression by biotic and abiotic stress is often mediated by integrating signaling molecules such as salicylic acid, jasmonic acid, and their derivatives [17]. Flavonoids, stilbenes and proanthocyanidins are collectively grouped in polyphenols, the name indicating both the compounds with a second aromatic ring and those arising from the polymerization of flavonoidic/catechin units. The main structure in flavonoids is the flavan nucleus consisting of 15 carbon atoms which are arranged into three rings with two benzene rings (A & B) connected by an oxygen containing pyran ring (C). Various flavonoids differ in the level of oxidation and saturation of the C ring and accordingly are classified into flavanones, flavones, flavanals, flavanols and anthocyanidins. The individual compounds in a particular class of flavonoids differ in the sunstitution pattern of the A and B rings [18]. Due to the heterogeneous structures of these phenolic acids which range from low molecular weight single aromatic ring structure to high molecular weight polymeric compounds, they can be broadly classified into simple and complex phenolics (**Figure 1**).

4. Metabolic pathways for the formation of phenolic secondary metabolites

As discussed above both primary and secondary metabolites share some of the precursor compounds and there is delicate balance in the production of these two metabolites. Accordingly carbon fluxes are diverted between the pathways.

Phenolics are formed by three different biosynthetic pathways:

I. The **shikimate** is also termed as chorizmate or succinylbenzoate pathway, This produces the phenyl propanoid derivatives (C_6–C_3);

II. The **acetate** is also termed as malonate or polyketide pathway, this produces the phenyl propanoids with side-chain-elongates including the large group of flavonoids (C_6–C_3–C_6) and some quinones; and

III. The **acetate** is also named as mevalonate pathway which produces the aromatic terpenoids mainly monoterpenes.

The addition of hydroxyl groups to the phenyl ring plays the major role, which involved in the biosynthesis of phenolic acids [19].

I. Shkimate/Phenylpropanoids Pathway

The precursor compound for phenyl propanoid pathway is phenylalanine which can form various types of phenolics which range in the number of aromatic rings from 1 to 6 which also differ in the substitution pattern (**Figure 2**). Hydroxycinnamic acids (HCAs) (C_6–C_3) which include caffeic, ferulic, *p*-coumaric and sinapic acids varies in the degrees of hydroxylation and methylation at C6 position. The cleavage of aliphatic side chain of P-coumaric acid can lead to the formation of hydroxybenzoic acids like salicylic, vanillic, gallic and syringic acids.
The synthesis of chorismic acid from the precursors phosphoenol pyruvate and erythrose-4-phosphate acts as a precursor for the synthesis of cinnamic acid derivatives. Various derivatives of benzoic acids are formed from chorismic acid via oxidative and non oxidative pathways and the precursor for protocatechuic acid is isochorismic acid.
The condensation of 3 C_2 residues with an activated hydroxycinnamic acid products are two classes of metabolites with a second aromatic ring linked to the phenylpropanoid moiety, the flavonoids (C_6–C_3–C_6) and the stilbenes (C_6–C_2–C_6).

II. Acetate/Malonate or Polyketide Pathway

The acetate-malonate pathway is the fatty acids both those 1° metabolites which arise universally and the more infrequent compounds with a limited distribution. This pathway also makes an important contribution to plant aliphatic and aromatic compounds; these are biosynthesised through the formation of polyketides.
Acetyl coenzyme-A is the precursor of the acetate-malonate pathway. This is a metabolite of very importance in both 1° and 2° metabolism. The metabolic pool of acetyl CoA is incessantly replenished by glycolysis and the catabolism of fatty acids and amino acids, which depletes for the synthesis of fatty acids, steroids, polyketides, terpenoids, aromatic compounds and acetyl esters and amides (**Figure 3**).

III. Acetate - Mevalonate Pathway

Mevalonic acid is the 1° precursor of steroids with phenolic ring biosynthesised by plants. It is consequent from acetyl CoA through the transitional formation of acetoacetyl CoA and 3-hydroxy-3-methylglutaryl CoA (HMG CoA) these reactions being catalyzed by acetyl CoA acetyltransferase and HMG CoA synthase respectively (**Figure 4**).

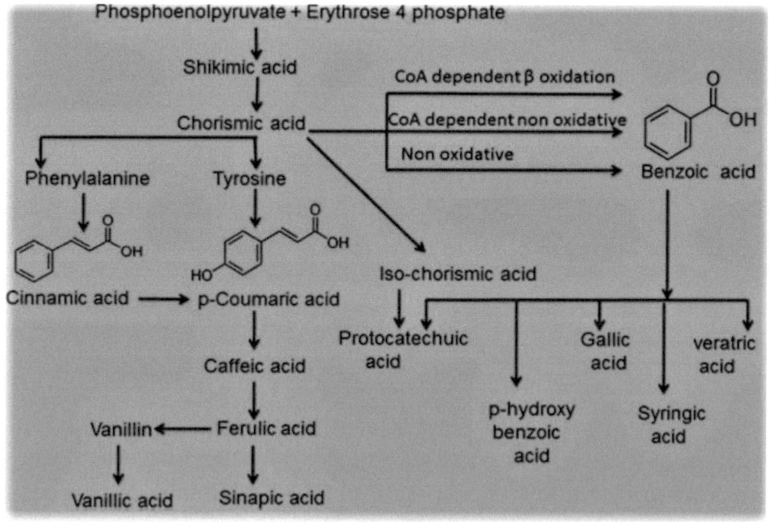

Figure 2.
Synthesis of plant phenolic compounds by shikimate pathway.

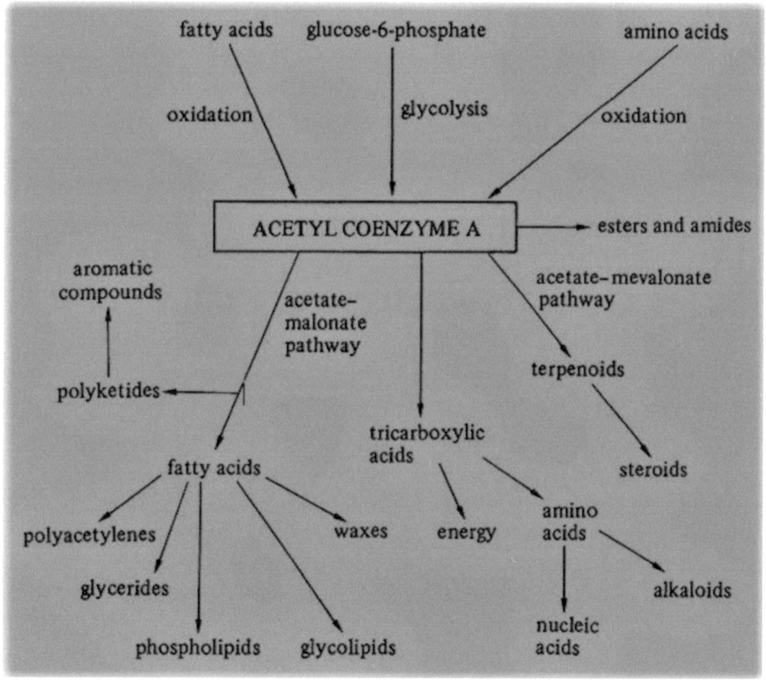

Figure 3.
The acetyl coenzyme a metabolie process.

5. Distribution of carbon fluxes between primary and secondary metabolism during stress conditions and regulation

The influence of exogenous/endogenous biotic and abiotic stresses influence the plant to make a trade off between growth & reproduction with defense mechanism intended to protect the plant. This feature in plants attracts special focus in plant ecophysiology. Primary metabolism which provides carbon skletons for the

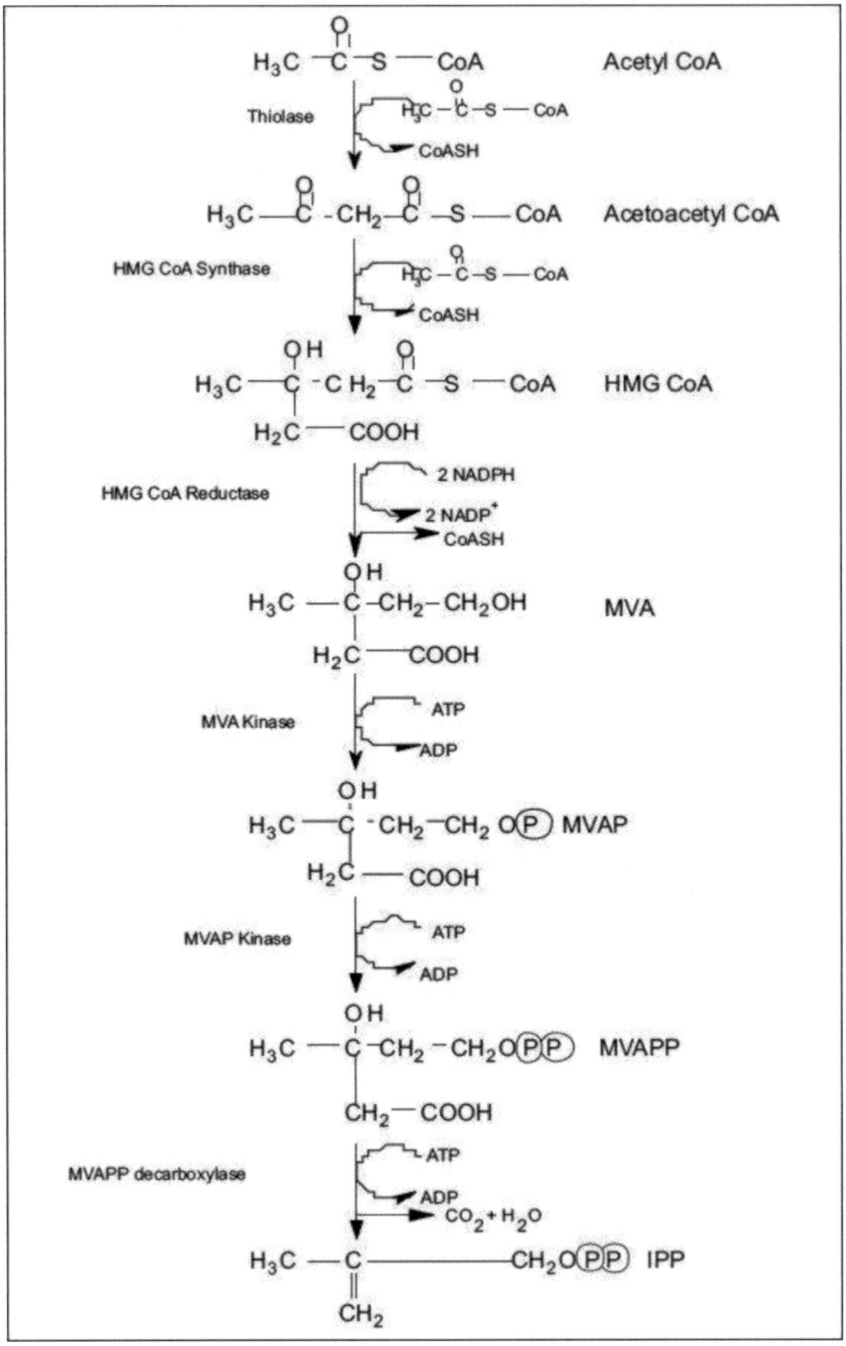

Figure 4.
The acetate/mevalonate pathway for the formation of IPP, the basic five-carbon unit of terpenoid biosynthesis. Synthesis of each IPP unit needs three molecules of acetyl-CoA.

synthesis of PSMs including phenolics requires large amounts of available resources. The stress condition leads to several biochemical and molecular mechanism triggering the adaptive response. As a part of this strategy cells try to divert the available carbon fluxes between primary and secondary metabolism and other limited resources. A considerable quantity of photosynthates are diverted to the production of phenolics and other PSMs. Lattanzio *et al.*, (2015) [20] proposed that there is a link between primary and secondary metabolism couples the accumulation of proline, a stress metabolite with energy transfer towards phenyl propanoid biosynthesis via the oxidative pentose phosphate pathway. Accordingly, some of the transduction pathways that involve a) proline redox cycle, b) pentose phosphate pathway are biased towards PSMs synthesis.

Phosphoenol pyruvate (PEP), a metabolite from glycolytic pathway is shared by four different metabolic routes leading to the formation of 1) TCA cycle, ATP generation and amino acid synthesis, 2) Methyl erythritol 4-phosphate pathway for the formation of isoprene units, 3) Shikimate-phenyl propanoid pathway for the formation of phenolic compounds, 4) Another anaplerotic route to refill PEP with PEP carboxylase. The stress conditions which favors phenolic formation increases the gene expression of shikimate dehydrogenase, phenyl ammonia lyase, chalcone synthase and PEP carboxylase specific enzymes involved in their production [21].

Biotic stress induced by *Amphibolis michoacaensis* induces gall formation in *Quercus castanea* wherein there is upregulation of phenolic related genes Phenylalanine ammonia lyase (PAL) at the intermediate and late growth stages; Phenyl propanoid genes at the intermediate stage and lignin genes at late stage. There is differential regulation of molecular switches related to secondary metabolites production during different stages of gall formation [22].

Infestation of rice by brown planthopper (BPH) in rice is observed to increase expression of *OsPAL6* and *OsPAL8* for the synthesis of phenylalanine ammonia lyase (PAL) through direct up-regulated by OsMYB30, an R2R3 MYB transcription factors [23].

Plants possess an effective immune system to combat most microbial attackers . There is an analogous immune system in plants like in animals to combat the microbial infestation. Salicylic acid is one of the hormone which is triggered in response to biotic stress. The immune response elicited leads to massive transcriptional reprogramming, cell wall strengthening, production of secondary metabolites and antimicrobial proteins [24].

Another mode of immune response to stress conditions is to influence epigenetic modifications in the development of stress memory. This is particularly of significance in high temperature heat shock stress. These modifications in turn can activate heat shock responsive genes and transcriptional factors by providing conceptual frame work for understanding molecular mechanisms behind the 'transcriptional stress memory' as potential memory tools in the regulation of plant heat stress response (HSR) [25].

6. Phenolics as antioxidants

Aerobic metabolism induces the formation of reactive oxygen and nitrogen species (RONS) radicals whose levels are expected to increase in various types of stress conditions. Plant phenolics are powerful antioxidants that can mediate scavenging of harmful reactive oxygen species (ROS). The antioxidant activity of phenolics is based on number of hydroxyl groups, the presence of alkyl chains and the number of unpaired electrons. Phenolic acids form stable phenoxyl radicals in reaction with radical molecules. Ellagic acid is a powerful antioxidant as it is high in hydrogen bonds so that they can act as electron acceptors and hydrogen donors. Hydroxycinnamic acids (HCAs) are more effective antioxidants than hydroxybenzoic acids (HBAs). Some of the compounds like ferulic acid not only act as antioxidants but also inhibit enzymes involved in free radical generation and activate other scavenging enzymes [26, 27]. The type of antioxidants and their quantity is dependent on the type of endogenous or exogenous stress and to exploit these antioxidants for human use and commercialisation induction can be achieved by exposing the plants to the selected exogenous stress conditions.

7. Conclusions

(Poly)phenols are plant secondary metabolites that constitute one of the most common and widespread groups of substances in plants. Plants adapt themselves their phenolic patterns to a changing environment through the emergence of new genes brought about by gene duplication and mutation and subsequent recruitment for adaptation to specific functions. Central metabolism requires high levels of limited plant resources, and during intense growth, the synthesis of phenolic metabolites may be substrate- and/or energy-limited. On the other hand, either abiotic or biotic stresses divert substantial amounts of substrates from primary metabolism into secondary defensive product formation, and this may lead to constraints on growth. The allocation pattern of a plant defines its ecological roles and is therefore an important factor in understanding plant distribution and adaptation. On the other hand, as far as the development of a new strategy to enable the production of useful secondary metabolites on a commercial scale is concerned, any progress made in the basic understanding of metabolic pathways and regulatory mechanisms may be addressed to exploit the plant cell and tissue culture potentials to produce food additives, such as antioxidant phenolics for specific recommendation for industrial and pharmaceuticals applications.

Acknowledgements

The authors are grateful to the Centre for Biotechnology, Institute of Science and Technology (IST), JNTUH, Hyderabad, India.

Authors' contributions

The author read and approved the final manuscript.

Funding

Not applicable.

Conflict of interest

The authors declare no conflict of interest.

Consent for publication

Authors declare that we has consent to publish the manuscript contents.

Ethics approval and consent to participate

Not applicable.

Availability of data and materials

Not applicable.

Abbreviations

ASA	Ascorbic acid
BA	Benzoic acid
BPH	Brown plant hopper
CA	Cinnamic acid
CGA	Chlorogenic acid
CHS	Chalcone synthase
HBA	Hydroxybenzoic acid
HCA	Hydroxycinnamic acid
HMG CoA	3-hydroxy-3-methylglutaryl CoA
HSR	Heat stress response
$NADP^+$	Nicotinamide adenine dinucleotide phosphate
ROS	Reactive oxygen species
RONS	Reactive oxygen and nitrogen species
SA	Salicylic acid
PCs	Phenolic compounds
PAL	Phenylalanine ammonia lyase
PEP	Phosphoenol pyruvate
PSM	Plant secondary metabolite
UV	Ultraviolet

Author details

Venkanna Banothu and Addepally Uma*
Department of Biotechnology, Centre for Biotechnology (CBT), Institute of Science and Technology (IST), Jawaharlal Nehru Technological University Hyderabad (JNTUH), Kukatpally, Hyderabad, Telangana, India

*Address all correspondence to: vedavathi1@jntuh.ac.in

References

[1] Molnar C and Gair J. Concepts if Biology-1st Canadian edition. 2015; Creative Commons Attribution 4.0 International License.

[2] Gull A, Ahmad A, Lone and Ul Islam Wani N. Biotic and Abiotic stresses in Plants. Intech open. 2019; 85832. DOI: 10.5772/intechopen.85832.

[3] Tasiu Isah. Stress and defense responses in plant secondary metabolites production. Biological Research. 2019; 52: 39. DOI: 10.1186/s40659-019-0246-3.

[4] Erb M & Kliebenstein DJ. Plant Secondary Metabolites as Defenses, Regulator, and Primary Metabolites: The Blurred Functional Trichotomy. Plant Physiology. 2020; 184 (1): 39-52. DOI: 10.1104/pp.20.00433.

[5] Ramakrishna A & Ravishankar GA. Influences of abiotic stress signals on secondary metabolites in plants. Plant Signal Behav. 2011; 6(11): 1720-31. DOI: 10.4161/psb.6.11.17613.

[6] Edreva A, Velikova V, Tsonev T, Dagnon S, Gurel A, Aktas L and Gesheva E. Stress-protective role of secondary metabolites: diversity of functions and mechanisms. Gen Appl Plant Physiology. 2008; 34(1-2): 67-78.

[7] Rejeb IB, Pastor V, Mauch-Mani B. Plant responses to simultaneous biotic and abiotic stress: molecular mechanisms. Plants. 2014; 3(4): 458-75. DOI: 10.3390/plant s3040 458.

[8] Chinnusamy V, Schumaker K, Zhu JK. Molecular genetics perspectives on cross-talk and specificity in abiotic stress signaling in plants. J Exp Bot. 2004; 55: 225-236.

[9] Thirumurugan D, Cholarajan A, Raja S.S.S & Vijayakumar R. An Introductory chapter: Secondary Metabolites. 2018. DOI: 10.5772/intechopen.79766.

[10] Sharma A, Shahzad B, Rehman A, Bhardwaj R, Landi M and Zheng B. Response of Phenylpropanoid Pathway and the Role of Polyphenols in Plants under Abiotic Stress. Molecules. 2019; 24: 2452. DOI: 10.3390/molecules 24132452.

[11] Cheynier V, Comte G, Davies KM, Lattanzio V, Martens S. Plant phenolics: recent advances on their biosynthesis, genetics, and ecophysiology. Plant Physiol Biochem. 2013; 72: 1-20. DOI: 10.1016/j.plaphy.2013.05.009.

[12] Samec D, Karalija E, Sola I, Vujcic Bok V, Salopek-Sondi B. The Role of Polyphenols in Abiotic Stress Response: The Influence of Molecular Structure. Plants. 2021; 10: 118. DOI: 10.3390/plants10010118.

[13] Vincenzo Lattanzio. Phenolic Compounds: Introduction. Ramawat K. G & Merillon J.M. (eds.), Natural Products. 2013. p 1543-1580. DOI: 10.1007/978-3-642-22144-6_57.

[14] Carrington Y, Guo J, Le CH, Fillo A, Kwon J, Tran LT and Ehlting J. Evolution of a secondary metabolic pathway from primary metabolism: shikimate and quinate biosynthesis in plants. The Plant Journal. 2018; 95 (5): 823-833. DOI: 10.1111/tpj.13990.

[15] Kroymann Juergen. Natural diversity and adaptation in plant secondarymetabolism. Curr Opin Plant Biol. 2011; 14(3): 246-251. DOI: 10.1016/j.pbi.2011.03.021.

[16] do Nascimento NC & Fett-Neto AG. Plant secondary metabolism and challenges in modifying its operation: an overview. In: Fett-Neto AG (ed) Plant secondary metabolism engineering–methods and application,

methods in molecular biology, vol 643. Humana Press, New York. 2010; p 1-13.

[17] Gould KS & Lister C. Flavonoid functions in plants. In: Andersen ØM, Markham KR (eds) Flavonoids – chemistry, biochemistry and applications. CRC Taylor & Francis, Boca Raton. 2006; 397-411.

[18] Panche A.N, Diwan A.D and Chandra S. R. Flavonoids: an overview. Journal of Nutritional Science. 2016; e47 (5): 1-15. DOI: 10.1017/jns.2016.41.

[19] Tzin V, Galili G. The biosynthetic pathways for shikimate and aromatic amino acids in *Arabidopsis thaliana*. Arabidopsis Book. 2010; 8: e0132.

[20] Caretto S, Linsalata V, Colella G, Mita G, Lattanzio V. Carbon Fluxes between Primary Metabolism and Phenolic Pathway in Plant Tissues under Stress. International Journal of Molecular Sciences. 2015; 16: 26378-26394. DOI: 10.3390/ijms161125967.

[21] Dizengremel P, Vaultier MN, Le Thiec D, Cabané M, Bagard M, Gérant D, Gérard J, Dghim AA, Richet N, Afif D, Pireaux JC, Hasenfratz-Sauder MP, Jolivet Y. Phosphoenolpyruvate is at the crossroads of leaf metabolic responses to ozone stress. New Phytologist. 2012; 195: 512-517.

[22] Betancourt EK, Soto PH, Anaya MR, Cortés NC and Oyama K. Differential expression of genes associated with phenolic compounds in galls of *Quercus castanea* induced by *Amphibolips michoacaensis*. Journal of plant interactions. 2019, 14(1): 177-186. DOI: 10.1080/17429145.2019.1603404.

[23] He J, Liu Y, Yuan D, Duan M, Liu Y, Shen Z, Yang C, Qiu Z, Liu D, Wen P, Huang J, Fan D, Xiao S, Xin Y, Chen X, Jiang L, Wang H, Yuan L and Wan J. An R2R3 MYB transcription factor confers brown planthopper resistance by regulating the phenylalanine ammonia-lyase pathway in rice. PNAS. 2020; 117 (1): 271-277. DOI: 10.1073/pnas.1902771116.

[24] Van Butselaar T, Van den Ackerveken G. Salicylic Acid Steers the Growth-Immunity Tradeoff. Trends in Plant Sci. 2020; 25(6):566-576. DOI: 10.1016/j.tplants.2020.02.002.

[25] Rai KK, Pandey N, Rai SP. Salicylic acid and nitric oxide signaling in plant heat stress. Physiol Plant. 2020; 168(2): 241-255. DOI: 10.1111/ppl.12958.

[26] Alfei S, Marengo B, Zuccari G. Oxidative stress, antioxidant capabilities, and bioavailability: Ellagic acid or urolithins? Antioxidants. 2020; 9: 707.

[27] Hernández-Ruiz J, Arnao MB. Relationship of Melatonin and Salicylic Acid in Biotic/Abiotic Plant Stress Responses. Agronomy. 2018; 8: 33.

Chapter 7

Medicinal Plants and Phenolic Compounds

Asma Nisar

Abstract

Medicinal plants were shown to play a significant role in curing many diseases of ancient times. The plant kingdom is truly a goldmine of potential drug compounds. Several earlier reviews and research studies summarized that the products from natural sources have contributed significantly to the discovery of drugs and health benefits for people. Moreover, it is believed that natural products are less deadly than synthetic medicines because of their plant origins. Medicinal plants are significant in the role of curing a variety of diseases and the properties that they possess for curing are related to the existence of phenolic compounds, flavonoids, anthocyanins and other phytochemicals. This chapter covers the effects of phenolic compound on plants and the importance of phenolic compounds for human health for prevention of various oxidative stress associated diseases.

Keywords: Medicinal Plants, Phenolic Compounds, flavonoids, stilbenes, lignans and phenolic acid

1. Introduction

Phenolics are a type of secondary metabolite that can be found almost all over in plants. They are an aromatic molecule with a benzene ring (C6) and one or more hydroxyl groups that belong to a broad and diversified group. In general, phenolics are classified according to the number of carbon atoms in the molecule. Three different biosynthetic pathways produce phenolics: (i) the shikimate/chorizmate or succinylbenzoate pathway, which produces phenylpropanoid derivatives (C6–C3); (ii) the acetate/malonate or polyketide pathway, which produces side-chain-elongated phenylpropanoids, including the large group of flavonoids (C6–C3–C6) and some quinones; and (iii) the acetate/mevalonate pathway, which produces the aromatic terpenoids, mostly monoterpenes, by dehydrogenation reactions [1].

The content of a certain phenolic in plant tissue varies depending on the season and stage of growth and development. Trauma, wounding and pathogen infection are only a few of the internal and external variables that alter phenolic production and accumulation. Furthermore, light increases the production of phenolics in chloroplasts and their accumulation in vacuoles. In some plant species, photoinhibition, as well as nutrient stressors such as nitrogen, phosphate, potassium, sulphur, magnesium, boron, and iron deficiency, cause the synthesis of phenylpropanoid chemicals [1].

The distribution of phenolics in plants is not consistent at the tissue, cellular, and subcellular levels. Plant cell walls contain insoluble phenolics, while plant cell vacuoles contain soluble phenolics. Certain polyphenols, such as quercetin, can be found in all plant products, including vegetables, fruit, cereals, tea, wine, fruit juices, infusions, and so on, whereas isoflavones and flavanones are found only in specific foods. Polyphenols are found in most foods in complex combinations. Higher levels of phenolics compounds found in outer layers of plants than inner layers. Plant polyphenol content is influenced by a variety of factors, including ripeness at harvest, environmental factors, processing, and storage. Environmental and edaphic factors, such as soil type, sun exposure, and rainfall, have a significant impact on the polyphenolic content of foods. The quantities and amounts of different polyphenols are greatly influenced by the degree of ripeness. In general, phenolic acid content declines as ripening progresses, although anthocyanin concentrations increase. Many polyphenols, particularly phenolic acids, are directly engaged in plants' responses to many types of stress: they aid in the healing of damaged areas by lignification, have antimicrobial capabilities, and their concentrations may rise following infection. Storage is another element that has a direct impact on the polyphenol content in foods. Polyphenolic content in foods changes during storage, according to studies, due to the simple oxidation of these polyphenols. Oxidation reactions result in the creation of more or less polymerised compounds, which affect food quality, especially colour and organoleptic qualities. Such alterations can be useful, as with black tea, or damaging, as with fruit browning. When wheat flour is stored, it loses a significant amount of phenolic acids. In terms of quality, flour after six months of storage had the same phenolic acids, although their concentrations were 70% lower than when it was fresh. Cold storage, on the other hand, has only a minor impact on the polyphenol content of apples, onions or pears. Cooking has a significant impact on polyphenol concentrations. After boiling for 15 minutes, onions and tomatoes lose between 75 and 80 percent of their initial quercetin content, 65 percent after cooking in a microwave oven, and 30 percent after frying [2].

2. Phenolics in plant defence

Phenolics perform a dual function in the plant's environment, repelling and attracting various organisms. They act as inhibitors, natural animal toxicants, and pesticides against invading organisms, such as herbivores, phytophagous insects, nematodes, fungal and bacterial pathogens. On the plant surface, simple phenolic acids, complex tannins, and phenolic resins deter birds by interfering with the gut microflora and impairing their digestive ability. Low-molecular-weight phenylpropanol derivatives attract symbiotic microbes, pollinators, and animals that disperse fruit [3].

Phenolics have long been recognised in animals as phytoestrogens and as allelochemicals for competitive weeds and plants. Allelochemicals that are widely effective include volatile terpenoids, toxic water-soluble hydroquinones, hydroxybenzoates, hydroxycinnamates, and 5-hydroxynapthoquinones. Numerous simple and complex phenolic compounds accumulate in plant tissues and function as phytoalexins, phytoanticipins, and nematicides against soil-borne pathogens and phytophagous insects. Thus, phenolic compounds have been proposed as useful alternatives to chemical control of agricultural crop pathogens for some time. The majority of polyphenols have been shown to have a negative effect on microbes. Plants accumulate phytoalexins in response to pathogen attacks, such as hydroxycoumarins and hydroxycinnamate conjugates. The synthesis, release, and accumulation of phenolic compounds—in particular, salicylic acid are critical for a variety of plant defence strategies against microbial invaders. Phenolics are synthesised when plant pattern recognition receptors recognise potential pathogens via conserved pathogen-associated molecular patterns (PAMPs), resulting in PAMP-triggered immunity. As a result, the pathogen's progress is slowed significantly before it takes complete control of the plant [1].

3. Classification of phenolic compounds

Polyphenols are classified according to the number of phenol rings they contain and the structural elements that connect these rings. The major classes of polyphenols are phenolic acids, flavonoids, stilbenes, and lignans. **Figure 1** depicts the various polyphenol groups and their chemical structures and **Figure 2** shows the structure of polyphenols.

3.1 Phenolic acids

Phenolic acids, alternatively referred to as phenol carboxylic acids, are aromatic acids composed of a phenolic ring and a carboxyl functional group. As a result, these compounds contain an aromatic ring, a hydroxyl group, and a carboxyl group. Salicylic acid is one of the most basic phenolic acids. Additionally, hydroxycinnamic and hydroxybenzoic acids are important naturally occurring phenolic acids. Hydroxycinnamic acids are derived from molecules of non-phenolic cinnamic acid, whereas hydroxybenzoic acids are derived from molecules of non-phenolic benzoic acid.

Naturally occurring phenolic acids are found in a variety of horse grams, dried fruits, the mushroom species Basidiomycetes, and human urine. Phenolic acids include protocatechuic acid (PCA), vanillic acid, p-hydroxybenzoic acid (PHBA), caffeic acid, ferulic acid, sinapinic acid, p-coumaric acid and syringic acid.

Phenolic compounds are phytochemicals found in cereals that are beneficial to health. Despite their antioxidant properties, phenolic compounds continue to

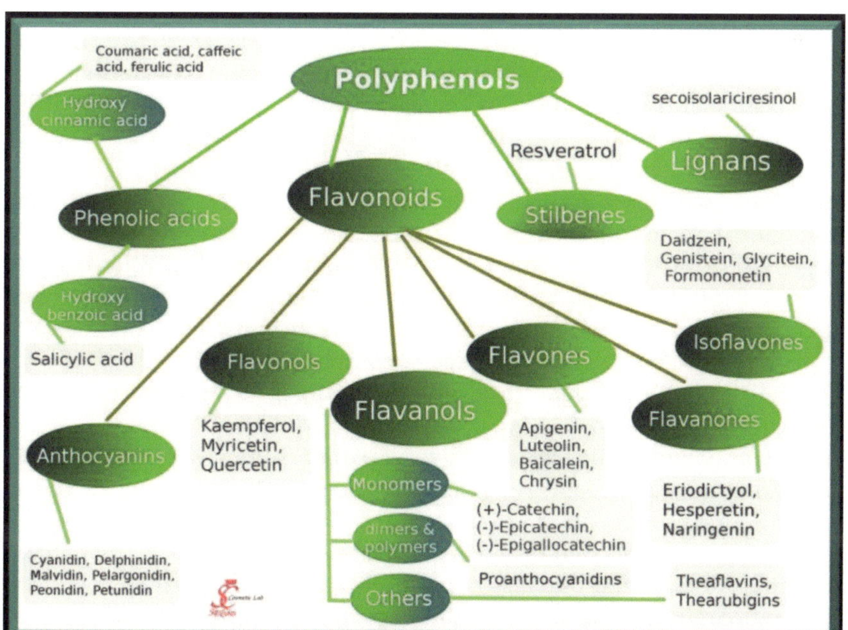

Figure 1.
The different groups of polyphenols.

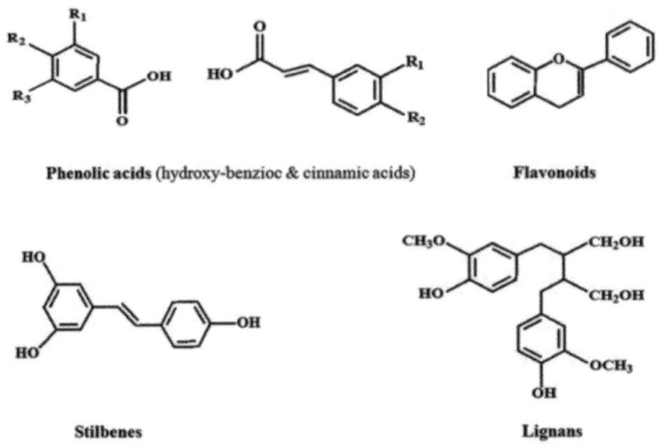

Figure 2.
Structure of polyphenols.

garner considerable attention. The phenolic acids and flavonoids are the two most abundant types of phenolic compounds found in whole grains. In cereals, phenolic acids are the most abundant. The gut microbiota is widely accepted as a factor in the biotransformation of phytochemicals, including phenolic acids, resulting in the formation of food-derived metabolites that are excreted in the urine. Phenolic acids are easily absorbed through the intestinal tract's walls, which is beneficial for human health because they act as antioxidants, preventing cellular damage caused by free radical oxidation reactions. If humans consume them on a regular basis, they may also help to maintain anti-inflammatory conditions in the body [4].

3.2 Lignans

Lignans are bioactive, non-caloric, non-nutrient phenolic plant compounds found in abundance in flax and sesame seeds and lesser amounts in grains, other seeds, fruits, and vegetables. Enterolignans (occasionally referred to as mammalian lignans) are metabolites of food lignans produced by intestinal bacteria in humans. They have been identified in urine and plasma from humans. Their insignificant estrogenic and other biochemical properties suggest that they may have nutritional value in preventing cardiovascular and other chronic diseases [5].

Monolignols, which are derived from hydroxycinnamic acids (p-coumaric, sinapic and ferulic acids), are either dimerized to form lignans or polymerised to form larger lignin structures in the cell wall. These structurally diverse compounds play a role in plant defence (as antioxidants, phytoalexins, biocides and others), protecting plants from diseases and pests and possibly assisting in plant growth control. Lignans and lignins are two distinct compounds that should not be confused. Lignans are stereospecific dimers of these cinnamic alcohols (monolignols) bonded at carbon 8 (C8- C8) [6].

Lignans (monolignol dimers) are found in plants either free or bound to sugars. There are numerous diglucosides of pinoresinol, secoisolariciresinol, and syringaresinol. 9–12 Sesame seeds contain sesaminol triglucoside and sesaminol diglucoside. Secoisolariciresinol occurs in flax as a diglucoside and is a component of an ester-linked complex or oligomer that also contains 3-hydroxyl-3-methylglutaric acid, a number of cinnamic acid glycosides (most commonly ferulic or p-coumaric acid), and the flavonoid herbacetin [7].

Lariciresinol, pinoresinol, matairesinol and secoisolariciresinol are the most abundant plant lignans found in foods. Numerous other lignans are found in a variety of foods, including medioresinol (found in sesame seeds, lemons and rye), syringaresinol (found in grains), sesamin, and sesamolin, a lignan precursor (in sesame seeds) [8]. Additionally, arctigenin, cyclolariciresinol (isolariciresinol), 7′-hydroxymatairesinol,b and 7-hydroxysecoisolariciresinol are found in foods but are rarely quantified. (Some cyclolariciresinol occurs naturally, while some are formed during the extraction and analysis of lariciresinol under acidic conditions). Lignans have no known nutritional value. While lignans are not classified as dietary fibres, they do share some chemical properties with lignin, an insoluble fibre [9].

Lignins are large plant polymers composed of the hydroxycinnamic alcohols, p-coumaryl, coniferyl, and sinapyl. They are racemic (non-stereospecific) polymers that contain monolignol units at C8 and four additional sites (C5-C5, C5-C8, C5-O-C4, C8-O-C4). Lignins are found in all higher plants' vessels and secondary tissues. They are found in a wide variety of foods, but are especially prevalent in cereal brans. Lignins are considered to be a type of insoluble dietary fibre from a nutritional standpoint. Lignins are necessary for plants because they strengthen the cell walls, aid in water transport, prevent the degradation of polysaccharides found in the cell walls, aid in the resistance of plants to pathogens and other threats, and provide texture in edible plants. Foods contain a small amount of lignan, typically less than 2 mg/100 g. The exceptions are flaxseed27 (335 mg/100 g) and sesame seeds (373 mg/100 g), which contain a hundred times the amount of lignan found in other foods. They are found in a wide variety of plant families, though the types and amounts vary considerably between them. Whole grains (particularly the bran layer) and seeds (in the seed coat) contain lignans. Several grains, including barley, flax, buckwheat, millet, rye, sesame seeds, oats, and wheat, contain a significant amount of lignans. Additionally, nuts and legumes are good sources. Although in smaller quantities than in grains, lignans are also found in fruits and vegetables such as asparagus, kiwi fruit, grapes, lemons, pineapple, oranges, and wine, as well as in coffee and tea [8].

In comparison to plants, animal foods contain almost no lignans. The enterolignans enterodiol and enterolactone are occasionally found in animal foods (milk products) as a byproduct of intestinal bacterial metabolism in the animals' guts, but these are exceptions. Little research has been conducted on the effects of storage and processing on lignans in the majority of foods, although it is known that the lignan content of flaxseed and sesame seed processing does not appear to change significantly [10].

3.3 Stilbene

Compounds of Stilbene in Plants While phenolic compounds are critical mediators of plants' adaptation and survival responses to acute and chronic stress, polyphenols also regulate cell growth, differentiation, pollen fertility, and nodulation, and thus appear to be essential for plant health. For instance, stilbenes are naturally occurring phenolic defence compounds found in a variety of plant species that exhibit antimicrobial and antioxidant activity against phytopathogens and ozone or ultraviolet stress Stilbene compounds are found in a wide variety of plant species, including wine grapes, peanuts, sorghum, and a variety of tree species [11]. Additionally, commercial sources of stilbenes include a number of plants cultivated in Asia as folk medicines, including *Polygonum cuspidatum*, Rheum undulatum, *Rhodomyrtus tomentosa*, *Melaleuca leucadendron*, and *Euphorbia lagascae*, whereas pterostilbene is found primarily in bilberries, blueberries, and some other berries. Grape pomaces, winemaking residues, and other grape juice solids contain high levels of polyphenols and are also an excellent source of a variety of stilbene compounds, not just resveratrol. Conifer tree bark waste contains significant amounts of stilbene compounds such as piceatannol, pinosylvin, and trans-resveratrol (t-Res). As a result, this massive amount of industrial byproducts represents an extremely attractive and affordable source of stilbenes with commercial applications. Genetic tools are a very promising method for producing specific stilbenes such as pterostilbene in plants via coexpression of stilbene synthase and O-methyltransferase. These stilbenes may be particularly well suited for pharmacological applications. The enzyme stilbene synthase (STS) is required for the biosynthesis of stilbenic compounds. STS appears to have evolved independently from chalcone synthases (CHSs) in stilbene-producing plants. Interestingly, different STS genes express differently in different tissues and developmental stages. Thus, it has been reported that STS genes were expressed at a lower level in young grape leaves than in mature leaves, whereas the transcript levels of eight STS genes increased dramatically in the berry skins of Cabernet Sauvignon and Norton grape cultivars following veraison, peaking at harvest. Although pine trees' heartwood contains a high concentration of pinosylvin, young seedlings accumulate significant amounts of the compound in response to stress induction (fungal or UV light) [12].

3.4 Flavonoids

Flavonoids are a large class of polyphenolic compounds with a benzo—pyrone structure that is abundant in plants. They are produced through the phenylpropanoid pathway. According to available data, secondary phenolic metabolites, including flavonoids, are responsible for a variety of pharmacological activities [13].

In plants, animals, and bacteria, flavonoids perform a variety of biological functions. Flavonoids have long been known to be synthesised in specific locations in plants and are responsible for the colour and aroma of flowers, as well as the ability of fruits to attract pollinators and thus aid in seed and spore germination, as well as the growth and development of seedlings. Flavonoids protect plants from biotic and abiotic stresses and act as one-of-a-kind UV filters. They also act as signal molecules, allopathic compounds, phytoalexins, detoxifying agents, and antimicrobial defensive compounds. Flavonoids protect plants from frost and drought and may also play a role in heat acclimatisation and freezing tolerance.

3.4.1 Family of flavonoids

Flavonoids are phytonutrients that belong to the polyphenol class. According to the Global Healing Center, polyphenols have historically been used in Chinese and Ayurvedic medicine and are associated with skin protection, brain function, blood sugar and blood pressure regulation, as well as antioxidant and anti-inflammatory activity.

Flavonoids are classified into several major groups, including flavanols, anthocyanidins, flavones, flavonones, flavonols, and isoflavones. There are additional subgroups within the flavanol subgroup. Each of these subgroups and each flavonoid type has a unique set of actions, benefits, and source foods.

a. **Flavones**: Lutein and apigenin are two examples. Celery, parsley, various herbs, and hot peppers are all good sources of flavones. Flavones have been linked to an array of antioxidant properties and a delay in the metabolization of drugs.

b. **Anthocyanidins**: Malvidin, pelargondin, peoidin, and cyanidin are examples. Red, purple, and blue berries; pomegranates; plums; red wine; and red and purple grapes are all good sources of anthocyanidins. Anthocyanidins are associated with cardiovascular health, antioxidant activity, and aid in the prevention of obesity and diabetes.

c. **Flavonones:** Hesperetin, eriodictyol, and naringenin are examples of flavonones. Citrus fruits contain a high concentration of flavonones. They are associated with cardiovascular health, relaxation, and anti-inflammatory and antioxidant activity in general.

d. **Isoflavones**: Genistein, glycitein, and daidzein are all members of this subgroup. Soybeans and soy products, as well as legumes, contain a high concentration of isoflavones. They are phytoestrogens, which means they are chemicals that mimic the oestrogen hormone. They may be beneficial in reducing the risk of hormonal cancers such as breast, endometrial, and prostate cancer, though current research findings are inconsistent. Isoflavones have been shown in various studies to act as antioxidants and as oxidants, leaving their effect on cancer unclear. Additionally, they are being studied as a possible treatment for menopausal symptoms.

e. **Flavonols:** Quercetin and kaempferol are members of this widely distributed subgroup of flavonoids. Onions, leeks, Brussels sprouts, kale, broccoli, tea, berries, beans, and apples all contain them. Quercetin is an antihistamine that may help alleviate symptoms of hay fever and hives. Additionally, it is well-known for its anti-inflammatory properties. Kaempferol and other flavonols have been linked to significant anti-inflammatory and antioxidant activity, which may help prevent chronic disease.

f. **Flavanols**: Flavanols are classified into three types: monomers (more commonly referred to as catechins), dimers, and polymers. Teas, cocoa, grapes, apples, berries, fava beans, and red wine all contain flavanols. Catechins are abundant in green and white teas, while dimers, which have been linked to cholesterol reduction, are found in black tea. Scientists believe catechins may be beneficial in alleviating the symptoms of chronic fatigue syndrome. Additionally, catechins have been linked to cardiovascular and neurological health [14].

4. Phenolic compounds' effects on human health

Epidemiological studies have repeatedly demonstrated an inverse relationship between the risk of chronic human diseases and polyphenol-rich diet consumption. Polyphenols contain phenolic groups that can accept an electron to form relatively stable phenoxyl radicals, interfering with chain oxidation reactions in cellular components. It is well established that foods and beverages high in polyphenols may enhance plasma antioxidant capacity. This increase in plasma's antioxidative capacity following consumption of polyphenol-rich foods can be explained by the presence of reducing polyphenols and their metabolites, their effects on the concentrations of other reducing agents (polyphenols' sparing effects on other endogenous antioxidants), or their effect on the absorption of pro-oxidative food components. Antioxidant consumption has been linked to decreased levels of oxidative damage to lymphocytic DNA. Similar findings have been made with polyphenol-rich foods and beverages, indicating polyphenols' protective properties. There is mounting evidence that polyphenols, as antioxidants, may protect cell constituents from oxidative damage and thus reduce the risk of developing various degenerative diseases associated with oxidative stress. **Figure 3** shows the different pharmacological functions of polyphenols.

The potential health benefits of dietary phenolics are dependent on their absorption and metabolism, which are determined by their structure, which includes their conjugation with other phenolics, their degree of glycosilation/acylation, their molecular size, and their solubility. These steps occur at various points throughout the small intestine's passage into the circulatory system and subsequent portal vein transport to the liver. Because polyphenol metabolites are rapidly eliminated from plasma, daily consumption of plant products is necessary to maintain adequate metabolite concentrations in the blood. However, it is important to keep in mind that the polyphenols found in the most abundant amounts in the daily diet are not necessarily the ones with the highest bioavailability. For example, hydroxycinnamic acids are found in high concentrations in foods, but their intestinal absorption is reduced by esterification. Additionally, differences in cell wall structures, glycoside distribution within cells, and phenolic compound binding to the food matrix can affect phenolic compound bioavailability. Epidemiological evidence to date indicates that polyphenols perform critical functions such as inhibiting pathogens and decay microorganisms, preventing triglyceride deposition, lowering the incidence of non-communicable diseases such as cardiovascular disease, diabetes, cancer, and stroke, and exerting anti-inflammatory and anti-allergic effects via processes involving reactive oxygen species. These protective effects are partially attributed to phenolic secondary metabolites. Initially, it was believed that the protective effect of dietary phenolics was due to their antioxidant properties, which resulted in a decrease in the body's free radical levels. However, there is emerging evidence that the metabolites of dietary phenolics, which are found in the circulatory system at concentrations ranging from nmol/L to low mmol/L, exert modulatory effects in cells via their selective actions on various components of intracellular signalling cascades critical for cellular functions such as growth, proliferation, and apoptosis. Polyphenols are thought to exert their antioxidant capacity in a variety of ways, depending on the hydroxylation state of their aromatic rings, including (i) radical scavenging, (ii) chelation and stabilisation of divalent cations, and (iii) modulation of endogenous antioxidant enzymes. Phenolic acids, hydrolysable tannins, and flavonoids have anti-carcinogenic and anti-mutagenic properties because they act as

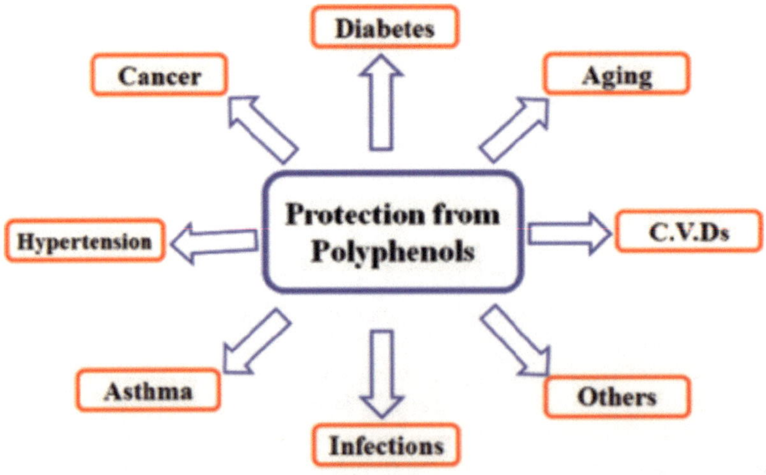

Figure 3.
Different pharmacological functions of polyphenols.

antioxidants for DNA, inactivating carcinogens, inhibiting pro-carcinogen activation enzymes, and activating xenobiotic detoxification enzymes. Flavonoids and L-ascorbic acid, in particular, have a synergistic protective effect against oxidative DNA damage in lymphocytes. Chlorogenic and caffeic acids are both antioxidants in vitro and may inhibit the formation of mutagenic and carcinogenic N-nitroso compounds in vitro. Flavonoids, catechins, and their derivatives are being investigated as potential therapeutic agents in studies of degenerative diseases and brain ageing processes, and may act as neuroprotective agents in progressive neurodegenerative disorders such as Parkinson's and Alzheimer's disease. Consumption of flavonoids results in a decrease in LDL oxidation. Resveratrol, also known as trans-3,5,4′-trihydroxystilbene, is the most well-known health-promoting molecule found in grapes and red wine. It has been studied for its effects on genes, as well as the heart, breast, prostate, uterus, and immune system. Additionally, recent research indicates that resveratrol supports healthy nerves and critical brain functions, such as cognitive processes. Tannins, more commonly referred to as tannic acid, have been implicated in experimental animals in reducing feed intake, growth rate, feed efficiency, net metabolizable energy, and protein digestibility. As a result, foods high in tannins, such as betel nuts and herbal teas, are regarded as having a low nutritional value. Numerous studies, however, indicated that the primary effect of tannins was not an inhibition of food consumption or digestion, but rather a decrease in the efficiency of converting absorbed nutrients to new body substances. Tannins' anticarcinogenic and antimutagenic properties may be related to their antioxidative capacity, which is critical for cellular oxidative damage protection, including lipid peroxidation. Tannins have also been reported to have additional physiological effects, including the acceleration of blood clotting, the reduction of blood pressure, the reduction of serum lipid levels, the induction of liver necrosis, and the modulation of immune responses. Polyphenols (phenolic acid, stilbenes, tannins, isoflavones, and catechins found in green tea) have been shown to inhibit the reproduction and growth of a variety of fungi, yeasts, viruses, and bacteria, including Salmonella, Clostridium, Bacillus, or *Chlamydia pneumoniae*, *Vibrio cholerae*, and enterotoxigenic *E. coli* (ETEC). Because phenolics act as a natural defence mechanism against microbial infections, they can be used in food processing to extend the shelf life of certain foods, such as catfish fillets [4].

5. Conclusion

Phenolic compounds and other bionutrients are abundant in medicinal plants. Phenolic compounds have several biotechnological applications in different industries. Their exploitation is mainly due to their antioxidant, antimicrobial, colouring, among other properties, specially explored for food preservation, by the food and packing industries, cosmetic and also the textile industry. Polyphenolic extracts are attractive ingredients for cosmetics and pharmacy due to their beneficial biological properties. Numerous studies conducted over the last two–three decades have demonstrated that these Phenolic compounds play a vital role in preventing chronic diseases such as cancer, diabetes, and coronary heart disease, among others. Dietary fibre, antioxidants, anticancer, detoxifying agents, immunity-enhancing agents, and neuropharmacological agents are the major classes of Phenolic compounds with disease-preventive properties. Each of these functional agents are composed of a diverse array of chemicals with varying degrees of potency. There is, however, considerable room for additional systematic research aimed at identifying these Phenolic compounds in medicinal plants and evaluating their potential to protect against a variety of diseases.

Author details

Asma Nisar
Faculty of Engineering, Universiti Malaysia Sabah, Malaysia

*Address all correspondence to: asmanisarr@gmail.com

References

[1] A. Bhattacharya, P. Sood, and V. Citovsky, "The roles of plant phenolics in defence and communication during Agrobacterium and Rhizobium infection," *Mol. Plant Pathol.*, p. no-no, May 2010.

[2] K. B. Pandey and S. I. Rizvi, "Plant Polyphenols as Dietary Antioxidants in Human Health and Disease," *Oxid. Med. Cell. Longev.*, vol. 2, no. 5, pp. 270-278, 2009.

[3] P. A. Ndakidemi and F. D. Dakora, "Legume seed flavonoids and nitrogenous metabolites as signals and protectants in early seedling development," *Funct. Plant Biol.*, vol. 30, no. 7, p. 729, 2003.

[4] T. Ozcan, A. Akpinar-Bayizit, L. Yilmaz-Ersan, and B. Delikanli, "Phenolics in Human Health," *Int. J. Chem. Eng. Appl.*, vol. 5, no. 5, pp. 393-396, Oct. 2014.

[5] P. Penttinen *et al.*, "Diet-Derived Polyphenol Metabolite Enterolactone Is a Tissue-Specific Estrogen Receptor Activator," *Endocrinology*, vol. 148, no. 10, pp. 4875-4886, Oct. 2007.

[6] T. Umezawa, "Diversity in lignan biosynthesis," *Phytochem. Rev.*, vol. 2, no. 3, pp. 371-390, Jan. 2003.

[7] X. Li, J.-P. Yuan, S.-P. Xu, J.-H. Wang, and X. Liu, "Separation and determination of secoisolariciresinol diglucoside oligomers and their hydrolysates in the flaxseed extract by high-performance liquid chromatography," *J. Chromatogr. A*, vol. 1185, no. 2, pp. 223-232, Mar. 2008.

[8] J. Peterson, J. Dwyer, H. Adlercreutz, A. Scalbert, P. Jacques, and M. L. McCullough, "Dietary lignans: physiology and potential for cardiovascular disease risk reduction," *Nutr. Rev.*, vol. 68, no. 10, pp. 571-603, Sep. 2010.

[9] L. B. Davin, M. Jourdes, A. M. Patten, K.-W. Kim, D. G. Vassão, and N. G. Lewis, "Dissection of lignin macromolecular configuration and assembly: Comparison to related biochemical processes in allyl/propenyl phenol and lignan biosynthesis," *Nat. Prod. Rep.*, vol. 25, no. 6, p. 1015, 2008.

[10] S. W. Lee, M. K. Jeung, M. H. Park, S. Y. Lee, and J. Lee, "Effects of roasting conditions of sesame seeds on the oxidative stability of pressed oil during thermal oxidation," *Food Chem.*, vol. 118, no. 3, pp. 681-685, Feb. 2010.

[11] C. Parage *et al.*, "Structural, Functional, and Evolutionary Analysis of the Unusually Large Stilbene Synthase Gene Family in Grapevine," *Plant Physiol.*, vol. 160, no. 3, pp. 1407-1419, Nov. 2012.

[12] M. Reinisalo, A. Kårlund, A. Koskela, K. Kaarniranta, and R. O. Karjalainen, "Polyphenol Stilbenes: Molecular Mechanisms of Defence against Oxidative Stress and Aging-Related Diseases," *Oxid. Med. Cell. Longev.*, vol. 2015, pp. 1-24, 2015.

[13] S. Kumar and A. K. Pandey, "Chemistry and Biological Activities of Flavonoids: An Overview," *Sci. World J.*, vol. 2013, pp. 1-16, 2013.

[14] A. N. Panche, A. D. Diwan, and S. R. Chandra, "Flavonoids: an overview," *J. Nutr. Sci.*, vol. 5, p. e47, Dec. 2016.

Section 3

Environmental Sustainability and Industrial Applications of Phenolic Compounds

Chapter 8

Novel Non-Thermal Processing Technologies: Impact on Food Phenolic Compounds during Processing

Josephine Ampofo and Michael Ngadi

Abstract

In recent times, food consumption has advanced beyond simply meeting growth and development needs to include the supply of ingredients that can protect against diseases. Among such non-nutritive ingredients are phenolic compounds. These are benzene-ringed secondary metabolites produced in plants upon exposure to environmental stress. Previous studies have linked phenolic compounds to bioactive benefits (e.g., antioxidative, anti-inflammatory, and anti-cancer) with these bioactivities dependent on their biochemical structure and concentrations of individual phenolic compounds present in the food system. However, majority of plant foods are thermally processed into ready-to-eat forms, with these processing methods potentially altering the structure and subsequent bioactivities of endogenous phenolic compounds. Thus, the aim of this chapter is to highlight on emerging non-thermal novel technologies (such as pulsed electric field, radiation, ultrasonication, high hydrostatic pressure processing and high pressure carbon dioxide processing) that can be exploited by the food industry to preserve/enhance bioactivities of phenolic compounds during processing.

Keywords: Phenolic compounds, Bioactivity, Non-thermal processing, Functional food

1. Introduction

In recent times, food consumption has advanced beyond simply meeting growth and development needs to include the supply of ingredients that can offer protection against diseases. The demand for such foods can be attributed to proven research data and advocacy by nutrition regulating bodies of the direct relationship between food composition and risks of diseases [1]. Food components that offer protection against disease development are known as bioactive compounds, with majority of these compounds reported as secondary metabolites. Secondary metabolites are non-nutritive compounds produced in plants as protection agents against oxidative stress upon exposure to above-threshold environmental conditions [2]. Among such non-nutritive secondary metabolites are phenolic compounds.

Phenolic compounds are benzene-ringed metabolites, with at least one phenol unit and one or more hydroxyl substituents [3]. Literature has reported about

10, 000 different classes of phenolic compounds, with these classes presented within three main groups including phenolic acids (e.g., hydroxycinnamic and hydroxybenzoic acids), flavonoids (e.g., anthocyanin, proanthocyanidins, flavonols etc.,), stilbenes (e.g., resveratrol and piceatannol etc.,), tannins (e.g., hydrolysable and condensed tannins), lignin, lignans and coumarins [4]. Different in-vitro and in-vivo studies have demonstrated bioactive capacity of phenolic compounds through their antioxidant, anti-inflammatory, anticancer, antidiabetic, cardiovascular protection and anti-cholesterol health effects. However, these reported bioactive properties are dependent on the type, concentration and biochemical structure of phenolic compounds present in a food system. Structurally, the bioactive capacity of phenolic compounds is dependent on factors such as the number and position of hydroxyl groups on the aromatic ring, hydrogen atoms of the adjacent hydroxyl groups (*o*-diphenol) present in the A, B and C rings of flavonoids, and the presence of double bonds in the benzene ring and oxo functional group (C=O) [5].

Nevertheless, majority of plant foods are subjected to different processing methods prior to consumption, with these processing methods causing changes in the biochemical stability and subsequent bioactivity of phenolic compounds present in the food. Naturally, phenolic compounds are present in foods as free or glycosylated (i.e., bound to protein and carbohydrate molecules), and are released from the food matrix during processing [6]. It is no doubt that, majority of traditional and industrial food processing methods are thermal intensive, with literature reporting a decrease in their concentrations of phenolic compounds and their subsequent bioactivities during transformation into ready-to-eat food products. Therefore, the food industry is continuously searching for alternative non-thermal techniques that can help retain/increase concentrations and bioactivities of phenolic compounds during processing. In this chapter, we focused on current non-thermal food processing technologies such as ultrasonication, high hydrostatic pressure processing, radiation, high pressure carbon dioxide processing and pulsed electric field. We seek to bring to light our understanding of their principles of operation, as well as how these novel non-thermal technologies influence yield and bioactivities of phenolic compounds during processing of some plant-based food products.

2. Thermal food processing and its effects on food phenolic compounds

Majority of plant foods are thermally processed into their ready-to-eat and stable forms. Notable examples of thermal processing techniques include roasting, microwaving, boiling, steaming and drying. However, research has demonstrated changes in food bioactive compositions such as phenolics during thermal processing. For instance, boiled broccoli showed reduced concentrations of caffeic acid (2.2 mg/100 g), quercetin (10 mg/100 g) and antioxidant capacity compared to their raw forms (caffeic aicd-6.6 mg/100 g and quercetin – 23.5 mg/100 g) [7]. Similarly, roasted coffee beans showed reduced phenolic levels and antioxidant capacity. These changes were attributed to oxidation by the triggering of non-enzymatic trans-glycosylation of phenolics into melanoidin as a result of Maillard reaction. Additionally, oxidation of phenolic compounds into their less bioactive forms such as quinones, and (−)-catechin into (+)-catechin has been reported as a result of thermal processes such as roasting [8]. Due to these bioactive reductions, there is a collective effort by researchers to develop non or medium thermal techniques that can help enhance/maintain phenolic levels along food processing.

3. Novel non-thermal food processing

It is no doubt that, the key purpose of agro processing is to convert raw food materials into their stable ready-to-eat forms through the inactivation of endogenous enzymes and microorganisms. Compared to thermal processing, non-thermal processing involves the application of no or minimal heat to a food system in order to inactivate spoilage enzymes and microbes, without compromising its nutrients, bioactive and organoleptic characteristics [9]. At household and industrial levels, the main advantage of non-thermal processing over thermal processing, is its capacity to exhibit thermal processing benefits while conserving the freshness and phytochemical composition of the food system under low energy consumption. Thus, non-thermal techniques can be considered as green alternatives for production of consumer-acceptable foods with rich composition of bioactive compounds such as phenolics. Examples of novel non-thermal processing techniques currently being applied in the food industry include high hydrostatic pressure processing, ultrasonication (acoustic emissions), radiation, pulsed electric field and high-pressure carbon dioxide processing. Therefore, the objective of this section is to discuss the underlining principles of key non-thermal processing techniques, with the main focus on how they influence concentrations and bioactivities of phenolic compounds during the conversion of food systems into ready-to-eat forms.

3.1 High hydrostatic pressure (HHP) processing

Compared to thermal food processing techniques, HHP processing is advantageous to the food industry due to its limited effects on nutritional and organoleptic qualities, as well as its negating effect on food size and geometry [10]. Basically, HHP involves the exposure of food systems to 100–1,000 MPa through a pressure transmitting medium (i.e., water or any other food grade solvent) at room or mild temperatures [11]. Depending on the density of the solvent and food components, temperature during HHP processing can increase by 3°C at every 100 MPa [12]. Therefore, the efficiency of HHP is dependent on parameters such as temperature, pressure level, pressure holding time, liquid/solid ratio, solvent type and solvent concentration [13]. When a food system is exposed to HHP, covalent bonds linking tightly packed cellular structural walls and membranes are broken, leading to cellular increment in porosity and permeability. Based on this principle, it can be postulated that, HHP processing can lead to (1) structural defragmentation for enhanced extraction of free and bound phenolic compounds (2) liberation of endogenous enzymes such as polyphenol oxidase and peroxidase which can cause degradation of phenolic compounds into *o*-quinones with subsequent reduction in their bioactivities. Thus, the pressure level and temperature applied during HHP are very crucial in order to avoid the generation of increasing temperatures and excess structural damage, which can negatively influence the yield and stability of phenolic compounds during processing. However, according to Heinz and Buckow [14] the low compressibility of covalent bonds compared to weak energy bonds, enables the native structures of low molecular weight molecules (such as phenolic compounds, vitamins and minerals) to be preserved during HHP processing, compared to macromolecules such as starch and protein.

Over the years, the food industry has mainly exploited HHP on commercial scale for preservation of plant foods (e.g., vegetables, juice and beverages) and muscle foods (e.g., meat, seafood and fish). Literature has reported mixed data on impacts of HHP on food phenolic compounds, with some reporting incremental changes and others showing an opposite trend. Patras et al. [15] studied the effect of HHP (i.e., 400 and 600 MPa for 15 min) on phenolic composition of strawberry puree. According to their study, strawberry puree treated with 400 MPa showed higher concentrations of phenolic compounds compared to the puree treated with 600 MPa and the control, with phenolic composition of the 600 MPa treated strawberry puree insignificantly different from the control. In this same study, the concentrations of anthocyanins and other phenolic compounds in blackberry pressure treated puree were increased by 108% with 200 MPa for 5 to 5 min, whereas blackberry puree treated with 400 MPa for 5 to 15 min showed phenolic increments by 92%, compared to their control counterparts. A similar trend was observed with apple treated HHP samples, where a pressure of 600 MPa induced a 75% loss of phenolic compounds after pasteurization, compared to their 400 MPa treated samples which showed significant retention of phenolic compounds [16]. Irrespective of this initial trend, Keenan and his colleagues interestingly observed an opposite trend when the HPP treated apple samples were stored for 14 and 21 days, where the apples treated with 600 MPa significantly retained their composition of phenolic compounds compared to their 400 MPa alternatives. In another study, Huang et al. [17] observed increased antioxidant activity and levels of phenolic compounds (i.e., +/− catechin, chlorogenic acid, neochlorogenic acid, epicatechin, ferulic acid and *p*-coumaric acid) with HHP treated apricot nectar. Similarly, manuka honey treated with HPP at 600 MPa (10 min, 25°C) showed increased concentration of phenolic compounds after 12 weeks of storage [18]. In another research investigated by Shen et al. [18], HHP (400, 500 and 600 MPa) treated jujube (*Ziziphus jujuba* Mill.) showed improved concentrations of phenolic compounds with increasing pressure, with the highest concentrations of total phenolic content (7.9%) and flavonoids (18.4%) obtained with 600 MPa.

Overall, reports showing increased concentrations of phenolic compounds can be linked to the structural damage of cell walls and membranes by HHP, thus increasing porosity and facilitating the release of bound phenolic compounds from structural carbohydrates and proteins. Also, reports showing decreased concentrations of phenolic compounds at high pressure levels were attributed by authors to the presence and activation rates of endogenous polyphenol oxidases and peroxidases during HHP as discussed earlier in this section. Thus, in order to retain/enhance the composition of food phenolic compounds and their accompanied bioactivities, HHP should be regulated between room temperature or slightly above, in order to avoid or limit the formation of oxidative undesirable products accompanied with high thermal treatments. Further data on how phenolic composition of foods changed upon exposure to HPP are displayed in **Table 1**.

3.2 Pulsed electric field (PEF) processing

PEF has gained much attention in the food industry due to its appreciable preservation capacity in comparison with high thermal pasteurization. According to Picart and Cheftel [24], PEF is advantageous over high thermal processing due to its capacity in inactivating microorganisms, while still preserving the nutritional and sensory quality attributes (e.g., colour, flavour and texture) of the food. Majority of the literature on PEF has been reported with food preservation, with dearth reports on its effects with phenolic compounds. Primarily, the principle of PEF involves the application of short pulses of high electric fields through a product placed between a set of electrodes [24]. When a food material is exposed to PEF, the generated electrical fields are able to create electroporation across the cell membranes, thus leading to cell membrane porosity as a result of structural disintegration [25]. The efficiency of PEF depends on factors such as electrical field strength, exposure time, applied energy density, pulse width and shape, pulse frequency and applied

Food type	HHP parameters	Effect on phenolics	Reference
Rough rice sprout	0.1–100 MPa	Increases total phenolic compounds after 24 h sprouting	[19]
Corn cob	600 MPa; 15 and 60 min, 20–60°C	+ 20% ferulic acid concentration at 15 min and 60°C	[20]
Watercress	0.1–600 MPa; 1.5–33.5 min; 20°C	Increased yield of phenolic acids, flavonoids and total phenolic content at 600 MPa and 1.5–7.8 min	[21]
Aronia berry puree	200–600 MPa; 2.5–5 min; 21–33°C	+ 3–13% total phenolic content; + 6–17% total anthocyanin content	[22]
Purple-skinned pelota pears (peels and pulp)	100, 300 and 600 MPa; 5 min; 17–34°C	+ 8.7% piscidic acid; + 55.9% hydroxybenzoic acid glycosides in peels at 350 MPa/5 min and 100 MPa/5 min, respectively. + 133.2% isorhamnetin glycosides in peels	[23]
Red-skinned Sanguinos prickly pears (peels and pulp)	100, 300 and 600 MPa; 5 min; 17–34°C	+ 90.6% retention of total piscidic acids and a general decrease in total phenolic compounds	[23]

Table 1.
Effect of high hydrostatic pressure (HHP) on phenolic composition of selected food systems.

food characteristics (e.g., pH and electrical conductivity) as discussed by Mañas and Pagán [26].

With respect to phenolic compounds, PEF can be used to exploit their concentrations and bioactivities from two approaches: (1) since phenolic compounds are mainly stored in the tightly packed cell wall and membranes, structural damage can enhance extraction and subsequent concentrations of phenolic compounds from storage cells compared to their native forms; (2) creation of structural porosity can induce oxidative stress in the food system, thus leading to the stimulation of biosynthesis pathways responsible for the production of antioxidants such as phenolic compounds [2]. For example, red cabbage treated with PEF showed a 2.15 times enhanced anthocyanin yield compared to the non-PEF treatment [27]. Luengo et al. [28] also reported an increased yield of total phenolic compounds with PEF treated tomatoes and grapes. Similarly, concentrations of phenolic compounds after treating borage (*Borago officinalis*) leaves with PEF were significantly enhanced, according to the work of Segovia et al. [29], with the authors also observing enhanced antioxidant capacities. In another study reported by Liu et al. [30–31], PEF treated onions showed 2.2 and 2.7 times increased total phenolics and flavonoid levels, compared to their control forms. Results from this study also showed enhanced antioxidant capacities with PEF treated onions. Their observation with enhanced antioxidant capacities can be attributed to the increased yield of phenolic compounds and the possibility of extracting other groups of bound phenolic compounds that were otherwise trapped in the cellular walls of the control samples. Regarding extraction efficiency and yield of phenolic compounds from plant foods, lots of positive literature has been reported with PEF applications and shown in **Table 2**.

PEF has also been associated with increased yield of fruit juice phenolic compounds and bioactivity. For juice application, the most exploited potential effects of PEF include colour, pH, acidity, soluble solids, concentration of phenolic compounds and activity of polyphenol oxidase (i.e., the enzyme responsible for juice browning through the degradation of phenolic compounds into *o*-quinones) [1].

Food type	PEF parameters	Effect on phenolics	Reference
Orange, pomelo and lemon fruits	3 kV/cm	+ 50% increased extraction yield	[32]
Blackberries (*Rubus fruticosus*)	PEF of 13.3 kV/cm; 10 μs pause after each 100 pulse	Sixfold higher compared to high voltage electric discharge (HVED)	[13]
Basil leaves	2, 3 and 4 kV/cm; 1, 2 and 3 min	The highest of 115.203 mg GAE/ 100 g at 3 kV/cm for 2 min	[25]

Table 2.
Effect of pulsed electric field (PEF) on phenolic composition of selected food systems.

PEF treated tomato juice showed higher concentrations of chlorogenic acid and quercetin, compared to their thermal treated alternatives [33]. In another study by Agcam et al. [34], concentrations of caffeic acid were enhanced with PEF treated tomato juice over storage time, whereas the concentrations of *p*-coumaric acid reduced over storage. The authors attributed their findings to the increased activity of hydroxylase, leading to the conversion of *p*-coumaric acid into caffeic acid along the storage period. Simultaneous to the tomato juice, the authors further reported the highest concentrations of total phenolic compounds (443.42 mg/GAE) with PEF treated orange juice, compared to their thermal treated forms (439.07 mg/GAE). Other authors such as Boussetta et al. [35] also postulated positive correlations between PEF and maximization of phenolic compounds during winemaking. Confirming their postulation is the work of Puértolas et al. [36], where an electric field of 5 kV/cm increased the yield of total phenolic compounds of wines developed from treated Cabernet Sauvignon grapes. A similar interesting observation that proved the relationship between pulse type and exposure time is the study of Delsart et al. [37], where PEF (4 kV/cm at 1 ms and 0.7 kV/cm at 200 ms) exposed Cabernet Sauvignon grapes showed improved extraction kinetic effects on vacuolar and parietal tannins, respectively.

Besides extraction and fruit juices, PEF has also been investigated with whole foods. Wiktor et al. [38] investigated different PEF intensities (1.85, 3 and 5 kV/cm) and pulse numbers (10, 50 and 100 exponential shaped pulses) on accumulation of phenolic compounds and antioxidant capacity in apple (var. Ligol). After their research, Wiktor and his colleagues [38] observed the highest total phenolic content and antioxidant capacity with apple tissues treated with 1.85 kV/cm at 10 pulse. Similar to their study, Soliva-Fortuny et al. [39] observed the highest yield of total phenolic content (13%) and flavone-3-ol (92%) with PEF treated apple, compared to control. Despites these positive results, it's also important to highlight that, some studies have reported reduced levels of phenolic compounds in foods treated with PEF. For instance, the findings of Odriozola-serrano et al. [33] and Aguilar-Rosas et al. [40–41] demonstrated significant reductions of phenolic compounds in strawberry and apple juices, respectively, upon treatment with PEF, with these authors correlating their observations to oxidative activities of polyphenol oxidase. In a normal unfractured cell, endogenous enzymes such as polyphenol oxidase are tightly held in packed membranes. However, when the cell is exposed to PEF, the induction of cellular porosity will lead to their release from tightly packed membranes, thus initiating oxidative reactions in the presence of oxygen and substrates such as phenolic compounds. However, these explanations are based on theoretical assumptions. Deeper studies investigating the presence and catalysis of polyphenol oxidase and peroxidase in PEF treated food systems should be conducted, in order to provide scientific proof of their activities, as well as threshold levels of PEF parameters to be applied towards the control/limitation of polyphenol oxidase/peroxidase activities.

3.3 Ultrasound (US)

In recent years, ultrasound (US) has gained attention in the food industry for pasteurization and preservation purposes. Unlike HHP, US is not only limited to inactivation of microorganism's, but also includes the deactivation of enzymes [1]. For food processing purposes, US can be divided into two categories (1) low intensity US (lower than 1 W/cm^2), which is a non-destructive approach used to measure the structure, composition and flow rate of a food (2) high intensity US (between 10 and 1000 W/cm^2), which involves the use of high frequencies to cause structural damage to the tissues and membranes of the exposed food system [42]. Among these two approaches, high intensity US is the most applied in the food industry with respect to phenolic compounds for (a) enhancing extraction yield of phenolic compounds from food materials (b) induction of oxidative stress in a food system by causing tissue cavitation and porosity, towards the stimulation of biosynthetic pathways (i.e., phenylpropanoid and shikimate pathways) responsible for the production of phenolic compounds as defense agents against induced oxidative stress [2].

According to Toma et al. [42] when US is applied to a food system, there is the formation of cavitation bubbles which creates a pressure zone change up to 400 km/h, leading to increased porosity, rupturing or removal of cell membranes for enhanced mass transfer from the cells interior upon collapsing. Yu et al. [43] treated Romaine lettuce (*Lactuca sativa* var. longifolia) with US (25 kHz, 2 kW, 1–3 min) and observed increased total phenolic content, compared to the control. From this same research, storage studies showed 22.5% increased phenolic concentrations after 60 h of storage with US treated Romaine lettuce (1 min treatment time) than their control forms. In another study involving black cumin (*Nigella sativa*), pretreatment with US (30, 60 and 90 W; 25 kHz; 30, 45 and 60 min) showed increasing total phenolic concentrations (ranges of 93.21 to 106.6 ppm) with increasing US conditions. The authors also observed enhanced antioxidant capacity with US treated black cumin and attributed this trend to the 5% increased total phenolic content in US treated samples compared to the control **Table 3** [47]. gives a summary of phenolic composition changes in some foods after US exposure.

Similarly, common bean (*Phaseolus vulgaris*) sprouts treated with US (180 and 360 W; 40 kHz; 30, 45 and 60 min) showed increased accumulation

Food type	US parameters	Effect on phenolics	Reference
Tomato (*Lycopersicon esculentum*) fruit	25 kHz; 1 kW; 1, 2, 3 and 4 min	+ 17.05% total phenolic content with 1 min US time. All US treatments showed higher antioxidant capacities than control	[44]
Black currant fruit (*Ribes nigrum* L.)	150 W; 40 kHz; amplitude-10, 40 and 70%; 3, 6 and 10 min	+ 4% total phenolic content (10 min; 70% amplitude) compared to control; + 20% total anthocyanin content (10 min; 70% amplitude) compared to control	[45]
Orange juice	24 kHz; 1, 10, 20 and 30 min	+ 63 and 64% levels of total phenolics and flavonoids with US (30 min) compared to control	[46]

Table 3.
Effect of ultrasound (US) on phenolic composition of selected food systems.

trend of phenolic compounds and antioxidant capacities with increasing US treatment parameters [2]. From this study, the highest level of total phenolic acids (216.7 mg/100 g), total flavonoids (203.5 mg/100 g), total anthocyanins (30.35 mg/100 g) and total antioxidant capacities (98%) were observed with 360 W (60 min) US treated common bean sprouts at 9 h of sprouting compared to the control. The authors further observed increased levels of oxidative stress markers (i.e., hydrogen peroxide, catalase and peroxidase) and activities of phenolic triggering enzymes (i.e., phenylalanine ammonia-lyse and tyrosine ammonia-lyase) with increasing US conditions. Thus, confirming with literature that US can improve the yield and bioactivity of food phenolic compounds through the induction of oxidative stress and triggering of the phenylpropanoid pathway. In another study, grape juice treated with ultrasound showed increased concentrations of phenolic compounds by 114.3%, compared to the control [48]. Naturally, phenolic compounds occur in foods as free or bound forms. According to Lieu and Le [48] the capacity of ultrasound to breakdown covalent bonds linking phenolic compounds bounded to cell wall components (i.e., carbohydrates and proteins) led to their release and observed increased yield. These explanations explain the report of Khan et al. [49], where orange peels treated with US (150 W) showed the highest yield of naringin (70.3 mg GAE/100 g), hesperidin (205.2 mg GAE/100 g) and total phenolic composition (275.8 mg GAE/100 g). Besides the principle of cavitation discussed previously in this section, another explanation for this increasing trend has been postulated by Khan et al. [1] to the addition of hydroxyl functional group to aromatic compounds such as phenolics during ultrasonication.

3.4 Radiation

Light is one of the most essential factors for plant growth and development. Plant foods grown in the field use sunlight for photosynthesis and production of bioactive metabolites such as phenolic compounds, whereas those grown commercially under controlled conditions use artificial light sources to meet their needs for photosynthesis and production of secondary metabolites such as phenolic compounds. Thus, changes in light quality, quantity, intensity and duration can be exploited to influence the final yield and bioactivities of phenolic compounds in different food systems [50]. According to Bantis et al. [51], five wavelength ranges including red (660–700 nm), far-red (700–750 nm), blue (495–400 nm), UV-A (400–315 nm) and UV-R8 (315–280 nm) has been described with respect to plant radiation. Among these wavelengths, blue, green, red and white are the most reported, with respect to accumulation and bioactivities of food phenolic compounds. Upon exposure to sunlight or artificial light, changes in the light parameters can induce oxidative stress, which are sensed by proteinaceous receptors on cell membranes, for subsequent triggering of metabolic pathways responsible for the production of secondary metabolites such as phenolic compounds [52]. For example, blue light was demonstrated by Qian et al. [52] to improve the biosynthesis of phenolic compounds through enhanced stimulation of malonyl CoA and coumaroyl (key substrates associated with the phenylpropanoid pathway for biosynthesis of phenolic compounds).

For controlled processing, artificial lighting technology such as light emitting diode (LED) has been proven by previous studies to be effective for accumulation of phenolic compounds in diverse human food crops, among other artificial light technologies such as high sodium pressure (HSP) and high-intensity discharge (HID). LED is the most preferred radiation technology among agroprocessors due to their efficient use of energy and capacity to produce food products rich in nutrients and bioactive secondary metabolites comparable to blue, green, red and far-red treated foods [53]. In the study of Alothman et al. [54], effect of UV-C (dose: 2.158 J/m^2) on flavonoid, total phenolics and vitamin C compositions were evaluated. From their result, total phenolic content and flavonoid levels were significantly increased with UV-C treated banana and guava, whereas no significant increments were observed with UV-C treated pineapple fruits and control. In another study, UV-C treated freshly cut mangoes showed increased concentrations of flavonoids and total phenolic compounds with increasing UV-C exposure time (10, 20 and 30 min) [55].

Also, black, red and white rice varieties were treated with gamma irradiation (10 kGy) and its influence on free and bound phenolic composites were investigated by Shao et al. [56]. According to their study, gamma irradiated white, red and black rice varieties produced the highest yield of bound phenolic compounds at 19.7, 40.2 and 59.5 mg GAE/100 g, respectively. The authors observed significant changes in bound phenolic compounds with all varieties of gamma irradiated rice, compared to free phenolic compounds. Bound phenolic compounds in cereal grains are covalently bonded to other cell wall fragments, thus the observation of Shao et al. [56] can be attributed to the capacity of gamma irradiation to break the covalent bonds, cause depolymerization of higher molecular weight phenolic compounds into smaller molecular weight phenolic compounds such as gallic and ferulic acids. Azad et al. [57] studied the impact of different LEDs such as blue (450–495 nm), green (510–550 nm) and fluorescent lamps on accumulation of phenolic compounds in soybean (*Glycine max* L.) sprouts at different sprouting stages (3, 4, 5, 6 and 7 days). According to their results, total phenolic content and isoflavones were maximum with blue light (at days 5 and 6), compared to the green and fluorescent light treated sprouts. Furthermore, the antioxidant capacity (DPPH) observed with blue light treated soybean sprouts were significantly higher than the other investigated LED treatments. Blue light treated soybean sprouts exhibited DPPH capacity of 75%, whereas the green and fluorescent treated sprouts showed 69 and 58%, respectively. Further literature demonstrating the effect of light on different food systems are displayed in **Table 4**.

3.5 High pressure carbon dioxide (HPCD) processing

Due to the connection between thermal pasteurization and degradation of organoleptic and nutritional qualities, the food industry is continuously searching for alternative techniques that can provide the advantages of thermal processing without compromise on nutritional and sensory attributes. High pressure carbon dioxide (HPCD) processing is a nonthermal mechanism applied for food pasteurization by using pressurized CO_2 between 0.1 MPa and 50 MPa [62]. Some advantages of HPCD include low cost, non-toxicity, non-inflammability and renewability of CO_2 [63]. Factors that influence the efficiency of HPCD processing include food structure, CO_2 physical state, time, pressure and temperature. Until now, research has provided data on HPCD with temperature ranges between 25 and 100°C, it is important to mention that pressure and temperature ranges of >28 MPa and > 60°C are recommended for food applications and fractionation of bioactive metabolites such as phenolics compounds [64]. Although previous authors have tried to explain the principle of HPCD, there still exist a non-well-defined theory. However, a common explanation among all reviewers is a reflection of the potential disintegration of structural cellular membranes and subsequent leakage of cytoplasmic composites during HPCD exposure [65]. Based on this explanation, it can be postulated that HPCD can modify phenolic composition of plant food systems through (1) cell wall and membrane deformation, thus enhancing the release of bound phenolic compounds from structural molecules (2) combination of CO_2 and low temperatures will help prevent oxidation of phenolic compounds through inactivation of polyphenol oxidase (PPO).

Food type	Light parameters	Effect on phenolics	Reference
Blueberry (*Vaccinium corymbosum* L.) leaves	Red, 661 nm-24 μmol/m^2/s Blue, 417 nm-6 μmol /m^2/s 12, 24, 48 h	Increased total phenolic content and antioxidant capacity with blue light at 12 h, compared to the control. Level of monomeric anthocyanins were improved up to 24 h with blue light, after which there was a decreasing trend. Red light deceased monomeric anthocyanins than with control and blue light.	[58]
Green and purple basil (*Ocimum basilicum*) plants	UV-B (16.0 μmol / m^2/s) 1H2D; 1 h/d for 2 days 2H2D; 2 h/d for 2 days 1H5D; 1 h/d for 5 days 2H5D; 2 h/d for 5 days Control; no UV-B	+ 80–169% total flavonoid content with green basil UV-B treatments, compared to control. Increased antioxidant capacity with all UV-B treatments. + 37–79% total flavonoid content and antioxidant capacity with purple basil UV-B treatment (2H5D) than control.	[59]
Habanero pepper (*Capsicum chinense*)	Combined Blue (48 W/m^2, 0, 1.5 and 3 min) and UV-C (11.3 W/m^2) Time: 0, 0.5 and 1 min	Increased levels of total phenolics, flavonoids and antioxidant capacity with all combined treatments. Optimum levels were obtained with blue light (3 min) + UV-C (0.5 min).	[60]
Date palm mazafati (*Phoenix dactylifera* L.)	Gamma radiation; 0, 0.5, 1 and 2.5 kGy	Optimum concentration of total phenolic compounds and antioxidant capacity were observed with 2.5 kGy than control and other treatments.	[61]

Table 4.
Effect of radiation on phenolic composition of selected food systems.

Gasperi et al. [66] investigated HPCD (10 MPa, 36°C and 10 min) effects on phenolic composition of carrot juice, with their results showing retention of phenolic compounds. In another study by Del Pozo-Insfran et al. [67], although phenolic composition of HPCD treated muscadine grape juice remained unchanged after processing, their antioxidant capacity was significantly increased after 1 week of storage whereas their control form exhibited reduced antioxidant capacity. This observation could be attributed to the groups of phenolic compounds released upon HPCD processing and the capacity of HPCD to inactivate endogenous PPO, than the control. Thus, helping retain the concentrations and bioactivity of native phenolic compounds in muscadine juice during HPCD processing. Ferrentino et al. [68] reported no changes in total phenolic levels with HPCD treated red grapes. Similarly, apple juice subjected to high thermal processing and HPCD were evaluated for changes in phenolic compositions by Noci et al. [69]. According to the authors, high thermal treated apple juice showed significant reductions in concentrations of total phenolic compounds, whereas the HPCD treatments exhibited insignificant phenolic reductions compared to the control and thermal treatments. In synchrony with these studies are the works of Guo et al. [70] and Agcam et al. [34], which investigated the influence of ultra-high temperature and HPCD processing on phenolic compositions of lychee and litchi juices, respectively. According to the findings of Guo et al. [70], ultra-high temperature processing significantly reduced concentrations of caffeic acid, epicatechin, 4-methylcatechol and rutin in treated lychee juice, whereas HPCD retained concentrations of the aforementioned phenolic compounds after treatment. Comparable to their findings, HPCD treated litchi juice showed preserved concentrations of total phenolic content, rutin, (−)-epicatechin, chlorogenic acid, and antioxidant capacity, compared to the control whereas their ultra-high temperature treated counterparts exhibited significant losses of these investigated phenolic compounds [34].

Dearth research data is available on the influence of HPCD on phenolic composition of food products during processing. Thus, in-depth studies are crucial in order to stimulate their efficient applications in the food industry. However, available literature on HPCD processing seems to provide a trend of phenolics preservation rather than increments or degradation. This trend may be attributed to the low solubility of phenolic compounds in CO_2. Therefore, posing a challenge for their efficient extraction from food matrices.

4. Conclusion and future perspectives

Consumption of phenolic-dense foods has been attributed to positive health benefits such as antioxidative, anticancer and inflammatory effects. It is no doubt that, food processing conditions can either increase or decrease their phenolic compositions and subsequent bioactivities. The wide application of thermal techniques along plant food processing has been associated with high cost, reduced concentrations of phenolic compounds and their subsequent bioactivities. Thus, to encourage the food industry to produce phenolic-dense functional foods, there is an increasing demand to exploit alternative non-thermal processing techniques that can help enhance/maintain phenolic levels and bioactivities. In this quest, literature has identified pulsed electric field, high hydrostatic pressure (HHP) processing, ultrasound, radiation and high-pressure carbon dioxide (HPCD) processing as novel non-thermal techniques that can be exploited to enhance/retain bioactivities of food phenolic compounds. Arguably, these novel non-thermal processing methods will not only help enhance/retain food phenolic compositions but will also help the industry to improve food nutritional value, shelf-life and sensory attributes. However, in order to scale-up their applications, key technical queries such as (a) besides sustainability, will the non-thermal method oxidize phenolics and other bioactive compounds (b) what key parameters of the non-thermal technique should be optimized in order to maximize the yield and bioactivities of phenolic compounds, case by case (c) how does the bioactive composition of different food groups respond upon exposure to each non-thermal method case by case (d) are there any groups of hazardous compounds liberated during non-thermal processing? If yes, what are the health effects of these compounds and how do they influence the nutritional and sensory attributes of the treated food.

Author details

Josephine Ampofo[1*] and Michael Ngadi[2]

1 Department of Sciences, Université Sainte Anne, Pointe-de-l'Église, Nova Scotia, Canada

2 Department of Bioresource Engineering, MacDonald Campus, McGill University, Sainte Anne-de-Bellevue, Quebec, Canada

*Address all correspondence to: josephine.ampofo@mail.mcgill.ca

References

[1] Khan MK, Ahmad K, Hassan S, Imran M, Ahmad N, Xu, C. Effect of novel technologies on polyphenols during food processing. Innovative Food Science and Emerging Technologies. 2018;45:361-381.

[2] Ampofo JO, Ngadi M. Ultrasonic assisted phenolic elicitation and antioxidant potential of common bean (*Phaseolus vulgaris*) sprouts. Ultrasonics Sonochemistry. 2020;64:104974.

[3] Gan RY, Lui WY, Wu K, Chan CL, Dai SH, Sui ZQ, Corke H. Bioactive compounds and bioactivities of germinated edible seeds and sprouts: An updated review. Trends in Food Science and Technology. 2017;59:1-14.

[4] Akyol H, Riciputi Y, Capanoglu E, Caboni MF, Verardo, V. (2016). Phenolic compounds in the potato and its by-products: An overview. International Journal of Molecular Sciences. 2016;17:835.

[5] Anantharaju PG, Gowda PC, Vimalambike MG, Madhunapantula SV. An overview on the role of dietary phenolics for the treatment of cancers. Nutrition Journal. 2016;15:1-16.

[6] Aquino-Bolanõs EN, García-Díaz YD, Chavez-Servia JL, Carrillo-Rodríguez JC, Guzman AM, Heredia-García E. (2016). Anthocyanin, polyphenol, and flavonoid contents and antioxidant activity in Mexican common bean (*Phaseolus vulgaris* L.) landraces. Emirates Journal of Food and Agriculture. 2016;28:581-588.

[7] Moreira ASP, Nunes FM, Simoes C, Maciel E, Domingues P, Domingues MRM, Coimbra MA. Transglycosylation reactions, a main mechanism of phenolics incorporation in coffee melanoidins: inhibition by Maillard reaction. Food Chemistry. 2017;227:422-31.

[8] Pellegrini N, Chiavaro E, Gardana C, Mazzeo T, Contino D, Gallo M, Riso P, Fogliano V, Porrini M. Effect of Different Cooking Methods on Color, Phytochemical Concentration, and Antioxidant Capacity of Raw and Frozen Brassica Vegetables. Journal of Agricultural Food Chemistry. 2010;58:4310-4321.

[9] Versteeg K. Next generation of innovative food processing technologies. Benefits and Challenges. In: Kai K, Pablo J, Geoffrey S, editors. Innovative Food Processing Technologies. Extraction, separation, component modification and process intensification. Elsevier; 2016. p. 455-464.

[10] Toepfl S, Mathys A, Heinz V, Knorr D. Review: potential of high hydrostatic pressure and pulsed electric fields for energy efficient and environmentally friendly food processing. Food Review International. 2006;22:405-23.

[11] Li F, Chen G, Zhang B, Fu X. Current applications and new opportunities for the thermal and non-thermal processing technologies to generate berry product or extracts with high nutraceutical contents. Food Research International. 2017;100:19-30.

[12] Fellows P. Food Processing Technologies: Principles and Practices. 3rd ed. Boca Raton: Elsevier; 2009. 928 p.

[13] Barba FJ, Terefe NS, Buckow R, Knorr D, Orlien V. New opportunities and perspectives of high-pressure treatment to improve health and safety attributes of foods. A review. Food Research International. 2015;77:725-42

[14] Heinz V, Buckow R. Food preservation by high pressure. Journal für Verbraucherschutz und Lebensmittelsicherheit. 2010;5:73-81.

[15] Patras A, Brunton NP, Da Pieve S, Butler F. Impact of high pressure processing on total antioxidant activity, phenolic, ascorbic acid, anthocyanin content and colour of strawberry and blackberry purées. Innovative Food Science and Emerging Technologies. 2009;10:308-313.

[16] Keenan DF, Rößle C, Gormley R, Butler F, Brunton NP. Effect of high hydrostatic pressure and thermal processing on the nutritional quality and enzyme activity of fruit smoothies. LWT-Food Science and Technology. 2012;45:50-57.

[17] Huang W, Bi X, Zhang X, Liao, X, Hu X, Wu J. Comparative study of enzymes, phenolics, carotenoids and color of apricot nectars treated by high hydrostatic pressure and high temperature short time. Innovative Food Science and Emerging Technologies. 2013;18:74-82.

[18] Fauzi AN, Farid MM. High pressure processed Manuka honey: change in nutritional and rheological properties over 1-year storage. Journal of Food Processing and Preservation. 2016;41:e13085–e13094.

[19] Shen J, Gou Q, Zhang Z, Wang M. Effects of high hydrostatic pressure on the quality and shelf-life of jujube (*Ziziphus jujuba* 4Mill.) pulp. Innovative Food Science and Emerging Technology. 2016;36:166-172.

[20] Kim MY, Lee SH, Jang GY, Li M, Lee YR, Lee J, Jeong HS. Changes of phenolic-acids and vitamin E profiles on germinated rough rice (*Oryza sativa* L.) treated by high hydrostatic pressure. Food Chemistry. 2017;217:106-111.

[21] Pérez-Rodríguez N, Torrado Agrasar A, Domínguez JM. High hydrostatic pressure as pretreatment and adjuvant for the enzymatic release of ferulic acid from corn cob. Process Biochemistry. 2017;58:204-210.

[22] Pinela J, Prieto MA, Barros L, Carvalho AM, Oliveira B, Saraiva JA, Ferreira ICFR. Cold extraction of phenolic compounds from watercress by high hydrostatic pressure: Process modelling and optimization. Separation and Purification Technology. 2018;192:501-512.

[23] Yuan B, Danao MGC, Stratton JE, Weier SA, Weller CL, Lu M. High pressure processing (HPP) of aronia berry purée: Effects on physicochemical properties, microbial counts, bioactive compounds, and antioxidant capacities. Innovative Food Science and Emerging Technologies. 2018;47:249-255.

[24] Picart L, Cheftel C. Pulsed electric fields. In: Zeuthen P, Bøgh-Sørensen L, editors. Food Preservation Techniques. Boca Raton: CRC Press; 2003. p.360-427.

[25] Hendrawan Y, Sabrinauly S, Hawa LC, Argo BD, Rachmawati M. Analysis of the phenol and flavonoid content from basil leaves (*Ocimum americanum* L) extract using pulsed electric field (PEF) pre-treatment. Agricultural Engineering International: CIGR Journal. 2019;21:149-158.

[26] Mañas P, Pagán R. Microbial inactivation by new technologies of food preservation. Journal of Applied Microbiology. 2005;98:1387-1399.

[27] Gachovska T, Cassada D, Subbiah J, Hanna M, Thippareddi H, Snow,D. Enhanced anthocyanin extraction from red cabbage using pulsed electric field processing. Journal of Food Sciences. 2010;75:E323-3-E329.

[28] Luengo E, Álvarez I, Raso J. Improving the pressing extraction of polyphenols of orange peel by pulsed electric fields. Innovative Food Science and Emerging Technologies. 2013;17: 79-84.

[29] Segovia J, Luengo E, Corral-Perez J, Raso J, Almajano M. Improvements in

the aqueous extraction of polyphenols from borage (*Borago officinalis* L.) leaves by pulsed electric fields: Pulsed electric fields (PEF) applications. Industrial Crops Production. 2014;65:390-396.

[30] Liu C, Grimi N, Lebovka N, Vorobiev E (2018a) Effects of preliminary treatment by pulsed electric fields and convective air-drying on characteristics of fried potato. Innovative Food Science and Emerging Technologies. 2018a;47:454-460.

[31] Liu ZW, Zeng XA, Ngadi MO (2018b) Enhanced extraction of phenolic compounds from onion by pulsed electric field (PEF). Journal of Food Processing and Preservation. 2018b;42:1-8.

[32] El Kantar S, Boussetta N, Lebovka N, Foucart F, Rajha HN, Maroun RG, Louka N, Vorobiev E. Pulsed electric field treatment of citrus fruits: Improvement of juice and polyphenols extraction. Innovative Food Science and Emerging Technologies. 2018;46:153-161.

[33] Odriozola-Serrano I, Soliva-Fortuny R, MartÃn-Belloso O. Phenolic acids, flavonoids, vitamin C and antioxidant capacity of strawberry juices processed by high intensity pulsed electric fields or heat treatments. European Food Research and Technology. 2008;228:1-10.

[34] Agcam E, Akyıldız A, Evrendilek GA. Comparison of phenolic compounds of orange juice processed by pulsed electric fields (PEF) and conventional thermal pasteurization. Food Chemistry. 2014;143:354-361.

[35] Boussetta N, Vorobiev E, Le L, Cordin-Falcimaigne A, Lanoisellé JL. (2012). Application of electrical treatments in alcoholic solvent for polyphenols extraction from grape seeds. LWT - Food Science and Technology. 2012;46:127-134.

[36] Puertolas E, Lopez N, Condon S, Alvarez I, Raso J. Potential applications of PEF to improve red wine quality. Trends in Food Science and Technology. 2010;21:247-255.

[37] Delsart C, Cholet C, Ghidossi R, Grimi N, Gontier E, Gény, L, et al. (2014). Effects of pulsed electric fields on Cabernet Sauvignon grape berries and on the characteristics of wines. Food and Bioprocess Technology. 2014;7:424-436.

[38] Wiktor A, Sledz M, Nowacka M, Rybak K, Chudoba T, Lojkowski W, Witrowa-Rajchert D. The impact of pulsed electric field treatment on selected bioactive compound content and color of plant tissue. Innovative Food Science and Emerging Technologies. 2015;30:69-78

[39] Soliva-Fortuny R, Vendrell-Pacheco M, Martin-Belloso O, Elez–Martinez P. Effect of pulsed electric fields on the antioxidant potential of apples stored at different temperatures. Postharvest Biology and Technology. 2017;132:195-201.

[40] Aguilar-Rosas S, Ballinas-Casarrubias M, Nevarez-Moorillon G, Martin-Belloso O, Ortega-Rivas E. Thermal and pulsed electric fields pasteurization of apple juice: Effects on physicochemical properties and flavour compounds. Journal of Food Engineering. 2007a;83:41-46.

[41] Aguilar-Rosas S, Ballinas-Casarrubias M, Nevarez-Moorillon G, Martin-Belloso O, Ortega-Rivas E. Thermal and pulsed electric fields pasteurization of apple juice: Effects on physicochemical properties and flavour compounds. Journal of Food Engineering. 2007a;83:41-46.

[42] Toma M, Vinatoru M, Paniwnyk L, Mason TJ (2001) Investigation of the effects of ultrasound on vegetal tissues

during solvent extraction. Ultrasonics Sonochemistry. 2001;8:137142.

[43] Yu J, Engeseth NJ, Feng H. High intensity ultrasound as an abiotic elicitor-Effects on antioxidant capacity and overall quality of romaine lettuce. Food Bioprocessing Technology. 2016;9:262-273.

[44] Lu C, Ding J, Park HK, Feng H. High intensity ultrasound as a physical elicitor affects secondary metabolites and antioxidant capacity of tomato fruits. Food Control. 2020;113:1-8

[45] Oancea S, Ghincevici D, Ketney O. The effect of ultrasonic pretreatment and sample preparation on the extraction yield of antioxidant compounds and activity of black currant fruits. Acta Chimica Slovencia. 2014;62:242-248.

[46] Guerrouj K, Sánchez-Rubio M, Taboada-Rodríguez A, Cava-Roda RM, Marín-Iniesta F. Sonication at mild temperatures enhances bioactive compounds and microbiological quality of orange juice. Food and Bioproducts Processing. 2016;99:20-28.

[47] Moghimi M, Farzaneh V, Bakhshabadi H. The effect of ultrasound pretreatment on some selected physicochemical properties of black cumin (*Nigella sativa*). Nutrire. 2018;43:1-8.

[48] Lieu LN, Le VVM. (2010). Application of ultrasound in grape mash treatment in juice processing. Ultrasonics Sonochemistry. 2010:17:273-279.

[49] Khan MK, Abert-Vian M, Fabiano-Tixier AS, Dangles O, Chemat F. (2010). Ultrasound assisted extraction of polyphenols (flavanone glycosides) from orange (*Citrus sinensis* L.) peel. Food Chemistry. 2010;119:851-858.

[50] Alvarado AM, Aguirre-Becerra H, Vázquez-Hernández MC, Magaña-Lopez E, Parola-Contreras I, Caicedo-Lopez LH, Contreras-Medina LM, Garcia-Trejo JF, Guevara-Gonzalez RG, Feregrino-Perez AA (2019) Influence of elicitors and eustressors on the production of plant secondary metabolites. In: Natural bio-active compounds. Springer, pp 333-388

[51] Bantis F, Smirnakou S, Ouzounis T, Koukounaras A, Ntagkas N, Radoglou K. Current status and recent achievements in the field of horticulture with the use of light-emitting diodes (LEDs). Scientia Horticulturae. 2018;235:437-451

[52] Qian H, Liu T, Deng M, Miao H, Cai C, Shen W, Wang Q. Effects of light quality on main health-promoting compounds and antioxidant capacity of Chinese kale sprouts. Food Chemistry. 2016;196:1232-1238.

[53] Aguirre-Becerra H, Vazquez-Hernandez MC, de la O, DS, Alvarado-Mariana A, Guevara-Gonzalez RG, Garcia-Trejo JF, Feregrino-Perez AA. Role of Stress and Defense in Plant Secondary Metabolites Production. In: Pal D, Nayak AK, editors. Bioactive Natural Products for Pharmaceutical Applications.1st ed. Switzerland: Springer Nature; 2021. p. 151-195.

[54] Alothman M, Bhat R, Karim A. (2009a). Effects of radiation processing on phytochemicals and antioxidants in plant produce. Trends in Food Science and Technology. 2009a;20:201-212.

[55] González-Aguilar GA, Villegas-Ochoa MA, Martínez-Téllez M, Gardea A, Ayala- Zavala JF. Improving antioxidant capacity of fresh-cut mangoes treated with UV-C. Journal of Food Science. 2007;72:S197–S202.

[56] Shao Y, Tang F, Xu F, Wang Y, Bao, J. Effects of γ-irradiation on phenolics

content, antioxidant activity and physicochemical properties of whole grain rice. Radiation Physics and Chemistry. 2013;85:227-233.

[57] Azad MOK, Kim WW, Park CH, Cho DH. Effect of artificial LED light and far infrared irradiation on phenolic compound, isoflavones and antioxidant capacity in soybean (*Glycine max* L.) sprout. Foods. 2018;7:1-10.

[58] Routray W, Orsat V, Lefsrud M. Effect of postharvest LED application on phenolic and antioxidant components of blueberry leaves. Chemical Engineering. 2018;2:1-9.

[59] Dou H, Niu G, Gu M (2019) Pre-harvest UV-B radiation and photosynthetic photon flux density interactively affect plant photosynthesis, growth, and secondary metabolites accumulation in basil (*Ocimum basilicum*) plants. Agronomy. 2019;9:434

[60] Pérez-Ambrocio A, Guerrero-Beltrán J, Aparicio-Fernández X, Ávila-Sosa R, Hernández-Carranza P, Cid-Pérez S, Ochoa-Velasco C. Effect of blue and ultraviolet-C light irradiation on bioactive compounds and antioxidant capacity of habanero pepper (*Capsicum chinense*) during refrigeration storage. Postharvest Biology and Technology. 2018;135:19-26.

[61] Ghadi FE, Ghara AR, Ghanbari T. (2015). Effect of gamma irradiation on the total phenolic content and free radical-scavenging activity of Iranian date palm Mazafati (*Phoenix dactylifera* L.). International Journal of Latest Research in Science and Technology. 2015;4:149-153.

[62] Alfaia A, Alfaia CM, Patarata L, Fernandes MJ, Fernandes MH, Elias M, Ribeiroa MH, Fraquezabet MJ. Binomial effects of high isostatic pressure and time on the microbiological, sensory characteristics and lipid composition stability of vacuum packed dry fermented sausages "chouriço". Innovative Food Science and Emerging Technologies. 2015;32:37-44.

[63] Briongos H, Illera AE, Sanz MT, Melgosa R, Beltrán S, Solaesa AG. (2016). Effect of high pressure carbon dioxide processing on pectin methyl esterase activity and other orange juice properties. LWT - Food Science and Technology. 2016;74: 411-419.

[64] Rock C, Guner S, Yang W, Gu L, Percival S, Salcido E. Enhanced antioxidant capacity of fresh blueberries by pulsed light treatment. Journal of Food Research. 2015;4:89-97

[65] Sevenich R, Mathys A. Continuous versus discontinuous ultra-high-pressure systems for food sterilization with focus on ultra-high-pressure homogenization and high-pressure thermal sterilization: A review. Comprehensive Review in Food Science and Food Safety. 2018;17:646-62.

[66] Gasperi F, Aprea E, Biasioli F, Carlin S, Endrizzi I, Pirretti G, Spilimbergo S. Effects of supercritical CO_2 and N_2O pasteurization on the quality of fresh apple juice. Food Chemistry. 2009;115:129-136.

[67] Del Pozo-Insfran D, Balaban MO, Talcott ST. Microbial stability, phytochemical retention, and organoleptic attributes of dense phase CO_2 processed muscadine grape juice. Journal of Agricultural and Food Chemistry. 2006;54:5468-5473.

[68] Ferrentino G, Plaza M, Ramirez-Rodrigues M, Ferrari G, Balaban M. Effects of dense phase carbon dioxide pasteurization on the physical and quality attributes of a red grapefruit juice. Journal of Food Science. 2009;74:E333-E341.

[69] Noci F, Riener J, Walkling-Ribeiro M, Cronin D, Morgan D, Lyng J.

Ultraviolet irradiation and pulsed electric fields (PEF) in a hurdle strategy for the preservation of fresh apple juice. Journal of Food Engineering. 2008;85:141-146.

[70] Guo M, Wu J, Xu Y, Xiao G, Zhang M, Chen Y. (2011). Effects on microbial inactivation and quality attributes in frozen lychee juice treated by supercritical carbon dioxide. European Food Research and Technology. 2011;232:803-811.

Chapter 9

Phenolic Compounds in the Built Environment

Elham H. Fini, Shakiba Ayat and Farideh Pahlavan

Abstract

This chapter examines source and application of phenolic compounds in the built environment as well as their environmental fate and treatment methods. We further describe the role of phenolic compounds in delaying aging and degradation of outdoor construction elements when exposed to intense solar radiation. In this chapter both plant-based and synthetic sources of phenolic compounds and their fate in the environment were examined. In addition, merits of select sources of phenolic compounds to resist ultraviolet radiation in composites as well as delaying degradation were studied. This chapter further provides insights pertaining to the underlying molecular interactions which afford phenol's role as an anti-aging additive for outdoor construction elements. This in turn provides a solution to promote bio-economy and enhance sustainability in the built environment.

Keywords: phenols, bio-mass, construction, aging, radiation, sustainability

1. Introduction to phenolic compounds

1.1 Structures and classification

There are more than ten thousand phenolic structures identified in nature, ranging from simple aromatic rings to complex polymerized compounds, making the phenolic compounds one of the main and largest groups of secondary metabolites of plants [1, 2].

The polyphenol structure (composed of several hydroxyl groups on aromatic rings) has been identified in higher plants in abundance, and to a lesser degree in edible plants [3]. Depending on the extent of their distribution in nature, phenolic compounds have been classified as being shortly distributed, widely distributed, or as polymers [4]. The types of phenolic compounds that are present or available in all plants are considered widely distributed. Examples include flavonoids and/or flavonoid derivatives, coumarins, and a wide range of phenolic acids including benzoic acid and cinnamic acid (**Figure 1**).

Phenolic compounds that are shortly or less widely distributed have limited presence in plants and include simple phenols, pyrocatechol, hydroquinone, and resorcinol (**Figure 2**).

The polymer class of phenolic compounds contains macromolecules such as tannin and lignin, illustrated in **Figure 3**.

Another method of classification is according to the size of a phenolic compound's carbon chain, dividing the compounds into 16 major classes: simple phenols

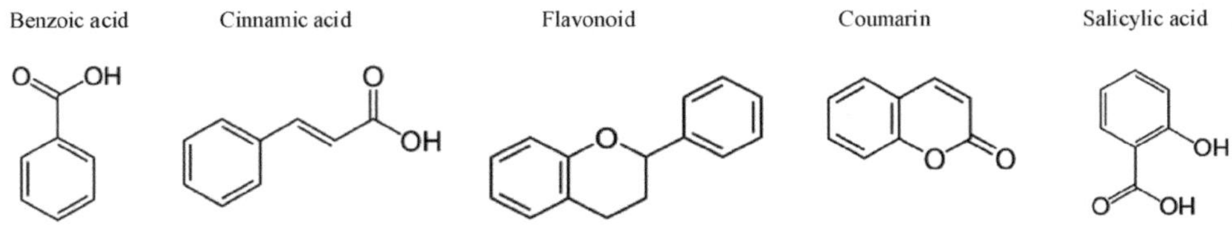

Figure 1.
Examples of widely distributed phenolics.

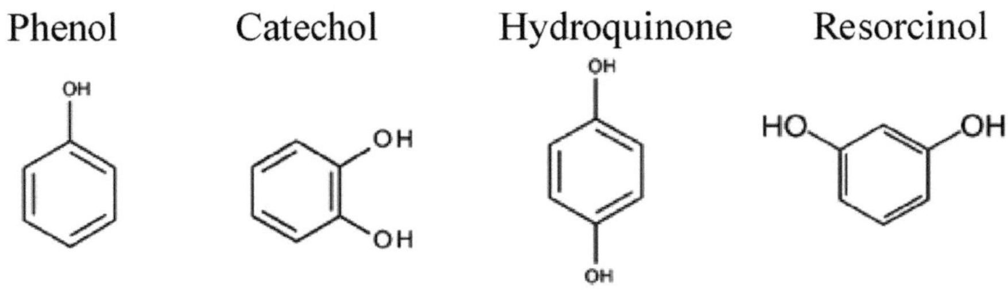

Figure 2.
Examples of shortly distributed phenolics.

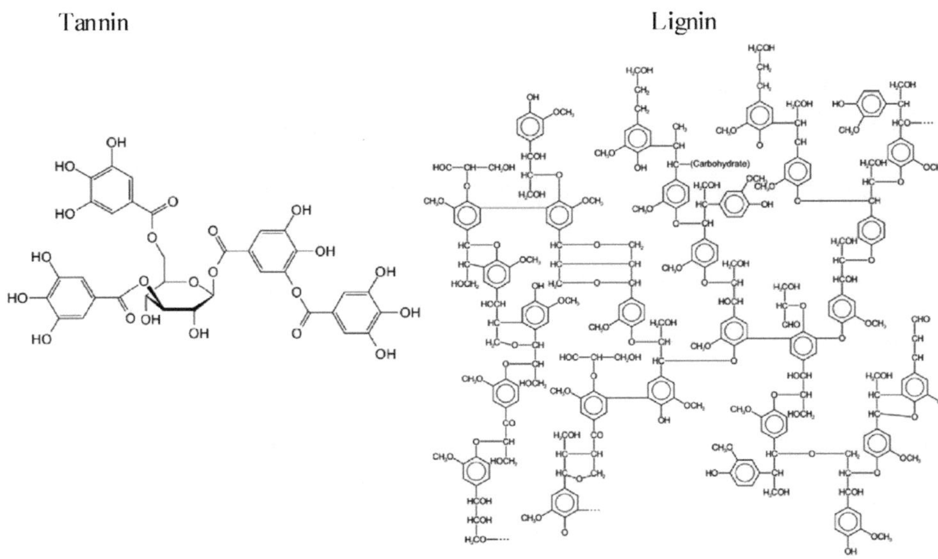

Figure 3.
Examples of polyphenolic compounds.

(C_6), benzoquinones (C_6), phenolic acids (C_6-C_1), acetophenones (C_6-C_2), phenylacetic acids (C_6-C_2), hydroxycinnamic acids (C_6-C_3), phenylpropenes (C_6-C_3), coumarins and isocoumarins (C_6-C_3), chromones (C_6-C_3), naphthoquinones (C_6-C_4), xanthones (C_6-C_1- C_6), stilbenes (C_6-C_2-C_6), anthraquinones (C_6-C_2- C_6), flavonoids (C_6-C_3-C_6), lignins ((C_6-C_3)$_n$), lignans and neolignans ((C_6-C_3)$_2$) [5].

A broader designation into flavonoids and non-flavonoids has traditionally been used; it was brought on based on the plethora of natural flavonoids and the diversity of C_6-C_3-C_6 structural offshoots [6]. The nonflavonoids group is classified according to the number of carbons that they have and comprises the following subgroups: phenolic acids, stilbenes, lignans, and others [5, 6]. **Figure 4** shows the main groups of plant phenolics.

1.2 Production and functions

The results of plants' photosynthesis can be classified as primary or secondary metabolites. Primary metabolites are usually described as substances that are essential chemical units of living plant cells. These fundamental substances are cellulose, hemicelluloses, polysaccharide, and lignin [7]. Plants synthesize a vast number of smaller molecules that are secondary metabolites. The secondary metabolites are formed by evolution to defend plants against harmful attacks by herbivores, pathogens, insects, and parasitic species [8].

Polyphenols, as secondary metabolites, are involved in functions related to reproduction, growth, defense, and pigmentation in plants, acting against pathogens, parasites, and predators. They also exhibit a great capacity to minimize the harmful effects of UV radiation, which may alter the regular metabolism in plants [2, 3]. Phenolic compounds are second only to cellulose in making up the bulk of organic matter, with phenolics (mainly lignin) accounting for about 40% of the organic carbon in the biosphere. The phenolic secondary metabolites are produced through the shikimic and malonic acid pathways, as shown in **Figure 5** (adapted from [9]).

Since the synthesis of phenols can proceed by different pathways, phenols are a diverse metabolic group, and their chemical diversity is matched by their varied roles in plants (**Table 1**). Phenols' roles include these actions: function in mechanical support; protect the plant from harmful ultraviolet solar radiation and excessive water loss; attract pollinators and seed dispersers; serve as signals that induce defensive reactions to biotic or abiotic stresses; suppress the growth of nearby

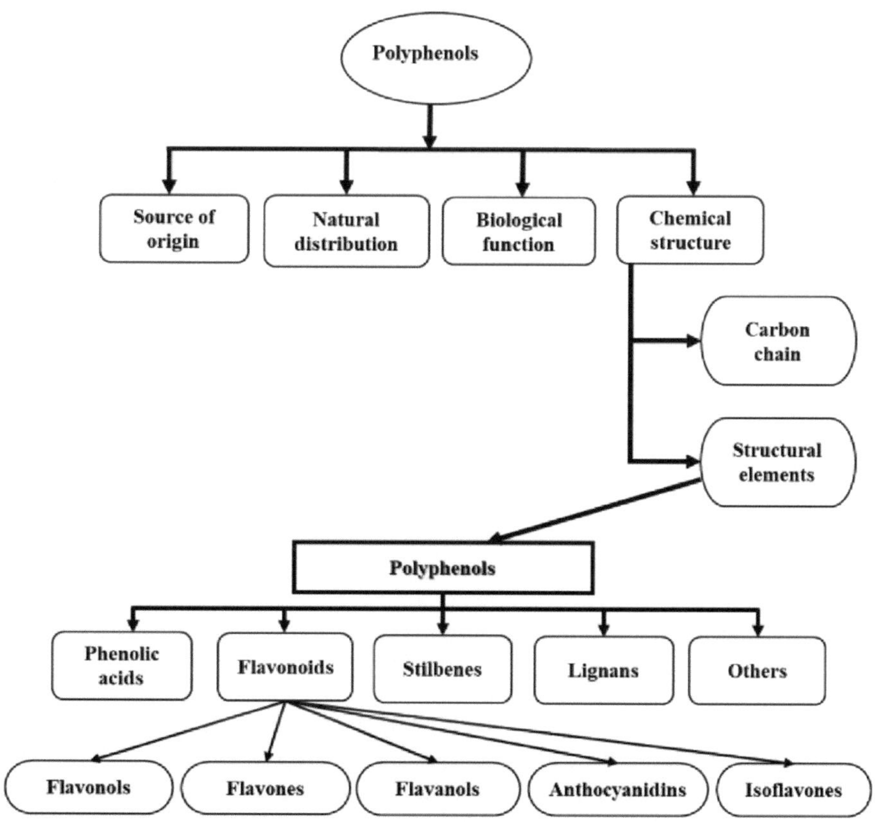

Figure 4.
Main groups of polyphenolic compounds based on plants' biological functions (adapted from [5]).

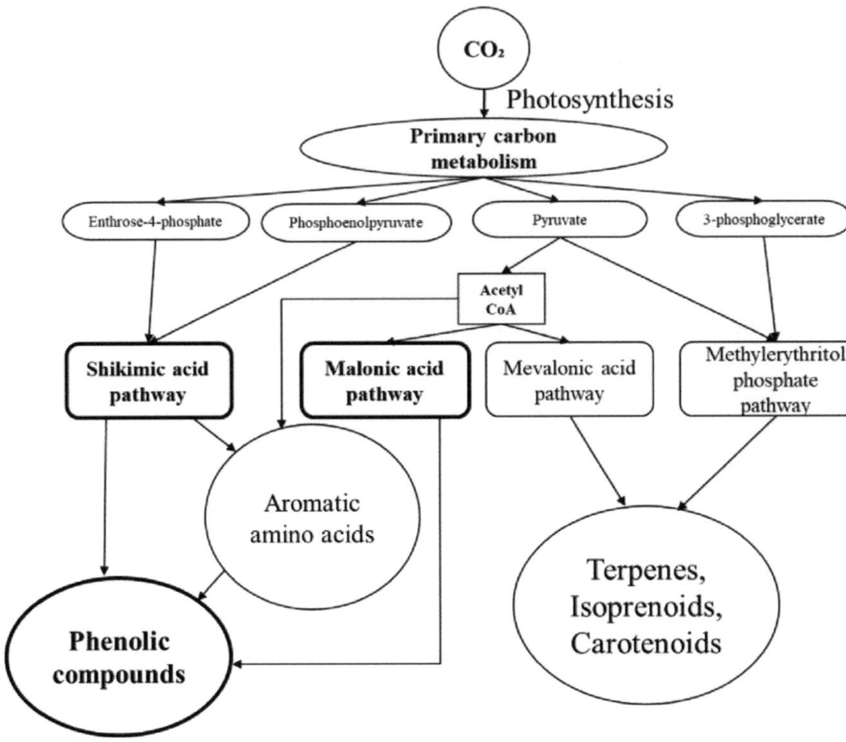

Figure 5.
Main plant pathways for production of phenolic compounds (adapted from [9]).

Compound	Biological functions
Coumarins	Antibiotics, discourages herbivores
Cutin	External barrier to water and gas diffusion in aerial parts
Flavonoids	Antimicrobial, signals, pigments, UV protection
Lignan	Antibiotics, discourages herbivores
Lignin	Strengthen cell walls
Suberin	External and internal barrier to water and gas diffusion in roots
Stilbenes	Antibiotics, fungicides
Tannins	Fungicides, discourages herbivores

Table 1.
The most prominent biological functions of phenolic groups.

competing plants (i.e., allelopathy); be attractive substances to accelerate pollination; provide coloring for camouflage and defense against herbivores; act as antibacterial and antifungal agents against pathogens; and provide protection against herbivores by repulsive taste or smell.

Phenolic acids constitute about one-third of the phenolic compounds in the human diet and are characterized by a remarkable antioxidant activity [10]. Phenolic acids can be divided into two groups: benzoic acids and their derivatives; and cinnamic acids and their derivatives. The benzoic acids are the simplest phenolic acids found in nature. Cinnamic acids are rarely found in their free form in plants and are generally in the form of esters.

1.3 Characteristics and reactions

Phenolics function as antioxidants in a number of ways. Phenolic hydroxyl groups are good hydrogen donors: hydrogen-donating antioxidants can react with reactive oxygen and reactive nitrogen species [11, 12] in a termination reaction, which breaks the cycle of generation of new radicals (reactions 1–5, adapted from [13] where, φ: phenolic antioxidant, •: free radical species, R-C-R: organic molecule).

$$e_{a}q^{-} + N_2O + H_2O \rightarrow HO\bullet + N_2 + HO^{-} \quad (1)$$

$$\varphi + HO\bullet \rightarrow \varphi\bullet + H_2O \quad (2)$$

$$\varphi + H\bullet \rightarrow H\varphi\bullet \quad (3)$$

$$R - C - R + HO\bullet \rightarrow R - C\bullet - R + H_2O \quad (4)$$

$$R - C\bullet - R + \varphi \rightleftharpoons [R - C - R \cdot \cdots \varphi]\bullet \quad (5)$$

$$[R - C - R \cdot \cdots \varphi]\bullet \rightarrow R - C - R + \varphi\bullet \quad (6)$$

The antioxidant molecule (φ) reacts with the initial reactive species and forms the antioxidant radical (φ•). The interaction between φ• and organic molecules produces an intermediate radical specie which has much greater chemical stability than the initial radical (reaction 5). The phenolics have the unique ability to produce stabilized free radicals due to delocalization of electrons between hydroxyl groups and the π-electrons of the benzene ring. These long-lasting radicals modify the oxidation processes and interrupt the free radical attack on other organic molecules (reaction 6) [14]. Similarly, the phenolic compounds can chelate metal ions and effectively stop the metal ions from producing free radicals. Another property which attributes to phenolic compounds antioxidant capability [10].

The phenolic compounds of plant origin act as antioxidants due to their redox properties, allowing them to act as reducing agents, hydrogen donors, free-radical quenchers, and metal chelators [15, 16]. In an organism, an oxidative process can be responsible for the generation of free radicals that attack the cells; the oxidative and nitrosative stress leads to serious diseases such as cancer, cardiovascular diseases, atherosclerosis, neurological disorders, hypertension, and diabetes mellitus [17, 18]. The principal function of antioxidants is to delay the oxidation of other molecules by inhibiting the initiation or propagation of oxidizing chain reactions by free radicals, consequently reducing oxidative damage [19]. Phenolic compounds can slow the oxidative degradation of lipids due to their antioxidant properties. Accordingly, the food industry is showing increasing interest in application of phenolic-rich plant materials, such as crude extracts and oils of herbs, fruits and spices, to improve the quality and nutritional value of foods [20].

Phenolic compounds have also been associated with other bioactivities important for maintaining good health, such as anti-inflammatory, antimicrobial, and anti-proliferative activities [21, 22]. In addition to the pharmacological interest in these compounds, their biological activities have also been explored in other industry sectors such as food [23, 24], cosmetics [22, 25], packaging, and textiles [18, 26, 27].

2. Sources of phenolic compounds

About 370 million years ago, nature first produced phenols in woody plants. With the development of phenylpropanoid metabolic pathways, lignin (a major constituent of wood) was synthesized; this was a key evolutionary advancement for land plants. Lignin made the cell walls stiffer and stronger, making plants to bear up better, grow to larger size, and develop greater variety of tissues. Furthermore, an extensive range of phenolic compounds are produced through phenylpropanoid metabolic reactions. These phenols are responsible for pigmentation (leaves, fruits, and flowers), defense mechanisms, and signaling in plants. Also, the Natural organic matter (NOM) is mainly generated from plant phenolics and has a major role in ecosystem [28].

All plants including lower forms such as ferns, algae, and lichens can produce phenolics compounds. No animal species (with the exception of a marine sponge) however, has been found to be able to produce phenolic compounds in their bodies. As bacteria are unable to degrade polyphenolic compound, phenolics have also been found in coal, shale oil, and petroleum. The water pumped up along with petroleum in oil-production fields often contains elevated levels of phenolics. Phenolics are also found in water pumped from coal beds to extract methane from the coal beds; much of this water's acute toxicity is due to these aromatic and phenolic compounds.

Most phenolic compounds in plants are condensed tannins (proanthocyanidins) and act as flavors and pigments. These compounds are the cause of intense colors of flowers, fruits, and leaves. Flavonoids (anthocyanins) are responsible for autumnal foliage colors in plants. Condensed tannins are also cause the characteristic astringent tastes of many fruits and wines. An abundance of flavorful phenolics produces the identifying tastes of many fruits, leaves, and roots such as grapes, teas, cranberries, grapefruit, coffee, cinnamon, ginger, and vanilla.

Two classes of naturally occurring phenolic-based materials – humic and fulvic substances – are especially noteworthy with respect to ecosystem-level function. Humic and fulvic substances are products of plants, derived from partially degraded lignin, tannins, and other phenolics. These phenolic acids are negatively charged and rapidly form complexes with metal ions. The complex of humic acid and bivalent cations is almost insoluble in water. Even trace concentrations of humic cmpounds (below 1–5 mg/l) can effectively trap and reduce the toxicities of metals such as Cu, Zn, and Cd. Humic substances can also bound to chlorinated pesticides and reduce their bio-availability to aquatic organisms since the pesticide–humic complex would not be able to pass through cellular membranes [28].

In addition to plants' natural debris, phenolic compounds are released or condensed through industrial endeavors. Anthropogenic sources of phenolic compounds are described below for the food, cosmetics, packaging, and textiles industries.

In the food industry (agroindustry), phenolic compounds are widely distributed through plants, including vegetables, fruits, legumes, herbs, and spices. Raw fruits and vegetables are good sources of polyphenols. Being seasonally produced, they are often industrially processed and stored. Consequently, significant amounts of by-products (peel, pulp, seeds, stones, stems) are manufactured that contain valuable bioactive compounds such as flavonols, flavanols, anthocyanins, and phenolic acids such as ferulic acid, vanillic acid, and caffeic acid. Cereals (maize, wheat, rice, barley, sorghum, oats, and rye) and their by-products (like bran) are rich in a variety of phytochemical compounds such as phenolic compounds, carotenoids, vitamin E, γ-oryzanols, dietary fibers, and β-glucans. The phenolic compounds in legumes (chickpeas, beans, lentils, and peas) and their by-products are mainly represented by tannins, phenolic acids, and flavonoids. Polyphenols are present in beverages such as coffee, tea, wine, and beer and also in the by-products created during their production (e.g., coffee silverskin, spent coffee grains, grape pomace, brewers' spent grain). The agro-industrial residues of grapes are mostly solid byproducts such as stalks, pomace, and the liquid filtrate. These residues are composed of water, proteins, lipids, carbohydrates, vitamins, minerals, and compounds with important biological properties such as fiber, vitamin C, and phenolic compounds such as tannins, phenolic acids, anthocyanins, and resveratrol [29]. Olive oil and its by-products (olive leaves, olive mill wastewater and pomace) contain secoiridoids, phenyl alcohols, flavonoids, lignans, and phenolic acids. Cocoa and cocoa-derived products include flavanols such as epicatechin, catechin, and procyanidins. Many herbs and spices (coriander, thyme, sage, and rosemary) and waste extracts obtained from essential-oil production are good sources phenolic acids. Additionally, the antioxidant and antimicrobial activity of phenolic compounds in plant extracts have been crucial to the application of these compounds as preservatives, thus increasing the shelf life of several foodstuffs [18].

The cosmetic industry has been exploring natural additives as alternatives to artificial ones [25, 30]. Phenolic compounds have shown great potential for use as bioactive ingredients in skincare and beauty products [22]. Due to the presence of chromophores in their structure, these compounds can absorb ultraviolet radiation and protect the skin. Protection from UV-light has been measured for some phenolics such as quercetin, resveratrol, and hydroxycinnamic acids; presenting a sun protection factor (SPF) of 7–30 [30]. Hydroxycinnamic acids, such as p-coumaric and protocatechuic acids, can increase stability of lotions (up to 6 months) while maintaining their anti-inflammatory and antimicrobial activity, working as multifunctional ingredients [25].

In the packaging industry, some phenolics have been used to improve packaging formulations with antioxidant and antimicrobial proprieties [27, 31], as well as adding interesting colors and boosting antioxidant activity.

In the textile industry, textile production and dyeing is an important source of diverse chemical pollutants, since high amounts of water are contaminated with a heavy load of chemicals. In addition, allergic reactions have been associated with synthetic dyes [26, 32]. Consequently, there has been increased interest in the use of phenolic compounds as natural dyes with high biodegradability [26, 32]. In addition to being less harmful to the environment, natural dyes obtained from oak (Quercus sp.) bark and from red, black and green tea extracts, showed UV protection when applied to Tussah silk and cotton, respectively [26, 32]. Antimicrobial activity against *E. coli* and *S. aureus* was also achieved with the use of natural dye from oak bark [18, 26].

In the pulp and paper industry, bark and knotwood are the most economically available wood residues; they are collected in stems and transported to sawmills and pulp mills, where bark is removed from the stems and knotwood is separated from wood chips. Many parts of cut trees such as leaves, branches, bark, roots, and stumps contain valuable phenolics but are considered waste material. Recently, more bark material is being used in energy, pulp, and paper production, and stumps have been collected for energy production due to restrictions on tree harvesting [33, 34].

3. Environmental fate and treatment of phenolic compounds

As previously stated, phenolic compounds are widely distributed through plants, and consequently, their dispensation in the environment, from either naturally discarded plant matter or decaying dead flora, is part of a balanced ecosystem. Plant phenolics have crucial impact on soil and fresh-water ecosystems mainly due to the large volume and variety of phenolic compounds in plants, the resistance of phenolics to degradation by bacteria; the low to average solubility of phenolic compounds in water; the conjugated system of C=C bonds, aromatic ring and hydroxyl group which translate into ease of absorbing certain wavelengths of light; and finally, their chemical energy content, available to potential decomposers.

However, many phenolic compounds are harmful to human health, causing necrosis, digestive problems, and liver and kidney damage in small doses and through chronic consumption of polluted water. Even in concentrations as low as 1 mg/l (1 ppm), phenolics can cause fish death in surface waters. At concentrations less than 1 mg/l, they are also toxic to other aquatic species and destroy the natural ecosystems. Phenolic compounds are phytotoxins, so their presence in soil would inhibit seed germination and the growth of local plants [35].

Phenolic compounds are present in the effluents of various industries such as oil refining, petrochemical production, pharmaceuticals production, resin manufacturing, and in pulp, paper, and wood products [36, 37]. Phenolic substances are widely used chemicals in several industries: as preservatives in the wood, lumber, and composites industries; as coke and coal gasifiers in coking operations; as antioxidants, flavorings, and other uses in the food industry; and in chemical production plants for the extraction and refining of minerals and metals, as well as the synthesis of organic chemicals, polymers, and plastics. Agricultural uses include pesticides, herbicides, and fungicides. Some of these phenolic substances are non-biodegradable and persistent in the environment and natural waters. Consequently, they will appear in treated water, as they usually cannot be treated through conventional methods and traditional treatment plants [36].

Nitrophenols and chlorinated phenols are usually labeled as priority pollutants since they are persistent and accumulative in nature [35, 38]. Phenol, cresols, and dimethyl-phenols have been considered as lesser environmental hazards because of their relative ease of biodegradation in activated sludge systems. However, as leachate from oil-shale waste, as well as in coal and coke leftover dumps, these short-chain phenols have become major pollutants of groundwaters [35, 39].

Phenolic common derivatives such as Bisphenol A (BPA), chlorophenols (CPs), and phenolic endocrine disrupting compounds (EDCs) are often listed at the top of lists of environmental pollutants. Phenolic compounds and their halogenated derivatives can also produce dioxin compounds [40]. Dioxins are notorious for their persistency in the environment and their high toxicity, so much so that they usually are at the top of "dirty dozen" lists.

Thus, the treatment of phenolic-rich agricultural and industrial wastes before discharge into the environment is a high priority. In the following sections, the environmental effects of phenolic compounds and their treatment techniques are discussed in more detail.

3.1 Phenolic compounds in air

Phenol is formed naturally in the atmosphere as a result of chemical reactions that occur in condensed water vapor that forms clouds. Natural background levels of phenolics in air are expected to be low, at about 1 ng/m [3] [38, 41]. Some phenolic compounds are typically emitted in ambient air by biomass burning, namely vanillin and acetosyringone [42]. Most phenolic compounds can be classified as volatile organic compounds (VOCs). VOCs are generally referred to as the highly reactive and/or toxic organics emitted from anthropogenic and natural sources, due to their high volatility in normal atmospheric conditions [43].

In industry, phenol is produced by extraction method from coal tar (tar being sourced from plants which contain high quantities of cumene). Phenol can be synthesized by a reaction between chlorobenzene and sodium hydroxide, or by oxidation of toluene, or by synthesis from benzene and propylene [41]. Phenol and its derivatives are used in several branches of industry: chemical production of alkylphenols, cresols, xylenols, phenolic resins, aniline and other compounds [44]; in oil and coal processing and metallurgy [45]. Phenol is also used in the production of pesticides, explosives, dyes, and textiles. Therefore, phenolic compounds are detected in industrial centers, especially near factories that incorporate wood in high concentrations. Phenol also enters the environment from vehicle exhaust and the use of disinfectants.

A large group of phenolic air pollutants are nitrophenols. The formation of nitrophenols happens under UV radiation from sunlight and in the atmosphere. The reaction of phenol, nitrite ions, and hydroxyl radicals leads to the formation of 2-nitrophenol and other nitrated compounds. Nitrophenols in the atmosphere are usually found in low concentrations (in the ng/dm [3] range). However, strong air pollution caused by industrial emissions leads to concentrations of nitrophenols up to 320 ng/dm [3, 41].

Nitrophenolic compounds mostly originate from anthropogenic emissions, such as automobile traffic, herbicide and insecticide use, coal combustion, and biomass burning [46–48]. Among these activities, biomass burning, traffic emissions, and coal combustion are regarded as the main sources, especially in urban areas [47–48]. Nitrophenols are a primary component of brown carbon (BrC), and the absorption properties of nitrated phenols in near-UV light can affect solar radiation and disturb atmospheric photochemistry, air quality, and regional climate [47, 49].

Incineration of phenolic waste contributes to phenolic pollutants in the air. Although the emissions from incinerators usually make up only a small proportion of air pollution, the secondary pollutants (heavy metals and polychlorinated dibenzo-p-dioxins and dibenzofurans) in the exhaust gases might pose greater health risks for the local populations [46, 50].

3.2 Phenolic compounds in water

The existence of phenolic compounds in water can be attributed to natural and anthropogenic activities. Natural sources of phenolic compounds in water include decomposition of dead plants and animals (which is called natural organic matter). Phenolic compounds are also synthesized by microorganisms and plants in the aquatic environment. Industrial, domestic, agricultural, and municipal activities account for the anthropogenic sources of phenolic pollutants in surface and ground waters [4, 51]. Phenolics are found to be one of the most common contaminants of wastewater streams from manufacturers of petrochemicals, polymeric resins, or pharmaceuticals, along with coal-conversion plants and chemical industries. Due to their high aqueous solubility and weak adsorption to soils, phenolic compounds are widespread water pollutants. They are characterized by low biodegradability, making them difficult to remove from the environment by naturally occurring processes. Consequently, they can be found in drinking water reservoirs and underground aquifers. These compounds tend to accumulate in nature and living tissues, causing severe health problems for many species and disturbing the ecosystems. These toxic and non-biodegradable organic compounds cannot be effectively removed from industrial wastewaters by common treatment technologies, as will be discussed in Section 3.4.

The industries with the highest phenolic concentrations in their effluent discharge can be categorized into two classes: industries processing natural ingredients such as agro-industries; and production plants manufacturing new products from synthesized chemicals.

Examples of agro-industries with the highest phenolic discharge in their wastewaters are olive-oil mills, vineyards, avocado-oil producers, soy processing plants, coffee and tea production, beer and liquor breweries, and fruit or fruit-juice processing plants. Section 2 contains more specifics on polyphenols in the food industry.

The chemical manufacturing sectors producing great volumes of phenolic waste are paper production, dye synthesis, and pesticide manufacturing, among others [52–55]. Two examples of chemical processing (pulp and paper milling) and manufacturing (polypropylene plastic production) are presented in more details here.

Pulp and paper production is one of the most freshwater-consuming industries in the world; consequently, it is one of the largest producers of wastewater. Hence, the toxic load of pulp and paper mills (PPMs) is extremely high, and they are considered a major source of pollution [56]. They have a high chemical load of about 700 different organic and inorganic compounds including phenols, sterols, dioxins, and furans. Not surprisingly, these wastewaters have been shown to have detrimental impacts on the environment such as endocrine disruption, oxidative stress, and genotoxicity [56].

Another industrial wastewater with a high phenolic load comes from the production of polypropylene. Polypropylene, a synthetic resin produced through the polymerization of propylene, is one of the important plastic resins; it is used in products that require toughness, flexibility, light weight, and heat resistance. Polypropylene is used in carpeting and upholstery, reusable containers, paper, adhesives, and electronics, among others; the annual worldwide production of polypropylene (PP) is estimated at several thousand tons [57]. Massive amounts of phenolic VOCs and substituted phenols are used in the production of polypropylene to improve the resin's thermo-oxidation properties. The presence of these substituted phenols in industrial wastewater and VOCs in the air has been shown to be in the hundreds of ppm range [57]. Such considerable quantities of toxic chemicals in production effluents can cause serious damage to the aquatic environment and the health of many species.

The technologies for removing phenols from industrial wastewater are classified as either conventional methods or advanced methods. Treatment methods and their advantages are discussed in more detail in Section 3.4.

3.3 Phenolic compounds in soil

Phenolic compounds are considered a major water pollutant group because of their high solubility in water. The existence of phenols in waste materials from industrial processes such as oil refineries, coking plants, wastewater treatment plants, petroleum-based processing, and phenol-resin-industry manufacturing plants has been well established [58]. Usually, the small amounts of phenolic compounds in soil come from natural sources. Since phenolic compounds are stored in the leaves, roots, and stems of plants, decomposition of dead leaves, roots, and plants transfers the phenolic compounds to the soil. Also, the root and leaf secretions of plants contain phenolic compounds, which are finally emitted into the soil either by the exudates or the degradation of the plants' material.

However, the presence of phenolics in municipal solid waste is an important source of pollution in soil. Disinfectants and cleaning products are the common hazardous compounds reported to occur in a highly organic and heterogeneous mix of household waste and are often deposited into a municipal landfill. Landfill leachate is composed of a complicated chemical mixture including these compounds: phthalates; phenolics; pesticides; aliphatic and aromatic hydrocarbons; fatty acids and carboxylic acids; volatile compounds such as benzene, toluene, ethylene, and xylene; polyaromatic hydrocarbons; and polychlorinated biphenyls [59, 60].

The degradation of phenolic compounds under anaerobic landfill conditions has been established in a few reports [60]. The removal of phenolic compounds under anaerobic conditions is achieved with de-chlorination. Anaerobic bacteria can convert all phenolic compounds to phenol which then, can be degraded to CH_4 and CO_2 under anaerobic conditions.

One of the major controlling parameters for the degradation of phenolic compounds is the redox condition. Oxygen strongly affects the natural degradation of these compounds, and a positive correlation between oxygen concentration and enhanced degradation has been established. Under aerobic conditions, the removal of mono- and di-chlorinated phenols occurs rapidly. However, aerobic degradation becomes less effective for more highly chlorinated compounds such as TeCP and PCP. Nitrophenols are swiftly transformed to amino groups under aerobic conditions first and then degraded to CH_4 and CO_2 under anaerobic conditions.

Land disposal of solid waste increases the risk of surface and groundwater contamination with landfill leachate. Therefore, attention to aerobic treatment of landfills not only increases the degradation of solid waste, but also promotes the decomposition of toxic compounds such as phenols produced during the degradation process. Increased degradation rates would reduce the transfer of pollutants to groundwater and is applicable to both active landfills and in the reclamation of old landfills [59].

3.4 Treatment methods

As previously mentioned, phenolic compounds enter aquatic environments from natural, industrial, domestic, and agricultural activities. Their presence may be due to the degradation or decomposition of natural organic matter (NOM) present in the water, a natural part of the carbon cycle. However, the disposal of industrial and domestic wastes into water bodies and through runoff from agricultural lands requires awareness, action, and remediation.

Standard treatment methods are unable to reduce the concentration and toxicity of phenol-rich wastewaters, mainly due to the high solubility of phenolic compounds and their toxicity to bacteria in activated sludge and anaerobic digesters; therefore, additional and alternate methods have been studied and implemented. The physical treatment processes include decantation (liquid-phase and solid-phase extractions), filtration (using reverse osmosis with micro, ultra, and nano membranes) and adsorption (activated carbon and ion exchange). The most-applied chemical methods are incineration, electrochemical methods, and advanced oxidation processes (AOP) [61]. Advanced technologies for removal of phenols include electrochemical oxidation, photo-oxidation, ozonation, UV/H_2O_2, Fenton reaction, membrane processes, and enzymatic treatment [36].

The main treatment methods of industrial effluents with biological processes are subdivided into microbial (aerobic and anaerobic) and enzymatic [61]. The microbial method involves the deployment of bacteria, yeast, and fungi in breaking down the phenolics into harmless products such as carbon dioxide and water. It has the advantage of comparatively low operational costs. Generally, the aerobic process is used for the degradation of phenolics with minimal halogenic substituents. On the other hand, the anaerobic process can efficiently reduce chlorinated phenolic compounds. The anaerobic system produces methane in addition to carbon dioxide and water. A major advantage of the anaerobic system of degradation is the absence of aeration cost, recovery of methane, and minimum excess biomass generation. Enzymes can also be used effectively to selectively eliminate pollutants in water, since they catalyze specific reactions under modest temperature, pH, and ionic strengths. Enzymatic reactions occur at much faster rates compared to other types of reactions and enzymatic systems can work under conditions unfavorable or toxic to bacteria. This method receives a high level of consideration due to its high pollutant-removal efficiency, its operation in a wide range of temperature and pressure, and the formation of harmless end products [4].

Considering the high value of phenolic compounds found in abundance in industrial waste discharge, however, it seems prudent to reclaim and extract the phenolic portion from effluent discharge. In addition to avoiding the extra cost and energy in treatment plants to process the phenolic compounds, the high levels of phenolic wastes could be viewed as resources. Therefore, reclaiming phenolic compounds offers a more attractive and sustainable solution to lowering the phytotoxicity of industrial waste. Wastes are residues of high organic load that are derived during raw materials processing and result in liquid or solid form. The fact that these substances are removed from the production process as undesirable materials defines them as wastes [62]. Since they contain high concentrations of valuable phenolic compounds, various recovery methods are being developed. The more popular ones are solid–liquid extractions, soxhlet extractions, pressurized fluid extractions and supercritical fluid extractions, ultrasound-assisted extractions, microwave-assisted extractions, pulsed electric field extractions, and enzyme-assisted extractions [63].

4. Industrial applications of phenolic compounds

Phenol derivatives have been attracting interest for decades as essential ingredients in various end-use industries due to their unique properties such as durability, chemical resistance, adhesion strength, plasticizing effect, and clear coating. These characteristics increase the sustainability of the phenol-derivatives market due to their applications in a broad range of industries such as electronics, paints and coatings, adhesives, household appliances, composites, textiles and packaging, pharmaceutical drugs, wood products, agricultural products, and automotive products. Popular phenol derivatives include bisphenol A (BPA), chloro-phenols, alkylphenols, phenylalanine, caprolactam, and salicylic acid. The most common applications of phenolic compounds are in the paint and coating industry, wood processing, and bituminous construction; these uses are briefly reviewed below.

4.1 Paint and coating industry

Bisphenol A is used to produce epoxy resins and polycarbonate plastics. The most common epoxy resins (ERs) are produced from the ring-opening reaction of bisphenol and epichlorohydrin, followed by pre-polymerization of the produced diglycidyl ether through a reaction with bisphenol A. Epoxy resin is a type of thermosetting material that shows fascinating characteristics such as excellent adhesion properties, thermal stability, high heat and chemical resistance, and good mechanical strength [64–67]. These properties make epoxy resin suitable for different applications including coating [68, 69]. The significant adhesion strength of epoxy resin with various substrates, especially metal surfaces, is mainly due to high surface functional groups. Compounds with high molecular size provide adequate surface coverage in their role as inhibitors resulting in corrosion mitigation [70]. The adsorption of epoxy resin macromolecules at the interface of a metallic surface and the environment offers good surface covering and forms an oxygen-rich layer as a protective film against the aggressive atmosphere reaching the surface, leading to excellent corrosion resistance. Epoxy resin macromolecules can be effectively applied as anti-corrosive coating formulations for different metals and alloys in all kinds of electrolytic media due to their hydrophilic groups [71, 72]. Several literature studies have reported on the effectiveness of bisphenol-A-based epoxy resin to inhibit corrosion of E24 carbon steel in the acidic electrolyte and aluminum alloy in NaCl solution [73].

4.2 Wood processing

Phenolic compounds are the primary material used in the production of phenolic resins. Phenolic resins are the first class of synthetic polymers synthesized by the reaction between phenolic compounds and formaldehyde under acidic or basic conditions. Based on the formaldehyde/phenol ratio and the PH of the medium, the prepared resins are divided in two types: thermosetting phenol resins and thermoplastic phenol resins [74]. Phenol-formaldehyde resin (PF) is a thermosetting phenolic resin, known as resol, that is synthesized by electrophilic attack of the excess of formaldehyde to the aromatic ring of phenols under a basic condition. Thermosetting resins usually have strong mechanical properties, flame-retardant behavior, environmental resistance, and high bonding/adhesive strength, making them one of the promising adhesives used widely in wood composites [75, 76]. Phenol-formaldehyde resins are able to infiltrate the wood cell walls and improve their hardness [77–79]. Phenol-formaldehyde resins are the preferred thermoset adhesive for exterior wood composites such as manufactured plywood, oriented strand board (OSB) panels, laminated veneer lumber (LVL), medium-density fiberboard (MDF), and other structural wood products [80, 81]. Plywood is another popular wood-panel product manufactured as three layers of wood (veneers) assembled by an adhesive binder (such as phenol-formaldehyde for exterior-use plywood, or urea-formaldehyde for interior-use plywood) and then brought under heat and pressure [82–84]. The advantages of plywood over natural wood are dimensional stability, resistance to splitting, and decorative value, making this product suitable for exterior and interior construction.

4.3 Bituminous construction

Bitumen, or asphalt binder, is the adhesive material that binds mineral aggregates together in an asphalt mixture. Bitumen is a waterproof and highly viscous material that is produced through vacuum distillation of crude oil. Because of the natural organic source of bitumen, oxidative aging is an inevitable phenomenon when a bitumen mixture is exposed to atmospheric oxygen. The major consequences of irreversible oxidative aging are the hardening of asphalt and the consequent pavement embrittlement, leading to deterioration of the asphalt's rheological properties and performance [85, 86]. During oxidation, the introduction of free radicals is believed to form new polar functionalities such as a carbonyl (C=O) group and also to break hydrocarbon side chains, leading to a reduction in aliphatic content in the bitumen [87, 88]. These structural changes in the bitumen fragments lead to further molecular agglomerations and unfavorable hardening of the bitumen mixture.

One of the counteractions to the aging of bitumen is adding "antioxidant" modifiers. Antioxidants are aimed at delaying the aging process and improving the aging resistance of bitumen by scavenging the free radicals generated in the process of oxidation [89–91]. The antioxidant mechanism of phenolic compounds through neutralizing free radicals and breaking the oxidation chain reactions has been previously discussed. Another group of compounds that function as free-radical scavengers and antioxidants for base bitumens and polymer-modified bitumens are hindered phenols [89, 91]. The antioxidation effectiveness of lignins and some of their derivatives in bitumen have already been studied [92–94].

To improve the performance properties and extend the service life of bitumen in the construction industry, it is necessary to use bitumen modifiers. Modifiers consist of fragments that are compatible with the bituminous environment and its processing temperatures. Thermosetting plastics, namely epoxy resins and phenolic resins, are important additives in bitumen modification, showing excellent adhesive ability, fatigue performance, and resistance to deformation. Blending phenol-formaldehyde resins (known as bakelite) or phenol-cresol-formaldehyde resins with bitumen causes significant improvements in rheological properties including the resistance to cracking and rutting, softening point, viscosity, and stability, reducing distresses during the bitumen's service life [95–98]. Besides the antioxidant nature of lignin, this polyphenol is also added as a modifier and renewable alternative into bitumen binder to improve the high-temperature and low-temperature performance and the resistance to rutting and cracking [99]. Adding bio-oils with high concentrations of phenolic compounds to bitumen binder increases resistance to ultraviolet exposure and decreases propensity to aging [100]. Phenol-rich bio-oils are effective at rejuvenating and restoring the properties of aged bitumen [101]. The thermomechanical properties of sulfur-extended bitumen can be tuned by introducing phenolic compounds. Phenolic compounds can activate the sulfur interactions within bitumen, so that the effect of sulfur can be more remarkable in the bitumen matrix [102].

Phenolic resins are used as curing agents in sulfur-containing synthetic rubbers, namely nitrile-butadiene rubber (NBR) and styrene-butadiene rubber (SBR), to increase the crosslinking density (vulcanization) and rigidity of the molecular network of rubber [103, 104]. Some study results show that the phenolic resins in the bio-oils activate rubber particles through their adsorption to the rubber surface. Their curative effects improve the rubber-asphalt interactions and reduce the segregation between rubber and bitumen that commonly occurs in rubberized asphalt binder [105].

Author details

Elham H. Fini*, Shakiba Ayat and Farideh Pahlavan
School of Sustainable Engineering and the Built Environment, Arizona State University, Arizona, United States

*Address all correspondence to: efini@asu.edu

References

[1] Carbonell-Capella, J. M., Buniowska, M., Barba, F. J., Esteve, J., & Frigola, A., Analytical Methods for Determining Bioavailability and Bioaccessibility of Bioactive Compounds from Fruits and Vegetables: A Review. Comprehensive Reviews in Food Science and Food Safety, 2014, 13, 155–171.

[2] Zuluaga-Domínguez,C. M., Nieto-Veloza, A., Classification, Extraction, Detection and Preservation of Bioactive Compounds: A Global Overview , Bioactive Compounds, Nova science publishers, 2017, 978-1-53612-418-7.

[3] Manach, C., Scalbert, A., Morand, C., Remesy, C., Jimenez, L., Polyphenols: food sources and bioavailability. American Journal of Clinical Nutrition, 2004. 79, 727–747.

[4] Anku, W. W., Mamo, M. A., Govender, P. P., Phenolic Compounds in Water: Sources, Reactivity, Toxicity and Treatment Method. Phenolic Compounds -Natural Sources, Importance and Applications. 2017.

[5] Belščak-Cvitanović, A., Durgo, K., Huđek, A., Bačun-Družina, V., Komes, D., Overview of polyphenols and their properties. Polyphenols: Properties, Recovery, and Applications, 2018, Elsevier Inc.

[6] Andr´es-Lacueva, C., Medina-Remon, A., Llorach, R., Urpi-Sarda, M., Khan, N., Chiva-Blanch, G., Zamora-Ros, R., Maria Rotches-Ribalta, M., Lamuela-Ravent´os, R. M., Phenolic Compounds: Chemistry and Occurrence in Fruits and Vegetables. 2010 Blackwell Publishing. ISBN: 978-0-813-80320-3.

[7] Metsa¨muuronen, Sari, Sire´n, H., Bioactive phenolic compounds, metabolism and properties: a review on valuable chemical compounds in Scots pine and Norway spruce. Phytochem Rev, 2019, 18:623–664.

[8] Wink, M., Evolution of secondary metabolites from an ecological and molecular phylogenetic perspective. PhytChem, 2003, 64:3–19.

[9] Saltveit, M. E., Synthesis and Metabolism of Phenolic Compounds" Fruit and Vegetable Phytochemicals: Chemistry, Nutritional Value, and Stability. Eds: de la Rosa, L. A., Alvarez-Parrilla, E., González-Aguilar, G. A., 2010, Blackwell Publishing. ISBN: 978-0-813-80320-3

[10] Yang, C.S., Landau, J.M., Huang, M. T., Newmark, H.L., Inhibition of carcinogenesis by dietary polyphenolic compounds. Annual Review of Nutrition, 2001, 21, 381–406

[11] Pereira, D. M., Valentão, P., Pereira, J. A., Andrade, P. B., Phenolics: From Chemistry to Biology. Molecules 2009, 14, 2202-2211.

[12] Heim, K.E.; Tagliaferro, A.R.; Bobilya, D.J., Flavonoid antioxidants: chemistry, metabolism and structure-activity relationships. J. Nutrit. Biochem. 2002, 13, 572-584.

[13] Nauser, T., Gebicki, J. M., Fast reaction of carbon free radicals with flavonoids and other aromatic compounds", Archives of Biochemistry and Biophysics 674, 2019, 108107.

[14] Parr, A.J.; Bolwell, J.P., Phenols in the plant and in man. The potential for possible nutritional enhancement of the diet by modifying the phenols content or profile. J. Sci. Food Agric. 2002, 80, 985-1012.

[15] Frankel, E. N.; Meyer, A. S., The problems of using one-dimensional methods to evaluate multifunctional food and biological antioxidants review. J. Sci. Food Agric. 2000, 80, 1925–1941.

[16] Ganh~ao, R.; Estevez, M.; Kylli, P.; Heinonen, M.; Morcuende, D.,

Characterization of selected wild Mediterranean fruits and comparative efficacy as inhibitors of oxidative reactions in emulsified raw pork burger patties. J. Agric. Food Chem. 2010, 58, 8854–8861.

[17] Caleja, C., Ribeiro, A., Barreiro, A., Ferreira, M. F., Ferreira, I. C. F. R., Phenolic compounds as nutraceuticals or functional food ingredients. Curr. Pharm. Des., 2017, 23, 2787.

[18] Albuquerque, B.R., Heleno, S. A., Oliveira, M.B.P.P., Barros, L., Ferreira, I.C.F.R., Phenolic compounds: current industrial applications, limitations and future challenges. Food Funct., 2020.

[19] Rodríguez-Carpena, J.-G., Morcuende, D., Andrade, M.-J., Kylli, P., Estevez, M., Avocado (*Persea americana* Mill.) Phenolics, In Vitro Antioxidant and Antimicrobial Activities, and Inhibition of Lipid and Protein Oxidation in Porcine Patties, J. Agric. Food Chem. 2011, 59, 625–5635.

[20] Bastida, S.; Sanchez-Muniz, F. J.; Olivero, R.; Perez-Olleros, L.; Ruiz-Roso, B.; Jimenez-Colmenero, F., Antioxidant activity of carob fruit extracts in cooked pork meat systems during chilled and frozen storage. Food Chem. 2009,116, 748–754.

[21] Durazzo, A., Caiazzo, E., Lucarini, M., Cicala, C., Izzo, A. A., Novellino, E., Santini, A., Polyphenols: a concise overview on the chemistry, occurrence, and human health, Phytother. Res., 2019, 33, 2221–2243

[22] Soto, M. L., Falqué, E., Domínguez, H., Relevance of natural phenolics from grape 519and derivative products in the formulation of cosmetics, Cosmetics, 2015, 259–276.

[23] Carocho, M., Morales, P., Ferreira, I. C. F. R., Natural food additives: Quo vadis?, Trends in Food Sci. Technol., 2015, 45, 284–295.

[24] Mark, R., Lyu, X., Lee, J. J. L., Parra-Saldívar, R., Chen, W.N., Sustainable production of natural phenolics for functional food applications, J. Funct. Foods, 2019, 52857, 233–254.

[25] Taofiq, O., Heleno, S.A., Calhelha, R.C., Fernandes, I.P., Alves, M.J., Barros, L., González-Paramás, A.M., Ferreira, I. C. F. R., Barreiro, M.F., Phenolic acids, cinnamic acid, and ergosterol as cosmeceutical ingredients: Stabilization by microencapsulation to ensure sustained bioactivity, Microchem J., 2019, 147, 469–477.

[26] Jia, Y., Liu, B., Cheng, D., Li, J., Huang, F., Lu, Y., Dyeing characteristics and functionability of tussah silk fabric with oak bark extract, Text. Res. J., 2017, 87, 5351806–1817.

[27] Zeng, P., Chen, Z.X., Qin, Y.-R., Zhang, Y.-H., Wang, X.-P., Wang, J.-Y., Ning, Z.-X., Ruan, Q. J., Zhang, Y.-S., Preparation and characterization of a novel colorimetric indicator film based on gelatin/polyvinyl alcohol incorporating mulberry anthocyanin extracts for monitoring fish freshness, Food Res. Int., 2019, 126, 108604.

[28] Stewart, A. J., Stewart, R. F., Phenols. In reference module in earth systems and environmental science, Encyclopedia of Ecology,Ed by S.E. Jorgensen & B. Fath, 2008, p. 2682-2689.

[29] Sousa, E.C., Uchoa-Thomaz, A.M. A., Carioca, J.O.B., de Morais, S.M., Alessandro de Lima, A., Martins, C.G., Alexandrino, C.D., Ferreira, P.A.T., Rodrigues, A.L.M., Rodrigues, S.P., do Nascimento Silva, J., Rodrigues, L.L., Chemical composition and bioactive compounds of grape pomace (*Vitis vinifera* L.), Benitaka variety, grown in the semiarid region of Northeast Brazil. Food Science and Technology, Campinas, 34(1): 135-142, Jan.-Mar. 2014.

[30] Cherubim, D. J., Martins, C.V., Fariña, L., Lucca, R.A., Polyphenols as natural antioxidants in cosmetics applications, J. Cosmet. Dermatol., 2020, 19, 33–37.

[31] Wang, S., Xia, P., Wang, S., Liang, J., Sun, Y., Yue, P., Gao, X., Packaging films formulated with gelatin and anthocyanins nanocomplexes: physical properties, antioxidant activity and its application for olive oil protection, Food Hydrocoll., 2019, 96, 617–624.

[32] Bonet-Aracil, M., Díaz-García, Á. P., Bou-Belda, E., Sebastiá, N., Montoro, A., Rodrigo, R., UV protection from cotton fabrics dyed with different tea extracts. Dyes Pigm., 2016, 134, 448–452.

[33] Arisandi, R., Marsoem, S.N., Lukmandaru, G., The Contents of Phenolics and Cell Wall Component of *Eucalyptus pellita* F. Mull Stemwood and Bark. Wood Research, 64 (3): 2019, 411-422

[34] Aspé, E., Fernández, K., Comparison of Phenolic Extracts Obtained of *Pinus radiata* Bark from Pulp and Paper Industry and Sawmill. Maderas. Ciencia y tecnología 13(3): 243-252, 2011.

[35] Kahru, A., Maloverjan, A., Sillakz, H., Pollumaa, L., The Toxicity and Fate of Phenolic Pollutants in the Contaminated Soils Associated with the Oil Shale Industry. Environ Sci & Poilut Res 9 Special Issue 1, 2002, 27 – 33.

[36] Cordova Villegas, L.G., Mashhadi, N., Chen, M., Mukherjee, D., Taylor, K. E., Biswas, N., A Short Review of Techniques for Phenol Removal from Wastewater. Curr Pollution Rep (2016) 2:157–167.

[37] Sun X., Wang, C., Li, Y., Wang, W., We, J., Treatment of phenolic wastewater by combined UF and NF/RO processes. Desalination. 2015; 355:68–74.

[38] Boyd, E.M., Killham, K., Meharg, A. A., Toxicity of mono-, di- and tri-chlorophenols to lux marked terrestrial bacteria, Burkholderia species Rasc c2 and *Pseudomonas fluorescens*. Chemosphere 43, 2001, 157-166.

[39] Licha, T., Herfort, M., Sauter, M., Do phenol plumes exist? In: Seller KP, Sohlich S (Eds): New Approaches Characterizing Ground- water Flow. Proc. Of the XXXI International Association of Hydrogeologists Congress. Munich. Balkema 2001.

[40] Ahmed, O.H., Altarawneh, M., Al-Harahshehd, M., Jianga, Z.-T., Dlugogorski, B.Z., Formation of phenoxy-type Environmental Persistent Free Radicals (EPFRs) from dissociative adsorption of phenol on Cu/Fe and their partial oxides. Chemosphere 240, 2020,124921.

[41] Michałowicz, J., Duda, W., Phenols —sources and toxicity. Polish Journal of Environmental Studies, 2007,16(3), 347–362.

[42] Vione, D., Albinetb, A., Barsottia, F., Mekicc, M., Jiangc, B., Minero, C., Marcello Brigantee, A., Sasho Gligorovsk, S., Formation of substances with humic-like fluorescence properties, upon photoinduced oligomerization of typical phenolic compounds emitted by biomass burning. Atmospheric Environment 206, 2019, 197–207.

[43] Tsai,W.-T., Current Status of Air Toxics Management and Its Strategies for Controlling Emissions in Taiwan. Toxics, 2016, Volume 4, Issue 2, 10.3390/toxics4020008.

[44] El-Naas, M.H., Mousa, H.A., Gamal, M.E., Microbial Degradation of Chlorophenols. In: Singh S. (eds) Microbe-Induced Degradation of

Pesticides. Environmental Science and Engineering. 2017, Springer, Cham.

[45] Chlebek, D., Hupert-Kocurek, K., Endophytic Bacteria in the Phytodegradation of Persistent Organic Pollutants. Chemistry 2019.

[46] Yang, Y., Li, X., Shen, R., Liu, Z., Ji, D., Seasonal variation and sources of derivatized phenols in atmospheric fine particulate matter in north china plain. Journal of Environmental Sciences, v89, 2020, p 136-144.

[47] Li, M., Wang, X., Lua, C., Lia, R., Zhanga, J., Donga, S., Yanga, L., Xuea, L., Chend, J., Wang, W., Nitrated phenols and the phenolic precursors in the atmosphere in urban Jinan, China", Science of the Total Environment, 714, 2020, 136760.

[48] Wang, X., Gu, R., Wang, L., Xu, W., Zhang, Y., Chen, B., Li, W., Xue, L., Chen, J., Wang, W., Emissions of fine particulate nitrated phenols from the burning of five common types of biomass." Environ. Pollut. 2017, 230, 405–412.

[49] Asmi, E., Kondratyev, V., Brus, D., Laurila, T., Lihavainen, H., Backman, J., Vakkari, V., Aurela,M., Hatakka, J., Viisanen, Y., Uttal, T., Ivakhov, V., Makshtas, A., Aerosol size distribution seasonal characteristics measured in Tiksi, Russian Arctic. Atmos. Chem. Phys. 2016,16, 1271–1287.

[50] Wang, N., Hu, X., Lu, S., Ma, S., Kang, L., Liao, S., Yu, Y., 2019. Interrelationship of anthropo-genic activity and parabens in fish from Taihu Lake during 2009-2017. Environ.Pollut. 252, 1002–1009

[51] Juretic, D., Kusic, H., Dionysiou, D. D., Bozic, A.L., Environmental aspects of photooxidative treatment of phenolic compounds, Journalof Hazardous Materials, 2013, 08, 60.

[52] Guo, P.-C., Qiua, H.-B., Yanga, C.-W., Zhanga, X., Shaoa, X.-Y, Laia, Y.-L., Sheng, G.-P., Highly efficient removal and detoxification of phenolic compounds using persulfate activated by MnOx@OMC: Synergistic mechanism and kinetic analysis. Journal of Hazardous Materials 402, 2021, 123846.

[53] Caetano, M., Valderrama, C., Farran, A., Cortina, J.L., Phenol removal from aqueous solution by adsorption and ion exchange mechanisms onto polymeric resins. J. Colloid Interface Sci. 2009, 338, 402–409.

[54] Liu, C., Min, Y., Zhang, A.-Y., Si, Y., Chen, J.-J., Yu, H.-Q., Electrochemical treatment of phenol-containing wastewater by facet-tailored TiO2: Efficiency, characteristics and mechanisms. Water Res. 2019,165, 114980.

[55] Tian, H., Xu, X., Qu, J., Li, H., Hu, Y., Huang, L., He, W., Li, B., Biodegradation of phenolic compounds in high saline wastewater by biofilms adhering on aerated membranes. J. Hazard. Mater. 2020, 392, 122463.

[56] Carvalho Neves, L., de Souzaa, J.B., de Souza Vidala, C.M., Herbertb, L.T., de Souzaa, K.V., Martinsa, K.G., Young, B.J., Phytotoxicity indexes and removal of color, COD, phenols and ISA from pulp and paper mill wastewater post-treated by UV/H2O2 and photo-Fenton. Ecotoxicology and Environmental Safety 202, 2020, 110939.

[57] Hernandez-Fernandez, J., Lopez-Martinez, J., Barcelo, D., Quantification and elimination of substituted synthetic phenols and volatile organic compounds in the wastewater treatment plant during the production of industrial scale polypropylene. Chemosphere 263, 2021, 128027.

[58] Liu, Y.J., Kuschk, P., Zhang, A.N., Wang, XC., Characterization of phenol

degradation by Acinetobacter sp. XA05 and Sphingomonassp. Chemistry and Ecology 2009, 25(2): 107–117.

[59] Yazıcı, R., Sekman, E., Top, S., Varank, G., Bilgili, M.S., Degradation of phenolic compounds in aerobic and anaerobic landfills: a pilot scale study. Waste Management & Research, 2011.

[60] Yiping, X., Yiqi, Z., Donghong, W., Shaohua, C., Junxin, L., Zijian, W., Occurrence and removal of organic micropollutants in the treatment of landfill leachate by combined anaerobic-membrane bio-reactor technology. Journal of Environmental Sciences 2008, 20: 1281–1287.

[61] Brandão, Y.B., de Oliveira, J.G.C., Benachour, M., Phenolic Wastewaters: Definition, Sources and Treatment Processes, Phenolic Compounds - Natural Sources, Importance and Applications, 2017.

[62] Galanakis, C.M., 2012. Recovery of high added-value components from food wastes: conventional, emerging technologies and commercialized applications. Trends Food Sci. Technol. 26, 68–87.

[63] Ruiz-Ruiz, J.C., Aldana, G. del C. E., Cruz, A.I.C., Segura-Campos, M.R., Antioxidant Activity of Polyphenols Extracted from Hop Used in Craft Beer, Biotechnological Progress and Beverage Consumption, 2020 Elsevier Inc.

[64] Qiu, S.; Zhou, Y.; Zhou, X.; Zhang, T.; Wang, C.; Yuen, R. K.; Hu, W.; Hu, Y., Air-stable Polyphosphazene-functionalized few-layer black phosphorene for flame Retardancy of epoxy resins. *Small* **2019,** *15* (10), 1805175.

[65] Xu, Y.-J.; Wang, J.; Tan, Y.; Qi, M.; Chen, L.; Wang, Y.-Z., A novel and feasible approach for one-pack flame-retardant epoxy resin with long pot life and fast curing. *Chemical Engineering Journal* **2018,** *337*, 30-39.

[66] Buonocore, G. G.; Schiavo, L.; Attianese, I.; Borriello, A., Hyperbranched polymers as modifiers of epoxy adhesives. *Composites Part B: Engineering* **2013,** *53*, 187-192.

[67] Guo, W.; Yu, B.; Yuan, Y.; Song, L.; Hu, Y., In situ preparation of reduced graphene oxide/DOPO-based phosphonamidate hybrids towards high-performance epoxy nanocomposites. *Composites Part B: Engineering* **2017,** *123*, 154-164.

[68] Zhou, H.; Chen, R.; Liu, Q.; Liu, J.; Yu, J.; Wang, C.; Zhang, M.; Liu, P.; Wang, J., Fabrication of ZnO/epoxy resin superhydrophobic coating on AZ31 magnesium alloy. *Chemical Engineering Journal* **2019,** *368*, 261-272.

[69] Wu, Y.; Song, N.; Wang, W.; Zhao, Y., Synthesis of graphene/epoxy resin composite via 1, 8-diaminooctane by ultrasonication approach for corrosion protection. *Ultrasonics sonochemistry* **2018,** *42*, 464-470.

[70] Umoren, S. A.; Eduok, U. M., Application of carbohydrate polymers as corrosion inhibitors for metal substrates in different media: a review. *Carbohydrate polymers* **2016,** *140*, 314-341.

[71] Guo, L.; Gu, C.; Feng, J.; Guo, Y.; Jin, Y.; Tu, J., Hydrophobic epoxy resin coating with ionic liquid conversion pretreatment on magnesium alloy for promoting corrosion resistance. *Journal of Materials Science & Technology* **2020,** *37*, 9-18.

[72] Atta, A. M.; A Ahmed, M.; A Al-Lohedan, H.; El-Faham, A., Multi-Functional cardanol triazine schiff base polyimine additives for self-healing and super-hydrophobic epoxy of steel coating. *Coatings* **2020,** *10* (4), 327.

[73] Hsissou, R.; Benhiba, F.; Abbout, S.; Dagdag, O.; Benkhaya, S.; Berisha, A.; Erramli, H.; Elharfi, A., Trifunctional epoxy polymer as corrosion inhibition material for carbon steel in 1.0 M HCl: MD simulations, DFT and complexation computations. *Inorganic Chemistry Communications* **2020,** *115*, 107858.

[74] Xu, Y.; Guo, L.; Zhang, H.; Zhai, H.; Ren, H., Research status, industrial application demand and prospects of phenolic resin. *RSC advances* **2019,** *9* (50), 28924-28935.

[75] Hunt, C. G.; Jakes, J. E.; Grigsby, W. In *Evaluation of adhesive penetration of wood fibre by nanoindentation and microscopy*, BIOCOMP 2010: 10th Pacific Rim Bio-Based Composites Symposium: 5-8 October 2010...[Sl: Sn, 2010]: p. 216-226., 2010; pp 216-226.

[76] Pilato, L., *Phenolic resins: a century of progress*. Springer: 2010; Vol. 11.

[77] Liu, C.; Zhang, Y.; Wang, S.; Meng, Y.; Hosseinaei, O., Micromechanical properties of the interphase in cellulose nanofiber-reinforced phenol formaldehyde bondlines. *BioResources* **2014,** *9* (3), 5529-5541.

[78] Liang, K.; Du, G. B.; Hosseinaei, O.; Wang, S. Q.; Wang, H. In *Mechanical properties of secondary wall and compound corner middle lamella near the phenol-formaldehyde (PF) adhesive bond line measured by nanoindentation*, Advanced Materials Research, Trans Tech Publ: 2011; pp 1746-1751.

[79] Atta-Obeng, E.; Via, B.; Fasina, O.; Auad, M.; Jiang, W., Cellulose reinforcement of phenol formaldehyde: Characterization and chemometric elucidation. *International Journal of Composite Materials* **2013.**

[80] Franke, T.; Herold, N.; Buchelt, B.; Pfriem, A., The potential of phenol–formaldehyde as plasticizing agent for moulding applications of wood veneer: two-dimensional and three-dimensional moulding. *European Journal of Wood and Wood Products* **2018,** *76* (5), 1409-1416.

[81] Furuno, T.; Saiki, H., Immamura, Y., Kajita, H., 2004. The modification of wood by treatment with low molecular weight phenol-formaldehyde resin: A properties enhancement with neutralized phenolic-resin and resin penetration into wood cell walls. *Wood Sci. Technol 37*, 349-361.

[82] Mao, A.; Xu, W.; Xi, E.; Li, Q.; Wan, H., Evaluation of phenol-formaldehyde resins modified and blended with pyrolysis bio-oil for plywood. *Forest Products Journal* **2018,** *68* (2), 113-119.

[83] Widyorini, R.; Syamsudin, T. S.; Alamsyah, E. M.; Purwasasmita, B. S., Bonding strength of plywood bonded using phenol formaldehyde mixed with wood bark powder nanofiller. *J Indian Ac Wood Sci* **2020,** *17* (1), 21-33.

[84] Gao, W.; Du, G., Curing kinetics of nano cupric oxide (CuO)-modified PF resin as wood adhesive: effect of surfactant. *Journal of Adhesion Science and Technology* **2013,** *27* (22), 2421-2432.

[85] Lu, X.; Isacsson, U., Effect of ageing on bitumen chemistry and rheology. *Construction and Building materials* **2002,** *16* (1), 15-22.

[86] Qin, Q.; Schabron, J. F.; Boysen, R. B.; Farrar, M. J., Field aging effect on chemistry and rheology of asphalt binders and rheological predictions for field aging. *Fuel* **2014,** *121*, 86-94.

[87] Zeng, W.; Wu, S.; Pang, L.; Chen, H.; Hu, J.; Sun, Y.; Chen, Z., Research on Ultra Violet (UV) aging depth of asphalts. *Construction and Building Materials* **2018,** *160*, 620-627.

[88] Hu, J.; Wu, S.; Liu, Q.; Hernández, M. I. G.; Wang, Z.; Nie, S.; Zhang, G., Effect of ultraviolet radiation in different wavebands on bitumen.

Construction and Building Materials **2018,** *159*, 479-485.

[89] Dessouky, S.; Contreras, D.; Sanchez, J.; Papagiannakis, A.; Abbas, A., Influence of hindered phenol additives on the rheology of aged polymer-modified bitumen. *Construction and building materials* **2013,** *38*, 214-223.

[90] Ouyang, C.; Wang, S.; Zhang, Y.; Zhang, Y., Improving the aging resistance of styrene–butadiene–styrene tri-block copolymer modified asphalt by addition of antioxidants. *Polymer degradation and stability* **2006,** *91* (4), 795-804.

[91] Li, Y.; Li, L.; Zhang, Y.; Zhao, S.; Xie, L.; Yao, S., Improving the aging resistance of styrene-butadiene-styrene tri-block copolymer and application in polymer-modified asphalt. *Journal of applied polymer science* **2010,** *116* (2), 754-761.

[92] Bishara, S.; Robertson, R.; Mohoney, D. In Lignin as an antioxidant: A Limited study on asphalts frequency used on Kansas roads, 42nd Annual Peterson Asphalt Research Conference, Cheyenne, WY, 2005.

[93] Xu, G.; Wang, H.; Zhu, H., Rheological properties and anti-aging performance of asphalt binder modified with wood lignin. Construction and Building Materials 2017, 151, 801-808.

[94] Pan, T., A first-principles based chemophysical environment for studying lignins as an asphalt antioxidant. Construction and Building Materials 2012, 36, 654-664.

[95] Saha, S. K.; Suman, S., Characterization of bakelite-modified bitumen. Innovative Infrastructure Solutions 2017, 2 (1), 3.

[96] Çubuk, M.; Gürü, M.; Çubuk, M. K.; Arslan, D., Rheological properties and performance evaluation of phenol formaldehyde modified bitumen. Journal of materials in civil engineering 2014, 26 (6), 04014015.

[97] Gunka, V.; Demchuk, Y.; Sidun, I.; Miroshnichenko, D.; Nyakuma, B. B.; Pyshyev, S., Application of phenol-cresol-formaldehyde resin as an adhesion promoter for bitumen and asphalt concrete. Road Materials and Pavement Design 2020, 1-13.

[98] Demchuk, Y.; Sidun, I.; Gunka, V.; Pyshyev, S.; Solodkyy, S., Effect of phenol-cresol-formaldehyde resin on adhesive and physico-mechanical properties of road bitumen. Chemistry & Chemical Technology 2018, 12 (4), 456-461.

[99] Xie, S.; Li, Q.; Karki, P.; Zhou, F.; Yuan, J. S., Lignin as renewable and superior asphalt binder modifier. ACS sustainable chemistry & engineering 2017, 5 (4), 2817-2823.

[100] Hosseinnezhad, S.; Hung, A. M.; Mousavi, M.; Sharma, B. K.; Fini, E., Resistance mechanisms of biomodified binders against ultraviolet exposure. ACS Sustainable Chemistry & Engineering 2020, 8 (6), 2390-2398.

[101] Li, C.; Rajib, A.; Sarker, M.; Liu, R.; Fini, E. H.; Cai, J., Balancing the Aromatic and Ketone Content of Bio-oils as Rejuvenators to Enhance Their Efficacy in Restoring Properties of Aged Bitumen. ACS Sustainable Chemistry & Engineering 2021.

[102] Mousavi, M.; Fini, E. H., Phenolic compounds to amplify the effect of sulfur on bitumen's thermomechanical properties. Fuel 2021, 287, 119532.

[103] Lattimer, R. P.; Kinsey, R. A.; Layer, R. W.; Rhee, C., The mechanism of phenolic resin vulcanization of unsaturated elastomers. Rubber chemistry and technology 1989, 62 (1), 107-123.

[104] Derakhshandeh, B.; Shojaei, A.; Faghihi, M., Effects of rubber curing ingredients and phenolic-resin on mechanical, thermal, and morphological characteristics of rubber/phenolic-resin blends. Journal of applied polymer science 2008, 108 (6), 3808-3821.

[105] Kabir, S. F.; Mousavi, M.; Fini, E. H., Selective adsorption of bio-oils' molecules onto rubber surface and its effects on stability of rubberized asphalt. Journal of Cleaner Production 2020, 252, 119856.

Chapter 10

Physiological Function of Phenolic Compounds in Plant Defense System

Vibhakar Chowdhary, Sheena Alooparampil, Rohan V. Pandya and Jigna G. Tank

Abstract

Plants respond to various abiotic and biotic stress conditions through accumulation of phenolic compounds. The specificity of these phenolic compounds accumulation depends on the type of stress condition and the response of plant species. Light stress induces biosynthesis of phenolic acids and flavonoids in plants. Temperature stress initially induces biosynthesis of osmoprotective compounds and then later stimulates synthesis of antioxidant enzymes and antioxidant compounds such as flavonoids, tannins and phenolic acids in plant cells. Salinity causes oxidative stress in plants by inducing production of reactive oxygen species. To resist against oxidative stress plants produce polyphenols, flavonoids, anthocyanins, phenolic acids and phenolic terpenes. Plants biosynthesize phenols and flavonoids during heavy metal stress.to scavenge the harmful reactive oxygen species and to detoxify the hydrogen peroxide. Plants accumulate phenols at the infection sites to slow down the growth of microbial pathogens and restrict them at infected site. Plants also accumulates salicylic acid and H_2O_2 at the infection site to induce the systemic acquired resistance (SAR) against microbial pathogens. Plants accumulate phenolic compounds which act as inhibitor or toxicant to harmful nematodes, insects and herbivores. Hence, phenols regulate crucial physiological functions in plants to resist against different stress conditions.

Keywords: plant defense, salinity, drought, microbial pathogens, insects, herbivores, phenols, flavonoids, tannins, terpenes

1. Introduction

Plants have developed various metabolic pathways which respond to different abiotic and biotic stress conditions specifically through biosynthesis of secondary metabolites. These metabolic pathways are linked with the primary metabolic pathways which are the integral part of growth regulating programmes in plants. During stress, plants reduce their growth and divert the primary metabolism towards biosynthesis of secondary metabolites. It specifically controls the expression level of genes through ontogeny and circadian clock phenomenon which are transcription factors responsible for regulation of growth and accumulation of various secondary metabolites in plants [1–6]. The transportation and accumulation of secondary metabolites regulates defense and development processes in plants based on the developmental stage, type of tissue or organ, and specific stress condition. Among various plant metabolites, phenolic compounds are the natural secondary metabolites that are biosynthesized in plants through metabolic pathways such as pentose phosphate, shikimate, and phenylpropanoid pathway [7–9]. These pathways are used by plants to produce either monomeric phenolic compounds such as flavanoids, phenolic acids and phenylpropanoids or polymeric phenolic compounds like tannins, lignins, lignans, and melanins. Phenolic compounds possess structural diversity due to their specific function in plant growth and defense mechanism. Some phenolic compounds are widely available in many plant species while others are specifically available only in certain plants species [10]. These phenolic compounds not only help in regulating various types of physiological functions in plants during growth and development but are also involved in plant defense mechanisms. They are known to have defensive function against abiotic and biotic stress conditions. Abiotic stress includes stress generated due to environmental changes such as high or low light and temperature, ultraviolet (UV) radiation, deficiency of nutrients, drought or flood like conditions. Biotic stress includes infection from microbial pathogen, attack by herbivorous organisms, increased production of oxidative species and free radicals in cells. The capability to synthesize specific phenolic compounds in response to biotic or abiotic stress is developed in plants through adaptive evolutionary phenomenon. Due to different environmental challenges plants have developed diversity in synthesizing various phenolic compounds [11].

For example, there are remarkable accumulation of flavanoids and isoflavones when plants experience low temperature stress, nutrients deficiency, exposure to UV radiation, microbial infection or injured through herbivores attack [12–14]. Anthocyanins accumulation was observed in flowers and fruits to attract pollinators for pollination. Anthocyanins also accumulate in young leaves to protect them from herbivorous insects and photodamage to regulate normal growth of plants [15]. Flavanoids are observed in guard cells of plants to protect tissue from UV radiation. They also accumulate to reduce the reactive oxidative stress generated through UV-B radiation [16]. Accumulation of phenols is observed in plants when plant experiences toxic metal stress from soil [17, 18]. Phenolic compounds help plant to develop resistance against microbial pathogens by inducing position explicit oversensitive response to protect spread of infection [8]. Proanthocyanidins, gallotannins and ellagitannins accumulation was observed in plants when infected with viruses, fungi or herbivores during early development stages of plant [8]. Secretion of t-cinnamic acid was observed from barley roots when it was infected by fungal pathogen fusarium [19]. Secretion of rosmarinic acid was observed in roots of *Ocimum basilicum* when it was infected with fungal pathogen *Pythium ultimum* [20]. Nematicide iridoid glycosides accumulation was observed in roots of plant *Plantago lanceolata* when it was infected with nematodes [21].

2. Plant defense against light stress

Plants accumulate phenolic acids and flavonoids in the vacuoles of mesophyll and epidermal cells during the light stress through photosynthetic apparatus and metabolism [22–24]. Falcone Ferreyra et al. [25] observed that when maize plants are exposed to UV-B radiation expression of genes P1, B and PL1 increases which induces biosynthesis of transcription regulators anthocyanin and 3-deoxyflavanoid which in turn regulates the activity of protein ZmFLS1 for converting the dihydroflavonols, dihydroquercetin and dihydrokaempferol to flavonols, quercetin and kaempferol respectively. Radyukina et al. [26] observed the accumulation of flavonoids, and anthocyanins in plants exposed to light and salinity stress. They suggested that flavonoids protect plants from UV-B radiation and anthocyanins protect from salinity stress. Manukyan [27] observed high accumulation of total phenol in *Melissa officinalis*, *Nepeta cataria* and *Salvia officinalis* plants after exposure to low UV-B radiation. Ma et al. [28] observed in *Salvia miltiorrhiza,* that UV radiation increases concentration of rosmarinic acid and lithospermic acid in plant. They suggested that methyl jasmonate induces transcripts of genes accountable for biosynthesis of enzymes tyrosine aminotransferase, cinnamic acid 4-hydroxylase, 4-hydroxyphenylpyruvate reductase and phenylalanine ammonia lyase (PAL) which in turn regulates the biosynthesis of rosmarinic acid and lithospermic acid. Ghasemzadeh et al. [29] observed that the accumulation of specific phenolic compounds in sweet basil leaves was dependent on the intensity of UV-B radiation. They suggested that phenolic compounds are synthesized in plants as a response towards the generated reactive oxygen species due to UV light damage. They observed that phenolic acids such as cinnamic acid, gallic acid, quercetin, ferulic acid, catechin, rutin, luteolin and kaempferol which are precursors for biosynthetic pathway of flavonoids are synthesized earlier in leaves through phenylpropanoid metabolism using PAL and chalcone synthase enzymes. Jang et al. [30] observed in plant *Salvia plebeian* that under sunlight the level of rosmarinic acid reduces whereas level of homoplantaginin and luteolin-7-glucoside increases. Csepregi et al. [31] observed that the accumulation of flavonols, quercetin and kaempferol derivatives increases in leaves of *Arabidopsis thaliana* when it is exposed to low UV-B light. León-Chan et al. [32] observed that the low temperature and UV-B radiation causes degradation of chlorophyll and accumulation of carotenoids, chlorogenic acid, flavonoids apigenin-7-O-glucoside and luteolin-7-O-glucoside in bell pepper plant leaves. They specifically observed that UV-B radiation increases flavonoids concentration in leaves whereas combination of low temperature and UV-B radiation increases chlorogenic acid concentration in leaves. They also observed that the luteolin-7-O-glucoside is involved in quenching of the reactive oxygen species developed due to low temperature and UV-B radiation stress. Peng et al. [33] observed that flavone O-glycosides are modulated by flavone 7-Oglucosyltransferase and flavone 5-O-glucosyltransferase during light stress. They suggested that allelic variation provides UV-B tolerance to plants in nature. Zhou et al. [34] also observed that flavonol accumulation is upregulated by UV-B irradiation in rice plants. Lobiuc et al. [35] suggested that the phytochemical content of basil green cultivar was high in red light whereas phytochemical content of basil red cultivar was high in blue light when exposed to different proportions of blue and red light. They observed that accumulation of rosmarinic acid, caffeic acid and anthocyanin increased when exposed to blue light as compared to white light. Chen et al. [36] suggested that the downregulation of genes *SmDXR, SmDXS2, SmGGPPS, SmCPS, SmHMGR* and *CYP76AH1* decreases tanshinone IIA content in *Salvia miltiorrhiza*. They also suggested that rosmarinic acid content increases when *Salvia miltiorrhiza* is exposed to UV light or combination of red and blue light. Taulavuori et al. [37] observed accumulation of phenolic compounds (chicoric acid and chlorogenic acid derivatives) in leaves of *Ocimum basilicum* and flavonoids (luteolin-glycoside derivatives, isorhamnetin diglycoside, apigenin derivatives) in plants of *Rumex sanguineus* after exposure to blue and blue-violet light. Stagnari et al. [38] observed that exposure of basil plants to colored light reduces the level of rosmarinic acid and caftaric acid in leaves whereas increased caffeic acid level in leaves. Nadeem et al. [39] observed that yellow light increases rosmarinic acid and chicoric acid in callus of basil whereas green light increases rosmarinic acid, eugenol and chicoric acid in callus of basil. They suggested that change in phytochemical content of callus of basil was due to the accumulation of reactive oxygen species by the metabolic action of CYP450 enzyme.

3. Plant defense against temperature stress

During high and low temperature stress, photosynthesis metabolism is inhibited and production of reactive oxygen species is stimulated which in turn damages the cells [40, 41]. To combat with this stress plants accumulate osmoprotective compounds such as soluble sugars, proline and glycine betaine which provides protection from oxidative damage [42]. Plants also biosynthesize antioxidant enzymes and substances to defense against oxidative stress [43]. Plants accumulate antioxidant metabolites such as phenolics, terpenes or alkaloids during temperature stress and develop stress resistance ability [44–47]. During temperature stress activity of enzyme phenylalanine ammonia lyase increases which results in accumulation of phenolic compounds in plant cells. Rivero et al. [48] has suggested that during heat and cold stress there is remarkable accumulation of soluble phenolics in watermelon and tomato. Kasuga et al. [49] suggested that cold induced phenols accumulation in plant cells decreases the freezing point, maintains water potential and protects from cell disruption. Weidner et al. [50] observed increased content of tannins and soluble phenols in roots of grapevine after cold treatment. Amarowicz et al. [51] observed increased concentration of gallic acid, ferulic acid and caffeic acid in grapevines during cold stress. Isshiki et al. [52] observed accumulation of farinose flavonoids on aerial part of primula during the freezing cold stress. Rana and Bhushan [53] have suggested that temperature stress induces biosynthesis of phenolic compounds in plants and provides tolerance against cold stress. Commisso et al. [54] suggested that phenolic compounds protect cytoskeleton of microfilaments from reactive oxygen species. Chalker-Scott and Fuchigami [55] suggested that cellular injury and stress tolerance capacity in plants is increased by accumulation of phenolic compounds and then its incorporation in to the cell wall of cells in the form of either suberin or lignin.

4. Plant defense against drought stress

During drought stress plants produce reactive oxygen species (hydrogen peroxide H_2O_2, singlet oxygen O, superoxide anion O^{2-}, and hydroxyl radical OH) which may cause protein degradation, cell mortality, membrane damage, lipid peroxidation and deoxy ribose nucleic acid (DNA) damage [56, 57]. In order, to prevent this damage, plants have detoxification system to neutralize the deleterious effect of reactive oxygen species which is regulated either by enzymes (superoxide dismutase (SOD), ascorbate peroxidase (APX), catalase (CAT), peroxidase (POD)) or by antioxidant molecules (phenols, vitamin C, carotenoids, tocopherol and glutathione) [58]. In plants overproduction of reactive oxygen species during stress is balanced through production of phenolic compounds and flavonoids using phenylpropanoid pathway [59]. Akula and Ravishankar [60] observed accumulation of flavonoids in leaves of willow plant during drought stress. Similarly, Nakabayashi et al. [61] observed increase in accumulation of anthocyanin and flavonoids in leaves of *Arabidopsis* in response to drought stress.

The biosynthesis and accumulation of phenolic compounds during drought stress is regulated by enzymes of phenylpropanoid pathway. Initially, phenylalanine ammonia lyase (PAL) diverts the central carbon flux of primary metabolism towards synthesis of phenolic compounds. Increase in PAL activity indicates beginning of plant antioxidant defense mechanism and is regulated by feedback inhibition process through increase in accumulation of its own product cinnamic acid [62]. The variations in the transcription level of genes encoding for phenylalanine ammonia lyase (PAL) regulates the activity of the enzyme and in turn specific phenolic compounds are synthesized in response to biotic or abiotic stress. Chalcone synthase is an enzyme which shows high activity during drought stress. It is a key enzyme in flavonoid synthesis pathway which acts on the CoA-ester of cinnamic acid to form chalcone. The chalcone is converted to flavanone by chalcone flavanone isomerase (CHI) enzyme through isomerization which is a precursor for synthesis of numerous flavonoid compounds [59]. Hura et al. [63] observed accumulation of ferrulic acid and high activity of PAL enzyme in leaves of maize under water stress conditions. Even Phimchan et al. [64] observed high PAL activity and ferrulic acid accumulation in fruits of capsicum during drought stress. Nakabayashi et al. [61] observed high activity of another enzyme chalcone synthase in response to drought stress in *Arabidopsis*. Gharibi et al. and Siracusa et al. [65, 66] have observed high accumulation of phenolic compounds in vegetables, fruits and cereals under drought stress. Sarker and Oba [67] observed high accumulation of flavonoids in leaves of *Amaranthus tricolor* during drought stress. Brunetti et al. [68] suggested that the high metabolic plasticity and accumulation of flavonoids in leaves of *Moringa oleifera* has provided ability to the plant to survive in water deficit conditions.

5. Plant defense against salinity stress

Salinity stress induces production of reactive oxygen species in plants which in turn causes oxidative stress. To resist against oxidative stress plants produce antioxidative metabolites such as polyphenols, flavonoids, anthocyanins, proanthocyanidins, phenolic acids and phenolic terpenes which quench the singlet oxygen, neutralize or absorb free radicals, decompose peroxides [45–47]. Yang et al. [69] suggested that accumulation of specific phenolic compounds in plants during salinity stress also depends on the type of plant species. Parida et al. [70] suggested that there was significant increase in polyphenols content in plants of *Aegiceras corniculatum* after 250 mM Nacl treatment. Ksouri et al. [71] suggested that there was significant increase in polyphenols in *jerba* plants after treatment with 100 mM and 400 mM NaCl. Hanen et al. [72] suggested that the phenol content in leaf of plant *Cynara cardunculus* increases in response to 50 mM NaCl treatment. Lim et al. [73] suggested that the accumulation of phenolic compounds in response to salinity stress in *Fagopyrum esculentum* (*Fagopyrum esculentum*) plants is due to the increased content of compounds such as vitexin, isoorientin, rutin, and orientin. Petridis et al. [74] suggested that the salinity stress stimulated the biosynthesis of phenols and oleuropein in leaves of olive plants. Borgognone et al. [75] observed that salinity stress increases the concentration of total phenols and flavonoids in leaves of artichoke and cardoon plants.

Another mechanism acquired by plants to resist against salinity stress is through salicylic acid which is an endogenous growth regulator and signaling molecule. It is a phenolic phytohormone which controls stress by decreasing H_2O_2 level and reducing oxidative damage in plants [76]. It enhances growth, development and productivity in plants during stress conditions [77]. Many research studies have suggested the function of salicylic acid in increasing salinity tolerance in plants. Jini and Joseph and Khan et al. [78, 79] had suggested that salicylic acid strengthens the salinity tolerance in plants such as *Medicago sativa*, *Vicia faba*, *Brassica juncea* and *Vigna radiate* (*Vigna radiate*). Jayakannan et al. [80] observed that exogenous salicylic treatment increased water content and growth of shoots in *Arabidopsis* plants growing under saline conditions. Various studies of mutant plants have suggested the function of salicylic acid in providing salinity tolerance to plants [81–85]. Various studies on exogenous application of salicylic acid to salinity stressed plants have also confirmed that salicylic acid alleviates the toxic effect of salt and increases the resistance of plants against salinity [86–91].

6. Plant defense against heavy metals

Ciriakova [92] has suggested that plants take up heavy metals through their roots which get accumulated inside the cell wall by apoplastic system. These heavy metals cause harm to plants by hindering the biochemical metabolisms such as cell division and elongation, photosynthesis, nitrogen metabolism, respiration, mineral nutrient utilization and water transportation [92, 93]. They inactivate essential enzymes by binding to their active sites, induce biosynthesis of reactive oxygen species, and exchange metal ions from biomolecules [94]. Plants biosynthesize phenols and flavonoids to scavenge the harmful reactive oxygen species which donates their electron to peroxidase enzymes to detoxify hydrogen peroxide produced under heavy metal stress conditions [95]. Shemet and Fedenko [96] observed accumulation of phenolic compounds in roots of maize under cadmium stress. Ali et al. [97] observed high activity of enzymes responsible for biosynthesis of phenols and flavonoids in roots of *Panax ginseng* exposed to copper sulphate. Kováčik et al. [98] observed in *Matricaria chamomilla* plants that when plants were exposed to nickel activity of pholyphenol oxidase enzyme decrease and there was increase in total phenol content of leaf rosettes. There was remarkable increase in activity of phenylalanine ammonia lyase (PAL) and shikimate dehydrogenase enzymes with accumulation of chlorogenic acid, protocatechuic acid and caffeic acid. Pawlak-Sprada et al. [99] suggested from transcriptional analysis of lupine and soyabean roots exposed to cadmium and lead that heavy metal stress induces phenylpropanoid pathway in plants. Márquez-García et al. [100] observed in *Erica andevalensis* plants that when plants are exposed to cadmium, the concentration of rutin, cinnamic acid derivatives and epigallocatechin increases. He suggested that excess cadmium exposure decreases the concentration of phenolic in plants to reduce the deleterious effect of produced phenoxyl radicals. Malčovská et al. [101] suggested that the production of phenolic compounds increases in plant cells when plants are under heavy metal stress as phenols are reactive oxygen species scavengers and metal chelators. Kisa et al. [102] observed in *Zea mays* leaves that when plants are exposed to cadmium and lead, the phenolic compounds increased in leaves were chlorogenic acid and rutin whereas there was decrease of caffeic acid and ferulic acid.

7. Plant defense against microbial pathogens

The plant defense mechanism occurs in two stages, in first response there is rapid accumulation of phenols at the infection site which slowdowns the growth of pathogen. In second response it biosynthesizes specific stress related substances (simple phenols, phenolic phytoalexins, hydroxycinnamic acids etc.) which restrict the pathogen at the infected site. The step by step process of plant defense mechanism includes host cell death, necrosis, accumulation of phenolic compounds, modification of cell wall through phenolic compounds deposition or development of barriers, and at last synthesis of specific toxic compounds to eliminate the pathogens [103]. Pathogenic microbes are recognized by plant cell membrane proteins which are known as pattern recognition receptors (PRRs). They recognize conserved pathogen associated molecular patterns (PAMP) of microorganisms and gives signal to synthesize specific phenolic compounds, through defense mechanism known as PAMP induced immunity [104–110]. Plants induce multicomponent defense response after pathogen attack which includes reprogramming of genetic resources, expression of large number of defense related genes, and encoding of enzymes that catalyze defense metabolites (phytoalexins). This physiological process is regulated by transcriptional factors responsible for accumulation of specific phytoalexins in plants. On the other hand, salicylic acid also plays crucial role in resisting pathogen attack in plants. During pathogenic infection there is remarkable accumulation of pathogenesis related (PR) protein at the location distant from the infection site. Simultaneously, there is accumulation of salicylic acid and H_2O_2 at the infection site to regulate the systemic acquired resistance (SAR) in plant. It is being observed that exogenous application of salicylic acid induces systemic acquired resistance (SAR) in plants and provides resistance against pathogens [10].

Plants possesses innate immunity against pathogenic bacterial species. They have developed metabolic mechanism to resist against pathogenic bacterial through accumulation of phenolic compounds. Postel and Kemmerling [111] suggested that plants recognize the bacterial pathogens through pathogen associated molecular patterns (PAMPs). Mikulic Petkovsek et al. [112] observed accumulation of hydroxycinnamic acid, gallic acid, quercetins and catechin in walnut husk plant infected by *Xanthomonas arboricola* bacteria. Cho and Lee [113] observed accumulation of sakuranetin in rice plants infected by *Xanthomonas oryzae* and *Burkholderia glumae*. Wang et al. [114] suggested that polyphenols inhibit bacterial species such as *Escherichia coli*, *Klebsiella pneumonia*, *Staphylococcus aureus*, *Salmonella choleraesuis*, *Bacillus subtilis*, *Serratia marcescens* and *Pseudomonas aeruginosa* by altering the properties and permeability of plasma membrane of cell and generation of reactive oxygen species.

Previous studies by various scientists have suggested that phenolic compounds eliminate fungal pathogens by altering the permeability of cell membrane, altering the integrity of cell wall, suppression of enzymes activity, formation of free radicals, inhibition of certain protein biosynthesis, damage of DNA and suppressing the expression of virulence genes [115–118]. The mode of action of flavonoids against fungal pathogens include damage of cytoplasmic membrane, distraction of cell wall, induction of cell death process, inhibition of enzyme activities, chelating of metal ions, binding with extracellular or soluble proteins, inhibition of efflux pump activity [12]. Gallego-Giraldo et al. [119, 120] suggested that the suppression of liginin biosynthesis genes (HCT) leads to the accumulation of salicylic acid which in turn increases transcription level of some pathogenesis related genes to improve immunity of plants. Widodo et al. [121] suggested that coumarins inhibit growth of fungi by altering the thickness of mitochondrial matrix, inducing apoptosis or inducing cell wall perforation which leads to release of cytoplasm from cell. Rahman [122] observed accumulation of furanocoumarin in celery and parsnip roots after *Sclerotinia sclerotiorum* infection. Al-Barwani and Eltayeb [123] observed antifungal activity of psoralen and furanocoumarin against fungi *Alternaria brassicicola*, *Sclerotinia sclerotiorum* and *Cercospora carotae*. Al-Amiery et al. [124] observed antifungal activity of coumarins against *Aspergillus niger* and *Candida albicans*. Serpa et al. [125] suggested that the flavone compound baicalein inhibits the infection caused by *Candida albicans* by inhibiting the activity of efflux pump and inducing apoptosis process. Zuzarte et al. [126] suggested that the chalcone carvacrol disrupts the cytoplasmic membrane of cell and induces apoptosis process in various *Candida* species. Belofsky et al. [127] suggested that the isoflavone sedonan A isolated from plant *Dalea formosa* prevents from infection caused by *Candida albicans* and *Cadida glabrata* by inhibiting the activity of intracellular transcription targets and efflux pumps. Sherwood and Bonello [128] suggested that lignin has potent antifungal activity against fungi *Diplodia pinea* under *in vitro* conditions. Anttila et al. [129] suggested that the tannins extract isolated from cone and bark of conifer plants has toxic effect on four soft rot fungi, three white rot fungi and eight brown rot fungi. Dos Santos et al. [130] observed antifungal activity of *Accacia mearnsii* tannin extract against *Aspergillus niger* and *Candida* sp. Wang et al. [114] observed that the ester derivatives of monoterpenes carvacrol and thymol were toxic against the phytopathogenic fungi in *in vitro* conditions. Rashed et al. [131] observed the toxic effect of *Ammi visnaga* seed extract against fungi *Rhizoctonia solani* was due to the presence of coumarins. Marques et al. [132] observed accumulation of phenolic compounds and lignin at the infected site during early stage to prevent the penetration of *Sporisorium scitamineum* fungi in other parts of sugarcane plant. Ogawa and Yazaki [133] suggested that the inhibitory mode of action of tannins is the inhibition of the activity of extracellular enzymes, inhibition of oxidative phosphorylation, or prevention of nutrient availability from substrate by protein insolubilization or metal complex formation.

Kumar and Pandey [134] suggested that Phenolic compounds suppress the viral infection in plants and represses the replication of viruses through mode of actions such as damage of protein, DNA or ribose nucleic acid (RNA), inhibition of viral enzyme activities. Zakaryan et al. [135] suggested that flavonoids suppresses the viral infection by distraction of viral RNA translation, inhibition of viral DNA replication, inhibition of viral protein synthesis, inhibition of transcription factors responsible for viral enzymes and genome synthesis and interfering with viral structural protein. Shokoohinia et al. [136] suggested that coumarins inhibit viral replication in cells by inhibition of enzymes such as protease, integrase and reverse transcriptase. Dunkić et al. [137] observed that the monoterpenes carvacrol and thymol present in essential oil of *Satureja montana* L. ssp. Variegate has antiviral activity against cucumber mosaic virus and tobacco mosaic virus. Hu et al. [138] observed antiviral activity of different phenolic compounds isolated from *Arundina graminifolia* against tobacco mosaic virus. Zhao et al. [139] suggested that the two flavonoids (fistula flavonoid B and C) isolated from bark and stem of plant *Cassia fistula* has antiviral activity against tobacco mosaic virus. Li et al. [140] identified phenolic compound gramniphenol which exhibited antiviral activity against tobacco mosaic virus. Liu et al. [141] observed antiviral potential of two coumarins (6-hydroxy-5-methoxy-7-methyl-3-(40-methoxyphenyl)-coumarin and 6-hydroxy-7-methyl-3-(40-methoxyphenyl)-coumarin) isolated from leaves of *Nicotiana tabacum* against tobacco mosaic virus.

8. Plant defense against insects, nematodes and herbivorous organisms

Plants have to face various pathogenic attacks in natural environment. To resist against these pathogens plants have adjusted their physiological metabolism and developed metabolic pathways which synthesize wide range of phenolic compounds. These phenolic compounds are used either to attract or repell different organism as per plants benefit. They protect plants by acting as inhibitors and toxicants against insects, nematodes and herbivorous animals which feeds on them [142–145]. Maxwell et al. [146] suggested that phenolic pigment (gossypol) found in cotton plants has toxic effect on *Heliothis zea*, *Heliothis virescens* and various other insect pests. Feeny [147] suggested that the tannins have inhibitory effect on the growth of *Opheropthera brumata* larvae. Levin [148] suggested that the phenolic quinone hypericin secreated by glans on leaves, sepals or petals of Lypericum spp. is toxic foe insects and mammals. He also suggested that the presence of gossypol in leaves and flowers of plants can inhibit grazing by mammals and infection by tobacco budworm or bollworm. Hedin et al. [149] suggested that some flavonoids present in cotton plants are feeding inhibitors for boll weevil, *Anthonomus grandis*.

Luczynski et al. [150] suggested that the concentration of catechol increases in leaves of strawberry when infected by spotted spider mites. Byers [151] suggested that the bark beetle *Scolytus multistriatus* does not consume *Carya ovate* due to the presence of phenolic compound juglone which is not palatable to them Accumulation of anthocyanins provides red, blue or purple color to leaves, flowers or fruits which protects plant from the herbivorous animals and insect pathogens. These pigments developed in leaves are either not palatable for animals to eat or they are not visible to animals due to lack of red visualization receptor. Insect pathogens avoid red leaves and they always colonize in green leaves. Better chemical defense, worst nutritional value and induced adverse effect in insects is observed in plants having red leaves. Hence autumn colors of leaves is an adaptive mechanism of plants to reduce the pathogen attacks [152–157]. Rehman et al. [158] suggested that catechol binds to the digestive system of mites and inactivates its digestive enzymes. Fürstenberg-Hägg et al. [159] suggested that wheat cultivars rich in phenolic content are not consumed by cereal aphids *Rhopalosiphum padi*.

9. Conclusions

Phenolic compounds regulate crucial physiological functions in plants to provide resistance against various biotic and abiotic stress conditions. To protect against UV radiation plants synthesize phenolic acids and flavonoids to scavenge the reactive oxygen species generated due to light stress. During temperature stress activity of phenylalanine ammonia lyase enzyme increases which results in accumulation of phenols in plants. The accumulation of phenols during drought stress is regulated by the activity of either phenylalanine ammonia lyase (PAL) or chalcone synthase. Phenylalanine ammonia lyase (PAL) activity accumulates phenolic acids which are used as precursors for biosynthesis of specific phenolic compounds. Chalcone synthase activity accumulates numerous flavonoid compounds in plants during water deficiency. During salinity stress plants accumulate polyphenols, flavonoids, anthocyanins, phenolic acids and terpenes to resist against the oxidative stress. Plants also accumulate salicylic acid during salinity stress to decrease the level of H_2O_2 and reduce the oxidative damage. Plants synthesize phenols and flavonoids to scavenge the reactive oxygen species produced during heavy metal stress. Plants accumulate phenolic compounds at infection site to reduce growth and penetration of microbial pathogens in other tissues and organs. It recognizes microbial pathogens and induces defense response at genetic level to biosynthesize defense metabolites. Plants also accumulates salicylic acid and H_2O_2 at infection site to regulate systemic acquired resistance. Plants accumulate phenolic compounds in organs which acts as inhibitors or toxicants for nematodes, insects and herbivores.

10. Future prospectives

The biosynthesis of phenolic compounds in plants during abiotic and biotic stress increases adaptation of plants in harsh environment. Hence, it is necessary to understand the molecular mechanism regulating biosynthesis and accumulation of specific phenolic compounds during particular stress condition. There should be genetic level studies on regulation of transcription factors responsible for biosynthesis of specific phenolic compounds during each stress. There should be progressive studies on interactive biology between phenolic compounds and salicylic acid to understand the crosstalk between them during salinity stress, oxidative damage and microbial pathogen attack.

Acknowledgements

Authors are whole heartedly thankful to Department of Biosciences, Saurashtra University for providing all the necessary facilities.

Conflict of interest

Authors declare that there is no conflict of interest.

Abbreviations

UV	ultra violet
H_2O_2	hydrogen peroxide
ROS	reactive oxygen species
SOD	superoxide dismutase
APX	ascorbate peroxidase
CAT	catalase
POD	peroxidase
PAL	phenylalanine ammonia lyase
CHI	chalcone flavanone isomerase
SA	salicylic acid
PRRs	pattern recognition receptors
PAMP	pathogen associated molecular patterns
PR	pathogenesis related
SAR	systemic acquired resistance
DNA	deoxy ribose nucleic acid
RNA	ribose nucleic acid

Author details

Vibhakar Chowdhary[1], Sheena Alooparampil[1], Rohan V. Pandya[2] and Jigna G. Tank[1*]

1 Department of Biosciences, Saurashtra University, Rajkot, Gujarat, India

2 Department of Microbiology and Biotechnology, Atmiya University, Rajkot, Gujarat, India

*Address all correspondence to: jignagtank@gmail.com

References

[1] Ornston LN, Yeh WK. Origins of metabolic diversity: Evolutionary divergence by sequence repetition. Proceedings of the National Academy of Sciences of the United States of America. 1979;**76**(8):3996-4000

[2] Wink M. Biochemistry of Plant Secondary Metabolism. UK/Sheffield, Boca Raton: Sheffield Academic Press/ CRC Press; 1999

[3] Lehfeldt C, Shirley AM, Meyer K, Ruegger MO, Cusumano JC, Viitanen PV, et al. Cloning of the SNG1 gene of Arabidopsis reveals a role for a serine carboxypeptidase-like protein as an acyltransferase in secondary metabolism. The Plant Cell. 2000;**12**(8): 1295-1306

[4] Tauber E, Last KS, Olive PJ, Kyriacou CP. Clock gene evolution and functional divergence. Journal of Biological Rhythms. 2004;**19**(5):445-458

[5] Broun P. Transcriptional control of flavonoid biosynthesis: A complex network of conserved regulators involved in multiple aspects of differentiation in Arabidopsis. Current Opinion in Plant Biology. 2005;**8**(3): 272-279

[6] do Nascimento NC, Fett-Neto AG. Plant secondary metabolism and challenges in modifying its operation: An overview. In: Fett-Neto A, editor. Plant Secondary Metabolism Engineering. Methods in Molecular Biology (Methods and Protocols). Vol. 643. Totowa, NJ: Humana Press; 2010. pp. 1-13. DOI: 10.1007/978-1-60761-723-5_1

[7] Balasundram N, Sundram K, Samman S. Phenolic compounds in plants and agri-industrial by-products: Antioxidant activity, occurrence, and potential uses. Food Chemistry. 2006;**99**(1):191-203. DOI: 10.1016/j.foodchem.2005.07.042

[8] Cheynier V, Comte G, Davies KM, Lattanzio V, Martens S. Plant phenolics: Recent advances on their biosynthesis, genetics, and ecophysiology. Plant Physiology and Biochemistry. 2013;**72**:1-20. DOI: 10.1016/j.plaphy.2013.05.009

[9] Heleno SA, Martins A, Queiroz MJ, Ferreira IC. Bioactivity of phenolic acids: Metabolites versus parent compounds: A review. Food Chemistry. 2015;**173**:501-513. DOI: 10.1016/j.foodchem.2014.10.057

[10] Kumar S, Abedin MM, Singh AK, Das S. Role of phenolic compounds in plant-defensive mechanisms. In: Plant Phenolics in Sustainable Agriculture. Singapore: Springer; 2020. pp. 517-532. DOI: 10.1007/978-981-15-4890-1_22

[11] Caputi L, Malnoy M, Goremykin V, Nikiforova S, Martens S. A genome-wide phylogenetic reconstruction of family 1 UDP-glycosyltransferases revealed the expansion of the family during the adaptation of plants to life on land. The Plant Journal. 2012;**69**(6): 1030-1042

[12] Mierziak J, Kostyn K, Kulma A. Flavonoids as important molecules of plant interactions with the environment. Molecules. 2014;**19**(10):16240-16265

[13] Schulz E, Tohge T, Zuther E, Fernie AR, Hincha DK. Flavonoids are determinants of freezing tolerance and cold acclimation in Arabidopsis thaliana. Scientific Reports. 2016;**6**(1):1. DOI: 10.1038/srep34027

[14] Rodríguez-Calzada T, Qian M, Strid Å, Neugart S, Schreiner M, Torres-Pacheco I, et al. Effect of UV-B radiation on morphology, phenolic compound production, gene expression, and subsequent drought stress responses in chili pepper (*Capsicum annuum* L.). Plant Physiology and Biochemistry. 2019;**134**:94-102. DOI: 10.1016/j.plaphy.2018.06.025

[15] Karageorgou P, Manetas Y. The importance of being red when young: Anthocyanins and the protection of young leaves of *Quercus coccifera* from insect herbivory and excess light. Tree Physiology. 2006;**26**(5):613-621. DOI: 10.1093/treephys/26.5.613

[16] Agati G, Tattini M. Multiple functional roles of flavonoids in photoprotection. New Phytologist. 2010;**186**(4):786-793

[17] Michalak A. Phenolic compounds and their antioxidant activity in plants growing under heavy metal stress. Polish Journal of Environmental Studies. 2006;**15**(4):523-530. DOI: 10.1016/j.fitote.2011.01.018

[18] Mandal SM, Chakraborty D, Dey S. Phenolic acids act as signaling molecules in plant-microbe symbioses. Plant Signaling and Behavior. 2010;**5**(4): 359-368

[19] Lanoue A, Burlat V, Schurr U, Röse US. Induced root-secreted phenolic compounds as a belowground plant defense. Plant Signaling and Behavior. 2010;**5**(8):1037-1038. DOI: 10.4161/psb.5.8. 12337

[20] Bais HP, Walker TS, Stermitz FR, Hufbauer RA, Vivanco JM. Enantiomeric-dependent phytotoxic and antimicrobial activity of (±)-catechin. A rhizosecreted racemic mixture from spotted knapweed. Plant Physiology. 2002;**128**(4):1173-1179

[21] Wurst S, Wagenaar R, Biere A, Van der Putten WH. Microorganisms and nematodes increase levels of secondary metabolites in roots and root exudates of *Plantago lanceolata*. Plant and Soil. 2010;**329**(1):117-126. DOI: 10.1007/s11104-009-0139-2

[22] Szymańska R, Ślesak I, Orzechowska A, Kruk J. Physiological and biochemical responses to high light and temperature stress in plants. Environmental and Experimental Botany. 2017;**139**:165-177. DOI: 10.1016/j.envexpbot.2017.05.002

[23] Tattini M, Galardi C, Pinelli P, Massai R, Remorini D, Agati G. Differential accumulation of flavonoids and hydroxycinnamates in leaves of *Ligustrum vulgare* under excess light and drought stress. New Phytologist. 2004;**163**(3):547-561. DOI: 10.1111/j. 1469-8137.2004.01126.x

[24] Conéjéro G, Noirot M, Talamond P, Verdeil JL. Spectral analysis combined with advanced linear unmixing allows for histolocalization of phenolics in leaves of coffee trees. Frontiers in Plant Science. 2014;**5**:39. DOI: 10.3389/fpls.2014.00039

[25] Falcone Ferreyra ML, Rius S, Emiliani J, Pourcel L, Feller A, Morohashi K, et al. Cloning and characterization of a UV-B-inducible maize flavonol synthase. The Plant Journal. 2010;**62**(1):77-91. DOI: 10.1111/j.1365-313X.2010.04133.x

[26] Radyukina NL, Toaima VI, Zaripova NR. The involvement of low-molecular antioxidants in cross-adaptation of medicine plants to successive action of UV-B radiation and salinity. Russian Journal of Plant Physiology. 2012;**59**(1):71-78. DOI: 10.1134/s1021443712010165

[27] Manukyan A. Effects of PAR and UV-B radiation on herbal yield, bioactive compounds and their antioxidant capacity of some medicinal plants under controlled environmental conditions. Photochemistry and Photobiology. 2013;**89**(2):406-414. DOI: 10.1111/j.1751-1097.2012.01242.x

[28] Ma P, Liu J, Zhang C, Liang Z. Regulation of water-soluble phenolic acid biosynthesis in *Salvia miltiorrhiza* Bunge. Applied Biochemistry and Biotechnology. 2013;**170**(6):1253-1262. DOI: 10.1007/s12010-013-0265-4

[29] Ghasemzadeh A, Jaafar HZ, Rahmat A. Antioxidant activities, total phenolics and flavonoids content in two varieties of Malaysia young ginger (*Zingiber officinale* Roscoe). Molecules. 2010;**15**(6):4324-4333

[30] Jang HJ, Lee SJ, Kim CY, Hwang JT, Choi JH, Park JH, et al. Effect of sunlight radiation on the growth and chemical constituents of *Salvia plebeia* R. Br. Molecules. 2017;**22**(8):1279. DOI: 10.3390/molecules22081279

[31] Csepregi K, Coffey A, Cunningham N, Prinsen E, Hideg É, Jansen MA. Developmental age and UV-B exposure co-determine antioxidant capacity and flavonol accumulation in Arabidopsis leaves. Environmental and Experimental Botany. 2017;**140**:19-25. DOI: 10.1016/j.envexpbot.2017.05.009

[32] León-Chan RG, López-Meyer M, Osuna-Enciso T, Sañudo-Barajas JA, Heredia JB, León-Félix J. Low temperature and ultraviolet-B radiation affect chlorophyll content and induce the accumulation of UV-B-absorbing and antioxidant compounds in bell pepper (*Capsicum annuum*) plants. Environmental and Experimental Botany. 2017;**139**:143-151. DOI: 10.1016/j.envexpbot.2017.05.006

[33] Peng M, Shahzad R, Gul A, Subthain H, Shen S, Lei L, et al. Differentially evolved glucosyltransferases determine natural variation of rice flavone accumulation and UV-tolerance. Nature Communications. 2017;**8**(1):1-2. DOI: 10.1038/s41467-017-02168-x

[34] Zhou Z, Schenke D, Miao Y, Cai D. Investigation of the crosstalk between the flg22 and the UV-B-induced flavonol pathway in *Arabidopsis thaliana* seedlings. Plant Cell Environment. 2017;**40**(3):453-458. DOI: 10.1111/pce.12869

[35] Lobiuc A, Vasilache V, Oroian M, Stoleru T, Burducea M, Pintilie O, et al. Blue and red LED illumination improves growth and bioactive compounds contents in acyanic and cyanic *Ocimum basilicum* L. microgreens. Molecules. 2017;**22**(12):2111. DOI: 10.3390/molecules22122111

[36] Chen GJ, Lee MS, Lin MK, Ko CY, Chang WT. Blue light decreases tanshinone IIA content in *Salvia miltiorrhiza* hairy roots via genes regulation. Journal of Photochemistry and Photobiology B: Biology. 2018;**183**:164-171

[37] Taulavuori K, Pyysalo A, Taulavuori E, Julkunen-Tiitto R. Responses of phenolic acid and flavonoid synthesis to blue and blue-violet light depends on plant species. Environmental and Experimental Botany. 2018;**150**:183-187. DOI: 10.1016/j.envexpbot.2018.03.016

[38] Stagnari F, Di Mattia C, Galieni A, Santarelli V, D'Egidio S, Pagnani G, et al. Light quantity and quality supplies sharply affect growth, morphological, physiological and quality traits of basil. Industrial Crops and Products. 2018;**122**:277-289. DOI: 10.1016/j.indcrop.2018.05.073

[39] Nadeem M, Abbasi BH, Younas M, Ahmad W, Zahir A, Hano C. LED-enhanced biosynthesis of biologically active ingredients in callus cultures of *Ocimum basilicum*. Journal of Photochemistry and Photobiology B: Biology. 2019;**190**:172-178. DOI: 10.1016/j.jphotobiol.2018.09.011

[40] Asada K. Production and scavenging of reactive oxygen species in chloroplasts and their functions. Plant Physiology. 2006;**141**(2):391-396

[41] Hasanuzzaman M, Nahar K, Alam M, Roychowdhury R, Fujita M. Physiological, biochemical, and molecular mechanisms of heat stress

tolerance in plants. International Journal of Molecular Sciences. 2013;**14**(5): 9643-9684

[42] Sakamoto A, Murata N. Genetic engineering of glycinebetaine synthesis in plants: Current status and implications for enhancement of stress tolerance. Journal of Experimental Botany. 2000;**51**(342):81-88

[43] Balla K, Bencze S, Janda T, Veisz O. Analysis of heat stress tolerance in winter wheat. Acta Agronomica Hungarica. 2009;**57**(4):437-444

[44] Hoque, Tahsina S, AAM Sohag, DJ Burritt, and MA Hossain. Salicylic acid-mediated salt stress tolerance in plants. In Plant Phenolics in Sustainable Agriculture. Singapore: Springer; 2020. pp. 1-38

[45] Oh MM, Carey EE, Rajashekar CB. Environmental stresses induce health-promoting phytochemicals in lettuce. Plant Physiology and Biochemistry. 2009;**47**(7):578-583

[46] Lo Piero AR, Puglisi I, Rapisarda P, Petrone G. Anthocyanins accumulation and related gene expression in red orange fruit induced by low temperature storage. Journal of Agricultural and Food Chemistry. 2005;**53**(23):9083-9088

[47] Christie PJ, Alfenito MR, Walbot V. Impact of low-temperature stress on general phenylpropanoid and anthocyanin pathways: Enhancement of transcript abundance and anthocyanin pigmentation in maize seedlings. Planta. 1994;**194**(4):541-549

[48] Rivero RM, Ruiz JM, Garcıa PC, Lopez-Lefebre LR, Sánchez E, Romero L. Resistance to cold and heat stress: Accumulation of phenolic compounds in tomato and watermelon plants. Plant Science. 2001;**160**(2): 315-321

[49] Kasuga J, Hashidoko Y, Nishioka A, Yoshiba M, Arakawa K, Fujikawa S. Deep supercooling xylem parenchyma cells of katsura tree (*Cercidiphyllum japonicum*) contain flavonol glycosides exhibiting high anti-ice nucleation activity. Plant, Cell & Environment. 2008;**31**(9):1335-1348

[50] Weidner S, Karolak M, Karamac M, Kosinska A, Amarowicz R. Phenolic compounds and properties of antioxidants in grapevine roots [*Vitis vinifera* L.] under drought stress followed by recovery. Acta Societatis Botanicorum Poloniae. 2009;**78**(2): 97-103

[51] Amarowicz R, Weidner S, Wójtowicz I, Karmac M, Kosinska A, Rybarczyk A. Influence of low-temperature stress on changes in the composition of grapevine leaf phenolic compounds and their antioxidant properties. Functional Plant Science and Biotechnology. 2010;**4**:90-96

[52] Isshiki R, Galis I, Tanakamaru S. Farinose flavonoids are associated with high freezing tolerance in fairy primrose (*Primula malacoides*) plants. Journal of Integrative Plant Biology. 2014;**56**(2): 181-188

[53] Rana S, Bhushan S. Apple phenolics as nutraceuticals: Assessment, analysis and application. Journal of Food Science and Technology. 2016;**53**(4):1727-1738

[54] Commisso M, Toffali K, Strazzer P, Stocchero M, Ceoldo S, Baldan B, et al. Impact of phenylpropanoid compounds on heat stress tolerance in carrot cell cultures. Frontiers in Plant Science. 2016;**7**:1439

[55] Chalker-Scott L, Fuchigami LH. The role of phenolic compounds in plant stress responses. In: Low Temperature Stress Physiology in Crops. Boca Raton: CRC Press; 2018. pp. 67-80

[56] Zhang J, Kirkham MB. Drought-stress-induced changes in activities of superoxide dismutase, catalase, and

peroxidase in wheat species. Plant & Cell Physiology. 1994;**35**(5):785-791

[57] Shao HB, Chu LY, Lu ZH, Kang CM. Primary antioxidant free radical scavenging and redox signaling pathways in higher plant cells. International Journal of Biological Sciences. 2008;**4**(1):8

[58] Ashraf MA, Riaz M, Arif MS, Rasheed R, Iqbal M, Hussain I, et al. The role of non-enzymatic antioxidants in improving abiotic stress tolerance in plants. In: Plant Tolerance to Environmental Stress: Role of Phytoprotectants. Boca Raton: CRC Press; 2019. pp. 129-144.

[59] Kumar S, Bhushan B, Wakchaure GC, Meena KK, Kumar M, Meena NL, et al. Plant phenolics under water-deficit conditions: Biosynthesis, accumulation, and physiological roles in water stress alleviation. In: Plant Phenolics in Sustainable Agriculture. Singapore: Springer; 2020. pp. 451-465

[60] Akula R, Ravishankar GA. Influence of abiotic stress signals on secondary metabolites in plants. Plant Signaling & Behavior. 2011;**6**(11):1720-1731

[61] Nakabayashi R, Yonekura-Sakakibara K, Urano K, Suzuki M, Yamada Y, Nishizawa T, et al. Enhancement of oxidative and drought tolerance in Arabidopsis by over accumulation of antioxidant flavonoids. The Plant Journal. 2014;**77**(3):367-379

[62] Boudet A. Evolution and current status of research in phenolic compounds. Phytochemistry. 2007; **68**(22-24):2722-2735

[63] Hura T, Hura K, Grzesiak S. Contents of total phenolics and ferulic acid, and PAL activity during water potential changes in leaves of maize single-cross hybrids of different drought tolerance. Journal of Agronomy and Crop Science. 2008;**194**(2):104-112

[64] Phimchan P, Chanthai S, Bosland PW, Techawongstien S. Enzymatic changes in phenylalanine ammonia-lyase, cinnamic-4-hydroxylase, capsaicin synthase, and peroxidase activities in Capsicum under drought stress. Journal of Agricultural and Food Chemistry. 2014;**62**(29): 7057-7062

[65] Gharibi S, Tabatabaei BES, Saeidi G, Goli SAH. Effect of drought stress on total phenolic, lipid peroxidation, and antioxidant activity of Achillea species. Applied Biochemistry and Biotechnology. 2016;**178**(4):796-809

[66] Siracusa L, Gresta F, Sperlinga E, Ruberto G. Effect of sowing time and soil water content on grain yield and phenolic profile of four buckwheat (*Fagopyrum esculentum* Moench.) varieties in a Mediterranean environment. Journal of Food Composition and Analysis. 2017;**62**:1-7

[67] Sarker U, Oba S. Drought stress enhances nutritional and bioactive compounds, phenolic acids and antioxidant capacity of Amaranthus leafy vegetable. BMC Plant Biology. 2018;**18**(1):258

[68] Brunetti C, Loreto F, Ferrini F, Gori A, Guidi L, Remorini D, et al. Metabolic plasticity in the hygrophyte *Moringa oleifera* exposed to water stress. Tree Physiology. 2018;**38**(11):1640-1165

[69] Yang L, Wen K, Ruan X, Zhao Y, Wei F, Wang Q. Response of plant secondary metabolites to environmental factors. Molecules. 2018;**23**:276

[70] Parida AK, Das AB, Sanada Y, Mohanty P. Effects of salinity on biochemical components of the mangrove, *Aegiceras corniculatum*. Aquatic Botany. 2004;**80**:77-87

[71] Ksouri R, Megdiche W, Debez A. Salinity effects on polyphenol content and antioxidant activities in leaves of

the halophyte *Cakile maritima*. Plant Physiology and Biochemistry. 2007; **45**:244-249

[72] Hanen F, Ksouri R, Megdiche W, Trabelsi N, Boulaaba M, Abdelly C. Effect of salinity on growth, leaf phenolic content and antioxidant scavenging activity in *Cynara cardunculus* L. In: Abdelli C, Öztürk M, Ashraf M, Grignon YC, editors. Biosaline Agriculture and High Salinity Tolerance. Basel: Birkhauser Verlag; 2008. pp. 335-343

[73] Lim JH, Park KJ, Kim BK, Jeong JW, Kim HJ. Effect of salinity stress on phenolic compounds and carotenoids in buckwheat (*Fagopyrum esculentum* M.) sprout. Food Chemistry. 2012; **135**:1065-1070

[74] Petridis A, Therios I, Samouris G, Tananaki C. Salinity-induced changes in phenolic compounds in leaves and roots of four olive cultivars (*Olea europaea* L.) and their relationship to antioxidant activity. Environmental and Experimental Botany. 2012;**79**:37-43

[75] Borgognone D, Cardarelli M, Rea E, Lucini L, Colla G. Salinity source-induced changes in yield, mineral composition, phenolic acids and flavonoids in leaves of artichoke and cardoon grown in floating system. Journal of the Science of Food and Agriculture. 2014;**94**:1231-1237

[76] Lee NY, Lee MJ, Kim YK, Park J, Park HK, Choi JS, et al. Effect of light emitting diode radiation on antioxidant activity of barley leaf. Journal of Korean Society for Applied Biological Chemistry. 2010;**53**:685-690

[77] Rajeshwari V, Bhuvaneshwari V. Salicylic acid induced salt stress tolerance in plants. International Journal of Plant Biology & Research. 2017;**5**:1067

[78] Jini D, Joseph B. Physiological mechanism of salicylic acid for alleviation of salt stress in rice. Rice Science. 2017;**24**:97-108

[79] Khan MIR, Fatma M, Per TS, Anjum NA, Khan NA. Salicylic acid-induced abiotic stress tolerance and underlying mechanisms in plants. Frontiers in Plant Science. 2015;**6**:462

[80] Jayakannan M, Bose J, Babourina O, Rengel Z, Shabala S. Salicylic acid improves salinity tolerance in Arabidopsis by restoring membrane potential and preventing salt-induced K^+ loss via a GORK channel. Journal of Experimental Botany. 2013;**64**(8): 2255-2268

[81] Borsani O, Valpuesta V, Botella MA. Evidence for a role of salicylic acid in the oxidative damage generated by NaCl and osmotic stress in Arabidopsis seedlings. Plant Physiology. 2001; **126**:1024-1030

[82] Cao Y, Zhang ZW, XueL W, Du JB, Shang J, Xu F, et al. Lack of salicylic acid in Arabidopsis protects plants against moderate salt stress. Zeitschrift für Naturforschung. 2009;**C64**:231-238

[83] Asensi-Fabado M, Munné-Bosch S. The aba3-1 mutant of *Arabidopsis thaliana* withstands moderate doses of salt stress by modulating leaf growth and salicylic acid levels. Journal of Plant Growth Regulation. 2011;**30**:456-466

[84] Miura K, Sato A, Ohta M, Furukawa J. Increased tolerance to salt stress in the phosphateaccumulating Arabidopsis mutants siz1 and pho2. Planta. 2011;**234**:1191-1199

[85] Hao L, Zhao Y, Jin D, Zhang L, Bi X, Chen H, et al. Salicylic acid-altering Arabidopsis mutants response to salt stress. Plant and Soil. 2012;**354**:81-95

[86] Ashraf M, Akram NA, Arteca RN, Foolad MR. The physiological, biochemical and molecular roles of brassinosteroids and salicylic acid in

plant processes and salt tolerance. Critical Reviews in Plant Sciences. 2010;**29**:162-190

[87] Hayat Q, Hayat S, Irfan M, Ahmad A. Effect of exogenous salicylic acid under changing environment: A review. Environmental and Experimental Botany. 2010;**68**:14-25

[88] Palma F, López-Gómez M, Tejera NA, Lluch C. Salicylic acid improves the salinity tolerance of *Medicago sativa* in symbiosis with *Sinorhizobium meliloti* by preventing nitrogen fixation inhibition. Plant Science. 2013;**208**:75-82

[89] Khan MIR, Asgher M, Khan NA. Alleviation of salt-induced photosynthesis and growth inhibition by salicylic acid involves glycinebetaine and ethylene in mungbean (*Vigna radiata* L.). Plant Physiology and Biochemistry. 2014;**80**:67-74

[90] Lee SY, Damodaran PN, Roh KS. Influence of salicylic acid on rubisco and rubisco activase in tobacco plant grown under sodium chloride in vitro. Saudi Journal of Biological Sciences. 2014;**21**(5):417-426

[91] Ardebili NO, Saadatmand S, Niknam V, Khavari-Nejad RA. The alleviating effects of selenium and salicylic acid in salinity exposed soybean. Acta Physiologiae Plantarum. 2014;**36**(12):3199-3205

[92] Ciriakova A. Heavy metals in the vascular plants of Tatra Mountains. Oecologia Montana. 2009;**18**(1-2):23-26

[93] Sytar O, Kumar A, Latowski D, Kuczynska P, Strzałka K, Prasad MNV. Heavy metalinduced oxidative damage, defense reactions, and detoxification mechanisms in plants. Acta Physiologiae Plantarum. 2013;**35**(4):985-999

[94] Schützendübel A, Polle A. Plant responses to abiotic stresses: Heavy metal-induced oxidative stress and protection by mycorrhization. Journal of Experimental Botany. 2002; **53**:1351-1365

[95] Sakihama Y, Yamasaki H. Lipid peroxidation induced by phenolics in conjunction with aluminum ions. Biologia Plantarum. 2002;**45**(2):249-254

[96] Shemet SA, Fedenko VS. Accumulation of phenolic compounds in maize seedlings under toxic cadmium influence. Fiziologiia i biokhimiia kul'turnykh rastenii. 2005;**37**(6):505

[97] Ali MB, Singh N, Shohael AM, Hahn EJ, Paek KY. Phenolics metabolism and lignin synthesis in root suspension cultures of *Panax ginseng* in response to copper stress. Plant Science. 2006; **171**(1):147-154

[98] Kováčik J, Klejdus B, Bačkor M, Repčák M. Phenylalanine ammonia-lyase activity and phenolic compounds accumulation in nitrogen-deficient *Matricaria chamomilla* leaf rosettes. Plant Science. 2007;**172**(2):393-399

[99] Pawlak-Sprada S, Arasimowicz-Jelonek M, Podgórska M, Deckert J. Activation of phenylpropanoid pathway in legume plants exposed to heavy metals. Part I. Effects of cadmium and lead on phenylalanine ammonia-lyase gene expression, enzyme activity and lignin content. Acta Biochimica Polonica. 2011;**58**(2):211-216

[100] Márquez-García B, Fernandez-Recamales M, Cordoba F. Effects of cadmium on phenolic composition and antioxidant activities of *Erica andevalensis*. Environmental and Experimental Botany. 2012;**75**(1):159-166

[101] Malčovská SM, Dučaiová Z, Maslaňáková I, Bačkor M. Effect of silicon on growth, photosynthesis, oxidative status and phenolic compounds of maize (*Zea mays* L.)

grown in cadmium excess. Water, Air, and Soil Pollution. 2014;**225**(8):2056

[102] Kısa D, Elmastaş M, Öztürk L, Kayır Ö. Responses of the phenolic compounds of *Zea mays* under heavy metal stress. Applied Biological Chemistry. 2016;**59**(6):813-820

[103] Nicholson RL, Hammerschmidt R. Phenolic compounds and their role in disease resistance. Annual Review of Phytopathology. 1992;**30**:369-389

[104] Bittel P, Robatzek S. Microbe-associated molecular patterns (MAMPs) probe plant immunity. Current Opinion in Plant Biology. 2007;**10**:335-341

[105] Hammerschmidt R, Hollosy SI. Phenols and the onset and expression of plant disease resistance. In: Daayf F, Lattanzio V, editors. Recent Advances in Polyphenol Research. Vol. 1. Oxford, UK: Wiley-Blackwell; 2008. pp. 211-227

[106] Heil M. Indirect defence via tritrophic interactions. The New Phytologist. 2008;**178**:41-61

[107] Me'traux J-P, Lamodie're E, Catinot J, Lamotte O, Garcion C. Salicylic acid and induced plant defenses. In: Daayf F, Lattanzio V, editors. Recent Advances in Polyphenol Research. Vol. 1. Oxford, UK: Wiley-Blackwell; 2008. pp. 202-210

[108] Zipfel C. Pattern-recognition receptors in plant innate immunity. Current Opinion in Immunology. 2008;**20**:10-16

[109] Nicaise V, Roux M, Zipfel C. Recent advances in PAMP-triggered immunity against bacteria: Pattern recognition receptors watch over and raise the alarm. Plant Physiology. 2009; **150**:1638-1647

[110] Wu J, Baldwin IT. Herbivory-induced signalling in plants: Perception and action. Plant, Cell & Environment. 2009;**32**:1161-1174

[111] Postel S, Kemmerling B. Plant systems for recognition of pathogen-associated molecular patterns. Seminars in Cell & Developmental Biology. 2009;**20**(9):1025-1031

[112] Mikulic-Petkovsek M, Slatnar A, Veberic R, Stampar F, Solar A. Phenolic response in green walnut husk after the infection with bacteria *Xanthomonas arboricola* pv. *juglandis*. Physiological and Molecular Plant Pathology. 2011;**76**(3-4):159-165

[113] Cho M, Lee S. Phenolic phytoalexins in rice: Biological functions and biosynthesis. International Journal of Molecular Sciences. 2015;**16**(12):29120-29133

[114] Wang L, Sun R, Zhang Q, Luo Q, Zeng S, Li X, et al. An update on polyphenol disposition via coupled metabolic pathways. In: Expert Opinion on Drug Metabolism & Toxicology. London: Ashley Publications; 2018. pp. 1-15

[115] Ansari MA, Fatima Z, Hameed S. Sesamol: A natural phenolic compound with promising anticandidal potential. Journal of Pathogens. 2014;**2014**:895193

[116] Upadhyay A, Mooyottu S, Yin H, Nair M, Bhattaram V, Venkitanarayanan K. Inhibiting microbial toxins using plant-derived compounds and plant extracts. Medicine. 2015;**2**:186-211

[117] Negritto MC, Valdez C, Sharma J, Rosenberg C, Selassie CR. Growth inhibition and DNA damage induced by X-phenols in yeast: A quantitative structure-activity relationship study. ACS Omega. 2017;**2**(12):8568-8579

[118] Fernandes KRP, Bittercourt PS, Souza ADL, Souza AQL, Silva FMA, Lima ES, et al. Phenolic compounds from *Virola venosa* (Myristicaceae) and evaluation of their antioxidant and

enzyme inhibition potential. Acta Amaz. 2019;**49**(1):48-53

[119] Gallego-Giraldo L, Escamilla-Trevino L, Jackson LA, Dixon RA. Salicylic acid mediates the reduced growth of lignin down-regulated plants. Proceedings of the National Academy of Sciences of the United States of America. 2011;**108**:20814-20819

[120] Gallego-Giraldo L, Jikumaru Y, Kamiya Y, Tang Y, Dixon RA. Selective lignin downregulation leads to constitutive defense response expression in alfalfa (*Medicago sativa* L.). The New Phytologist. 2011;**190**:627-639

[121] Widodo GP, Sukandar EY, Adynyana IK. Mechanism of action of coumarin against *C. albicans* by SEM/TEM analysis. ITB Journal of Science. 2012;**44A**:145-151

[122] Rahman AU. Studies in Natural Product Chemistry. Vol. 24. Amesterdam: Elsevier; 2000

[123] Al-Barwani FM, Eltayeb EA. Antifungal compounds from induced *Conium macultum* L. plants. Biochemical Systematics and Ecology. 2004;**32**:1097-1108

[124] Al-Amiery AA, Kadhum AA, Mohamad AB. Antifungal activities of new coumarins. Molecules. 2012; **17**(5):5713-5723

[125] Serpa R, Franca EJ, Furlaneto-Maia L, Andrade CG, Diniz A, Furlaneto MC. In vitro antifungal activity of the flavonoid baicalein against Candida species. Journal of Medical Microbiology. 2012;**61**:1704-1708

[126] Zuzarte M, Vale-Silva L, Goncalves MJ, Cavaleiro C, Vaz S, Canhoto J, et al. Antifungal activity of phenolic-rich *Lavandula multifida* L. essential oil. European Journal of Clinical Microbiology & Infectious Diseases. 2012;**31**(7):1359-1366

[127] Belofsky G, Kolaczkowski M, Adams E, Schreiber J, Eisenberg V, Coleman CM, et al. Fungal ABC transporter-associated activity of isoflavonoids from the root extract of *Dalea formosa*. Journal of Natural Products. 2013;**76**(5):915-925

[128] Sherwood P, Bonello P. Austrian pine phenolics are likely contributors to systemic induced resistance against *Diplodia pinea*. Tree Physiology. 2013;**33**(8):845-854

[129] Anttila A-K, Pirttilä AM, Häggman H, Harju A, Venäläinen M, Haapala A, et al. Condensed conifer tannins as antifungal agents in liquid culture. Holzforschung. 2013; **67**(7):825-832

[130] Dos Santos C, Vargas Á, Fronza N, Dos Santos JHZ. Structural, textural and morphological characteristics of tannins from *Acacia mearnsii* encapsulated using sol-gel methods: Applications as antimicrobial agents. Colloids and Surfaces B: Biointerfaces. 2017;**151**:26-33

[131] Rashed YM, Aseel DG, Hafez EE. Antifungal potential and defense gene induction in maize against Rhizoctonia root rot by seed extract of *Ammi visnaga* (L.) Lam. Phytopathologia Mediterranea. 2018;**57**(1):73-88

[132] Marques JPR, Hoy JW, Appezzato-da-Glória B, Viveros AFG, Vieira MLC, Baisakh N. Sugarcane cell wall-associated defense responses to infection by *Sporisorium scitamineum*. Frontiers in Plant Science. 2018;**9**:698

[133] Ogawa S, Yazaki Y. Tannins from *Acacia mearnsii* De wild. Bark: Tannin determination and biological activities. Molecules. 2018;**23**(4):E837

[134] Kumar S, Pandey AK. Chemistry and biological activities of flavonoids:

An overview. Scientific World Journal. 2013;**2013**:162750

[135] Zakaryan H, Arabyan E, Oo A, Zandi K. Flavonoids: Promising natural compounds against viral infections. Archives of Virology. 2017;**162**(9): 2539-2551

[136] Shokoohinia Y, Sajjadi SE, Gholamzadeh S, Fattahi A, Behbahani M. Antiviral and cytotoxic evaluation of coumarins from *Prangos ferulacea*. Pharmaceutical Biology. 2014;**52**(12):1543-1549

[137] Dunkić V, Bezić N, Vuko E, Cukrov D. Antiphytoviral activity of *Satureja montana* L. ssp. *variegata* (host) P. W. Ball essential oil and phenol compounds on CMV and TMV. Molecules. 2010;**15**(10):6713-6721

[138] Hu Q-F, Zhou B, Huang J-M, Gao X-M, Shu L-D, Yang G-Y, et al. Antiviral phenolic compounds from *Arundina gramnifolia*. Journal of Natural Products. 2013;**76**(2):292-296

[139] Zhao W, Zeng XY, Zhang T, Wang L, Yang GY, Chen YK, et al. Flavonoids from the bark and stems of *Cassia fistula* and their anti-tobacco mosaic virus activities. Phytochemistry Letters. 2013;**6**:179-182

[140] Li L, Xu W-X, Liu C-B, Zhang C-M, Zhao W, Shang S-Z, et al. A new antiviral phenolic compounds from *Arundina graminifolia*. Asian Journal of Chemistry. 2015;**27**:3525-3526

[141] Liu CB, Shen QP, Wang Y, Zhang FM. Coumarins from the leaves of *Nicotiana tabacum* and their anti-tobacco mosaic virus activities. Chemistry of Natural Compounds. 2016;**52**:992-995

[142] Fraenkel G. The raison d'e^tre of secondary plant substances. Science. 1959;**129**:1466-1470

[143] Cornell HV, Hawkins BA. Herbivore responses to plant secondary compounds: A test of phytochemical coevolution theory. The American Naturalist. 2003;**161**:507-522

[144] Lattanzio V, Lattanzio VMT, Cardinali A. Role of phenolics in the resistance mechanisms of plants against fungal pathogens and insects. In: Imperato F, editor. Phytochemistry: Advances in Research. Trivandrum: Research Signpost; 2006. pp. 23-67

[145] Bhattacharya A, Sood P, Citovsky V. The roles of plant phenolics in defence and communication during Agrobacterium and Rhizobium infection. Molecular Plant Pathology. 2010;**11**:705-719

[146] Maxwell FG, Lafever HN, Jenkins JN. Blister beetles on glandless Cotton1. Journal of Economic Entomology. 1965;**58**:792-793. DOI: 10.1093/jee/58.4.792

[147] Feeny P. Seasonal changes in oak leaf tannins and nutrients as a cause of spring feeding by winter moth caterpillars. Ecology. 1970;**51**:565-581. DOI: 10.2307/1934037

[148] Levin DA. Plant phenolics: An ecological perspective. The American Naturalist. 1971;**105**(942):157-181

[149] Hedin PA, Jenkins JN, Thompson AC, McCarty JC, Smith DH, Parrott WL, et al. Effect of bioregulators on flavonoids, insect resistance and yield of seed cotton. Journal of Agricultural and Food Chemistry. 1988;**36**:1055-1061

[150] Luczynski A, Isman MB, Raworth DA. Strawberry foliar phenolics and their relationship to development of the two spotted spider mite. Journal of Economic Entomology. 1990;**83**:557-563. DOI: 10.1093/jee/83.2.557

[151] Byers JA. Host-tree chemistry affecting colonization in bark beetles. Chemical Ecology of Insects. 1995; **2**:154-213

[152] Lee DW, Gould KS. Why leaves turn red. American Scientist. 2002; **90**:524-531

[153] Gould KS. Nature's Swiss army knife: The diverse protective roles of anthocyanins in leaves. Journal of Biomedicine & Biotechnology. 2004; **5**:314-320

[154] Archetti M. Decoupling vigour and quality in the autumn colours game: Weak individuals can signal, cheating can pay. Journal of Theoretical Biology. 2009;**256**:479-484

[155] Archetti M, Döring TF, Hagen SB, Hughes NM, Leather SR, Lee DW, et al. Unravelling the evolution of autumn colours: An interdisciplinary approach. Trends in Ecology & Evolution. 2009; **24**:166-173

[156] Nikiforou C, Manetas Y. Strength of winter leaf redness as an indicator of stress vulnerable individuals in *Pistacia lentiscus*. Flora. 2010;**205**:424-427

[157] Hughes NM. Winter leaf reddening in "evergreen" species. The New Phytologist. 2011;**190**:573-581

[158] Rehman F, Khan FA, Badruddin SMA. Role of phenolics in plant defense against insect herbivory. In: Chemistry of Phytopotentials: Health, Energy and Environmental Perspectives. Berlin/Heidelberg: Springer; 2012. pp. 309-313

[159] Fürstenberg-Hägg J, Zagrobelny M, Bak S. Plant defense against insect herbivores. International Journal of Molecular Sciences. 2013;**14**:10242-10297. DOI: 10.3390/ijms140510242

Chapter 11

Citrus Peels as a Source of Bioactive Compounds with Industrial and Therapeutic Applications

Doha Hussien Abou Baker, Eman Ahmed Ibrahim and Zeinab Abd El-Rhaman Salama

Abstract

Agriculture wastes are considered a good starting point to discover for new drugs all over the world. In this context, Agriculture wastes contain millions of compounds to be screened to find bioactive compounds responsible for the activity to be used in drugs. Citrus agriculture is one of the most important commercial and industrial agricultural activities in the world. The peel waste of Citrus species is a rich source of bioactive compounds such as essential oils, flavones, polyphenols, and pigment. Citrus peel has been widely used in the medicine industry. The waste peel of citrus consider a rich source of pharmacologically active metabolites with antioxidant activities.

Keywords: Citrus waste, Phenolic compounds, Liminoids, Antioxidant activity, Therapeutic Activity, Industrial uses

1. Introduction

Agriculture crops, fruits, vegetables, cereals, bean crops produce large amount of wastes or by-products. These huge amounts of wastes could be of significant value if properly utilized. They could be more valuable than the main products and hence an added value will maximize these wastes. The main uses of these wastes are as animal feed or as compost used in enhancing soil fertility and used instead of chemical fertilization. Some products wastes such as banana stem waste can be used as an fiber for hand made paper and several grades of recycled papers. Beet waste can produce natural colors. Papaya is used as medicine (papain), toothpaste and meat tenderizers. Pine apple core for natural sweetener, grape pomace is a source of tartaric acid and polyphenols a natural antioxidants. Resveratrol a compound found in grape pomace known for its beneficial cardiovascular effects. Citrus peel includes cellulose, hemicellulose, lignin, pectin (galacturonic acid), chlorophyll pigments and other low molecular weight compounds (eg limonene) [1]. Polyphenol from grape seeds is used for management of Alzheimer disease [2]. Imitation vanilla is a liquid concentrate comes from treated wood pulp by –products.

In Egypt the major cultivated fruit trees are citrus, which came after mango and grapes in its cultivated area. Citrus has been cultivated in Egypt since ancient times, and there are some types grown in different regions such as Baladi orange, sweet (sugar) orange, and blood orange mandarin, lime, lemon, grapefruit, sour orange, kumquat, shadouk, pummelo and citron. At present, the area of citrus cultivation has increased rapidly, as this area reached 204,095 hectares, representing about 29% of the total fruit area (700,854 hectares), while the total productive area reached about 175734 hectares, producing approximately 4.27 million metric tons [3]. In Egypt these wastes were a main sources of agricultural cultural waste. The main use of these wastes are as animal feed or as compost. However higher amounts of these wastes were burned in the field or throwed in water canals, causing hazards and environmental pollution [4]. Recycling of such residues is one suitable technology adopted in industrial and developed countries, because these wastes are high value product and their recovery may be economically attractive. Citrus juice production generates 15 million tons of waste annually in the world, including peels, seeds, and fruit pulp [5].

Recently there is increase in the use of plant byproducts. The availability, potentiality, no side effect and no cost of byproducts in comparison to modern therapeutic drugs for the treatment of dangerous diseases such as cancer, and Alzheimer makes them more attractive [2, 6, 7]. Citrus fruits are the biggest fruit sector production all over the world, at the same context, waste the dominant byproduct of Citrus processing industries [8]. These citrus fruit residues, which are generally discarded as waste in the environment, can act as potential nutraceutical resources. Due to their low cost and easy availability such wastes are capable of offering significant low-cost nutritional dietary supplements. The utilization of these bioactive rich citrus residues can provide an efficient, inexpensive, and environment friendly platform for the production of novel nutraceuticals or for the improvement of older ones.

Citrus by –products is a major source of phenolic compounds; flavonoids, [9]. These flavonoids belong to six classes and have different biological activities i.e. antioxidant, anticancer, antiviral and antinflammatory. Dimou et al. 2019 on their review concluded that by –product of fruits and vegetables have an important role to be used as functional activity in cosmotics, nutraceuticals and as functional foods either in their raw material for additive processes or as ingredients for a new products [10]. Citrus waste have limonoids and flavonoids as their anticancer constituents. The most abundant citrus flavonoids, generally known as the flavanones, include hesperidin, naringin, narirutin, and neohesperidin, and these compounds have been found to provide health benefits such as antioxidative, anticancer, antiinflammatory, and cardiovascular protective activities. Furthermore, the consumption of naringin and hesperidin reduced cholesterol levels in hamsters by 32 to 40% [11].

2. Materials and methods

The peel waste of citrus fruit after juice extraction was obtained from a local food processing company. Samples were extracted using a different polar solvent. Yield extract of peels by different solvent was determined phenol contents according to Singleton et al., [12] total flavonoid were assayed by method Zhishen et al. [13].

Extract Limonin from citrus pee were prepared and purified according to the procedures Tian et al., and Tian et al. [14, 15]. A kilogram of powder peels was extracted (Soxhlet) overnight with hexane at 25°C to remove the oil. The solvent was changed to acetone and methanol sequentially to extract the peels. The methanol fraction was evaporated by a rotary evaporator under vacuum at 60°C and the residue was partitioned with 1:1 methylene chloride-water using an ultrasonic sonicator. The methylene chloride fraction and the previous acetone fraction were combined and evaporated to dryness or purification of the limonoid aglycones. Limonin was purified by repeated crystallization in methylene chloride and isopropanol.

Extraction of the essential oils.a kilogram of citrus peel were macerated in 1Lof distilled water during 24 h before extraction. Peels were then submitted to Clevenger hydrodistillation for 3 h. The obtained the essential oils were dried over anhydrous sodium sulfate and after filtration stored at 4C. The yield of extraction was isolated according to Williams and Lasunzi [16].

3. Useful materials and compounds isolated from citrus peels

3.1 Useful materials, e.g. dietary fibers

Dietary fiber which is often classified as soluble fiber and insoluble fiber consists of a mixture of vegetable carbohydrate polymers, both oligosaccharides and polysaccharides, eg. Inulin, pectin, gums, cellulose, and resistant starch (**Figure 1**) [17]. Fewer sources of fiber. Apart from helping to avoid digestion, and absorption in the small intestine, fiber has one of the following functions, increases colon fermentation, lowers cholesterol levels and maintains insulin levels [17]. Healthy people prefer natural supplements for fear that synthetic ingredients could be a source of poisoning. A high fiber by-product that is high in fiber and bioactive constituents is a treat for food processors. Dietary fiber supplements can produce more economical diet with many health benefits. The average daily fiber requirement is 25 g per day for women and 38 g per day for men [18]. Most nutritionists suggest that 30% of our daily fiber intake should come from soluble fiber. Apart from health benefits, dietary fiber has several functional properties such as increased shelf life, water retention capacity, emulsion stability, oil retention capacity, viscosity or gelling, bile acid and binding capacity.

The wastes of whole grains, and fruits that are produced in large quantities every day can be used as value-added products. They provide fiber as well as bioactive constituents such as polyphenols and EOs and offer economic benefits

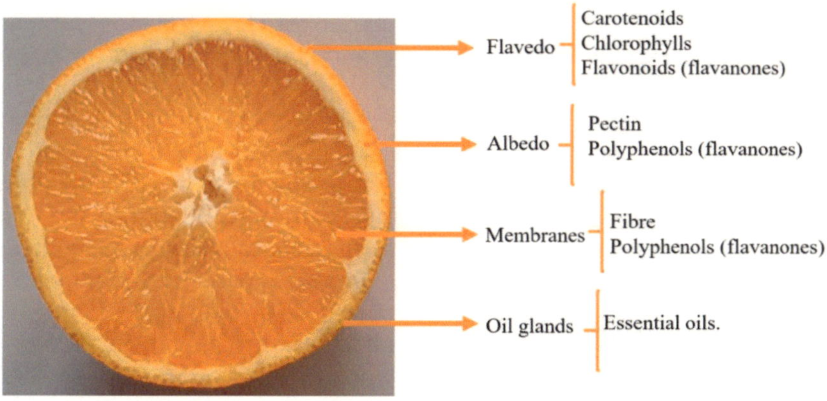

Figure 1.
Structure of peel from citrus sp.

to both producers and consumers. A typical example is the residue from the citrus waste treatment industry [19]. Garcia et al. reported that the addition of grain or fruit fiber, especially citrus fiber, can be used as a fat substitute in dry fermented sausages [20]. Citrus fiber, which has a bioactive function due to the presence of components such as polyphenols, can be used as an effective inhibitor of lipid oxidation in meat products, thereby increasing oxidation stability and extending its shelf life [21].

Citrus fiber can also be used to reduce residual nitrite levels [22]. Citrus waste can be seen as a potential source of pectin [23]. Fiber consumption is often associated with a lower risk of life-threatening chronic diseases such as gastrointestinal disease, intestinal disease, diabetes, cardiovascular disease, obesity, cancer and improved physiological functions including lowering blood cholesterol, weight loss due to glucose [24]. The effectiveness of citrus waste on low plasma liver cholesterol, serum triglyceride levels, total serum cholesterol, total liver lipids and liver cholesterol [23] has been proven by many epidemiologists. The waste fiber extracted from citrus fruits is involved in improving intestinal function and health [25]. Waste, cellulose and waste fiber from *C. hystrix* and *C. maxima* can be used as potential food fiber sources for food fortification due to their high physicochemical properties.

3.2 Phenolic compounds, e.g. flavonoids

Total phenolics from orange, mandarin, and lemon were 178.90, 169.54, and 61.22 mg GAE /100 peel, respectively [26] and flavonoid was 80.94 to 87.71 mg/rutin/100 g [27]. The main bioactive constituents known for their health benefits are phytochemicals, especially phenolic constituents found in vegetables and fruits. Studies report that phenols are not only present in edible parts of plants, but their presence has also been reported in inedible parts of plants with various biological effects. The mechanisms behind the contribution of phenolic constituents to improved health and prevention of related diseases include carcinogen inactivation, cell differentiation, maintenance of DNA repair, changes in estrogen metabolism and inhibition of N-nitrosamine formation. The main mechanisms for the antioxidant effects of phenolics in functional foods include metal chelating activity and free radical scavenging activity. It has been shown that reactive oxygen species such as superoxide radicals support human pathogenesis [28]. Phenols provide an effective way to prevent and treat free radical-mediated diseases such as cancer, neurodegenerative diseases, diabetes [29–31], the aging process [32] and cardiovascular dysfunction due to free radical scavenging and cooling ROS [33]. In addition, many of the antioxidants found in plants exhibit a variety of biological effects, including antiviral, antiallergic, anti-inflammatory, antibacterial, and antithrombotic effects [34].

Citrus considered one of the most popular fruit plants in the world, contains many active constituents that can protect health. In addition, it contains enough folic acid, vitamin C, pectin and potassium. Citrus species from various origins have been evaluated for their phytochemical composition and contribution to improved health [35], and it has been recognized that citrus species have promising biological properties, including anti-inflammatory, antiatherogenic, anti-tumor activity, anticoagulant and antioxidant activity [36].

Citrus waste has been shown to be rich in healthy constituents, including vitamin C, carotenoids, and polyphenol antioxidants [37]. Benamrouchea and Madania confirm that *C. sinensis* L. and *C. aurantium* L [38]. wastes are powerful antioxidants. In the last decade, Interesting phytochemicals such as 40-geranyloxyferulinic acid and boropic acid have been found to have valuable pharmacological effects such as chemoprophylactic, anti-inflammatory, neuroprotective and antipyloric agents. *C. sinensis* is richest sources of phytochemicals such as 40-geranyloxyferulinic acid and boropic acid [39].

Flavonoids are phenolic constituents with the structure of phenylbenzopyrone which are two benzene rings connected by a linear triangular carbon chain with a carbonyl group in position C. The prevention of serious chronic disease has attracted the attention of many researchers. Citrus flavonoids include one group of glycosides, namely naringin, and hesperidin [40]. Wang et al. mentioned that a 117 flavonoid were isolated from different *citrus species* by using LC–MS/MS. The flavonoids were identified as 39 polymethoxylated flavonoids (PMFs), 7 flavones, 10 C-O-glycosylflavonoids, 44 O-glycosylflavonoids, 10 C-glycosylflavonoids and 7 newly O-glycosylpolymethoxylated flavonoids, O-glycosylated flavonoids [41]. Citrus flavonoids have been shown to have health-related properties that include cancer-fighting, antiviral, and anti-inflammatory activities, reducing capillary fragility, and limiting human platelet aggregation [42]. The broad biochemical functions of flavonoids in citrus waste have recently been extensively studied. They increase the antioxidant capacity of serum against lipid peroxidation and reduce oxidative stress in the elderly [43]. These constituents have beneficial effects of anti-inflammatory, anti-tumor anti-diabetes, neuroprotective agent and anti-atherosclerosis [44–47]. HPLC analysis of citrus waste extract showed that hesperidin was present in all extracts in the highest concentrations [48]. The flavonoid glycosides naringin, didimine, pontsirin, narirutin [49]. Several reports highlighted the relationship between structure and antioxidant activity of the flavonoid subclass in citrus extracts. Johan found that bioactive compounds of flavonoids were extracted from orange peel. The compounds were polymethoxylated flavones, flavanone-O-trisaccharides, flavone-O-disaccharides, and, finally, flavone-C-glycosid. Flavonoids showed to have antioxidant properties [50].

Dry mandarin waste is used as a traditional Chinese medicine to cure various diseases including dyspepsia, bronchial asthma and cardiovascular disease [51]. Numerous scientific studies report that it is a rich source of many flavonoids, especially flavonoid glycosides, which play an important role in protecting against life-threatening diseases such as neurodegenerative disorders, atherogenesis and cancer [52–54].

3.3 Essential oils, e.g. liminoids

This citrus fruit is one of the most original oranges in Egypt. It is important to note that the citrus wastes variety offers an excellent EC yield. GC–MS analysis of citrus wastes essential oil EO identified many bioactive components. Terpenes form base-grade constituents and d-limonene. Interestingly, the resulting EO of orange waste showed remarkable antibacterial activity against C. acnes, which is a potential therapy for the treatment of acne. However, further research is needed to investigate the mechanism of their biological activity and its effect on C. acnes in order to exploit this EG on a commercial scale. Citrus EO is an important biologically active ingredient in orange wastes. It is collected intensively in the oil glands of orange wastes [55]. On average, it makes up about 1–3% of the skin weight of fresh orange wastes [56]. Citrus EO consists many different constituents depending on the citrus variety [57]. The ingredients also differ significantly depending on the method of extraction [58]. Citrus EO is widely used in the chemical, medical and food due to its pleasant aroma, antimicrobial activity, and antioxidant properties. The nature of EO is very attractive. Previous research has shown that orange EO has a broad spectrum of antimicrobial activity against yeasts, fungi, and bacteria and that activity mainly depends on the EO constituents [59].

Limonoids are a unique class of highly oxygenated tetracyclic triterpenoids, Members of the class limonoids have wide health-promoting and disease-preventing activities, including anticancer, antibacterial, antioxidant, larvicidal, antimalarial and antiviral activities, and thus has potential applications in nutriceuticals, pharmaceuticals and agriculture [60]. Kikuchi et al. mentioned that a new limonoids has been isolated from *Satsuma orange* and characterized as limonoids 1–5; 21,23-dihydro-21-hydroxy-23-oxonomilin (1),21,23-dihydro-23-methoxy-21-oxonomilin (2), 21,23-dihydro-21-hydroxy-23-oxonomilinic acid methylester (3), 21,23-dihydro-23-methoxy-21-oxolimonin (4), and 21,23-dihydro-21-oxolimonin (5), along with known compounds (6–12)(**Figure 2**) [61]. The most important citrus fruits are mandarin (*C. reticulata*), bergamot (*C. bergamia*), bitter orange (*C. aurantium*), lime (*C. aurantifolia*), sweet orange (*C. sinensis*), and lemon. (*C. limon*) [62]. The citrus limonoids are responsible for a wide variety of therapeutic properties such as antiviral, antifungal, antibacterial and antimalarial [63]. Senevirathne et al. and Miyake et al. reported on the occurance of limonoid in large amount in citrus juice and citrus tissue as water soluble glycosides and found in seeds as water insoluble a glycones [64, 65]. The latter is responsible for delaying the bitterness of citrus fruit. These limonoids are converted to the non bitter glycosides during maturation. These limonoids are similar to the limonoid found in Neem seeds and possess insecticidal activity.

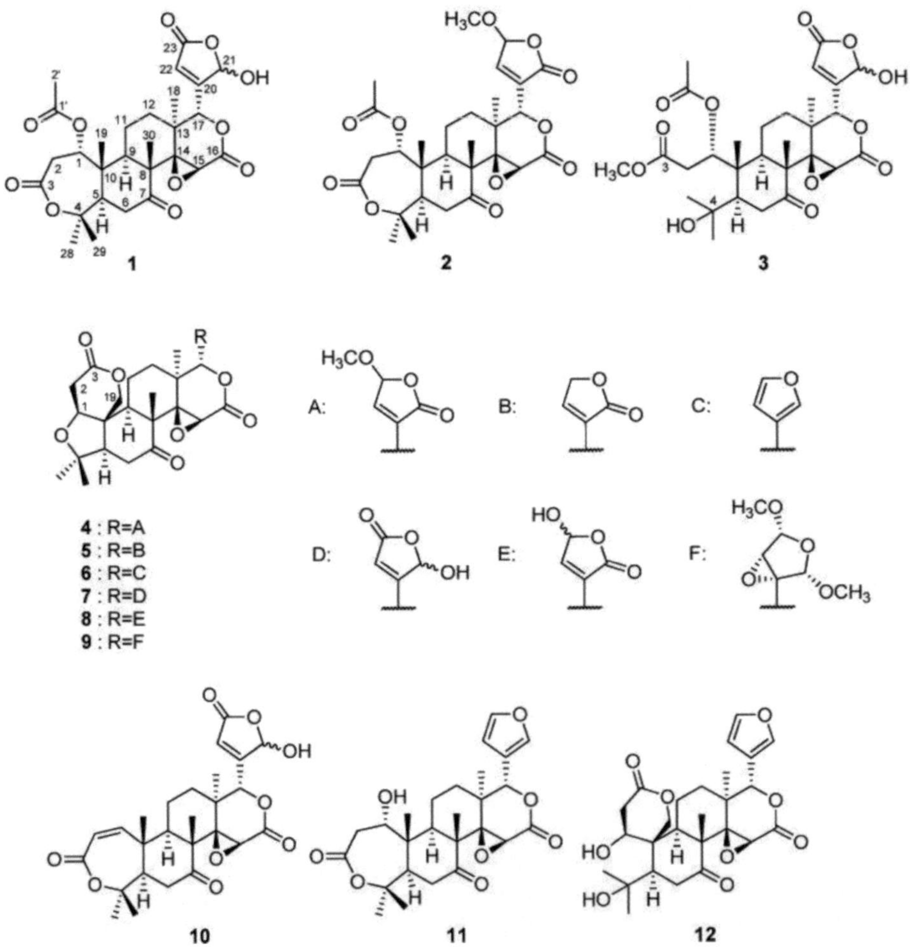

Figure 2.
New limonoids isolated from Satsuma orange.

The potential of citrus limonoids as anticancer agent was studied by Jacob et al. [66]. They concluded that limonin and nomilin topical application showed 60% reduction in tumer borden, however nomilin is less effective. Limonine glycoside and as aglycone administered in vitro to estrogen dependent and independent human breast cancer cell lines proved that the limonoids were equally potent like the standard drug tamoxifen for inhibiting the proliferation of estrogen dependant breast cancer cells, while more potent than tamoxifen for its activity against estrogen independent cancer cells.

4. Conclusions

Recent research on the functional properties of citrus wastes has added to our knowledge. Due to the low cost and availability of leftover fruit, which if not disposed of as environmental waste, should be seen as a potential source of nutrients capable of providing significant inexpensive nutritional supplements. Good use of citrus peels in the production of polystyrene. Polystyrene is one of the most common thermoplastic polymers used in the production of packaging materials and household and consumer goods. This unwanted manufacturing waste is rich in bioactive constituents and can be recycled as a value-added nutritional supplement that provides beneficial phenols, flavonoids, EOs, and fiber. They function as calorie-free fillers, enhance emulsions, increase water and oil retention, and can prevent a wide variety of ailments. Citrus waste extract holds promise as a source of bioactive constituents in the food industry. Some of the vital compounds extracted from citrus peels such as limonene, pectin, myrcene, and α-Pinene are used for flavor and good smell as safe food additives. Peels are a rich source of micro-nutrients and can be used as a source to improve the growth of agricultural crops and feed animals. According to literature, the biologically active compounds in the citrus peel can effectively prevent or inhibit diseases, enhance immune function, prevent cancer, and have antioxidant activity. In addition, the use of identified citrus waste will also help reduce pollution problems caused by poor residue disposal. Further research is needed to determine the bioavailability of this waste extract *in vivo.*

Acknowledgements

Replace the entirety of this text with acknowledgments. Usually, the acknowledgments section includes the names of people or institutions who in some way contributed to the work, but do not fit the criteria to be listed as the authors.

The authorship criteria are listed in our Authorship Policy: https://www.intechopen.com/page/authorship-policy.

This section of your manuscript may also include funding information.

Conflict of interest

The authors declare no conflict of interest.

Author details

Doha Hussien Abou Baker[1], Eman Ahmed Ibrahim[2]* and Zeinab Abd El-Rhaman Salama[2]

1 Medicinal and Aromatic Plant Department, National Research Centre (NRC), Dokki, Giza, Egypt

2 Plant Biochemistry Department, National Research Centre (NRC), Dokki, Giza, Egypt

*Address all correspondence to: eman_1975_11@yahoo.com

References

[1] Pathak, P. D., Mandavgane, S A. and Kulkarni, B. D. Fruit peel waste: characterization and its potential uses. Current Science, 2017. 113- 444-454. Doi: 10.18520/cs/v113/i03/444-45

[2] El-Gengaihi S, Emad M. Hassan, Abeer Y. Ibrahim, Faten M. Aboul Ella and Doha H. Abou Baker. Phenolic Compounds from Grape Wastes and their Impact in Neurodegenerative Disease. Journal of Chemical and Pharmaceutical Research. 2016a 8-207-217. Available online www.jocpr.com

[3] Abobatta, W.F. (2019). Citrus Varieties in Egypt: An Impression. International Research Journal of Applied Sciences. 2019, 1-63-66. pISSN: 2663-5577, eISSN: 2663-5585.

[4] Alnaimy, A., Gad A.E, Mustafa M.M., Atta M.A.A., Basuony, H.A.M. Using of citrus by-products in farm animals feeding. Sci. 2017;1(3):58-67. DOI: 10.15406/oajs.2017.01.00014.

[5] Kholaf, G. M., Gomaa E. G., Ziena H. M. Antimicrobial Activity of some Egyptian Citrus Peels Extracts. 2017 23- 872-883 DOI: 10.21608/ASEJAIQ JSAE.2017.4771

[6] Mossa, A.T.H., Ibrahim, F.M., Mohafrash, S.M., Abou Baker, D.H. and El Gengaihi, S. Protective effect of ethanolic extract of grape pomace against the adverse effects of cypermethrin on weanling female rats. Evidence-Based Complementary and Alternative Medicine. 2015.2015-1-10 doi: 10.1155/2015/381919

[7] El-Gengaihi, S. E., Hamed, M. A., Aboubaker, D. H., and Abdel-tawab, H. M..,. Flavonoids from sugar beet leaves as hepatoprotective agent. Int. J. Pharm., Sci., 2016b 8, -281-286. https://www.researchgate.net/publication/301490287

[8] Raimondo, M., Caracciolo, F., Cembalo, L., Chinnici G., Pecorino, B., and D'Amico, Mario. Making virtue out of necessity: Managing the citrus waste supply chain for bioeconomy applications. Sustainability. 2018, 10, 4821-2-19.doi:10.3390/su1012482.

[9] Rafiqa, S., Kaula, R., Sofia, S.A., Bashira, N., Nazirb, F., Nayik G.A. Citrus peel as a source of functional ingredient: A review. Journal of the Saudi Society of Agricultural Sciences. 2018.17- 351-358. Doi.org/10.1016/j.jssas.2016.07.006

[10] Dimou, C., Karantonis, H. C., Skalkos, D., Koutelidakis, A. E. alorization of Fruits by-products to Unconventional Sources of Additives, Oil, Biomolecules and Innovative Functional Foods. Current Pharmaceutical Biotechnology. 2019. 20- DOI : 10.2174/138920102066619 0405181537

[11] Kurowska, E.M., Manthey, J.A. Hypolipidemic effects and absorption of citrus polymethoxylated flavones in hamsters with diet-induced hyper-cholesterolemia. J Agric Food Chem. 2004;52:2879-2886. doi: 10.1021/jf035354z

[12] Singleton VL. and Rossi JA. Colorimetric of total phenolics with phosphomolibdic –phosphor tungstic acid reagents. American Journal of Enology and Viticulture. 1965. 16- 144-158.

[13] Zhishen J, Mengcheng T, and Jianming W. The determination of flavonoid contents in mulberry and their scavenging effects on superoxide radicals. Food Chemistry. 1999; 64: 555-559

[14] Tian, Q., Miller, E.G., Ahmad, H., Tang, L. and Patil B.S. Differential inhibition of human cancer cell proliferation by citrus limonoids. Nutr. Cancer. 2001.40-180-184.

[15] Tian, Q.G., Barbacci, D. Schwartz, S. and Patil B.S. Electron ionization of mass spectrometry. Rapid Commun. Mass. 2003.17-2517-2522

[16] Williams, I.R., and Lasunzi, I. Essential oil from Melaleuce dissi. Ora a potential source of high quality tea tree oil. Ind, Crop. Prod. 1994.211-217

[17] Fuentes-Zaragoza, E., Riquelme-Navarrete, M.J., Sanchez-Zapata, E., Perez-Alvarez, J.A. Resistant starch as functional ingredient: a review. Food Res. Int. , 2010. 43 (4), 931-942. doi.org/10.1016/j.foodres.2010.02.004

[18] Institute of Medicine, National Academy of Sciences. Dietary reference intakes: proposed definition of dietary fiber. Washington, DC: National Academy Press; 2001. doi.org/10. 17226/10161.

[19] Braddock, R.J. Handbook of Citrus By-products and Processing Technology. , 1999. John Wiley, New York.

[20] Garcia, M.L., Dominguez, R., Galvez, M., Gavlez, D., Casas, C., Selgas, M.D. Utilisation of cereal and fruit fibres in low fat dry fermented sausage. Meat Sci., 2002. 60, 227-236. Doi.org/10.1016/S0309-1740(01) 00125-5

[21] Sayago-Ayerdi, S.G., Brenes, A., Goni, I. Effect of grape antioxidant dietary fibre on the lipid oxidation of raw and cooked chicken hamburgers. LWT Food Sci. Technol. 2009. 42, 971-976. DOI : 10.1016/j.lwt.2008. 12.006

[22] Fernandez-Gines, J.M., Fernandez-Lopez, J., Sayas-Barbera, E., Sendra, E., Perez Alvarez, J.A., Effect of storage conditions on quality characteristics of bologna sausages made with citrus fibre. J. Food Sci. 2003. 68, 710-715. Doi:10. 1111/j.1365-2621.2003.tb05737.x

[23] Terpstra, A.H., Lapre, J.A., Vries, H.T., Beynen, A.C. The hypo cholesterolemic effect of lemon peels, lemon pectin, and the waste stream material of lemon peels in hybrid F1B hamsters. Eur. Nutr. 2002. 41 (1), 19-26. DOI: 10.1007/s0039402 00002.

[24] Figuerola, F., Hurtado, M.L., Estevez, A.M., Chiffelle, I., Asenjo, FFibre concentrates from apple pomace and citrus peel as potential fibre sources for food enrichment. Food Chem. 2005.91, 395– 401. doi.org/10.1016/j. foodchem.2004.04.036

[25] Chau, C.F., Sheu, F., Huang, Y.L., Su, L. H. Improvement in intestinal function and health by the peel fibre derived from Citrus sinensis L cv Liucheng. J. Sci. Food Agric., 2005. 85, 1211-1216.

[26] Al-Juhaimi, F .Y. Citrus fruit by- products as sources of bioactive compounds with antioxidant potential. Pak. J. Bot., 2004. 46(4): 1459-1462.

[27] Abd El-aal H.A., Halaweish F.T. Food preservative activity of phenolic compounds in organge peel extracts (Citrus sinesis L) Lucrări Ştiinţifice. Universitatea de Ştiinţe Agricole şi Medicină VeterinarăIaşi. 2010. 53- 233-240

[28] Aruoma, O.I. Methodological considerations for characterizing potential antioxidant actions of bioactive components in plant foods. Mutat. Res., 2003. 523-524, 9-20. Doi. org/10.1016/S0027-5107(02)00317-2

[29] Boynes, J.W. Role of oxidative stress in the development of complication in diabetes. Diabetes. 1991. 40, 405-411. Doi.org/10.2337/diab.40.4.405

[30] Huang, R.P., Golard, A., Hossain, M.Z., Huang, R., Liu, Y.G., Boynton, A.L. Hydrogen peroxide promotes transformation of rat liver non-neoplastic epithelial cells through activation of epidermal growth factor

receptor. Mol. Carninog., 2001. 30, 209-217. Doi.org/10.1002/mc.1030

[31] Perry, G., Raine, K.A., Nunomura, A., Watayc, T., Sayre, L.M., Smith, M.A. How important is oxidative damage? Lessons form Alzheimer's disease. Free Radic. Biol. Med., 2000. 28, 831-834.

[32] Hensley, K., Floyd, R.A.,. Reactive oxygen species and proteinoxidation in aging: a look back, a look ahead. Arch. Biochem. Biophys., 2002.397- 377-383. Doi.org/10.1006/abbi.2001.2630

[33] Hool, L.C. Reactive oxygen species in cardiac signaling: from mitochondria to plasma membrane ion channels. Clin. Exp. Pharmacol. Physiol., 2006. 33, 146-151. http://www.aups.org.au/Proceedings/36/55-61

[34] Cook, N.C., Sammon, S. Flavanoids-chemistry, metabolism, cardioprotective effects, and dietary sources. Nutr. Biochem. 1996.7-66-76. Doi.org/10.1016/S0955-2863(95)00168-9

[35] Guimarães, R., Barros, L., Barreira, J.C.M., Sousa, M.J., Carvalho, A.M., Ferreira, I.C.F.R. Targeting excessive free radicals with peels and juices of citrus fruits: grapefruit, lemon, lime and orange. Food Chem. Toxicol. 2009. 48 (1)-99-106. Doi.org/10.1016/j.fct.2009.09.022

[36] Montanari, A., Chen, J., Widmer, W., 1998. Citrus flavonoids: a review of past biological activity against disease. In: Manthey, J.A., Buslig, B.S. (Eds.), Flavonoids in the Living System. Plenum Press, New York, pp. 103-113.

[37] Anwar, F., Naseer, R., Bhanger, M.I., Ashraf, S., Talpur, F.N., Aladededune, F.A. Physicochemical characteristics of citrus seeds and oils from Pakistan. J. Am. Oil Chem. Soc., 2008. 85-321-330. Doi.org/10.1007/s11746-008-1204-3

[38] Benamrouchea, S.L., Madania, K. Phenolic contents and antioxidant activity of orange varieties (Citrus sinensis L. and Citrus aurantium L.) cultivated in Algeria: peels and leaves. Ind. Crops Prod., 2013 50, 723-730. doi.org/10.1016/j.indcrop.2013.07.048

[39] Genovese, S., Fiorito, S., Locatelli, M., Carlucci, G., Epifano, F. Analysis of biologically active oxyprenylated ferulic acid derivatives in Citrus fruits. Plant Foods Hum. Nutr., 2014. 69 (3)-255-260. DOI 10.1007/s11130-014-0427-8

[40] Li, S., Wang, H., Guo, L., Zhao, H., Ho, C.T. Chemistry and bioactivity of nobiletin and its metabolites. J. Funct. Foods. 2014. 6- 2-10. Doi.org/10.1016/j.jff.2013.12.011

[41] Wang, S., Yang, C., Tu, H., Zhou, J., Liu, X., Cheng, Y., Luo, J., Deng X., Zhang H., Xu, J. Characterization and metabolic diversity of flavonoids in citrus species. Scientific Reports. 2017. 7-1-10. doi:10.1038/s41598-017-10970-2.

[42] Benavente-Garcia, O., Castillo, J., Francisco, R., Ana-Ortuno, M., Delrio, J. Uses and properties of citrus flavonoids. J. Agric. Food Chem., 1997. 45, 4505-4515. Doi.org/10.1021/jf970373s

[43] Assini, J.M., Mulvihil, E.E., Sutherland, B.G. Naringenin prevents cholesterol-induced systemic inflammation, metabolic dysregulation, and atherosclerosis in Ldlr/mice. J. Lipid. Res., 2013.54-711-724. doi.org/10.1194/jlr.M032631

[44] Aruoma, O.I., Landes, B., Ramful-Baboolall, D. Functional benefits of citrus fruits in the management of diabetes. Prev. Medi., 2012. 54-12-16. doi.org/10.1016/j.ypmed.2012.02.012

[45] Hwang, S.L., Shih, P.H., Yen, G.C. Neuroprotective effects of citrus flavonoids. J. Agric. Food Chem., 2012. 60-877-885. doi.org/10.1021/jf204452y

[46] Park, E., Pezzuto, J.M. Flavonoids in cancer prevention. Anti Cancer Agents

Med. Chem., 2012. 12 (8)-836-851. doi.org/10.2174/187152012802650075

[47] Mulvihill, E.E., Huff, M.W. Citrus flflavonoids and the prevention of atherosclerosis. Cardiovasc. Hematol. Disord. Drug Targets. 2012. 12 (2)-84-91. doi.org/10.2174/18715291280390 1210

[48] Londono-Londono, J., de Lima, V.R., Lara, O., Gil, A., Pasa, T.B.C., Arango, G.J., Pineda, J.R.R. Clean recovery of antioxidant flavonoids from citrus peel: optimizing an aqueous ultrasound assisted extraction method. Food Chem., 2010 119- 81-87. Doi.org/10.1016/j.foodchem.2009.05.075

[49] Tomás-Barberán, F.A., Clifford, M.N. Flavanones, chalcones and dihydrochalcones– nature, occurrence and dietary burden. J. Sci. Food Agric., 2000.80-10731080.Doi.org/10.1002/(SICI)10970010(20000515)80:7<1024::AID-JSFA567>3.0.CO;2-S

[50] John, A. M. Fractionation of orange peel phenols in ultrafiltered molasses and mass balance studies of their antioxidant levels. J. Agric. Food Chem., 2004. 52-7586–7592

[51] China Pharmacopoeia Committee, 2010. Chinese Pharmacopoeia (II). Beijing, China. https://www.scirp.org/reference/ReferencesPapers.aspx?ReferenceID=287958

[52] Tripoli, E., Guardia, M.L., Giammanco, S., Majo, D.D., Giammanco, M. Citrus flavonoids: molecular structure, biological activity and nutritional properties: a review. Food Chemistry. 2007.104 (2)-466-479

[53] Benavente-Garcia, O., Castillo, J. Update on uses and properties of citrus flavonoids: new findings in anticancer, cardiovascular, and anti-inflammatory activity. J. Agric. Food Chem., 2008. 56, 6185– 6205. Doi.org/10.1021/jf8006568

[54] Hwang, S.L., Shih, P.H., Yen, G.C. Neuroprotective effects of citrus flavonoids. J. Agric. Food Chem., 2012. 60-877-885. doi.org/10.1021/jf204452y

[55] Boussaada, O., Skoula, M., Kokkalou, E., Chemli R. Chemical variability of flowers, leaves, and peels oils of four sour orange provenances. J. Essent. Oil Bear. Plants. 2007.10 - 453-464. Doi.org/10.1080/09720 60X.2007.10643579

[56] Njoroge, S.M., Koaze, H., Karanja, P.N., Sawamura M. Essential oil constituents of three varieties of Kenyan sweet oranges (Citrus sinensis). Flavour Fragrance J., 2005.20 - 80-85 . Doi.org/10.1002/ffj.1377

[57] Sharma, K., Mahato, N., Cho, M.H., Lee Y.R. Converting citrus wastes into value-added products: economic and environmently friendly approaches. Nutrition. 2017. 34 -29-46. Doi.org/10.1016/j.nut.2016.09.006

[58] González-Mas, M.C., Rambla, J.L., López-Gresa, M.P., Blázquez, M.A., Granel, A. l.Volatile compounds in citrus essential oils: a comprehensive review Front. Plant Sci., 2019.10- 12. doi.org/10.3389/fpls.2019.00012

[59] Reyes-Jurado, F., Navarro-Cruz, A.R., Ochoa-Velasco, C.E., Palou, E., López-Malo, A., Ávila-Sosa R. Essential oils in vapor phase as alternative antimicrobials: A review Crit. Rev. Food Sci. Nutr., 2019. 18 - 1-10. Doi.org/10.1080/10408398.2019.1586641

[60] Olatunji, T.L., Odebunmi, C. A.and Ademola A. E. Biological activities of limonoids in the Genus Khaya (Meliaceae): a review. Future Journal of Pharmaceutical Sciences. 2021.7-2-18. Doi.org/10.1186/s43094-021-00197-4

[61] Kikuchi, T., Ueno, Y., Hamada, Y., Furukawa, C., Fujimoto, T.Takeshi Yamada and Reiko Tanaka. New Limonoids from Peels of Satsuma

Orange (*Citrus reticulata*) Molecules . 2017. 22(6)-3 -10doi:10.3390/ molecules22060907

[62] Tranchida, P.Q., Bonaccorsi, I., Dugo, P., Mondello, L., Dugo G., Analysis of Citrus essential oils: state of the art and future perspectives. A review Flavour Fragr. J., 2012.27 - 98-123. Doi.org/10.1002/ffj.2089

[63] Manners, G, D. Citrus limonoids: analysis, bioactivity, and biomedical prospects J Agric Food Chem., 2007.55(21)-8285-8294. doi: 10.1021/ jf071797h

[64] Senevirathne, M.; Teon, Y.J.; Ha, J.H.; Kim, S.H. The role of essential fatty acids in human health. J Food En., 2009. 92- 157-163. Doi.org/10.1016/j. jfoodeng.2008.10.033

[65] Miyake, Y.; Yamamato, K.; Morimastsa, Y. and Osawa, T. Isolation of C-glucosyflavone from lemon peel & antioxidated activity of flavonoid compounds in lemon fruit. J Agric Food chem., 1997. 45-4619 -4623. Doi. org/10.1021/jf970498x

[66] Jacob, R., Hasegawa, S., and Manner, G. The Potential of Citrus Limonoids as Anticancer Agents Robert. erishables Handling Quarterly. 2000. 102-6-8. ucce.ucdavis.edu/files/ datastore/234-200

Chapter 12

Brazilian Caatinga: Phenolic Contents, Industrial and Therapeutic Applications

Elba Lúcia Cavalcanti de Amorim, Patrícia Cruz, Jorge Veras Filho, Italo Caio Silva, Uyara Costa, Jenifer Oliveira, Maria Santa Medeiros, Marcelino Diniz, Kivia Machado and Ana Caroline Xavier

Abstract

Phenolic compounds, mainly represented by flavonoids, tannins and coumarins, bioactive molecules with various applications, have antioxidant, photoprotective, antimicrobial, anti-inflammatory and even antitumor properties. The main mechanism of action of phenolic compounds is due to the transfer of electrons to free radicals, which leads to the interruption of oxidative reactions. The flora of the Brazilian caatinga is full of species with high concentrations of these compounds, which are possibilities for researching new pharmaceutical products and functional foods, and may even generate technological and economic impact, contributing directly or indirectly to the development of communities that are inserted in this context. This is extremely important, considering the large amount of ecotoxic residues resulting from the industrial chain, where it is necessary to use methods to reduce this impact on the environment, such as adsorption, oxidation, biotransformation, liquid–liquid partition and hybrid techniques. This shows the need to reuse this waste and even improve production processes in order to make the most of the content of these compounds with varied applications that sometimes end up being underused. This chapter brings some of the main species involved in this context, their contributions to health and possible applications at a technological, industrial and even sustainable level.

Keywords: Phenolic compounds, Folk Medicine, Ethnopharmacology, Industrial, Therapeutic

1. Introduction

Brazil is often mentioned as a country with great diversity, which takes into account its historical miscegenation considering the participation of native peoples, colonizers and immigrants from the most varied countries [1], relating this also with its biodiversity contained in its ecosystems that are divided into six biomes: Amazon, Atlantic Forest, Caatinga, Cerrado, Pantanal and Pampa [2]. This characteristic contributes to the modification of the human relationship with the environment, as well as in the extraction of its resources and its applications, contributing to the development, but having an impact on the preservation of the typical species of each of these environments. A problem that has attracted the attention of researchers, since natural alternatives can, on the one hand, minimize environmental damage, such as, for example, the use of natural pigments instead of synthetics or the reprocessing of waste from the production chain [3], however on the other hand, the extraction of these resources in an indiscriminate way can lead to important environmental imbalances, in addition to threatening species with extinction [4].

The use of plants, whether for technological, medicinal, food, commercial or even religious purposes, is directly interconnected with the knowledge shared through generations within the cultural apparatus of the people who cultivate or live in their surroundings, knowledge that can be registered and monitored through ethnobotanical, ethnogeographic and ethnopharmacological studies, mainly [5, 6]. Since their properties are related to the content of secondary metabolites carried by these plant species, chemical components that are synthesized by plants to defend itself against pathogens and predators or to favor germination, thus associated with the production and maintenance strategy of the species and that are called secondary because they are not associated with the growth, development and structure of species. Among these metabolites are the essential oils, alkaloids, quinones, saponins and phenolic compounds, which include tannins, flavonoids and coumarins [7, 8].

Phenolic compounds, which are substances that have an aromatic ring with one or more hydroxy substituents [9], in the plants that synthesize them, act as allelopathic, preventing other species from interfering with their growth and development, fighting pests, including microorganisms and parasites, promote coloring and characteristic odors, protect against ultraviolet radiation [10] and are potent antioxidants, a function that is one of the most explored when using these compounds for medicinal purposes [11], and is also the largest source of antioxidants in human food, present in leaves, fruits and teas. The application of these resources, however, can be optimized, in order to promote their use also at a technological level, considering their versatility [12]. The present study intends to address these applications, highlighting these compounds as a source of innovations and sustainable development.

2. Caatinga: initial considerations, geographic and cultural aspects

The biome Caatinga extends in the northeast of Brazil, covering the states of Ceará, Rio Grande do Norte, with greater distribution in the states of Paraíba and Pernambuco, in addition to the southeast of Piauí, west of Alagoas and Sergipe, north and center of Bahia and part of Minas Gerais, as illustrated in **Figure 1** [13]. Its name comes from the Tupi Guarani language that means "White Forest", due to the dry seasons, where only the trunks and shrubs, without leaves, remain in the environment. It is the only biome found exclusively in Brazil, one of the least studied and, consequently, one of the least protected, where only 2% of the region is inside of a protected area. For this reason, the Caatinga continues to be one target of deterioration and changes in its territory, caused by the non-responsible use of its natural resources [14].

The caatinga weather has extreme characteristics, with high solar radiation and low cloudiness, presenting the lowest average annual humidity, low evapotranspiration and irregular and sparse occurrence of rain through the year [15]. And despite being highly threatened, the caatinga is seen as a poor area, a common view about arid and semi-arid areas around the world. However, the caatinga exhibits a vast

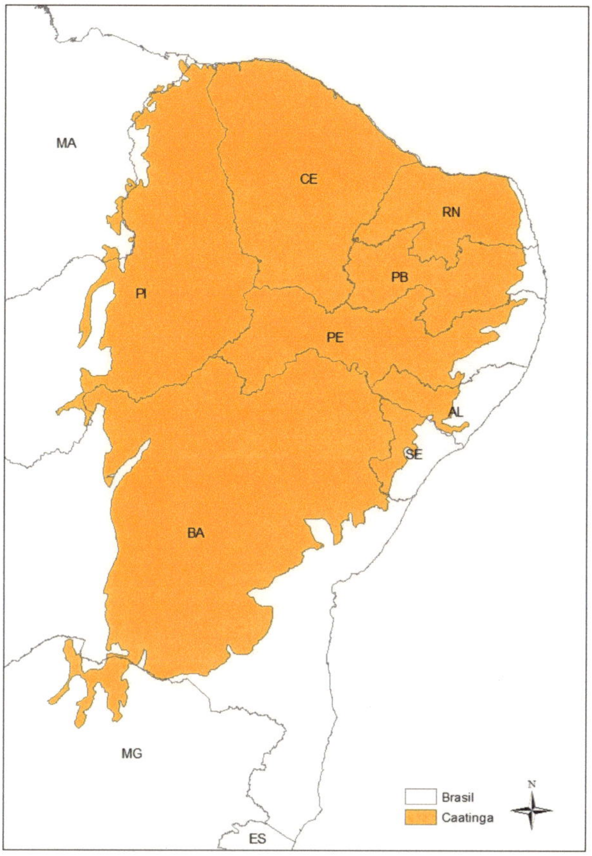

Figure 1.
Geographical location of the caatinga biome. Map available in the National Forest Information System of the Brazilian Forest Service.

biodiversity, adapted to the difficult climatic conditions and is also associated with the source of natural resources, such as wood and medicinal plants. Some of these characteristics can be seen in **Figure 2**, which shows an area of the caatinga in the state of Pernambuco [16]. In the northeastern semiarid, the application of these resources is quite popular, considering the socioeconomic component, since there is a large concentration of families dependent on subsistence agriculture and cattle raising that frequently applies these species as low cost medicinal and food alternatives [17].

The environment adversities, however, end up bringing a high concentration of secondary metabolites in the plant species of the caatinga, including the phenolic compounds, since these compounds respond to the stress caused by the environment, in order to promote survival [18], which can make these resources have different applicability and have a excellent application, as medicinal purpose, in the treatment of effluents, such as natural pigments, in leather tanning for cultural or commercial purposes, in the production of wines, in the optimization of functional foods, among other applications. It is necessary to bring attention, in this case, to the conscious use of these resources aiming to reduce the impact on the environment, with less degradation and predatory extraction, but under the guidance of using natural resources without bringing ecological imbalances that encompass both flora and fauna [19].

Ethnobiology and ethnoecology have already shown that the knowledge of local communities on use, management, including ethnic, biological and cultural

Figure 2.
Vegetation in the Brazilian caatinga, in the state of Pernambuco.

implications, is extremely important for the issue of conservation of natural resources [20]. Ethnobotanical surveys have been carried out in recent decades in areas of the caatinga in order to register their species, their importance related to botany, preservation and also the development of new medicines. So, studies with this purpose can be useful, by increasing the focus on these environments. Since ethnopharmacological studies have revealed that many people who live within the context of the caatinga use their species as the first source of healing, using these alternatives for the treatment of various illnesses, such as coughs and colds, wound healing, with antimicrobial, antiparasitic and pesticide purposes, and even chronic diseases such as hypertension and diabetes. With majoritarian use of the species bark, considering that it is the part that is available throughout the year, even in the dry periods, followed by the leaves and seeds and fruits. The most cited species in these studies are mentioned in Section 4 [21].

3. Phenolic compounds in caatinga plant species: methods for obtaining, characterizing and purifying

Phenolic compounds are a group of substances easily found in nature, present in several plant species [22]. In this group are flavonoids, tannins, coumarins and other phenolic acids that are essential for the development of plants, acting mainly in the protection of stress caused by the environment, such as insects, infections, UV radiation, among others [23]. They are metabolites with increasing pharmacological interest, among the activities attributed to them are anti-inflammatory, antioxidant and antibacterial [24]. The chemical structure of these chemical species has similarities as well as specificities.

Flavonoids, tricyclic compounds with an arrangement of 15 atoms, have activities both in the species where they occur and in a medicinal way, regarding protection against various pathogens, including fungi and bacteria, in addition to viruses and insects; photoprotection and antioxidant action, mainly, also acting as anti-inflammatories, through enzymatic modulation. Tannins are characterized by their astringency, which guarantees, in the species in which they are observed, defense against predators, beside antioxidant and antimicrobial activity, in addition of the ability to combine itself with macromolecules such as proteins to form insoluble complexes [25]. Their structures can be viewed in **Figure 3**.

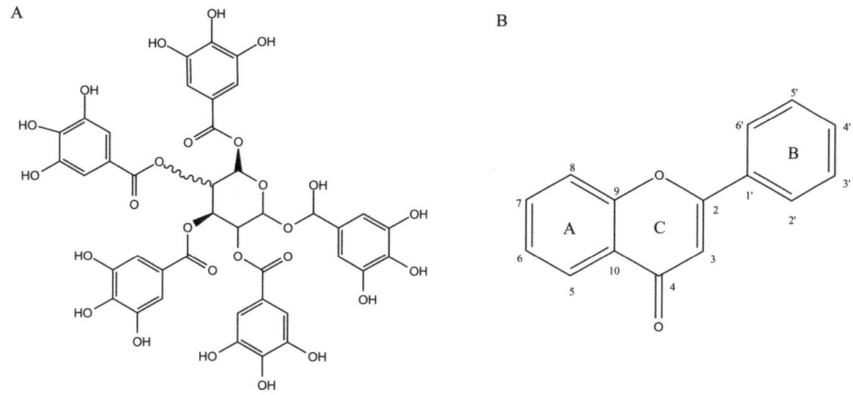

Figure 3.
Structural representation of tannins (A) and flavonoids (B).

Coumarins also have antioxidant effects and in plants they act as enzyme inhibitors, in the control of plant growth, in respiration, photosynthesis and in defense against infections and their biological applications are associated with their ability to make non-covalent interactions with protein structures, having a broad spectrum of biological activities, including the synthesis of a potent anticoagulant, warfarin. This is the most popular application of coumarins in medical sciences, which have served as a basis for new research that pursues the development of new synthetic alternatives with better response and fewer side effects, such as risks of bleeding when using these drugs [26].

The extraction is one of the first stages of studies with medicinal plants, being extremely important in the results obtained, as it interferes in the qualitative and quantitative tests of the metabolites, playing a fundamental role in the result of the processing of pharmaceutical and food products. This stage can suffer interference from several factors such as temperature, extraction time, solvent and part of the plant used, besides seasonal effects, considering that environmental aspects can bring modifications in the final chemistry composition of the species [27]. Before that, it is necessary to proceed the identification of the species to be studied by confirming botanical parameters, including macro and microscopic tests, considering the possibility of mistakes, due to morphological and even synonymic similarities of the popular names of these species [27]. From then on, a phytochemical study begins, which must go through extraction processes that can be identified in **Figure 4**.

In order to promote a higher yield of these resources extracted from plants, to guarantee their use at the technological level, whether for medicinal, food or in the production chain, in general, studies appear with the objective of optimizing, mainly, methods of extraction, addressing issues such as solvent, temperature, pH, quantity of material, standardization of new techniques and procedures or updating of protocols already in use [28]. The extraction time is one of the parameters that is often optimized, considering that the longer the extraction time, higher the difficulty to apply the method on a large scale, beside the potential material loss due to degradation, since these compounds, being antioxidants, also end up undergoing degradation by the action of light [29].

To assist in the quantification of phenolic compounds, it is possible to use simple spectrophotometer techniques, favoring efficiency of the process and also reducing costs at this stage, maintaining the quality of the analysis. For the research of flavonoids, the standard for performing the calibration curve is rutin. For tannins, tannic acid can be used as a standard and the coumarins, can be analyzed through the Borntrager reaction with a calibration curve using 1,2-benzopyrone as a standard [30].

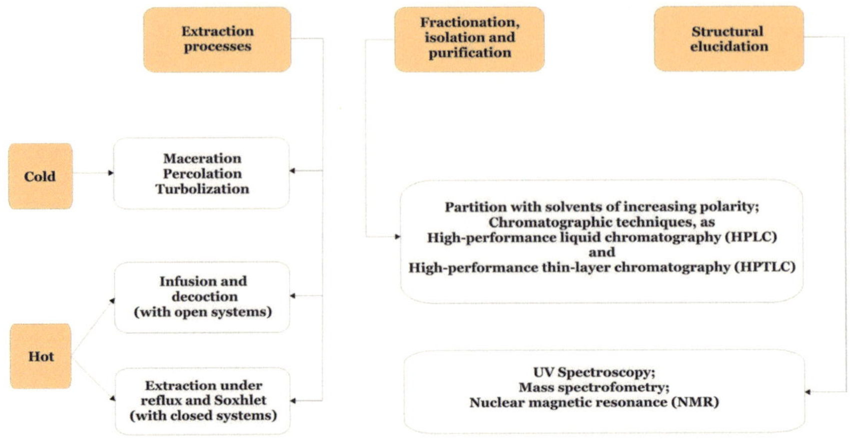

Figure 4.
Classical methods in the stages of studies with medicinal plants.

4. Medicinal and nutritional applications of phenolic compounds and related plant species

From the development of public policies that include natural products and medicinal plants in Brazil, these alternatives have become even more viable in primary health care. This has been observed since the year 2006, when the National Program of Medicinal Plants and Herbal Medicines was created, which provides an identification of the population with the treatments they use, considering that this knowledge comes, in most cases, from the community itself [31]. Several of the species that are used in traditional medicine in Brazil have a high concentration of phenolic compounds in their composition, with the associated mechanisms of action relating precisely to these chemical components and several of the species associated with these strategies are highlighted in ethnopharmacological studies [5].

Coumarins are also phenolic compounds and stand out for having a widely used representative, due to their anticoagulant potential, warfarin, an oral anticoagulant traditionally used in thrombophilic disorders which has its chemical representation indicated in **Figure 5**, where it is possible to identify the various phenolic groups present in its structure. In the caatinga biome, among the main representative species rich in coumarins, is the amburana (*Amburana cearensis*). This potential can serve as a basis, including for the study of synthetic alternatives, for the identification of optimized pharmaceutical products [26].

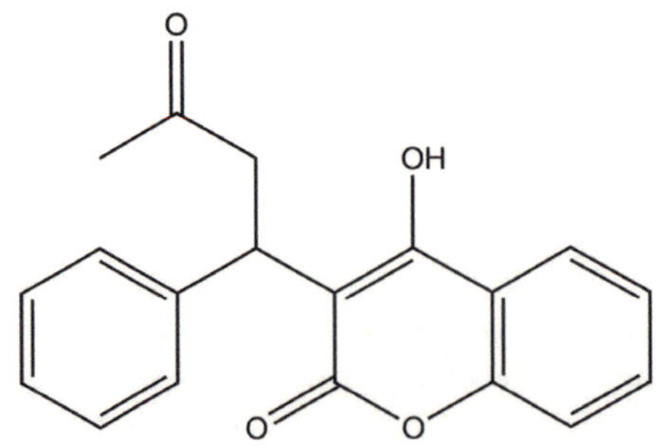

Figure 5.
Structural representation of a phenolic compound widely used in medicine as an anticoagulant (coumarin). Warfarin.

Some of the main species of the caatinga with a high content of phenolic compounds are shown in **Table 1**, some of them can also be seen in **Figure 6** with their respective representations in studies that use High-performance thin-layer chromatography (HPTLC) as part of the research. These cited species are frequently referenced in ethnopharmacological studies, especially those with a focus on medicinal activity, whether by antimicrobial, antioxidant, anti-inflammatory action, among others [17, 32].

In Nutrition, some medicinal plants can also be considered functional foods, and the beneficial effects occur, since the compounds can act simultaneously on different cellular targets. Among the bioactive compounds already identified

Scientific name	Popular name	Use/composition/mechanisms
Amburana cearensis	Amburana	Its composition comprises mainly flavonoids and coumarins. Making it have an excellent anti-inflammatory action, being also mentioned as having a possible bronchodilator action, not yet fully elucidated [17, 32, 33].
Anacardium occidentale	Cashew tree	It also has, as well as the species mentioned above, antimicrobial action, in wound healing and as an anti-inflammatory, mainly due to the presence of tannins and flavonoids. In addition, it also stands out for its efficiency as an antioxidant, which leads people to use the tea made from cashew barks also to prevent diseases [17, 32].
Anadenanthera colubrina	Angico	It stands out for its high tannin content, with efficient application against infections in general, especially those affecting the skin [17, 32].
Mimosa tenuiflora	Jurema preta	One of the most popular species, according to ethnopharmacological studies applied in the caatinga. It has a very high concentration of tannins, being useful as an antimicrobial, anti-inflammatory and wound healing [17, 32].
Momordica charantia	São Caetano melon	Its fruits, leaves and roots are used for the treatment of diabetes, colic and as a healing agent [34] also presents a gastroprotective effect, due to the presence of bioactive compounds [35]. It is characterized as a plant rich in nutrients and considered quite versatile. Its fruits contain a high number of vitamins, including those of the B complex and minerals [36].
Schinus terebinthifolius Raddi	Aroeira vermelha	Excellent alternative as anti-inflammatory, wound healing, antimicrobial and antioxidant. The main mechanisms of action are related to the capacity of scavenging free radicals (antioxidant activity), interference in microbial cell walls, making these species act as an antifungal, antibacterial and also antiparasitic, some flavonoids can interfere in the performance of proteins and enzymes by hydrophobic interactions, promoting modulations that leads to anti-inflammatory effect, mostly because of the great content of tannins and flavonoids, which also participate in wound healing process, promoting a cover that offers defense against contaminations helping in the hemostasis phase. At the same time, cell signaling can occur, influencing the inflammatory phase by interference in the work of macrophages, responsible for combating possible contaminants [37, 38].
Selaginella convoluta (Arn.) Spring	Jericho	Antimicrobial, antioxidant and anti-inflammatory effects, with similar mechanisms. In addition to being associated with use as an antidepressant and antinociceptive, mechanisms of which have not yet been fully elucidated, but which may also be related to interactions with protein sites and enzymes [39, 40].

Table 1.
Caatinga plants with high concentration of phenolic compounds/medicinal use.

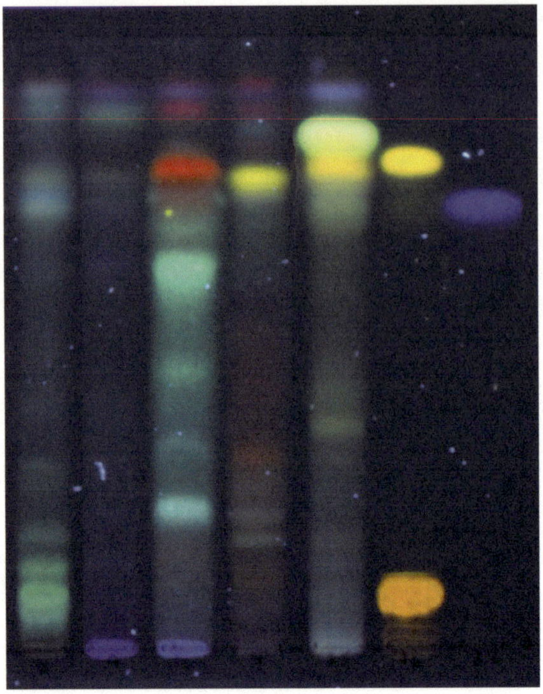

Figure 6.
High performance thin-layer chromatography (HPTLC) of the crude extract of Caatinga plants. 1 – Mimosa tenuiflora *(Jurema preta); 2* – Anacardium occidentale *(caju roxo); 3* – Myracrodruon. Urundeuva *(aroeira do sertão); 4* – Anadenanthera colubrina *(Angico); 5* – Amburana cearensis *(amburana); P1 – Rutine + quercetin; P2 – Gallic acid.*

are soluble and insoluble fibers, antioxidants (such as polyphenols, carotenoids, tocopherols, phytosterols, isoflavones, organosulfur compounds), plant steroids and phytoestrogens [41]. However, the prescription of medicinal plants must be performed with caution considering the possibility of side effects and interactions with drugs and nutrients, which can generate organic imbalances [42]. Among phytochemicals, there is a growing interest in the discovery and identification of phenolic compounds that occur naturally in plant species, with the aim of finding new and promising sources of antioxidants for human health [43]. In food, they are responsible for color, astringency, aroma and oxidative stability [44].

The main food sources of phenolic compounds are citrus fruits, such as lemon, orange and tangerine, in addition to other fruits such as cherry, grape, plum, pear, apple and papaya, being found in greater concentration in the pulp and in the fruit juice, in addition to green pepper, broccoli, red cabbage, onions, garlic and tomatoes, which are also excellent sources of these compounds [45]. In addition, we can emphasize some plants that belong to the caatinga biome that have a great antioxidant potential and also potential for nutritional use, such as *Moringa oleifera*, characterized as nutritious and with a wide variety of uses, almost all parts of it can be used, its leaves being a food source to combat malnutrition [46], in addition to containing considerable amounts of proteins and several micronutrients, among them vitamin A, vitamin C, potassium, iron and calcium. They are also a good source of phytonutrients, such as carotenoids, and tocopherols [47].

These plants are also classified as unconventional food plants (UFP), because although they are edible, they are commonly underutilized, neglected and even considered weeds. Although not widespread, UFPs are alternatives for food and income improvement for family farmers, and can also be grown in urban backyards, adding food value to meals and even giving an exotic touch to some dishes. They are characterized by rusticity, weather resistance, longevity and great adaptability to different climates and regions. They are, in general, less demanding in fertility and irrigation, easy in maintenance and can be planted using seeds or seedlings. With extremely diversified possibilities of flavors and very interesting nutritional characteristics, the UFPs contribute to the improvement of the local diet in the communities involved, and are also a true cultural rescue [48].

From that definition, these plants are also distinguished for being considered for localized, regional and/or seasonal consumption, with limited distribution and with no established production chain [48]. In addition, if carried out in a sustainable manner, it can be considered a form of land use with low impact on agriculture, associated with environmental conservation. Thus, the dissemination of studies carried out to the communities can help in a better direction of the use of these resources, where through them it can contribute to an improvement of health, as well as provide strategies to fight hunger in these localities, using these species in favor of the individuals who live in their surroundings [49].

5. Phenolic compounds with technological application and economic potential

Among the main technological applications for phenolic compounds, as mentioned, its use as a medicinal alternative stands out, where these and other metabolites, medicinal plants and herbal medicines can assist in self-care, making individuals as protagonists of their own care, and also, in reducing health costs. In the Brazilian health system, this alternatives are applied inside the so-called integrative and complementary practices, where they are present, ensured by the legislation, including the official pharmacopeia of the country, in which is possible to consult the monographs with techniques of extraction and basic information about some of the plants used popularly, always remembering that in order to be considered within these official means, these plants, as well as herbal medicines need to undergo studies that prove their efficacy and toxicity patterns, in order to guarantee safety in its use [50–52].

Within the context of medicines, phytotherapeutic drugs of wide use are also highlighted, which are rich in phenolic compounds, such as flavonoids, present in *Ginkgo biloba* L, where they contribute with antioxidant action; Passion fruit (*Passiflora spp*), from which drugs that act as anxiolytics are extracted, where in this case, the flavonoids act mainly as markers for indication of quality in the extraction, a function that is facilitated due to its fundamental structure, which usually has 15 atoms of carbon arranged in rings, forming a tricyclic compound. The tannic compounds are also noteworthy. Some of them present in the species *Hamamelis virginiana* L. that favor its hemostatic activity, since one of the main characteristics of tannins, in addition to promoting the antioxidant effect, is also able to bind to molecules and macromolecules, such as proteins, for example, being able to contribute to the stabilization in the treatment of injuries, of the most diverse types [27].

In addition, from the perspective of nutrition, there is a growing concern with the food supply capacity at global scale, as well as a high consumption of processed foods and fast food is identified, considering the fast pace of life that people in general maintain. Because of this, the incorporation of these biological assets, such as phenolic compounds and the integration of plant species that are, underutilized or neglected in food, can be an alternative to compose food products that promote health, at the same time that they become productive pathways for families and their financial support, from the moment that a resignification of these alternatives is promoted [53].

This positive health effect resulting from phenolic compounds application of nutrition, is what justifies the action of wines as potent antioxidants, since it is estimated that in its composition it contains more than 200 polyphenols, among flavonoids, tannins and anthocyanins, mainly. And although the grape is not a typical species of the caatinga, there is an increase in investments in the northeast region of Brazil, in which the caatinga is also part, in the so-called São Francisco valley. And this profile generates interaction with the environment, its modification and helps in economic and human development [54].

Another application of caatinga species, such as Babaçu (*Orbignya phalerata* (Mart.), for example, can also contribute to the production of biodiesel, being, therefore, an alternative to consider in the formulation of sustainable energy sources, in addition to having the possibility of using all parts of the species in this process, which also impacts sustainability demands, considering that all portions can be processed for application in the production of inputs and products, such as brushes and carpets, in food, from the preparation of vegetable oils, chocolates, foods and cakes [55].

The action of flavonoids as photoprotectors has also leveraged research for their cosmetic application. Considering that caatinga species are rich in these metabolites, they can be sources of pharmaceutical inputs or the purification and isolation of these compounds can serve as a basis for synthesizing new molecules that promote the potentiation of the photoprotective effect. Such activity is mainly due to its chemical structure with double conjugated bonds, which directly interferes with the absorption process in the region of ultraviolet A (UVA) and ultraviolet B (UVB) [56].

6. Reprocessing waste containing phenolic compounds in its composition as a sustainability initiative

The current scenario involves all the problems in which globalization and development is involved in the so-called fourth industrial revolution, where, in addition to the competitiveness of the market, there is an urgent need for sustainable proposals and measures within development systems, in order to provide growth with significantly less impact on the environment. From the 1990s, this paradigm shift is observed when we see that development needs sustainable attitudes, where the concept of Triple Bottom Line (TBL) begins to be considered, which understands the viability of companies' businesses according to the dynamics between economic, social and environmental aspects [57].

Based on this, the reprocessing of residues, mainly those resulting from agribusiness, which is one of the economic sections that most impact the environment, can be an alternative to maximize the efficiency of the production process, meeting the needs related to sustainability. It is clear, from studies published in recent years, that the use of these materials resulting from this reprocessing has high versatility, being able to meet, for example, demands of the food chain [58], and civil construction [59].

As an example of reprocessing, we can mention the contribution of one of the most endemic species of the caatinga, the cashew tree (*Anacardium occidentale* L.), which has several applications: medicinal, food and industrial in general. From this production process, the cashew nut is obtained, its fruit, which contributes economically both to the local product and to exports; as well as cashew nut shell liquid (CNSL), which is seen as a by-product and is still little used in Brazil, while in other countries, some of those who acquire this liquid from Brazil and with the appropriate technology employed, perform its reprocessing as becoming a product of high added value, used as resins and polymers, in addition to additives, surfactants, drugs, pesticides, among others, configuring itself as an alternative with high potential for profitability and which continues to be underutilized [60].

However, phenolic compounds need to be quantified so that they can be used with these purposes safely, considering that these compounds can also, if accumulated, generate complications for the environment, being necessary to maintain safe concentrations when developing new alternatives that contain them. Where it is observed that there is a need to reduce the concentration of phenolic compounds in wastewater in an increasing way, requiring increasingly efficient technologies in this process. In Italy, a law aims to guarantee the quality of fresh, coastal, brackish and marine waters from polluting waste discharge locations, stipulates that waste water must contain a limit value of total phenols before disposal 0.05 mg/L and that, if the disposal is done in freshwater, these waters, after disposal, must contain up to 0.01 mg/L of total phenols, being considered safe for the environment, not putting in risk the health and quality of marine organisms. The reuse of waste can be a tool to reduce the accumulation of these compounds in nature [57].

In Brazil, this concentration varies according to the destination that the wastewater will have and this is divided into four classes. Class 1: the water must be free of phenolic compounds, intended for domestic use without having undergone previous treatment; Class 2: water used for domestic use, irrigation and recreation, which has undergone previous treatment, can contain 0.001 mg/L; Class 3: it can also contain a concentration of 0.001 mg/L, where this water is destined for domestic use or for disposal in places where there is a need for environmental preservation, fauna and flora; Class 4: the concentration can reach 1 mg/L, for which this water can be used for purposes that demand less quality standards such as some domestic uses, industrial use, irrigation, among others. When it is necessary to treat these effluents with high concentrations of phenolic compounds, various techniques can be used. Some of them are described below [57].

Adsorption Methods are the most traditional for the treatment of wastewater with organic contaminants. For phenolic compounds, the continuous flow fixed bed technique is widely used, which generally consists of a cylinder containing an adsorbent inside, with activated carbon (most common compound), with inlet and outlet, through which it is fed by residual water. Factors that interfere with a good efficiency of the technique are the concentration of contaminants in the wastewater and the feed flow rate. For phenolic compounds, a good efficiency is when the concentration of these compounds is small in the wastewater in addition to a low flow rate [58]. The advantage of considerably reducing the concentration of phenolic compounds in wastewater, allowing for their safe disposal, concomitantly generates. As a disadvantage, the formation of solid waste formed by activated carbon with adsorbed phenolic components, considering that their improper disposal will also cause environmental damage [59].

Another traditional method is the Advanced Oxidation Processes (OAP's) also known as Fenton reaction. The reaction consists in the formation of hydroxyl radicals (HO^-) and Fe^{+3} as a product of the interaction between Fe^{+2} and hydrogen peroxide (H_2O_2). Hydroxyl radicals will react with organic compounds, such as phenols, giving rise to organic radicals that interact with oxygen in the environment, causing a series of degradation of these compounds, generating mainly carbon dioxide and water as products. With the advancement of technologies, the association with other techniques shows that there is a decrease in the reaction time, as this reaction occurs slowly, in addition to a significant increase in the formation of hydroxyl radicals, consequently increasing the degradation of organic compounds [60].

And then came the Eletro-Fenton and Foto-fenton techniques. Electro-fenton consists in the use of electrodes composed of transition metals such as manganese oxide (MnO_x) and nickel oxide (NiO_x) that act as catalysts, increasing efficiency and decreasing the reaction time, accelerating the process [61]. In the case of Foto-Fenton, its principle is the use of ultraviolet radiation not as a catalyst, but as a photolytic agent. In the Fenton reaction, the Fe^{+2} ion with hydrogen peroxide, forming Fe^{+3}. Ultraviolet radiation works by recovering the Fe^{+2} ions by reducing, by photolysis, the nox number of Fe^{+3}, which makes the system always have the Fe^{+3} ion as the producing agent of hydroxyl radicals in a continuous process. The advantage of this method is that both the ultraviolet radiation used can be either by lamps or by sunlight, which reduces costs [62].

Some microorganisms, such as Gammaproteobacteria, Actinobacteria, Betaproteobacteria and Alphaproteobacteria, can absorb certain types of phenolic compounds and use them as substrates for vital biochemical reactions. From this, the use of biomass from aerobic microorganisms immobilized on solid supports, usually membranes, through which they are subjected to direct contact with wastewater under aerobic conditions, a process called biofiltration, may be interesting alternatives [63].

The use of enzymes can also be a good alternative in the biopurification process of wastewater containing phenolic compounds, especially peroxidases, which are oxidoreductive enzymes capable of oxidizing aromatic compounds and, furthermore, when oxidation occurs in phenolic compounds in the presence of hydrogen peroxide, polymerization of these compounds occurs, forming insoluble precipitates that are easily removed by physical processes of solid–liquid separation (**Figure 7**). The peroxidases can be used pure obtained by commercialization or contained in crude extracts of plants such as horseradish, soybean, turnip, garlic, sweet potato, radish and sorghum for which it is a cheaper alternative and as functional as using purified enzymes [64].

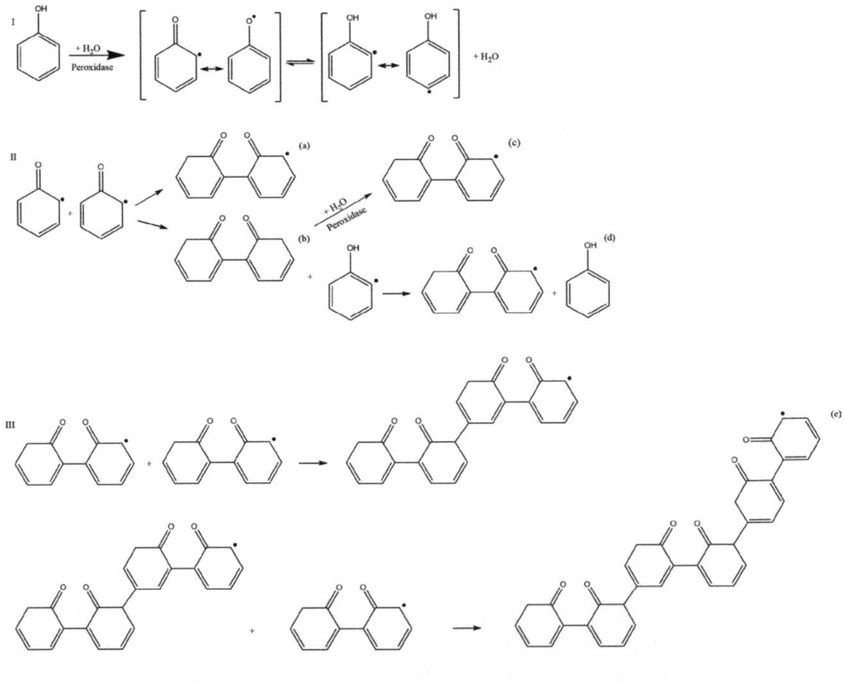

Figure 7.
Polymerization scheme of phenolic compounds by peroxidases, with the following sequence: I. phenol oxidation step with formation of phenoxy radical and water as product; II. Radical dimerization step; III. Polymerization itself; (a) dimer radical; (b) non-radical dimer; (c) radical dimer formation via the peroxidase pathway; (d) radical dimer formation by reaction with phenoxy radical; (e) insoluble polymer.

Liquid–liquid partition is a well-known and used way of extracting and purifying substances. It consists of a mix of two immiscible liquids in which one of the liquids has the solute that migrates to the other liquid phase by affinity according to its polarity. Usually, an aqueous phase and an organic phase are used that vary in more or less polar, so it is chosen according to the solute in question. The most commonly used traditional organic solvents are ethanol, methanol, acetone, ethyl acetate and hexane. Taking the characteristics of this technique into consideration, it was seen that it could be used for the removal of phenolic compounds from wastewater, where the choice of solvent occurs according to the physicochemical properties of the phenolic compounds [65].

However, these solvents have several disadvantages, as they are as toxic as the residues themselves, are flammable, generate atmospheric gases, and are non-biodegradable. With this in mind, the technique was linked to biopurification, replacing traditional solvents with solvents considered "clean" or "green". Called neoteric solvents, their use minimizes environmental impacts, reuse of the solvent itself, reducing process costs, in addition to increasing the efficiency of removal of residual compounds. Neoteric solvents encompass ionic liquids, eutectic solvents, biologically-based solvents and supercritical fluids [65].

Methods that combine more than one technique for the bio-depuration process, called hybrid technologies, can be a very promising alternative. Cavitation is a phenomenon in which tiny bubbles form within a system from the very intense agitation of molecules in a liquid by ultrasound. Cavitational reactors cause this agitation to generate a high energy content in the system. This technology by itself is not so interesting for the treatment of wastewater on an industrial scale, since it demands high costs and causes operational problems when referring to the dissipation of the generated energy. However, when combined with known bio-depuration methods and used as oxidative or catalytic (enzymatic) methods, it can be viable on a large scale. Basically, cavitation reactors will add energy to the medium, facilitating the formation of reactions [66].

Cavitation associated with hydrogen peroxide causes the formation of hydroxyl radicals without the need for the presence of Fe^{+2} in the system, but in well-controlled conditions of pH and temperature, as disorders in this regard can cause the radicals to interact with the hydrogen peroxide itself forming water. Some disadvantages of the photo-fenton technique, such as inhibition of the reach of ultraviolet radiation by interaction with non-interesting contaminants on the surface of the waste water, as well as limitations on mass transfer, can be reduced or eliminated when associated with cavitation, in addition to an increase in the production of free radicals, a technique called oxidative Sonophotocavitation [66].

The association of cavitation with Electro-Fenton, in addition to helping in the formation of free radicals, the movement of molecules helps the solution to remain mixed and acts to clean the electrodes, removing crusts formed in the process that hinder the exchange of energy between the electrodes and the solution. In combination with oxidative enzymes (peroxidases), cavitation helps to eliminate some disadvantages of the process, such as increasing the useful life and decreasing the inactivation of these enzymes, in addition to being able to act synergistically in the production of radicals, increasing the polymerization of phenolic compounds [66].

7. Future perspective, industrial, therapeutics and sustainability applications

It becomes important to envision the perspective of promoting the integration of academic studies and the industry for the joint development of strategies using the species mentioned, in order to bring together objectives and set concrete goals for possible application, thus contributing to the increasing number of alternatives such as these arrive at the final development, with economic gain, valorization of the local component, academic contribution and for the general community, culminating in added value in conjunction with sustainable development strategies. Which is a necessity imposed by our current reality [67].

For example, the products that have the grape, as its raw material, in addition to wines, are items of high added value, highlighting the participation of grape juice, these products can provide good responses at the commercial and economic level [60]. However, the large scale of production ends up generating waste, which can also be reprocessed for other applications, becoming sustainable alternatives for local development, including the generation of food alternatives, such as the production of flour, also rich in phenolic compounds, with good sensory acceptability and large-scale applicability [68, 69].

Within this context, a major transformation occurred in the so-called São Francisco Valley, in northeastern Brazil and also inserted within the extension where the caatinga grows. From irrigation strategies, this region became one of the main producers of grapes and wines in the country, also participating in exportation in this area, promoting the addition of economic value to that region, generating employment and development This demonstrates that with the correct investment, the region in which the caatinga is located can, with its own means, grow and develop, also in a sustainable way, since production can reuse all waste associated with the production chain [54].

And this theme can be highlighted considering the following situations: 1. The caatinga region can use these resources sustainably in order to guarantee the livelihood of the families that are inserted in this context. 2. The presence of finished products on the pharmaceutical market, either as medicines or as functional foods, makes it clear that these species can be applied for these purposes, adding value. 3. There is a need to expand studies that promote the visibility of this region in general, as there are still unexplored potentials [69, 70].

8. Conclusion

Phenolic compounds are present in several of the endemic species of this biome, characteristically brazilian, the Caatinga. The underutilization of these species impacts on a loss of applicability and potential profitability, which can, if applied, leverage the production processes in this location that comprises the Northeast region of Brazil, favoring sustainable development with added value not only to elaborated products, but to the very identity of the communities that depend on the processing of these alternatives. The applications can be diverse, either for use in food, as functional foods, or in traditional medicine, where it adds value even to the cultural factor related to families that pass this knowledge on plants to the generations away. They also participate in industrial and civil construction activities, among other areas. This appreciation also ends up increasing the identity of these communities and their peoples who contribute directly or indirectly to raising the use of plant species, making them feel represented. Therefore, initially, the greatest value to add with the best use of these alternatives is social improvement.

Author details

Elba Lúcia Cavalcanti de Amorim*, Patrícia Cruz, Jorge Veras Filho,
Italo Caio Silva, Uyara Costa, Jenifer Oliveira, Maria Santa Medeiros,
Marcelino Diniz, Kivia Machado and Ana Caroline Xavier
Department of Pharmacy, Federal University of Pernambuco, Pernambuco, Brazil

*Address all correspondence to: elba.amorim@ufpe.br

References

[1] Silva DM, Ribeiro ACD, Brum DV. Brazil's identity is diversity. Humanities & Innovation. 2018;5(6):57-69. [cited 2021 may]. Available from: https://revista.unitins.br/index.php/humanidadeseinovacao/article/view/675.

[2] Coutinho LM. Brazilian biomes. 1st. ed. São Paulo: text workshop; 2016.

[3] Neves CR, Gomes AVS, Rezende EJC. The return to natural dyes as a way to preserve the environment and its historical and cultural importance. DAT Journal. 2021;6(1):279-293.[cited 2021 may]. Available from: https://doi.org/10.29147/dat.v6i1.

[4] Demartelaere ACF, Preston HAF, Nascimento MNP, Gomes KKF, Silva MEA, Souza JB et al. Utilities and the economic importance of *Copernicia prunifera* for the state of Rio Grande do Norte: an endangered species. Brazilian Journal of Development. 2021;7(1):5065-5088. [cited 2021 may]. Available from:https://www.brazilianjournals.com/index.php/BRJD/article/view/23129.

[5] Gonçalves RN, Gonçalves JRSN, Buffon MCM. Methodology and sample sufficiency in ethnobotanical and ethnopharmacological studies: experience report. Academic Vision. 2018;18(4):14-24. [cited 2021 may]. Available from: https://revistas.ufpr.br/academica/article/view/56120.

[6] Diniz RF. Etno-knowledge and afrobrasilian tratitional cultures: farmacopeia, magic and material and symbolic reproduction of quilombola communities in the Jequitinhonha Valley-MG. GEOgraphia. 2019;21(47):13-28. [cited 2021 may]. Available from: https://periodicos.uff.br/geographia/article/view/28178.

[7] Sousa RF, Sousa JA. Secondary metabolics associated with water stress and their functions in plant tissues. Brazilian Journal of Environmental Management. 2017;11(1):01-08. [cited 2021 may]. Available from: https://www.gvaa.com.br/revista/index.php/RBGA/article/view/5008.

[8] Cruz JHA, Ferreira JLS, Simões APG, Cristino DL, Costa EID, Souza ERL et al. *Malva sylvestris*, *Vitis vinifera* and *Punica granatum*: a review of the contribution to the treatment of periodontitis. Archives of Health Investigation. 2018;7(11):486-491. [cited 2021 may]. Available from: https://www.archhealthinvestigation.com.br/ArcHI/article/view/3039.

[9] Malacrida CR, Motta S. Total phenolic compounds and anthocyanins in grape juice. Food Science and Technology. 2005;25(4):659-664. [cited 2021 may]. Available from: https://www.scielo.br/j/cta/a/Kp9JZy3xyrky5JTHN3yMPbM/abstract/?lang=pt.

[10] Vizzoto M, Krolow AC, Weber GEB. Secondary metabolites found in plants and their importance. Embrapa Temperate Climate-Documents (INFOTECA-E). 2010;1:1-16. [cited 2021 may]. Available from: https://www.infoteca.cnptia.embrapa.br/infoteca/handle/doc/886074.

[11] Verruck S, Prudencio ES, Silveira SM. Bioactive compounds with antioxidant and antimicrobial capacity in fruits. Journal of the Southern Brazilian Congress of Food Engineering. 2018;4(1):111-124. [cited 2021 may]. Available from: https://www.revistas.udesc.br/index.php/revistacsbea/article/view/13312.

[12] Arnoso BJM, Costa GF, Schmidt B. Bioavailability and classification of phenolic compounds. Nutrition Brazil. 2019;18(1):39-48. [cited 2021 may]. Available from: https://

portalatlanticaeditora.com.br/index.php/nutricaobrasil/article/view/1432.

[13] Alves, JJA, Araújo MA, Nascimento SS. Caatinga degradation: an ecogeographic investigation. Caatinga Journal. 2009;22(3): 126-135. [cited 2021 may]. Available from: https://periodicos.ufersa.edu.br/index.php/caatinga/article/view/560

[14] Leal IR, Tabarelli M, Silva JMC. Caatinga ecology and conservation. University Publisher UFPE. 2003. [cited 2021 may]. Available from: http://www.bibliotecaflorestal.ufv.br/bitstream/handle/123456789/9865/Livro_Ecologia-e--Conserva%C3%A7%C3%A3o-da-Caatinga_MMA.pdf?sequence=1.

[15] Souza LSB, Moura MSB, Sediyama GC, Silva TGF et al. Energy balance and biophysical control of evapotranspiration in the Caatinga under intense drought conditions. Brazilian Agricultural Research. 2015;50(8):627-636. [cited 2021 may]. Available from: https://www.scielo.br/j/pab/a/wzzwvkXpskxcsbGq9yMNRwm/abstract/?lang=pt.

[16] Queiroz LP, Rapini A, Giulietti AM. Towards broad knowledge of the biodiversity of the Brazilian semi-arid region. Ministry of Science and Technology. 2005. [cited 2021 may]. Available from: https://www.terrabrasilis.org.br/ecotecadigital/pdf/rumo-ao-amplo-conhecimento-da-biodiversidade-do-semi-arido-brasileiro.pdf.

[17] Albuquerque UP, Andrade LHC. Use of plant resources from the caatinga: the case of the agreste region in the state of Pernambuco (Northeast Brazil). Interscience. 2002;27(7):336-346. [cited 2021 may]. Available from: https://www.agencia.cnptia.embrapa.br/Repositorio/Recursos_vegetais_caatinga_000fmkfyjm102wyiv80kxlb36f8ucm41.pdf.

[18] Chaves TP. Seasonal variation in the production of secondary metabolites and in the antimicrobial activity of plant species from the Brazilian semiarid region (Doctoral thesis). State University of Paraíba; 2012. 78 p. [cited 2021 may]. Available from: http://tede.bc.uepb.edu.br/jspui/handle/tede/1879.

[19] Schober J. Preservation and rational use of the only exclusively national biome. Science and Culture. 2002;54(2):06-07. [cited 2021 may 30]. Available from: http://cienciaecultura.bvs.br/scielo.php?script=sci_arttext&pid=S0009-67252002000200003.

[20] Alves A, Sá JC, Araújo IS. The right to land as a strategy for the preservation of medicinal species and the strengthening of the native culture of the Fulni-ô Indians (Águas Belas–PE). Journal of Cultural and Contemporary Studies. 2012;7:182-208. [cited 2021 may]. Available from: https://www.revistadialogos.com.br/Dialogos_7/12_ALVES_JEMERSON_IDIARURY_Fulnio.pdf.

[21] Vallejo LR. Conservation unit: a theoretical discussion in light of the concepts of territory and public policies. Geographia. 2002;4(8):57-78. [cited 2021 may]. Available from: https://periodicos.uff.br/geographia/article/view/13433.

[22] Jatobá SUS, Cidade LCF, Vargas GM. Ecologism, environmentalism and political ecology: different visions of sustainability and territory. society and state. 2009; 24(1):47-87. [cited 2021 may]. Available from: https://www.scielo.br/j/se/a/CSrVxYphhYvHrgcZgRNF8WF/abstract/?lang=pt.

[23] Cunha LHO. Extractive reserves: an alternative for the production and conservation of biodiversity. Meeting of

the peoples of Vale do Ribeira. 2001:1-42. [cited 2021 may]. Available from: http://nupaub.fflch.usp.br/sites/nupaub.fflch.usp.br/files/color/resex.pdf.

[24] Svobodová A, Psotova J, Walterova D. Natural phenolics in the prevention of UV-induced skin damage. A review. Biomedical papers of the Medical Faculty of the University Palacký. 2003;147(2):137-145. [cited 2021 may]. Available from: https://pubmed.ncbi.nlm.nih.gov/15037894/

[25] Silva MLC, Costa RS, Santana AS, Koblitz MGB. Phenolic compounds, carotenoids and antioxidant activity in plant products. Seminar: Agricultural Sciences. 2010;31(3):669-682. [cited 2021 may]. Available from: https://www.redalyc.org/pdf/4457/445744097017.pdf

[26] Franco, DP. et al. The importance of cumarines for medicinal chemistry and the development of bioactive compounds in recent years. New Chemistry; 2021. 44(2)180-197, 2021. [cited 2021 may]. Available from: https://www.scielo.br/j/qn/a/NdnnhHYx8b8tJFSFYXms6nM/?lang=pt.

[27] Simões CMO, Schenkel EP, Mello JCP, Mentz LA, Petrovick PR. Organizadores. Farmacognosia: do produto natural ao medicamento. Porto Alegre : Artmed; 2017. 486 p.

[28] Azmir J, Zaidul ISM, Rahman MM, Sharif KM, Mohamed A, Sahena F et al. Techniques for extraction of bioactive compounds from plant materials: A review. Journal of Food Engineering. 2013;117:426-436. [cited 2021 may]. Available from: https://fac.ksu.edu.sa/sites/default/files/23-jfe_117_2013.pdf.

[29] Guerra AP; Santos Garcia VA; Silva C. Optimization of the extraction of phenolic compounds from mango peel (Tommy atkins) using an ultrasound-assisted process. e-Xacta. 2016; 9(1):103-110. [cited 2021 may]. Available from: https://revistas.unibh.br/dcet/article/view/1783.

[30] Amorim ELC, Castro VTNA, Melo JGD, Correa AJC, Peixoto Sobrinho TJS. Standard Operating Procedures (SOP) for the Spectrophotometric Determination of Phenolic Compounds Contained in Plant Samples. In: Isin Akyar. (Org.). Latest Research into Quality Control. 1ed.Rijeka: InTech, 2012; (3):47-66. [cited 2021 july]. Available from: https://www.intechopen.com/chapters/38925

[31] BRAZIL. Ministry of Health. Secretariat of Science, Technology and Strategic Inputs. Department of Pharmaceutical Assistance. National Policy on Medicinal Plants and Herbal Medicines. Brasília: Ministry of Health, 2006. 60 p. – (Series B. Basic Health Texts). [cited 2021 july]. Available from: https://bvsms.saude.gov.br/bvs/publicacoes/politica_nacional_fitoterapicos.pdf

[32] Araújo, TADS. (2012). Antioxidant activity of medicinal plants from the caatinga and atlantic forest: ethnobotanical and ecological aspects. [cited 2021 may]. Available from: https://repositorio.ufpe.br/bitstream/123456789/10367/1/tese%20THIAGO%20ARA%C3%9AJO%20eletronica.pdf.

[33] Lima, LR et al. Assessment of the antiedematogenic, antimicrobial and mutagenic activity of *Amburana cearensis* seeds (AC Smith). Brazilian Journal of Medicinal Plants, 15(3): 415-422, 2013. [cited 2021 May]. Available from: https://www.scielo.br/j/rbpm/a/wkZR6qz67SQXgJRgQM7Frwn/?lang=pt.

[34] Chen J, Tian R, Qiu M, Lu L, Zheng Y, Zhang Z. Trinorcucurbitane and cucurbitane triterpenoids from the roots of *Momordica charanti*a. Phytochemistry [Internet]. 2008 [cited

2021 May];69(Issue 4): 1043-8. 2008. Available from: https://doi.org/10.1016/j.phytochem.2007.10.020.

[35] Leite KL, Nunes-Pinheiro DCS, Campello CC. Gastroprotective effect of hexanic extract from aerial parts of *Momordica charantia*. Ciência Animal. 2005; 15(1):15-20. [cited 2021 july]. Available from: http://www.uece.br/cienciaanimal/dmdocuments/Artigo2.2005.1.pdf

[36] Bakare RI, Magbagbeola OA, Akinwande AI, Okunowo OW. Nutritional and Chemical evaluation of *Mormodica charantia*. J Med Plant Res [Internet]. 2010 [cited 2021 May]; 4 (21):2189-2193. Available from: https://doi.org/10.5897/JMPR10.274.

[37] Carvalho MG, Melo AGN, Aragão CFS, Raffin FN, Moura TFAL. *Schinus terebinthifolius* Raddi: chemical composition, biological properties and toxicity. Brazilian Journal of Medicinal Plants. 2013; 15(1): 158-169. [cited 2021 july]. Available from: https://www.scielo.br/j/rbpm/a/bNdsZSp6jMDqM6qVXxCHGgL/?lang=en&format=pdf

[38] Falcão MPMM, Oliveira TKB, Sarmento DA, Ó NPR, Gadelha NC. *Schinus terebinthifolius* Raddi (Aroeira) and its properties in Popular Medicine. Green Journal of Agroecology and Sustainable Development. 2015; 10 (5): 23-27. [cited 2021 july]. Available from: https://www.gvaa.com.br/revista/index.php/RVADS/article/view/3455

[39] Sá PGS, Guimarães A. L.; Oliveira A. P.; Siqueira-Filho JAS; Fontana AP, Damasceno PKF, Branco CRC, Branco A, Almeida JRGS. Fenóis totais, flavonoides totais e atividade antioxidante de *Selaginella convoluta* (Arn.) Spring (Selaginellaceae).Rev Ciênc Farm Básica Apl. v. 33(4), p.561-566, 2012. [cited 2021 july]. Available from: http://rcfba.fcfar.unesp.br/index.php/ojs/article/view/260/258

[40] Novaes TER, Novaes ASR. Potential and medicinal uses of the jericho, *Selaginella convoluta* (Arn.) Spring, in the Caatinga biome: a brief review. Research, Society and Development, [S. l.], v. 10, n. 1, p. e43810111989, 2021. DOI: 10.33448/rsd-v10i1.11989. Available from: https://rsdjournal.org/index.php/rsd/article/view/11989. Acesso em: 3 jul. 2021.

[41] Angelo PM, Jorge M. Compostos fenólicos em alimentos – Uma breve revisão. Rev. Inst. Adolfo Lutz. 2007;66(1)1-9. [cited 2021 may 30]. Available from: http://periodicos.ses.sp.bvs.br/pdf/rial/v66n1/v66n1a01.pdf

[42] Manach C, Scalbert A, Morand C, Rémésy C, Jiménez L. Polyphenols: Food source and bioavailability. The American Journal of Clinical Nutrition [Internet]. 2004 [cited 2021 May); 79 (5):727-747. [cited 2021 may 30]. Available from: http://10.1093/ajcn/79.5.727

[43] Liu, Y., Qi, Y., Chen, X., He, H., Liu, Z., Zhang, Z., Ren, Y, Ren, X. (2018). Phenolic compounds and antioxidant activity in red- and in green-fleshed kiwifruits. Food Research International [Internet]. 2019 [cited 2021 May];116:291-301. Available from: https://doi.org/10.1016/j.foodres.2018.08.038.

[44] Peleg H, Bodine KK, Noble AC. The influence of acid on adstringency of alum and phenolic compounds. Chem Senses [Internet]. 1998 [cited 2021 May]; 23(3):371-378. Available from: https://doi: 10.1093/chemse/23.3.371.

[45] Pimentel CVMB, Francki VM, Gollücke APB. Functional foods: introduction to the main bioactive substances in foods. São Paulo: Varela; 2005. 95p.

[46] Anwar F, Latif S, Ashraf M, Gilani AH. Moringa oleífera: a food plant with multiple medicinal uses.

Phytother Res, 2007 [cited 2021 May]; 21 (Issue 1):17-25. Available from: https://doi.org/10.1002/ptr.2023.

[47] Hekmat S, Morgan K, Soltani M, Gough R. Sensory evaluation of locally-grown fruit purees and inulin fibre on probiotic yogurt in Mwanza, Tanzania and the Microbial Analysis of Probiotic Yogurt Fortified with *Moringa oleifera*. J Health Popul Nutr. 2015;33(1):60-7. [cited 2021 may]. Available from: https://www.ncbi.nlm.nih.gov/pmc/articles/PMC4438649/.

[48] Nesbitt M, McBurney RPH, Broin M, Beentje HJ. Linking biodiversity, food and nutrition: the importance of plant identification and nomenclature. Journal of Food Composition and Analysis [Internet]. 2010 [cited 2021 May]; 23(Issue 6): 486-498. Available from: https://doi.org/10.1016/j.jfca.2009.03.001.

[49] Whallans RCM, Ana Paula F, LMS, HDMS, RMMH. Incorporation of phenolic compounds in food products: a review. Brazilian Journal of Development. 2021; 7(5):46470-46499. Available from: https://doi: 10.34117/bjdv7n5-184

[50] Castro MR, Figueiredo FF. Traditional knowledge, biodiversity, integrative and complementary practices: The use of medicinal plants in the SUS. Hygeia [Internet]. 2019 [cited 2021 May];15(31):56-70. Available from: http://www.seer.ufu.br/index.php/hygeia/article/view/4660.

[51] Tesser CD, Sousa IMC, Nascimento MC. Traditional and Complementary Medicine in Primary Health Care in Brazil. Saúde em debate [Internet]. 2018 [cited 2021 May]; 42(Spec 1):174-188. Available from: https://doi.org/10.1590/0103-11042018S112

[52] Lima CA, Santos AMVS, Messias RB, Costa FM, Barbosa DA, Silva CSO, et al. Integrative and complementary practices: use by community health agents in self-care. Rev Bras Enferm [Internet]. 2018;71(Suppl 6):2683-9. [Thematic Issue: Good practices in the care process as the centrality of the Nursing]. [cited 2021 May]. Available from: http://dx.doi.org/10.1590/0034-7167-2018-0078.

[53] Proença RPC. Food and globalization: some reflections. Science and Culture. [cited 2021 May]. 2010:62(4):43-47. Available from: http://cienciaecultura.bvs.br/scielo.php?script=sci_arttext&pid=S0009-67252010000400014

[54] Sá NC, Silva EMS, Bandeira AS. The culture of grapes and wine in the São Francisco Valley. RDE-Economic Development Magazine - RDE, Year XVII [Internet]. 2015 [cited 2021 May]: Special edition: 461 – 491. Available from: doi: http://dx.doi.org/10.21452/rde.v17nesp.4017.

[55] Souza SP, Seabra JE, Nogueira LAH. (2018). Feedstocks for biodiesel production: Brazilian and global perspectives. Biofuels, 9(4), 455-478. Available from: DOI: 10.1080 / 17597269.2017.1278931

[56] Alencar-Filho JMT, Sampaio PA, Pereira ECV, Oliveira-Júnior RG, Silva FS, Almeida JRGS, Rolim LA, Nunes XP, Araújo, ECC. (2016). Flavonoids as photoprotective agents: a systematic review. Journal of Medicinal Plants Research, 10(47), 848-864. [cited 2021 May]. Available from: https://academicjournals.org/journal/JMPR/article-abstract/C1278AB62177

[57] Guerra R. Ecotoxicological and chemical evaluation of phenolic compounds in industrial effluents. Chemosphere. 1(44): 1737-1747, 2001. [cited 2021 May]. Available from: https://pubmed.ncbi.nlm.nih.gov/11534905/

[58] El-Naas MH, Alhaija MA, Al-Zuhair S. Evaluation of na activated carbon packed bed for the adsorption of phenols from petrolium refinery wastwater. Environ Sci. Pollut Res. 1-10, 2017.

[59] Neves LCM., et al. Biofiltration Methods for the Removal of Phenolics Residues. Applied Biochemistry and Biotecnology. 1 (1): 129-132, 2006. Available from: https://doi.org/10.1385/ABAB:129:1:130

[60] Gernjak W. et al. Photo-Fenton treatment of water containing natural phenolic pollutants. Chemosphere, 50 (1): 71-78, 2003. Available from: DOI: 10.1016/s0045-6535 (02) 00403-4

[61] Abbas ZI, Abbas AS. Oxidative degradation of phenolic wastewater by electro-fenton process using MnO2-graphite electrode. Journal of Environmental Engineering. v.7, p.1-7, 2019.

[62] Cunha GMA, Evangelista-Neto AA, Medeiros GGD, Silva DN, Mota ALN, Chiavone-Filho O. Uso do processo Foto-Fenton no tratamento de águas produzidas em campos de petróleo. 4° PDPETRO, Campinas, SP. 21-24 de Outubro de 2007. [cited 2021 july]. Available from: http://www.portalabpg.org.br/PDPetro/4/resumos/4PDPETRO_6_2_0443-1.pdf

[63] Tian H., et al. Biodegradation phenolics compounds in high saline wastewater by biofilms adhering on aerated membranes. Journal of Hazardous Materials, 392 (1):1-11, 2020. Available from: https://doi.org/10.1016/j.jhazmat.2020.122463

[64] Diao M., et al. Biodepollution of wastwater containing phenolic compounds from leather industry by plant peroxidases. Biodegradation, 22(1):389-396, 2011. Available from: DOI: 10.1007/s10532-010-9410-8

[65] Cañadas R., et al. Overview of neoteric solvents as extractants in food industry: A focus on phenolic compounds separation from liquid streams. Food Research International, 136 (1): 1-17, 2020. Available from: https://doi.org/10.1016/j.foodres.2020.109558

[66] Gogate P. R. Treatment of wastewater streams containing phenolic compounds using hybrid techniques based on cavitation: A review of the current status and the day forward. Ultrasonics Sonochemistry, 15 (1):1-15, 2008. Available from: DOI: 10.1016 / j.ultsonch.2007.04.007

[67] Sá KM et al. Assessing the impact of the Brazilian policy on medicinal plants and herbal medicines in higher education in the health area. Ibero-American Journal of Education Studies, 1106-1131, 2018. [cited 2021 May]. Available from: https://periodicos.fclar.unesp.br/iberoamericana/article/view/11160.

[68] Nunes EO, Bonato-Júnior D, Pegoraro DDB. Evolution of phenolic compounds and ochratoxin a in the processing of isabel grape juice (*Vitis labrusca* L.). Scientific Initiation Seminar and Integrated Teaching, Research and Extension Seminar, 2013. [cited 2021 May], Available from: https://portalperiodicos.unoesc.edu.br/siepe/article/view/3259.

[69] Vaccari NFS, Soccol MCH, Ide GM. Phenolic compounds in wines and their antioxidant effects in disease prevention. Agricultural Sciences Journal [Internet]. 2009 [cited 2021 May]; 8 (Suppl 1): 71-83. Available from: https://www.bvs-vet.org.br/vetindex/periodicos/revista-de-ciencias-agroveterinarias/8-(2009)-1. ISSN 1676-9732

[70] Silva MAM, Frutuoso MNMA, Rodrigues SSFB, Nogueira RJMC.

Socio-environmental factors influenced by drought in the conservation of the caatinga. HOLOS [Internet]. 2016 [cited 2021 May]; 4:245-257. Available from: https://doi.org/10.15628/holos.2016.4175.

Chapter 13

Effect of Climate Change on Polyphenols Accumulation in Grapevine

Monis Hussain Shah, Rizwan Rafique, Tanzila Rafique, Mehwish Naseer, Uzman Khalil and Rehan Rafique

Abstract

Phenolics compounds in grapes contribute to berry and must color, organoleptic properties, nutritional value, antioxidant properties and provide protection against environmental challenges. Climate change has place mammoth challenges for the viticulture industry in different viticulture regions. Environmental variables determine to the greater extent, suitable grapes varieties for fresh as well as premium quality wine production. Grape berry composition is particularly affected by heat, drought, and intensity of solar irradiation. It is expected that climatic extremes will have an adverse effect on berry quality traits such as phenolic compounds in different grape cultivars. Polyphenols particularly anthocyanins decrease at elevated temperature, similarly flavanols levels increase with better exposure to solar radiation. Water availability is crucial for better vine growth and good production, however modest water stress particularly near veraison, upregulates the activity of key enzymes of the phenylpropanoid and flavonoid pathways. Therefore, it is important to know that how and when phenolic substance accumulate in berries and how various cultivars respond. This review elaborates the effect of weather conditions on biosynthesis of different phenolic compounds in grapes. Berry phenolic substances e.g., total phenolic compounds (TPC), total anthocyanins (TAC) and total flavonoid contents (TFC) synthesis is strongly regulated under the influence of environmental conditions during growing season. In this chapter we, shall focus on accumulation of phenolic compounds in grapevine in relation to climatic variations.

Keywords: Grapevine, berry phenolics, anthocyanins, temperature, CO_2, radiations, water

1. Introduction

1.1 Global climate change

Climate changes are the mammoth challenges that human race will face in coming decades as described by Intergovernmental Panel on Climate Change. The increase in release of greenhouse gases, particularly CO_2 is considered as the main cause of global warming. The concentration of CO_2 has increased from 280 ppm to 400 ppm subsequently of 0.5–1°C rise in an average temperature. It is expected that mean global temperature will rise by 0.2–0.3°C per decade hence rise of 1.2 to 5.8°C by the end of the twenty-first century. The increase in mean temperature in key viticulture regions was 1.6–1.8°C in Europe and 1.2–1.4°C across the globe during the growing seasons from 1950 to 2000 [1–4]. Similarly, a decrease in precipitation has been recorded in over southern Europe [5]. In addition to rising temperature, corresponding heat waves are becoming more common and frequent. Climate change is no doubt an inevitable challenge that must be dealt with serious policies in the upcoming decades. It is a major challenge that viticulture industry has to face in coming decades.

1.2 Climate a key determinant for viticulture

Climate is a limiting factor determining phenology, vegetative growth, physiological development, fruit production and consequently wine quality [6–8]. Geographical distribution of vineyards is determined by climatic factors. Weather parameters: temperatures, solar radiation, precipitation, and the inter-annual seasonal variability leads to annual changes in vine productivity [9–11]. Extreme weather events: hailstorms, excessive rainfall, late frost spells have been recognized as factors having detrimental impacts on grapevine productivity and quality [12].

1.3 Climate change impacts on viticulture

It is evident that climate change will have a negative impact on viticulture industry. Higher temperature during the active growing season will strongly affect grapevines because it is a major driver of development stages of grapevine [13]. Extreme heat stress during ripening period will abruptly reduce grapevine metabolism. It may result in higher sugar levels and lower acidity with potential increase in chances of wine spoilage [14] thereby lower production and quality. Furthermore, extreme heat and water stress, under future climates, may threaten final yields and productivity [15].

2. Grapevine phenolic compounds

Phenolic Compounds in grapes account for only a trivial proportion of the berry weight but contribute significantly to fresh fruit. All phenolic compounds have some common features as; an "aromatic ring" comprising of six carbon atoms having one or more hydroxyl (OH) groups or their derivatives as indicated in **Table 1**. They play an important role in color development, astringency, flavor and aroma to grapes. These compounds are the main substrates for grape juice and wine oxidation [16–18]. Their susceptibility to oxidation due to unsaturated double bonds and hydroxyl groups make phenolic compounds valuable antioxidants [19, 20]. Flavonoids and non-flavonoids phenolics are produced inside grape berries through biochemical pathway (**Figure 1**). Flavonoids accumulate mainly in the skin, seeds, and stem while neoflavanoids mostly accumulate in the mesocarp of the berry.

Phenolic profile of grapevines depends on, region, prevailing weather conditions, and site-specific viticultural practices [22–28]. Higher the total phenolic content more is antioxidant activity and it is a genotypic character [29–32]. Skin color (yellow, pink, red, blue-black and full black) is due to presence of anthocyanins. Anthocyanins are synthesized to protect the berries from the negative effect of adverse environmental conditions particularly ultraviolet radiation. Accumulation and degradation of already synthesized anthocyanins was noticed due to elevated temperatures during the ripening period [33, 34]. Therefore, in hotter regions the

Polyphenolic Compounds	Basic Chemical Structure	Examples
Anthocyanin	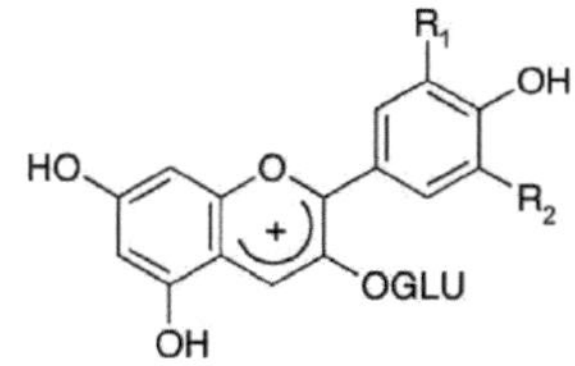	Cyanidin-3-GLUa **R1=OH, R2=H** **Delphinidin-3-GLU** R1=OH, R2=OH **Peonidin-3-GLU** R1=OCH3, R2=H **Malvidin-3-GLU** R1=OCH3,R2=OCH3 **Petunidin-3-GLU** R1=OCH3, R2=OH
Flavonols	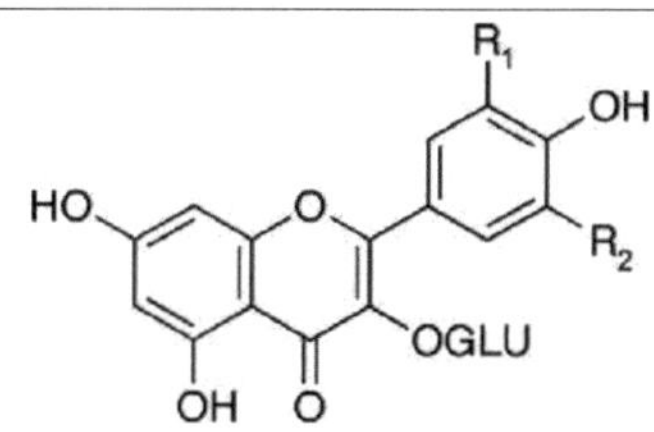	**Isorhamnetin-3-GLU** R1=OCH3, R2=H **Kampferol-3-GLU** R1=H, R2=H **Laricitrin-3-GLU** R1=OCH3, R2=OH **Myricetin-3-GLU** R1=OH, R2=OH **Quercetin-3-GLU** R1=OH, R2=H **Syringetin-3-GLU** R1=OCH3, R2=OCH3
Flavan-3-ols	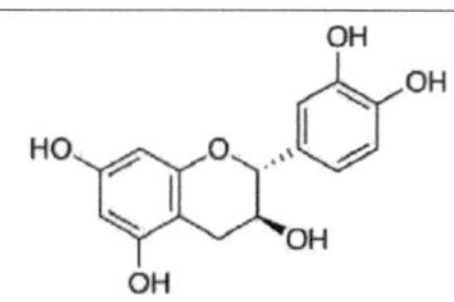 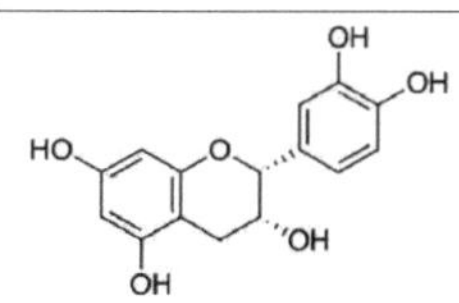	**Catechin** (Left) **Epicatechin** (Right)
Tannins	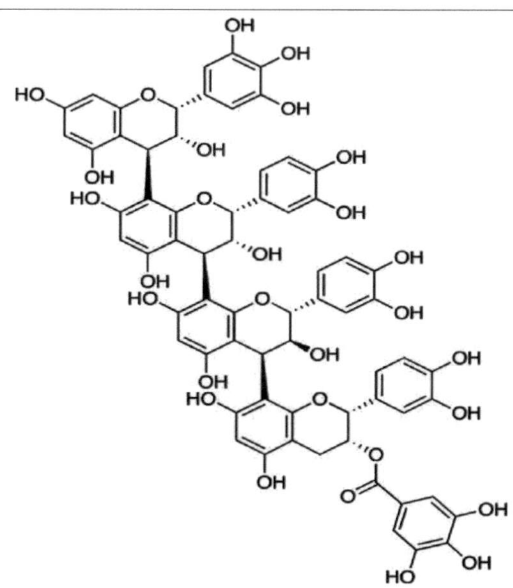	**Proanthocyanidin** tetramer having (from top to bottom) epigallocatechin, epicatechin, catechin, and epicatechin gallate
Hydroxycinnamic acid	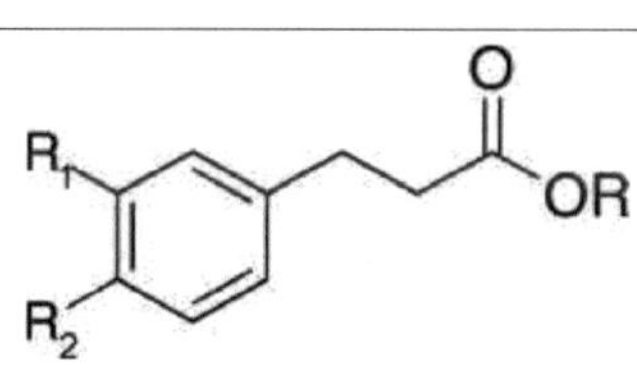	**Caffeic Acid** R1=OH, R2=OH **Cinnamic Acid** R1=H, R2=H **Coumaric Acid** R1=H, R2=OH **Ferulic Acid** R1=OCH3, R2=OH

Polyphenolic Compounds	Basic Chemical Structure	Examples
Hydroxybenzoic acid	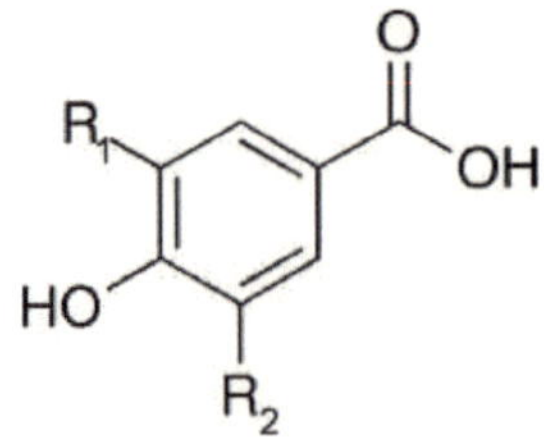	**Gallic acid** R1=OH, R2=OH **Protocatechuic Acid** R1=H, R2=OH **Syringic acid** R1=OCH3, R2=OCH3
Stilbenes	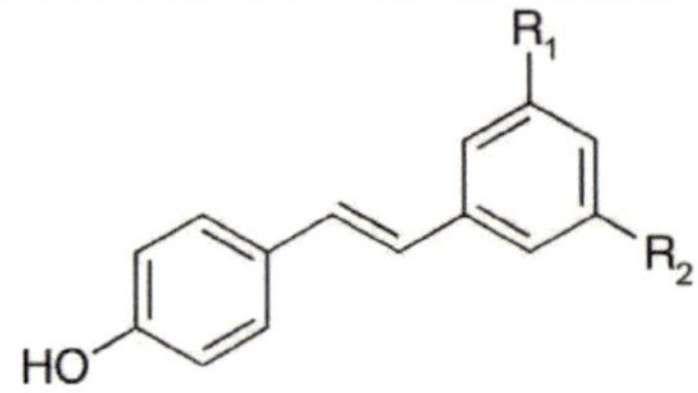	**Piceid** R1=OH, R2=GLU **Pterostilbene** R1=OCH3, R2=OCH3 **Resveratrol** R1=OH, R2=OH **Viniferins** resveratrol polymers

Table 1.
Different classes of polyphenolic compounds and their basic structures along with examples are given.

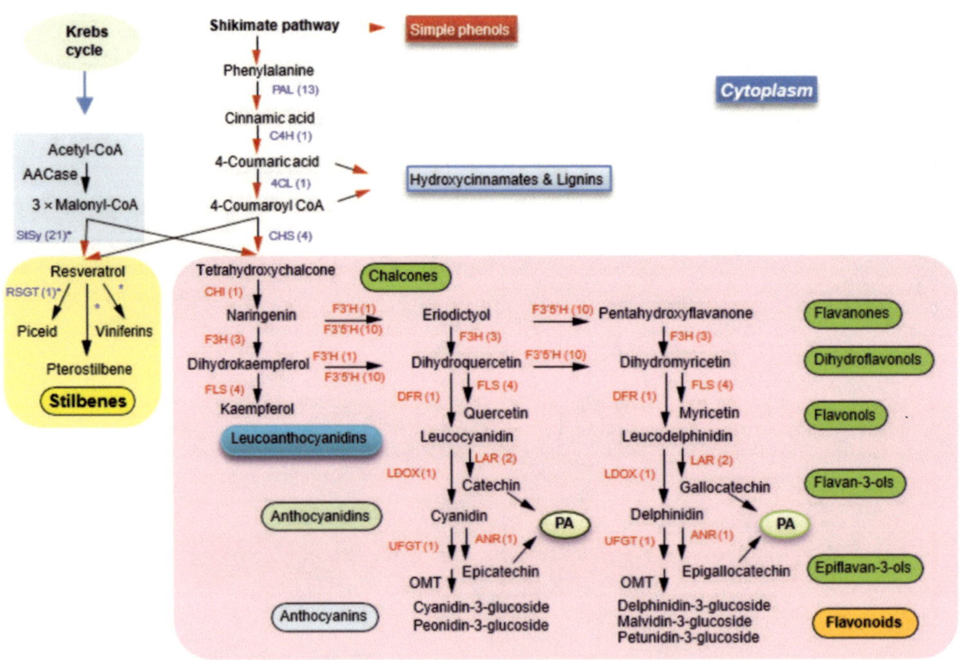

Figure 1.
Shikimate pathway for the biosynthesis of anthocyanins, Flavonols and flavonoids. (reproduced from the idea of Velasco et al. [21] with few modifications).

anthocyanin in red and black grapes skin is affected more, while climatic conditions in colder growing regions favor their biosynthesis. Grapevine varieties (var.) have particular anthocyanin fingerprints e.g., malvidin-3-Oglycoside is most abundant in var. 'Hasansky Sladky' while in var. 'Zilga' it is delphinidin-3-O-glycoside. Moreover, their biosynthesis varies from year to year due to annual seasonal climatic variability [35].

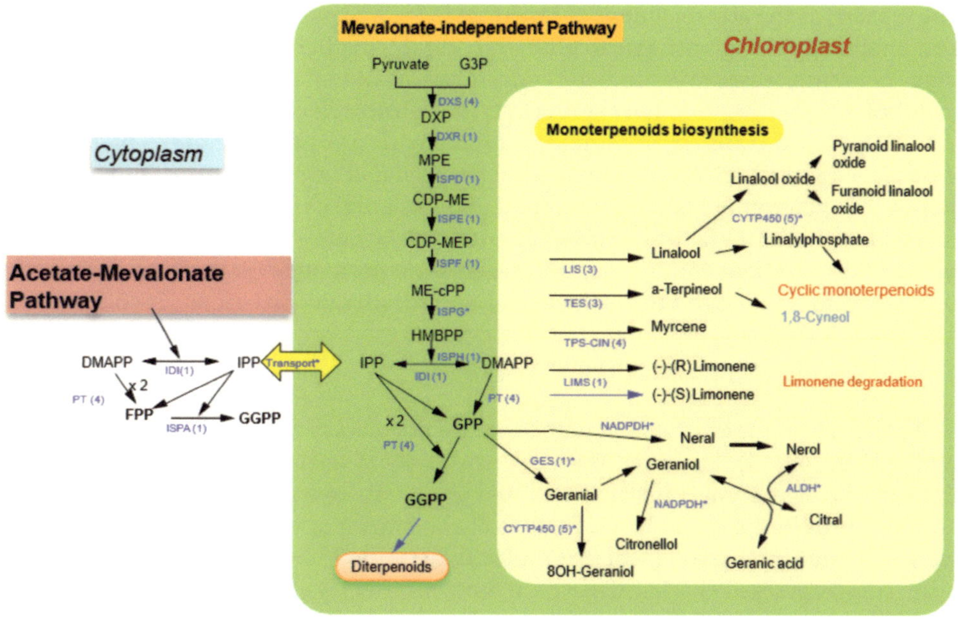

Figure 2.
Acetate Mevalnoate pathway for monoterpenoids biosynthesis. (reproduced from the idea of Velasco et al. [21] with few modifications).

2.1 Phenolic compounds biosynthesis in grapevine

Production of phenoloic compounds is regulated by transcription factors which regulate the activity of genes involved in phenolic biosynthetic pathways. Moreover, location, timing, and extent of the production of phenolic compounds is also dependent on these transcription factors [36, 37]. In addition to grape berries, some flavonoids are produced in leaves and are imported via the phloem [38, 39]. Shikimate and malonate pathways are the two main "assembly lines". The shikimate pathway (**Figure 1**) is the part of the biosynthesis chain of most plant phenolics, whereas the malonate pathway (**Figure 2**) is less important compared with Shikimate pathway in plants, but the malonate pathway is essential in fungi and bacteria.

3. Key climatic variables affecting grapes polyphenolic compounds

Secondary metabolites such polyphenols play significant ecological functions within the defense and signaling mechanisms in plants [40]. Different climatic variables such as air temperature, radiation, rainfall, relative humidity, wind, altitude, and topographic features play vital role in the polyphenol biosynthesis pathway in grapes. In this section, we shall review research studies focusing on key environmental variables.

3.1 Temperature

Temperature plays a significant role in vine phenology whereas increase in mean temperature prolonged the vegetative and reproductive cycle of grapevine and hence berry developmental and maturity stages are shifted in warmer months of the growing plant reproductive cycles [41]. Available historical records of harvest timings from different grape growing regions indicate an advance of 1–2 weeks during last few decades [42–47]. Although, some management practices may be the one reason for the advancement of ripening [48]. The conjugated effects of progressive phenology along with rise in temperatures during berry ripening with higher sugar contents, lesser organic acids concentration and altered berry composition of metabolites, such as phenolic compounds [41]. Research studies have encompassed the effect of wide ranging of temperature intensities; from moderate to high heat stress i.e., up to 35–45°C during day or night period at key berry development stages. The genotype, plant material and experimental constraints may affect the response of berry metabolism to temperature variations [49]. Although, difficult to fully relate with field conditions, controlled climate chamber experiments are also conducted to understand the influence of environmental variations [50–52].

3.1.1 Temperature impact on phenolic compounds

Effects of temperature on polyphenols are not always consistent as recently highlighted [49, 53]. However, there are unequivocal scientific evidence which indicate the deleterious effects of elevated temperature on the biosynthesis of anthocyanins in the grape berry. Studies of the impact of elevated temperature were validated at physiological and molecular studies [54–63]. It was noticed that heat stress repressed chief anthocyanin biosynthesis regulators, such as VviMYBA1 and downstream regulating genes such as VviCHI, VviUFGT, VviDFR, VviF3H2, and VviLDOX. However, not all of these research studies indicated unambiguous suppression or a strong correlation with lower anthocyanin accumulation. Various aspects of viticulture e.g., vines, cultivars, berry development stages, treatment intensities and sampling strategy take part in accumulation and production of anthocyanin. The effect of temperature on anthocyanin biosynthesis varies highly between different genotypes. For instance, when maximum temperature exceeds 35°C during berry ripening, it inhibits the color formation prominently e.g., in cv. Grenache than in cv. Carignan [64].

Previously, it has also been established that timings of temperature variations during day-night period have a strong influence on berry metabolites and lower temperature near berry ripening time particularly at night was related with improved coloration of grapes [55]. It was recently confirmed through experiments at molecular level that lower night temperatures increased anthocyanin accumulation and expression of related genes e.g., VviF3H1, VviUFGT, VviCHS3, and VviMYBA1 [65]. More pronouncing effects of lower night temperature were noticed near veraison stage in Corvina grapes. In a related study on Kyoho grapes, 3°C rise in temperature (27 to 30°C) during berry ripening caused less berry coloration and induced a significant decrease in transcript levels of anthocyanin regulating genes [66]. Similarly, in cv. Merlot, increase in day's temperature from 20 to 25°C during ripening caused decrease in anthocyanin levels by 37% [63]. In addition to repressing of anthocyanin regulating genes, high temperature may stimulate anthocyanin degradation due to the augmented activity of peroxidases [33]. It has been established that a peroxidase coding gene; VviPrx3 is up regulated, in berries when exposed to high temperature [67], and similar effects have been noticed in other plant species, such as Brunfelsia, litchi and strawberry [68–70]. A related increase in quantity of acylated and tri-hydroxylated anthocyanins has been observed in cvs. Merlot, Cabernet Sauvignon, Sangiovese and Malbec under higher temperature conditions [52, 61, 63, 71, 72], alongside overexpression of the acyltransferase gene Vvi3AT activity. Similarly for anthocyanins, elevated temperature impeded flavanol buildup while significantly augmented methoxylated (isorhamnetin & syringetin) and 3′, 4′, 5′-substituted (myricetin & syringetin) flavonols in cv. Merlot. More interestingly, rise in temperature may cause a disconnection of sugar-anthocyanin accumulation and biosynthesis, hence leading to a lower anthocyanin-sugar ratio which might be due to delayed anthocyanin biosynthesis or lesser anthocyanin accumulation during ripening phases. The extent of this thermal decoupling is highly cultivar dependent as indicated for cv. Grenache and cv. Carignan and can vary even among the clones of same cultivar as discussed for cv. Tempranillo [63, 64, 72–75].

The effect of temperature on tannins biosynthesis is yet not fully understood. However, it may be pointed that elevated temperature can enhance the production of tannin monomers, flavan-3- ols as highlighted by [76]. However, some other studies report non-significant effects on tannin production as tannins were not much affected by heat stress in cv. Sangiovese at veraison stage. More recently, scientists came up with similar results indicating no effect on flavan-3-ol or tannin levels. Although, significantly higher galloylation of flavan-3-ols levels were noticed in consistent with earlier findings. It was further indicated that an overexpression of UDP glucose-gallic acid-glucosyltransferase genes under elevated temperature. Moreover, heat stress also reserved the expression levels of members of STS biosynthetic pathway. However, lower temperature upregulated STS transcripts hence accelerated stilbene biosynthesis [49, 61, 76–79].

3.2 Radiation

Berry exposure to sunlight is generally associated with better berry quality attributes due to more total soluble solids (TSS), anthocyanins, and phenolics. On the other hand, it also lowers acidity and pH along with lower disease incidence due to favorable improved microclimate [49, 80–82].

3.2.1 Effect of radiations on phenolic compounds

Increased levels of phenolic compounds have been noticed in cvs. Pinot Noir, Riesling, Summer Black and Cabernet Sauvignon owing to better exposure to sunlight [83–86]. It also augmented the expression level of regulatory and structural phenylpropanoid genes as highlighted by recent studies [87–90]. Flavonoids particularly flavonol glucosides are the most light-responsive phenolic compounds ones whose levels increased with better exposure to sunlight. This positive effect was in consistent with their UV radiation-screening activity and their capability to reduce oxidative damage. Flavonoids were produced upon exposure to UV-B radiation as adaptive traits to reduce the radiation damage, as there exists a strong correlation between physiology and quercetin-3-O-glucoside & kaempferol-3-O-glucoside levels in UV-B radiation stressed vines [91–97]. Recently, a more comprehensive study elucidated that shoot removal and leaf thinning in cvs. Cabernet Sauvignon & Petit Verdot improved light exposure, hence it significantly augmented the flavonols kaempferol, quercetin and myricetin levels. However, little or no change was noticed for other flavonoid compounds. Similarly, higher levels of hydroxycinnamic acids and flavonol were noticed due to increased sun light exposure in cv. Cabernet Sauvignon [90, 98].

Several transcriptomic studies indicated that flavonol genes such as VviGT5, VviGT6 and VviFLS1 were induced more than other phenylpropanoids genes when exposed to UV radiation as observed in cv. Tempranillo berry skin. In return, lower expression level of VviFLS4 gene and its transcriptional regulator i.e., MYB12 was noticed under shade [92]. However, it has not yet been established that to what level UV light contributed to stimulate the synthesis of phenolic compounds. It can be deduced from literature that UV-B radiations are responsible for overexpression of key flavonoid genes [40, 99–104]. Recently, VviHY5 and VviHYH; the two bZIP TFs elongated hypocotyl 5 protein (HY5) orthologs were identified as the key components of UV-B reaction pathway along with mediated flavonol accumulation owing to high radiation exposure in grapevines [100, 105].

Anthocyanin accumulation increased significantly when grapes clusters were exposed to increased light, whereas shading decreased them. Recently, it was indicated that the UV-B radiation might prompted up-regulation of miR3627/4376 which facilitated anthocyanin accumulation [106, 107]. In a related *in vitro* study in which effect of berry exposure to light and temperature was studied it was inferred that elevated light increased anthocyanin levels in grapes [59]. The augmented anthocyanin levels found associated with the up regulation of correlated genes of anthocyanin biosynthesis pathway. Some other studies also endorse the stimulation of key anthocyanin genes e.g., TF VviMYBAa and VviUFGT under higher sun light exposure [66, 99]. Interestingly, UV-B radiation prompted the expression of VviMYBA1 gene while delaying the down regulation of VviMYBA6 and VviMYBA7 genes at later berry developmental stages [105]. Less light exposure modulated the quantity of di- to tri-hydroxylated anthocyanins more toward tri-hydroxylated anthocyanins as demonstrated through the down regulation of VviF3′ 5 ′Hs, somewhat similar but inconstant trends have been reported in cvs. Cabernet Sauvignon and Petite Verdot under warm climatic conditions [59, 90, 98, 99, 106]. However, low light conditions may increase non-acylated anthocyanins concentration as highlighted by [92, 107]. There is still need for further research to develop a better understanding.

3.3 Water

Water is an important constituent of plant structure and performs variety of functions in addition to transport of mineral nutrients from soil. It is a key component of photosynthetic pathway in plants. Moreover, water balance is necessary for quality table and wine grape production. Similarly, primary and secondary metabolite production is regulated by balanced water availability.

3.3.1 Impact of vine water status on phenolic compounds

Different primary and secondary metabolites are significantly influenced by drought stress in grapevines. Recent research has focused on probing the effects of water on berry physiology and quality attributes [40, 108, 109] and it has been noticed that drought stress may increase primary metabolites and polyphenols up to 85% and 60% respectively under different stress treatments. The impact of water deficit varies with intensity and duration of the stress conditions as well as berry developmental stage. Water deficit during the initial growth phases has more negative impact on final volume and yield at harvest as it reduces cell expansion, however rate of cell expansion is not affected much [110] while ripening phase, and it has little impact on berry size. Primary metabolites such as citric acid and glyceric acid synthesis was affected by both short and prolonged stress whereas polyphenols biosynthesis was accelerated only by the prolonged drought stress treatment.

Selective water deficit applications increased anthocyanin accumulation in grape skin along with the activation of genes of corresponding anthocyanin biosynthesis pathway [111]. For instance, in grape cv. Chardonnay, water stress increased the content of flavonols and decreases the expression of genes involved in biosynthesis of stilbene precursors [40].

It has been observed that modest water deficit i.e., predawn leaf water potential of 0.3 to −0.5 MPa is useful for better wine quality especially for red cultivars [112, 113] These positive effects may partially be attributed to increased solute concentration owing reduced berry volume under water deficit conditions. However, a higher buildup of secondary metabolites independent of change in berry volume has been reported [114]. More elaborative research findings at molecular level highlighted an upregulation of key enzymes of the phenylpropanoid and flavonoid pathways in response to water stress [40, 115–120]. But these beneficial effects were more noticeable when water deficit occurred throughout berry ripening phase [49].

In addition to an increase in the accumulation of phenylpropanoids and flavonoids due to water stress, an altered composition of anthocyanins has also been noticed owing to increased levels of tri-hydroxylated anthocyanins i.e., petunidin, delphinidin and malvidin [111, 121–124]. However, these observed changes in the anthocyanin profile of grapes due to water stress appear to be highly varietal dependent [125, 126] due to varying genotypic response associated to environmental variables. Similarly, an increase in proanthocyanidin concentration and proanthocyanidin polymerization along with higher catechin levels in grape berry skins have also been indicated by [127–129]. The increase in phenolic levels when water deficit occurred before veraison may be due to concentration effects [130, 131] however, several other scientists discussed increase in anthocyanin content at berry level [111, 114, 123, 132]. More focused research is needed to validate ribose, glyceric acid, citric acid, kaempferol-3-O-glucoside and quercetin-3-O-glucoside interactions as indicators of drought stress [133].

3.4 Impacts of elevated CO_2 concentration

Elevated atmospheric CO_2 is usually favorable for plant growth as it causes an increased photosynthetic carbon fixation hence more biomass and yield. Free Air Carbon enrichment (FACE) experiments on agronomic crops such as wheat, rice and soybean have outlined 12–14% increase in harvestable yield owing to elevated carbon dioxide (eCO_2) [134–136]. Although, there are limited studies on horticultural crops however, it has been indicated that eCO_2 increased total antioxidant capacity of fruits and vegetables, along with higher concentration of glucose, fructose, total soluble sugars, polyphenols compounds, flavonoids, ascorbic acid, and calcium [90]. Research studies on grapevine related to eCO_2 mainly focused on vegetative growth and photosynthetic responses while records on berry metabolism at physiological and molecular level are relatively scarce. However, most of the available records suggest an increase in photosynthetic activity hence better yield and biomass accumulation [94, 137–141]. Recently dependence of berry ripening rates on the carbon fixation was investigated however, only few quality attributes were found to be affected due to eCO_2 and that particularly; sugars, acids, and berry size [138, 142, 143]. Recently, it has been inferred in FACE experiment that eCO_2 did not negatively affected juice and wines quality [144]. Similarly, it had already been established that anthocyanins and proanthocyanidins were not affected by eCO_2 [137, 138, 142–145]. Moreover, in multi stress experiments on cv. Temperanillo cuttings where elevated temperature condition i.e., +4°C and CO_2 i.e., 700 ppm were simulated it was deduced that high CO_2 in combination with elevated temperature hastened berry ripening and decreased high temperature tempted anthocyanin–sugar decoupling in berries [146].

4. Conclusions

Polyphenols are the key secondary metabolite of grapes and have ample amounts of antioxidants. The production and biosynthesis of phenols is regulated by varying climatic conditions in addition to genotypic traits. Elevated temperature impairs phenolic biosynthesis pathways hence lesser accumulation, while lowers temperatures favor their production. On the other hand, excessive radiation may cause degradation of these compounds. Optimum sun light penetration is necessary for the activation of genes of phenolic biosynthesis pathways. Water balance is also important as modest water deficit near verasion can also promote their activity. For elevated carbon dioxide levels (eCO_2) despite limited studies, no major negative effects have been reported. However, there is need to study grapes phenolic compounds in relation to global climate change.

Author details

Monis Hussain Shah[1], Rizwan Rafique[2,3,4*], Tanzila Rafique[5], Mehwish Naseer[5], Uzman Khalil[4] and Rehan Rafique[6]

1 Horticultural Research Institute for Floriculture and Landscaping, Rawalpindi, Pakistan

2 Agriculture Department (Extension), Punjab, Pakistan

3 PMAS Arid Agriculture University, Rawalpindi, Pakistan

4 University of Florida, Gainesville, USA

5 Department of Botany, GC Women University Faisalabad, Pakistan

6 Foot and Mouth Disease Research Center, Lahore Cantonment, Pakistan

*Address all correspondence to: rizwanrafiqueao@gmail.com

References

[1] Quenol H, de Cortazar Atauri IG, Bois B, Sturman A, Bonnardot V, Le Roux R. Which climatic modeling to assess climate change impacts on vineyards? Oeno One. 2017; 51:91-97.

[2] Raza A, Razzaq A, Mehmood SS, Zou X, Zhang X, Lv Y, Xu J. Impact of climate change on crops adaptation and strategies to tackle its outcome: A review. Plants. 2019; 8:34.

[3] Fraga H, Malheiro AC, Moutinho-Pereira J, Santos JA. An overview of climate change impacts on European viticulture. Food and Energy Security. 2012; 1:94-110.

[4] Ramos MC, Jones GV, Yuste J. Phenology and grape ripening characteristics of cv Tempranillo within the Ribera del Duero designation of origin (Spain): influence of soil and plot characteristics. European Journal of Agronomy. 2015; 70:57-70.

[5] Christensen JH, Hewitson B, Busuioc A, Chen A, Gao X, Held I, Jones R, Kolli RK, Kwon WT, Laprise R, Magaña Rueda V. Regional climate projections. Chapter 11. 2007.

[6] Costa R, Fraga H, Fonseca A, García de Cortázar-Atauri I, Val MC, Carlos C, Reis S, Santos JA. Grapevine phenology of cv. Touriga Franca and Touriga Nacional in the Douro wine region: Modelling and climate change projections. Agronomy. 2019; 9:210.

[7] Van Leeuwen C, Friant P, Chone X, Tregoat O, Koundouras S, Dubourdieu D. Influence of climate, soil, and cultivar on terroir. American Journal of Enology and Viticulture. 2004; 55:207-217.

[8] Keller M. The science of grapevines. In Anatomy and Physiology, 1st ed.; Keller, M., Ed.; Elsevier Academic Press: London, UK, 2010; p. 400.

[9] Fraga H, Pinto JG, Santos JA. Climate change projections for chilling and heat forcing conditions in European vineyards and olive orchards: A multi-model assessment. Climatic Change. 2019; 152:179-193.

[10] Jones GV, Davis RE. Climate influences on grapevine phenology, grape composition, and wine production and quality for Bordeaux, France. American Journal of Enology and Viticulture. 2000; 51:249-261.

[11] Fraga H, Santos JA. Daily prediction of seasonal grapevine production in the Douro wine region based on favourable meteorological conditions. Australian Journal of grape and wine research. 2017; 23:296-304.

[12] Mosedale JR, Wilson RJ, Maclean IM. Climate change and crop exposure to adverse weather: changes to frost risk and grapevine flowering conditions. PloS one. 2015;10(10): e0141218.

[13] Parker M, Baldock G, Hayasaka Y, Mayr C, Williamson P, Francis IL, Krstic M, Herderich M, Johnson D. Seeing through smoke. Wine and Viticulture Journal. 2013; 28:42-46.

[14] De Orduna RM. Climate change associated effects on grape and wine quality and production. Food Research International. 2010; 43:1844-1855.

[15] Fraga H, de Cortazar Atauri IG, Santos JA. Viticultural irrigation demands under climate change scenarios in Portugal. Agricultural Water Management. 2018; 196:66-74.

[16] Macheix JJ, Sapis JC, Fleuriet A, Lee CY. Phenolic compounds and polyphenoloxidase in relation to browning in grapes and wines. Critical Reviews in Food Science & Nutrition.1991; 30:441-486.

[17] Singleton VL. Tannins and the qualities of wines. In Plant polyphenols 1992 (pp. 859-880). Springer, Boston, MA.

[18] Keller M. The science of grapevines. Academic press; 2020.

[19] Danilewicz JC. Review of reaction mechanisms of oxygen and proposed intermediate reduction products in wine: Central role of iron and copper. American Journal of Enology and Viticulture. 2003; 54:73-85.

[20] Danilewicz JC. Review of reaction mechanisms of oxygen and proposed intermediate reduction products in wine: Central role of iron and copper. American Journal of Enology and Viticulture. 2003; 54:73-85.

[21] Velasco R, Zharkikh A, Troggio M, Cartwright DA, Cestaro A, Pruss D, Pindo M, FitzGerald LM, Vezzulli S, Reid J, Malacarne G. A high quality draft consensus sequence of the genome of a heterozygous grapevine variety. PloS one. 2007; 2(12):e1326.

[22] Fernandes de Oliveira A, Mercenaro L, Nieddu G. Assessing thermal efficiency for berry anthocyanin accumulation in four different sites and field-growing conditions. InX International Symposium on Grapevine Physiology and Biotechnology 1188. 2016.p. 181-188.

[23] Zhu L, Zhang Y, Lu J. Phenolic contents and compositions in skins of red wine grape cultivars among various genetic backgrounds and originations. International Journal of Molecular Sciences. 2012; 13:3492-3510.

[24] Wojdyło A, Samoticha J, Nowicka P, Chmielewska J. Characterisation of (poly) phenolic constituents of two interspecific red hybrids of Rondo and Regent (Vitis vinifera) by LC–PDA–ESI-MS QTof. Food chemistry. 2018; 239:94-101.

[25] Samoticha J, Wojdyło A, Golis T. Phenolic composition, physicochemical properties and antioxidant activity of interspecific hybrids of grapes growing in Poland. Food chemistry. 2017; 215:263-273.

[26] Soubeyrand E, Basteau C, Hilbert G, van Leeuwen C, Delrot S, Gomes E. Nitrogen supply affects anthocyanin biosynthetic and regulatory genes in grapevine cv. Cabernet-Sauvignon berries. Phytochemistry. 2014; 103:38-49.

[27] Katalinic V, Mozina SS, Skroza D, Generalic I, Abramovic H, Milos M, Ljubenkov I, Piskernik S, Pezo I, Terpinc P, Boban M. Polyphenolic profile, antioxidant properties and antimicrobial activity of grape skin extracts of 14 Vitis vinifera varieties grown in Dalmatia (Croatia). Food chemistry. 2010; 119:715-723.

[28] Tarko T, Duda-Chodak A, Sroka P, Satora P, Jurasz E. Polish wines: Characteristics of cool-climate wines. Journal of food composition and analysis. 2010; 23:463-468.

[29] Yang J, Martinson TE, Liu RH. Phytochemical profiles and antioxidant activities of wine grapes. Food Chemistry. 2009; 116:332-339.

[30] Liang Z, Cheng L, Zhong GY, Liu RH. Antioxidant and antiproliferative activities of twenty-four Vitis vinifera grapes. PloS one. 2014; 9:e105146.

[31] Liang Z, Wu B, Fan P, Yang C, Duan W, Zheng X, Liu C, Li S. Anthocyanin composition and content in grape berry skin in Vitis germplasm. Food Chemistry. 2008;111:837-844.

[32] Ortega-Regules A, Romero-Cascales I, Lopez-Roca JM, Ros-García JM, Gomez-Plaza E. Anthocyanin fingerprint of grapes: environmental and genetic variations.

Journal of the Science of Food and Agriculture. 2006; 86:1460-1467.

[33] Mori K, Goto-Yamamoto N, Kitayama M, Hashizume K. Loss of anthocyanins in red-wine grape under high temperature. Journal of Experimental Botany. 2007; 58:1935-1945.

[34] Poudel PR, Mochioka R, Beppu K, Kataoka I. Influence of temperature on berry composition of interspecific hybrid wine grape 'Kadainou R-1'(Vitis ficifolia var. ganebu× V. vinifera 'Muscat of Alexandria'). Journal of the Japanese Society for Horticultural Science. 2009; 78:169-174.

[35] Maante-Kuljus M, Ratsep R, Moor U, Mainla L, Poldma P, Koort A, Karp K. Effect of vintage and viticultural practices on the phenolic content of hybrid winegrapes in very cool climate. Agriculture. 2020;10:169.

[36] Guo C, Guo R, Xu X, Gao M, Li X, Song J, Zheng Y, Wang X. Evolution and expression analysis of the grape (Vitis vinifera L.) WRKY gene family. Journal of Experimental Botany. 2014; 65:1513-1528.

[37] Hichri I, Barrieu F, Bogs J, Kappel C, Delrot S, Lauvergeat V. Recent advances in the transcriptional regulation of the flavonoid biosynthetic pathway. Journal of Experimental Botany. 201; 62:2465-2483.

[38] Buer CS, Muday GK, Djordjevic MA. Flavonoids are differentially taken up and transported long distances in Arabidopsis. Plant Physiology. 2007; 145:478-490.

[39] Zhao J, Pang Y, Dixon RA. The mysteries of proanthocyanidin transport and polymerization. Plant physiology. 2010; 153:437-443.

[40] Teixeira A, Eiras-Dias J, Castellarin S, Geros H. Berry phenolics of grapevine under challenging environments. International Journal of Molecular Sciences. 2013; 14:18711-18739. doi: 10.3390/ijms140918711

[41] Van Leeuwen C, Destrac-Irvine A. Modified grape composition under climate change conditions requires adaptations in the vineyard. OENO One. 2017; 51:147-154.

[42] Chuine I, Yiou P, Viovy N, Seguin B, Daux V, Ladurie EL. Grape ripening as a past climate indicator. Nature. 2004; 432:289-290.

[43] Duchene E, Schneider C. Grapevine and climatic changes: a glance at the situation in Alsace. Agronomy for Sustainable Development. 2005; 25:93-99.

[44] Jones GV, Duchene E, Tomasi D, Yuste J, Braslavska O, Schultz H, Martinez C, Boso S, Langellier F, Perruchot C, Guimberteau G. Changes in European winegrape phenology and relationships with climate. InXIV International GESCO Viticulture Congress, Geisenheim, Germany, 2005 p. 54-61.

[45] Petrie PR, Sadras VO. Advancement of grapevine maturity in Australia between 1993 and 2006: putative causes, magnitude of trends and viticultural consequences. Australian Journal of Grape and Wine Research. 2008; 14:33-45.

[46] Webb LB, Whetton PH, Bhend J, Darbyshire R, Briggs PR, Barlow EW. Earlier wine grape ripening driven by climate warming and declines in soil water content. Nature Climate Change. 2012; 2:259-264.

[47] Cook BI, Wolkovich EM. Climate change decouples drought from early wine grape harvests in France. Nature Climate Change. 2016; 6:715-719.

[48] Webb LB, Watterson I, Bhend J, Whetton PH, Barlow EW. Global

climate analogues for winegrowing regions in future periods: projections of temperature and precipitation. Australian Journal of Grape and Wine Research. 2013; 19:331

[49] Rienth M, Vigneron N, Darriet P, Sweetman C, Burbidge C, Bonghi C, Walker RP, Famiani F, Castellarin SD. Grape Berry Secondary Metabolites and Their Modulation by Abiotic Factors in a Climate Change Scenario–A Review. Frontiers in Plant Science. 2021; 12:262.

[50] Pillet J, Egert A, Pieri P, Lecourieux F, Kappel C, Charon J, Gomès E, Keller F, Delrot S, Lecourieux D. VvGOLS1 and VvHsfA2 are involved in the heat stress responses in grapevine berries. Plant and Cell Physiology. 2012; 53:1776-1792.

[51] Carbonell-Bejerano P, Santa María E, Torres-Pérez R, Royo C, Lijavetzky D, Bravo G, Aguirreolea J, Sanchez-Diaz M, Antolín MC, Martínez-Zapater JM. Thermotolerance responses in ripening berries of Vitis vinifera L. cv Muscat Hamburg. Plant and Cell Physiology. 2013; 54:1200-1216

[52] Puntel LA, Sawyer JE, Barker DW, Dietzel R, Poffenbarger H, Castellano MJ, Moore KJ, Thorburn P, Archontoulis SV. Modeling long-term corn yield response to nitrogen rate and crop rotation. Frontiers in plant science. 2016; 7:1630.

[53] Gouot JC, Smith JP, Holzapfel BP, Walker AR, Barril C. Grape berry flavonoids: A review of their biochemical responses to high and extreme high temperatures. Journal of Experimental Botany. 2019; 70:397-423.

[54] Buttrose MS, Hale CR, Kliewer WM. Effect of temperature on the composition of'Cabernet Sauvignon'berries. American Journal of Enology and Viticulture. 1971; 22:71-75.

[55] Kliewer WM, Torres RE. Effect of controlled day and night temperatures on grape coloration. American Journal of Enology and Viticulture. 1972; 23:71-77.

[56] Spayd SE, Tarara JM, Mee DL, Ferguson JC. Separation of sunlight and temperature effects on the composition of Vitis vinifera cv. Merlot berries. American journal of Enology and Viticulture. 2002; 53:171-182.

[57] Mori K, Sugaya S, Gemma H. Decreased anthocyanin biosynthesis in grape berries grown under elevated night temperature condition. Scientia Horticulturae. 2005;105: 319-330.

[58] Yamane T, Jeong ST, Goto-Yamamoto N, Koshita Y, Kobayashi S. Effects of temperature on anthocyanin biosynthesis in grape berry skins. American Journal of Enology and Viticulture. 2006; 57:54-59.

[59] Azuma A, Yakushiji H, Koshita Y, Kobayashi S. Flavonoid biosynthesis-related genes in grape skin are differentially regulated by temperature and light conditions. Planta. 2012; 236:1067-1080.

[60] Rienth M, Torregrosa L, Luchaire N, Chatbanyong R, Lecourieux D, Kelly M. Day and night heat stress trigger different transcriptomic responses in green and ripening grapevine (Vitis vinifera) fruit. BMC Plant Biology. 2014; 14:108.

[61] Pastore C, Dal Santo S, Zenoni S, Movahed N, Allegro G, Valentini G, Filippetti I, Tornielli GB. Whole plant temperature manipulation affects flavonoid metabolism and the transcriptome of grapevine berries. Frontiers in plant science. 2017; 8:929.

[62] Torres N, Hilbert G, Luquin J, Goicoechea N, Antolín MC. Flavonoid and amino acid profiling on *Vitis vinifera* L. cv Tempranillo subjected to deficit

irrigation under elevated temperatures. Journal of Food Composition and Analysis. 2017; 62:51-62.

[63] Yan Y, Song C, Falginella L, Castellarin SD. Day Temperature Has a Stronger Effect Than Night Temperature on Anthocyanin and Flavonol Accumulation in 'Merlot'(*Vitis vinifera* L.) Grapes During Ripening. Frontiers in Plant Science. 2020; 11:1095.'

[64] Fernandes de Oliveira A, Mercenaro L, Del Caro A, Pretti L, Nieddu G. Distinctive anthocyanin accumulation responses to temperature and natural UV radiation of two field-grown (*Vitis vinifera* L.) cultivars. Molecules. 2015; 20: 2061-2080.

[65] Gaiotti F, Pastore C, Filippetti I, Lovat L, Belfiore N, Tomasi D. Low night temperature at veraison enhances the accumulation of anthocyanins in Corvina grapes (Vitis vinifera L.). Scientific reports. 2018; 8:1-3.

[66] Shinomiya R, Fujishima H, Muramoto K, Shiraishi M. Impact of temperature and sunlight on the skin coloration of the 'Kyoho'table grape. Scientia Horticulturae. 2015; 193:77-83.

[67] Movahed N, Pastore C, Cellini A, Allegro G, Valentini G, Zenoni S, Cavallini E, Dinca E, Tornielli GB, Filippetti I. The grapevine VviPrx31 peroxidase as a candidate gene involved in anthocyanin degradation in ripening berries under high temperature. Journal of Plant Research. 2016; 129:513-526.

[68] Vaknin H, Bar-Akiva A, Ovadia R, Nissim-Levi A, Forer I, Weiss D, Oren-Shamir M. Active anthocyanin degradation in Brunfelsia calycina (yesterday–today–tomorrow) flowers. Planta. 2005; 222:19-26.

[69] Zhang Z, Pang X, Xuewu D, Ji Z, Jiang Y. Role of peroxidase in anthocyanin degradation in litchi fruit pericarp. Food Chemistry. 2005; 90: 47-52.

[70] Chisari M, Barbagallo RN, Spagna G. Characterization of polyphenol oxidase and peroxidase and influence on browning of cold stored strawberry fruit. Journal of Agriculture and Food Chemistry. 2007; 55: 3469-3476.

[71] Tarara JM, Lee J, Spayd SE, Scagel CF. Berry temperature and solar radiation alter acylation, proportion, and concentration of anthocyanin in Merlot grapes. American Journal of Enology and Viticulture. 2008; 59:235-247.

[72] De-Rosas I, Ponce MT, Malovini E, Deis L, Cavagnaro B, Cavagnaro P. Loss of anthocyanins and modification of the anthocyanin profiles in grape berries of Malbec and Bonarda grown under high temperature conditions. Plant Science. 2017; 258:137-145.

[73] Sadras VO, Petrie PR, Moran MA. Effects of elevated temperature in grapevine. II juice pH, titratable acidity and wine sensory attributes. Australian Journal of Grape and Wine Research. 2013; 19(:107-15.

[74] Arrizabalaga M, Morales F, Oyarzun M, Delrot S, Gomes E, Irigoyen JJ, Hilbert G, Pascual I. Tempranillo clones differ in the response of berry sugar and anthocyanin accumulation to elevated temperature. Plant Science. 2018; 267:74-83.

[75] Rienth M, Vigneron N, Darriet P, Sweetman C, Burbidge C, Bonghi C, Walker RP, Famiani F, Castellarin SD. Grape berry secondary metabolites and their modulation by abiotic factors in a climate change scenario–a review. Frontiers in Plant Science. 2021; 22:12:262.

[76] Cohen SD, Tarara JM, Gambetta GA, Matthews MA, Kennedy JA. Impact of

diurnal temperature variation on grape berry development, proanthocyanidin accumulation, and the expression of flavonoid pathway genes. Journal of Experimental Botany. 2012; 63: 2655-2665.

[77] Gouot JC, Smith JP, Holzapfel BP, Barril C. Impact of short temperature exposure of Vitis vinifera L. cv. Shiraz grapevine bunches on berry development, primary metabolism and tannin accumulation. Environmental and Experimental Botany. 2019;168:103866.

[78] Rienth M, Torregrosa L, Sarah G, Ardisson M, Brillouet JM, Romieu C. Temperature desynchronizes sugar and organic acid metabolism in ripening grapevine fruits and remodels their transcriptome. BMC Plant Biology. 2016: 16:164.

[79] Rienth M, Torregrosa L, Kelly MT, Luchaire N, Pellegrino A, Grimplet, J, et al. Is transcriptomic regulation of berry development more important at night than during the day? PLoS ONE. 2014; 9:e88844.

[80] Dokoozlian NK, Kliewer WM. Influence of light on grape berry growth and composition varies during fruit development. Journal of the American Society for Horticultural Science. 1996;121:869-874.

[81] Bergqvist J, Dokoozlian N, Ebisuda N. Sunlight exposure and temperature effects on berry growth and composition of Cabernet Sauvignon and Grenache in the Central San Joaquin Valley of California. American Journal of Enology and Viticulture. 2001; 52:1-7.

[82] Abeysinghe SK, Greer DH, Rogiers SY. The effect of light intensity and temperature on berry growth and sugar accumulation in Vitis Vinifera "Shiraz" under vineyard conditions. Vitis. 2019; 58:7-16.

[83] Song J, Smart R, Wang H, Dambergs B, Sparrow A, Qian MC. Effect of grape bunch sunlight exposure and UV radiation on phenolics and volatile composition of Vitis vinifera L. cv. Pinot noir wine. Food Chemistry. 2015; 173: 424-431.

[84] Xi X, Zha Q, Jiang A, Tian Y. Impact of cluster thinning on transcriptional regulation of anthocyanin biosynthesis-related genes in 'Summer Black'grapes. Plant Physiology and Biochemistry. 2016; 104:180-187.

[85] Brandt M, Scheidweiler M, Rauhut D, Patz CD, Will F, Zorn H, Stoll M. The influence of temperature and solar radiation on phenols in berry skin and maturity parameters of *Vitis vinifera* L. cv. Riesling: This article is published in cooperation with the 21th GIESCO International Meeting, June 23-28 2019, Thessaloniki, Greece. Guests editors: Stefanos Koundouras and Laurent Torregrosa. Oeno One. 2019;53(2).

[86] Blancquaert EH, Oberholster A, Ricardo-da-Silva JM, Deloire AJ. Grape flavonoid evolution and composition under altered light and temperature conditions in Cabernet Sauvignon (*Vitis vinifera* L.). Frontiers in Plant Science. 2019; 10:1062.

[87] Chorti E, Guidoni S, Ferrandino A, Novello V. Effect of different cluster sunlight exposure levels on ripening and anthocyanin accumulation in Nebbiolo grapes. American Journal of Enology and Viticulture. 2010; 61:23-30.

[88] Matsuyama S, Tanzawa F, Kobayashi H, Suzuki S, Takata R, Saito H. Leaf removal accelerated accumulation of delphinidin-based anthocyanins in 'Muscat Bailey A'[Vitis× labruscana (Bailey) and Vitis vinifera (Muscat Hamburg)] grape skin. Journal of the Japanese Society for Horticultural Science. 2013; CH-062.

[89] Wu BH, Cao YG, Guan L, Xin HP, Li JH, Li SH. Genome-wide transcriptional profiles of the berry skin of two red grape cultivars (Vitis vinifera) in which anthocyanin synthesis is sunlight-dependent or -independent. PLoS ONE. 2014: 9:e105959.

[90] Sun RZ, Cheng G, Li Q, He YN, Wang Y, Lan YB, et al. Light induced variation in phenolic compounds in cabernet sauvignon grapes (*Vitis vinifera* L.) involves extensive transcriptome reprogramming of biosynthetic enzymes, transcription factors, and phytohormonal regulators. Front. Plant Science. 2017; 8:547.

[91] Downey MO, Rochfort S. Simultaneous separation by reversed-phase high-performance liquid chromatography and mass spectral identification of anthocyanins and flavonols in Shiraz grape skin. Journal of Chromatography A. 2008; 1201:43-47.

[92] Matus JT, Loyola R, Vega A, Peña-Neira A, Bordeu E, Arce-Johnson P, Alcalde JA. Post-veraison sunlight exposure induces MYB-mediated transcriptional regulation of anthocyanin and flavonol synthesis in berry skins of Vitis vinifera. Journal of Experimental Botany. 2009; 60:853-867.

[93] Agati G, D'Onofrio C, Ducci E, Cuzzola A, Remorini D, Tuccio L, Lazzini F, Mattii G. Potential of a multiparametric optical sensor for determining in situ the maturity components of red and white Vitis vinifera wine grapes. Journal of Agricultural and Food Chemistry. 2013; 61:12211-12218.

[94] Moutinho-Pereira J, Goncalves B, Bacelar E, Cunha JB, Coutinho J, Correia CM. Effects of elevated CO2 on grapevine (*Vitis vinifera* L.): physiological and yield attributes. Vitis. 2009; :159-165.

[95] Reshef N, Walbaum N, Agam N, Fait A. Sunlight modulates fruit metabolic profile and shapes the spatial pattern of compound accumulation within the grape cluster. Frontiers in Plant Science. 2017; 8:70.

[96] Reshef N, Agam N, Fait A. Grape berry acclimation to excessive solar irradiance leads to repartitioning between major flavonoid groups. Journal of Agricultural and Food Chemistry. 2018; 66:3624-3625.

[97] Ibanez JG, Carreon-Alvarez A, Barcena-Soto M, Casillas N. Metals in alcoholic beverages: A review of sources, effects, concentrations, removal, speciation, and analysis. Journal of Food Composition and Analysis. 2008; 21:672-683.

[98] Torres N, Martínez-Lüscher J, Porte E, Kurtural SK. Optimal ranges and thresholds of grape berry solar radiation for flavonoid biosynthesis in warm climates. Frontier Plant Sciences. 2020a; 11:931.

[99] Koyama K, Ikeda H, Poudel PR, Goto-Yamamoto N. Light quality affects flavonoid biosynthesis in young berries of Cabernet Sauvignon grape. Phytochemistry. 2012; 78:54-64.

[100] Loyola R, Herrera D, Mas A, Wong DC, Höll J, Cavallini E, Amato A, Azuma A, Ziegler T, Aquea F, Castellarin SD. The photomorphogenic factors UV-B RECEPTOR 1, ELONGATED HYPOCOTYL 5, and HY5 HOMOLOGUE are part of the UV-B signalling pathway in grapevine and mediate flavonol accumulation in response to the environment. Journal of Experimental Botany. 2016; 67:5429-5445.

[101] Keller M and Torres-Martinez N. Does UV radiation affect winegrape composition? Acta Horticulturae. 2004; 640:313-319.

[102] Schreiner M, Martínez-Abaigar J, Glaab J, Jansen M. UV-B induced secondary plant metabolites: potential benefits for plant and human health. Optik and Photonik. 2014; 9:34-37.

[103] Carbonell-Bejerano P, Diago MP, Martínez-Abaigar J, Martínez-Zapater JM, Tardaguila J, Nunez-Olivera E. Solar ultraviolet radiation is necessary to enhance grapevine fruit ripening transcriptional and phenolic responses. BMC plant biology. 2014; 14:1-6.

[104] Liu L, Gregan S, Winefield C, Jordan B. From UVR 8 to flavonol synthase: UV-B-induced gene expression in Sauvignon blanc grape berry. Plant Cell Environment. 2015; 38:905-919.

[105] Matus JT, Cavallini E, Loyola R, Holl J, Finezzo L, Dal Santo S, Vialet S, Commisso M, Roman F, Schubert A, Alcalde JA. A group of grapevine MYBA transcription factors located in chromosome 14 control anthocyanin synthesis in vegetative organs with different specificities compared with the berry color locus. The Plant Journal. 2017; 91:220-236.

[106] Guan L, Dai Z, Wu BH, Wu J, Merlin I, Hilbert G, Renaud C, Gomes E, Edwards E, Li SH, Delrot S. Anthocyanin biosynthesis is differentially regulated by light in the skin and flesh of white-fleshed and teinturier grape berries. Planta. 2016 ; 243:23-41.

[107] Sunitha S, Loyola R, Alcalde JA, Arce-Johnson P, Matus JT, Rock CD. The role of UV-B light on small RNA activity during grapevine berry development. 2019; 9:769-787.

[108] Rienth M, Scholasch T. State-of-the-art of tools and methods to assess vine water status. Oeno One. 2019;53: 2403.

[109] Gambetta GA, Herrera JC, Dayer S, Feng Q, Hochberg U, Castellarin SD. The physiology of drought stress in grapevine: towards an integrative definition of drought tolerance. Journal of Experimental Botany. 2020; 71:4658-4676.

[110] Ojeda H, Deloire A, Carbonneau A. Influence of water deficits on grape berry growth. Vitis. 2001; 40:141-146.

[111] Castellarin SD. matthews, mA, Di Gaspero, G. & Gambetta, GA 2007. Water deficit accelerate ripening and induce changes in gene expression regulating flavonoid biosynthesis in grape berries. Planta.;227:101-112.

[112] Van Leeuwen C, Tregoat O, Chone X, Bois B, Pernet D, Gaudillere JP. Vine water status is a key factor in grape ripening and vintage quality for red Bordeaux wine. How can it be assessed for vineyard management purposes?. Oeno One. 2009; 43:121-134.

[113] Zufferey V, Spring JL, Verdenal T, Dienes A, Belcher S, Lorenzini F, Koestel C, Rosti J, Gindro K, Spangenberg J, Viret O. The influence of water stress on plant hydraulics, gas exchange, berry composition and quality of Pinot Noir wines in Switzerland. Oeno One. 2017; 51: 17-27.

[114] Ojeda H, Andary C, Kraeva E, Carbonneau A, Deloire A. Influence of pre-and postveraison water deficit on synthesis and concentration of skin phenolic compounds during berry growth of Vitis vinifera cv. Shiraz. American Journal of Enology and Viticulture. 2002; 53:261-267.

[115] Cramer GR, Ergul A, Grimplet J, Tillett RL, Tattersall EA, Bohlman MC, Vincent D, Sonderegger J, Evans J, Osborne C, Quilici D. Water and salinity stress in grapevines: early and late changes in transcript and metabolite profiles. Functional & integrative Genomics. 2007: 7:111-134.

[116] Deluc LG, Quilici DR, Decendit A, Grimplet J, Wheatley MD, Schlauch KA,

Mérillon JM, Cushman JC, Cramer GR. Water deficit alters differentially metabolic pathways affecting important flavor and quality traits in grape berries of Cabernet Sauvignon and Chardonnay. BMC genomics. 2009: 10:1-33.

[117] Savoi S, Wong DCJ, Arapitsas P, Miculan M, Bucchetti B, Peterlunger E. et al. Transcriptome and metabolite profiling reveals that prolonged drought modulates the phenylpropanoid and terpenoid pathway in white grapes (*Vitis vinifera* L.). BMC Plant Biology. 2016; 16:67.

[118] Savoi S, Wong DCJ, Degu A, Herrera JC, Bucchetti B, Peterlunger E. et al. Multi-omics and integrated network analyses reveal new insights into the systems relationships between metabolites, structural genes, and transcriptional regulators in developing grape berries (*Vitis vinifera* L.) exposed to water deficit. Frontier Plant Sciences. 2017; 8:1124.

[119] Zarrouk O, Brunetti C, Egipto R, Pinheiro C, Genebra T, Gori A. Grape ripening is regulated by deficit irrigation/elevated temperatures according to cluster position in the canopy. Frontier Plant Sciences 2016a; 7:1640.

[120] Zarrouk O, Costa JM, Francisco R, Lopes CM, Chaves MM. "Drought and water management in mediterranean vineyards," In: Geros H, Chaves MM, Gil HG, Delrot S, editors. Grapevine in a Changing Environment. 2016b; 38-67.

[121] Chorti E, Kyraleou M, Kallithraka S, Pavlidis M, Koundouras S, KOTSERIDIS Y. Irrigation and leaf removal effects on polyphenolic content of grapes and wines produced from cv.'Agiorgitiko'(*Vitis vinifera* L.). Notulae Botanicae Horti Agrobotanici Cluj-Napoca. 2016;44: 133-139.

[122] Koundouras S. Environmental and viticultural effects on grape composition and wine sensory properties. Elements. 2018; 14:173-178.

[123] Olle D, Guiraud JL, Souquet JM, Terrier N, Ageorges A, Cheynier V, Verries C. Effect of pre-and post-veraison water deficit on proanthocyanidin and anthocyanin accumulation during Shiraz berry development. Australian Journal of Grape and Wine Research. 2011;17:90-100.

[124] Cook MG, Zhang Y, Nelson CJ, Gambetta G, Kennedy JA, Kurtural SK. Anthocyanin composition of Merlot is ameliorated by light microclimate and irrigation in central California. American Journal of Enology and Viticulture. 2015; 66:266-278.

[125] Niculcea M, Lopez J, Sanchez-Díaz M, Carmen Antolín M. Involvement of berry hormonal content in the response to pre-and post-veraison water deficit in different grapevine (V itis vinifera L.) cultivars. Australian Journal of Grape and Wine Research. 2014; 20:281-291.

[126] Theodorou N, Nikolaou N, Zioziou E, Kyraleou M, Stamatina K, Kotseridis Y. Anthocyanin content and composition in four red winegrape cultivars (*Vitis vinifera* L.) under variable irrigation. OENO One. 2019; 53: 39-51.

[127] Kyraleou M, Kotseridis Y, Koundouras S, Chira K, Teissedre PL, Kallithraka S. Effect of irrigation regime on perceived astringency and proanthocyanidin composition of skins and seeds of Vitis vinifera L. cv. Syrah grapes under semiarid conditions. Food Chemistry. 2016; 203, 292-300.

[128] Caceres-Mella A, Talaverano MI, Villalobos-Gonzalez L, Ribalta-Pizarro C, Pastenes C. Controlled water deficit during ripening affects proanthocyanidin synthesis, concentration and composition in

Cabernet Sauvignon grape skins. Plant Physiology and Biochemistry. 2017; 117- 34-41.

[129] Zsofi Z, Villango S, Palfi Z, Toth E, Balo B. Texture characteristics of the grape berry skin and seed (Vitis vinifera L. cv. Kekfrankos) under postveraison water deficit. Scientia Horticulturae. 2014; 172:176-182.

[130] Santesteban LG, Miranda C, Royo JB. Regulated deficit irrigation effects on growth, yield, grape quality and individual anthocyanin composition in *Vitis vinifera* L. cv. "Tempranillo." Agricultural Water Management. 2011; 98: 1171-1179.

[131] Brillante L, Martinez-Luscher J, Kurtural SK. Applied water and mechanical canopy management affect berry and wine phenolic and aroma composition of grapevine (Vitis vinifera L., cv. Syrah) in Central California. Scientia Horticulturae. 2018; 227:261-271.

[132] Koundouras S, Hatzidimitriou E, Karamolegkou M, Dimopoulou E, Kallithraka S, Tsialtas JT. Irrigation and rootstock effects on the phenolic concentration and aroma potential of Vitis vinifera L. cv. Cabernet Sauvignon grapes. Journal of Agriculture and Food Chemistry. 2009; 57:7805-7813.

[133] Griesser M, Weingart G, Schoedl-Hummel K, Neumann N, Becker M, Varmuza K, Liebner F, Schuhmacher R, Forneck A. Severe drought stress is affecting selected primary metabolites, polyphenols, and volatile metabolites in grapevine leaves (Vitis vinifera cv. Pinot noir). Plant Physiology and Biochemistry. 2015; 88:17-26.

[134] Ainsworth EA, Long SP. What have we learned from 15 years of free-air CO2 enrichment (FACE)? A meta-analytic review of the responses of photosynthesis, canopy properties and plant production to rising CO2. New Phytologist. 2005;165: 351-371.

[135] Ainsworth EA, Rogers A. The response of photosynthesis and stomatal conductance to rising [CO2]: mechanisms and environmental interactions. Plant Cell Environment. 2007; 30: 258-270.

[136] Ainsworth EA. Rice production in a changing climate: a meta-analysis of responses to elevated carbon dioxide and elevated ozone concentration. Global Change Biology. 2008; 14:1642-1650.

[137] Goncalves B, Falco V, Moutinho-Pereira J, Bacelar E, Peixoto F, Correia C. Effects of elevated CO2 on grapevine (*Vitis vinifera* L.): volatile composition, phenolic content, and in vitro antioxidant activity of red wine. Journal of Agricuture and Food Chemistry. 2009; 57, 265-273.

[138] Kizildeniz T, Mekni I, Santesteban H, Pascual I, Morales F, Irigoyen JJ. Effects of climate change including elevated CO2 concentration, temperature and water deficit on growth, water status, and yield quality of grapevine (Vitis vinifera L.) cultivars. Agricultural Water Management. 2015; 159, 155-164.

[139] Edwards EJ, Unwin DJ, Sommer KJ, Downey MO, Mollah M. The response of commercially managed, field grown, grapevines (Vitis vinifera L.) to a simulated future climate consisting of elevated CO2 in combination with elevated air temperature. Acta Horticulturae. 2016; 1115: 103-110.

[140] Edwards EJ, Unwin D, Kilmister R, Treeb M. Multi-seasonal effects of warming and elevated CO2; on the physiology, growth and production of mature, field grown, Shiraz grapevines. OENO One. 2017; 51:127-132.

[141] Wohlfahrt Y, Smith JP, Tittmann S, Honermeier B, Stoll M. Primary productivity and physiological responses of Vitis vinifera L. cvs. under Free Air Carbon dioxide Enrichment (FACE). European Journal of Agronomy. 2018; 101:149-162.

[142] Martinez-Luscher J, Morales F, Sanchez-Diaz M, Delrot S, Aguirreolea J, Gomes E, et al. Climate change conditions (elevated CO2 and temperature) and UV-B radiation affect grapevine (Vitis vinifera cv. Tempranillo) leaf carbon assimilation, altering fruit ripening rates. Plant Science. 2015; 236:168-176.

[143] Bindi M, Fibbi L, Miglietta F. Free air CO2 enrichment (FACE) of grapevine (*Vitis vinifera* L.): II. Growth and quality of grape and wine in response to elevated CO2 concentrations. European Journal of Agronomy. 2001; 14, 145-155.

[144] Wohlfahrt Y, Patz CD, Schmidt D, Rauhut D, Honermeier B, Stoll M. Responses on Must and Wine Composition of *Vitis vinifera* L. cvs. Riesling and Cabernet Sauvignon under a Free Air CO2 Enrichment (FACE). Foods. 2021; 10:145.

[145] Salazar-Parra C, Aguirreolea J, Sánchez-Diaz M, Irigoyen JJ, Morales F. Climate change (elevated CO2, elevated temperature and moderate drought) triggers the antioxidant enzymes' response of grapevine cv. Tempranillo, avoiding oxidative damage. Physiologia Plantarum. 2012; 144:99-110.

[146] Arrizabalaga-Arriazu M, Morales F, Irigoyen JJ, Hilbert G, Pascual I. Growth performance and carbon partitioning of grapevine Tempranillo clones under simulated climate change scenarios: elevated CO2 and temperature. Journal of Plant Physiology. 2020; 252:153226.

Chapter 14

Mathematical Relationship Based on Experimental Data, for Corrosion Inhibition Mechanism of Phenolic Compounds Obtained from *Echium italicum* L.

Boudiba Sameh, Hanini Karima, Boudiba Louiza, Saouane Izzeddine and Benahmed Merzoug

Abstract

We highlight in this chapter the corrosion protection using phenolic extract. The building of mathematical models using experimental results obtained from the investigation of phenolic molecules or fractions extracted from *Echium italicum* L., used as corrosion inhibitors is one of the new trends in the study of steel protection. The evaluation of the corrosion inhibition of carbon steel (API 5 L-X60) in a solution 1 M of hydrochloric acid was performed using gravimetric method, potentiodynamic polarization and electrochemical impedance spectroscopy (EIS). The predicted mathematical relationships between the corrosion rate and the inhibitory efficiency in the presence of the butanolic extract of *Echium italicum* L. (BEEI), when increasing temperature proved a good agreement between experimental and mathematical studies.

Keywords: mathematical, relationship, *Echium italicum* L., phenolic compounds, corrosion inhibition

1. Introduction

Secondary metabolites, derived from natural plants, of which polyphenols constitute the grand part, have a wide range of activities [1]: biological (antibacterial, antioxidant [2, 3], anti-inflammatory [4], antidiabetic [5], anti-carcinogenic [6], etc.) and chemical (in connection with their chelating power with metals and their reducing properties generated by the hydroxyl functions of their aromatic rings). These molecules are used in several fields: as preservative food additives (constituting an alternative to the use of synthetic ones, such as buthylhydroxyanisol (BHA) and buthylhydroxytoluene (BHT), which have carcinogenic effects [7]; as flavoring in cosmetology [8]; as additives in electrolytic baths during the metals electrodeposition [9–11] and as corrosion inhibitors [12–14].

IntechOpen

Even today, these molecules have not revealed all their secrets. Our research aims to understand the power of polyphenols in the field of metal protection against corrosion.

The *Echium italicum L.* is an annual herbaceous plant, available in most of Europe, Asia [15–18], and Africa [19]. This plant widespread in Tebessa area (Algeria) [20], is used traditionally for several infections [15, 17, 18]. Furthermore, the *Echium italicum L.* contains different polyphenolic compounds, known as effective additives in electroplating baths [16, 17, 21].

We continue in this investigation to focus on the *effect* of plant extract obtained from the aerial parts of *Echium italicum L.*, on the corrosion inhibition of carbon steel (CS, API 5 L-X60) against 1 M hydrochloric acid solution, using weight loss, potentiodynamic polarization and electrochemical impedance measurements.

On the other hand, and in order to better understand the underlying reaction mechanisms of the corrosion phenomena to choose which inhibitors to use to combat it, the prediction of the implementation behavior of effective corrosion control measures is paramount. For this purpose, mathematical simulation was used as a powerful method [12, 22] to evaluate the kinetic parameters of both corrosion rate and inhibition efficiency.

2. Corrosion

Corrosion, from the Latin "corrodere" (which means "to attack") is one of the harmful global problems affecting several industrial fields such as maritime installations, petroleum, chemical, civil engineering, electrical, nuclear, sanitary, and other industries, without forgetting the environmental impact [23–25].

The corrosion of metals and their transformation into various compounds cause an alteration in their appearance, either on the surface or in-depth, thus reducing their effectiveness (parts breakage, the toxicity of the resulting metal oxides, etc.).

Several factors come into play in this phenomenon. They can be chemical (water, oxygen, salinity, acidity, etc.), physical (temperature, pressure, etc.), or biological (marine biological deposit of plants or animals, bacteria, etc.).

This phenomenon has Several definitions:

- The ISO 8044 standard defines corrosion as a physicochemical process, which leads to the deterioration of a material (metal or alloy), or degrades its functional properties following its interaction with an aggressive environment, making it unsuitable for supposed application [26].

- The National Association of Corrosion Engineers (NACE) outlines this phenomenon as the deterioration of a material, usually a metal, generated by its interaction with its environment [27].

3. Corrosion protection

Corrosion control is the set of measures that can protect metallic materials from the harmful effects of the aggressive environment. Many of these methods were reported in the literature [28, 29]. The first protection is the choice of pure metal or alloys resistant to these attacks [30].

After designing the equipment using the appropriate material, it must be protected against corrosion should be considered to avoid many problems and ensure a certain service life. For this, the preferred outlet must comply with environmental protection requirements and allow the recycling or disposal of the various components at the end of their use by applying the following choices [31]:

- Prevention by an adapted shape of the parts;
- Protection by coatings;
- Electrochemical protection;
- Protection by corrosion inhibitors.

All these solutions have drawbacks of efficiency over time, cost and environmental pollution, which is why other alternatives are exploited in research for the benefit of sustainable development that respects the environment. To this end, there have recently been some new alternatives products both environmentally friendly and less expensive as the use of inhibitors from naturals origin [32].

4. Corrosion inhibitors

Inhibitors have been successfully applied to prevent corrosion and damage in many and varied technical fields for a long time. These products have been frequently studied because they provide simple solutions to protect metals from corrosion in the aquatic environment. With the originality of being the only means of interference of the corrosive environment with steel, these compounds reduce the rate of corrosion of metals when added in appropriate amounts, without apparent change [9, 14].

5. Materials and methods

5.1 Materials

The vegetable material was collected during May 2017 from the East-North of Tebessa (Algeria). The extraction and purification of the *n*-butanolic extract obtained from *Echium italicum* L. (BEEI) were performed as reported in litérature [10, 33, 34].

The tested aggressive medium was a chlorhydric acid solution (1 M), and the investigated inhibitors were freshly prepared solutions of BEEI with different concentrations (from 100 to 500 ppm).

Before each mesurements, the substrates were abraded using emery paper with different grades (from 200 to 2000), cleaned with distilled water and then acetone. The used substrates were carbon steels with the following composition: (by weight %): C (0.26%), Mn (1.35%), P (0.03%), S (0.03%) and Fe (98.33%). Weight loss measurements were performed with specimens with dimensions of 1 cm x 1 cm x 1 cm. For the electrochemical experiments, only an exposed surface of 1 cm^2 was used. All measurements were conducted in an aerated area.

5.2 Methods

5.2.1 Electrochemical measurements

All electrochemical measurements were accomplished through a Voltalab (PGZ 301) potentiostat and controlled by software model (Voltamaster 4) under given conditions. The electrochemical characteristics of CS sample in uninhibited and inhibited solutions were realized in conventional three-electrode cell: CS as working electrode, platinum electrode as counter electrode and saturated calomel electrode (SCE) as a reference.

5.2.1.1 Potentiodynamic polarization

Potentiodynamic polarization curves were recorded after total immersion of the working electrode (CS) in 1 M HCl solution containing different concentrations of the tested inhibitor by changing the electrode potential from +250 to −250 mV vs. open circuit potential (OCP) with a scan rate of 1 mV/s. The linear Tafel segments of anodic and cathodic curves were extrapolated to corrosion potential (E_{corr}) to get corrosion current densities (i_{corr}). The (η_w%) at different concentrations of BEEI were calculated using the formula (1) [35]:

$$\eta_p(\%) = \frac{i^0_{corr} - i_{corr}}{i^0_{corr}} \times 100 \tag{1}$$

where i^0_{corr} and i_{corr} are the corrosion current density in the absence and presence of BEEI, respectively.

5.2.1.2 Electrochemical impedance spectroscopy

Electrochemical impedance spectroscopy (EIS) experiments were conducted in the frequency range of 100 kHz–10 mHz with a signal amplitude perturbation of 10 mV at open circuit potential (OCP) measured during 60 min of immersion in the tested solutions. The percentage of the inhibition efficiency (η_w%) was calculated using the polarization resistance by the following relationship [36]:

$$\eta_R = \frac{R_p - R^0_p}{R_p} \times 100 \tag{2}$$

where R^0_p and R_p are polarization resistances without and with inhibitor addition, respectively.

5.2.2 Gravimetric method

For these measurements, the prepared and pre-weighed CS substrates were totally immersed in beakers containing 1 M HCl without and with the addition of diverse concentrations of BEEI. The substrates were taken out after two hours, washed with 20% NaOH solution containing 200 g/l of zinc dust with a brush, rinsed severally with bidistilled water, dried with acetone, washed again with bidistilled water, dried and reweighted [34]. From the weight loss data, the corrosion rate (CR) was calculated according to Eq. (3) [37]:

$$CR = \frac{w}{A.t} \tag{3}$$

where w is the average weight loss, A is the total area of one CS sample, and t is the immersion time. The inhibition efficiency η_w can be calculated by Eq. (4) [36]:

$$\eta_w\% = \frac{CR^0 - CR}{CR^0} \times 100 \tag{4}$$

where CR^0 and CR are the values of corrosion rate in absence and presence of BEEI respectively.

5.2.3 Mathematical regression

A mathematical model is used to correlate variables by fitting an equation to experimental data. When using two variables, one of them is considered as explanatory, whereas the other is considered as a dependent. The linear regression of $\boldsymbol{y}$ is given as a function of $\boldsymbol{x}$ in the equation $\boldsymbol{y} = \boldsymbol{ax} + \boldsymbol{b}$ which minimize the value $\Delta(a,b)$ [38]:

$$\Delta(a,b) = \sum_{i=1}^{i=N} [y_i - (ax_i + b)]^2 \tag{5}$$

where a and b can be calculated as follow:

$$a = \frac{\sum_{i=1}^{i=N} (x_i - \overline{X})(y_i - \overline{Y})}{\sum_{i=1}^{i=N} (x_i - \overline{X})^2} \tag{6}$$

and

$$b = \overline{Y} - a\,\overline{X}. \tag{7}$$

6. Results and discussion

6.1 Electrochemical impedance spectroscopy

EIS was performed to estimate CS corrosion behavior in the presence of 500 ppm of BEEI at 298 K.

The inspection of Nyquist plots presented in **Figure 1** shows only one depressed capacitive imperfect semicircle at the higher frequency range, indicating that the corrosion reaction is controlled by the charges transfer process on a heterogeneous and irregular steel surface electrode [39–41].

The electrical equivalent circuit model (EEC) adjusted by fitting from the resulting impedance diagrams for Nyquist plots is reported in literature [12].

From **Table 1**, a noticeable increase in R_{ct} values and a decrease in C_{dl} at 500 ppm of BEEI due to the formation of protective diapers on the CS surface are observed. This phenomenon can be explained by the higher adsorption of phytochemical components above CS upon BEEI addition which ultimately reduces the charge transfer between the CS surface and the corrosion medium [42]. Consequently, we noticed that the η_R reached 74% at 500 mg.L^{-1}.

6.2 Potentiodynamic polarization measurement

To further evaluate the efficiency of the expected green inhibitor, the polarization technique was studied due to its excellent reliability. **Figure 2** shows the parameters given in **Table 2** and extracted from the CS electrode polarization curves for 500 ppm BEEI in the presence of 1 M HCl solution, at 293 K.

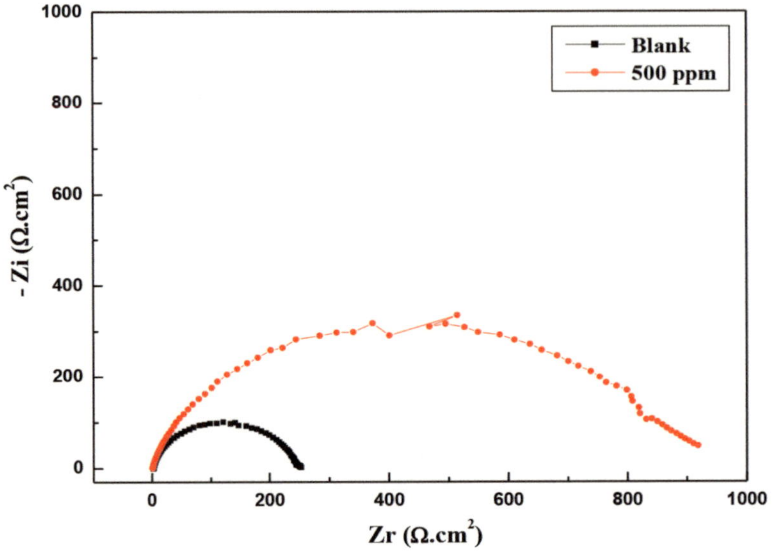

Figure 1.
Nyquist spectra for carbon steel in HCl solution in the absence and presence of 500 ppm of BEEI.

C (mg/l)	R_{ct} (Ωcm^2)	$Q \times 10^5$ ($\Omega^{-1}S^n cm^{-2}$)	*n*	C_{dl} ($\mu F cm^{-2}$)	η_R (%)
Blank	232.2	21.00	0.89	142.5	—
500	889.9	09.11	0.80	47.87	73.82

Table 1.
EIS parameters of CS in HCl solution in the absence and presence of 500 ppm of BEEI.

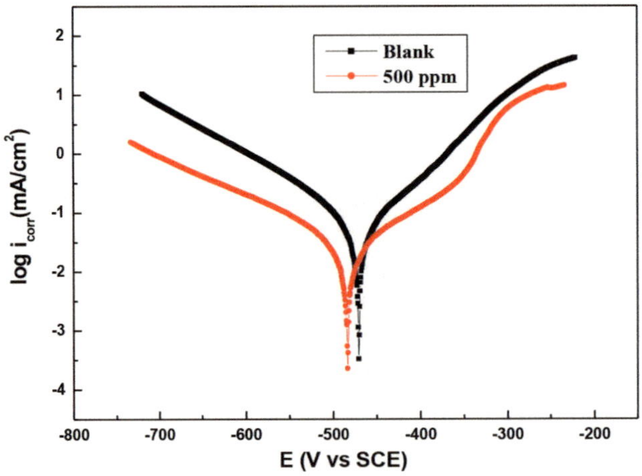

Figure 2.
Potentiodynamic polarization curves for carbon steel in HCl solution with and without inhibitor addition.

As can be seen in this figure, at 500 ppm of BEEI, the cathodic and anodic corrosion current densities will decrease. This comportment can be attributed to the adsorption of the inhibitor at the carbon steel interface [43] by reducing the dissolution of steel and delaying the hydrogen evolution reaction [44].

In our study, the maximum displacement in E_{corr} value for the optimum concentration of BEEI was −13.3 mV (< 85 mV), which exhibits that the inhibitor acts

C (mg/l)	$-E_{corr}$ (mV)	i_{corr} (mA cm^{-2})	β_a (mV dec^{-1})	$-\beta_c$ (mV dec^{-1})	η_p (%)
Blank	470.6	0.0861	72.6	117.9	—
500	483.9	0.0216	75.8	134.2	74.91

Table 2.
Potentiodynamic polarizations parameters of CS in HCl solution in the absence and presence of 500 ppm of BEEI.

Temperature (K)	C (ppm)	CR (g cm^{-2} h^{-1})	θ	η_w%
293	Blank	0.1232	—	—
	500	0.0334	0.7288	72.88
303	Blank	0.1681	—	—
	500	0.0548	0.6740	67.40
313	Blank	0.1716	—	—
	500	0.0682	0.6025	60.25

Table 3.
Corrosion parameters obtained from weight loss measurements of CS in 1 M HCl with different concentrations of BEEI at different temperatures.

as a mixed type [43, 45]. The highest inhibition efficiency was 75% at a concentration of 500 ppm.

6.3 Weight loss measurements

To study the behavior of the steel protection in the presence of an aggressive solution, the weight loss method was used. Based on this approach, the *CR*, the η_w and the degree of surface coverage θ were calculated and are very useful for discussing adsorption properties.

From **Table 3**, it can be observed that the *CR* values reduced progressively with increasing BEEI concentration, reaching 0.0334 g cm^{-2} h^{-1} for 500 ppm at 293 K. This result indicates that the addition of this inhibitor slows down the dissolution process of the CS. There was also a *CR* increase with increasing temperature.

This behavior can be attributed to the removal of some adsorbed molecules contained in the BEEI, through thermal energy-induced mechanical vibration [46]. As the inhibition efficiency is derived from the weight-loss method, the highest inhibition efficiency was 72.88% for 500 ppm of BEEI at 293 K.

6.3.1 Thermodynamic parameters

The good fitting for experimental data of gravimetric measurements at all measured temperatures supports the applicability of Freundlich adsorption model expressed by the subsequent Equation [47, 48]:

$$log\theta = log\,K_{ads} + \alpha\,logC \qquad (8)$$

where,

K_{ads} is the adsorption constant which could be taken as a measure of the strength of the adsorption forces between the inhibitor molecules and the metal surface [49].

α is the parameter taking into account the heterogeneity of the intermolecular interactions in the adsorbed layer and the steel surface.

K_{ads} represent the values calculated from the intercept lines of $log\theta$ versus Log C.

The standard adsorption free enthalpy ΔG°_{ads} can be calculated from the equation given below [50, 51]:

$$\Delta G^{\circ}_{ads} = -RTln(C_{H_2O} \times K_{ads}) \tag{9}$$

where

R is the gas constant,

T is the absolute temperature,

C_{H_2O} is the water concentration expressed in mg L^{-1} with an approximate value of 10^6. [52].

The standard adsorption enthalpy (ΔH°_{ads}) can be calculated by the relationship (10) [53]:

$$ln\, K_{ads} = \frac{-\Delta H^{\circ}_{ads}}{RT} + I \tag{10}$$

where

I is the integration constant.

The ΔH°_{ads} values were calculated from the intercept lines of $ln\, K_{ads}$ versus $1/T$.

The standard adsorption entropy (ΔS°_{ads}) was calculated using the equation below:

$$\Delta S^{\circ}_{ads} = \frac{\Delta H^{\circ}_{ads} - \Delta G^{\circ}_{ads}}{T} \tag{11}$$

The obtained results for K_{ads}, ΔG°_{ads}, ΔH°_{ads} and ΔS°_{ads} are presented in **Table 4**. From **Table 4**, we noticed the following:

1. A decrease in K_{ads} values with increasing temperatures, indicating that the adsorbent molecules contained in the BEEI adsorbed from the metal surface [54].

2. Negative ΔG°_{ads} values point out the spontaneity of the adsorption process. Moreover, the results obtained for ΔG°_{ads} adsorption values below -40 kJmol^{-1} confirm the physisorption mechanism [55].

3. The negative sign of ΔH°_{ads} indicates that the BEEI adsorption process on the steel surface is exothermic [56, 57].

4. The positive sign of ΔS°_{ads}, indicates that the adsorption process is accompanied by increased disorder due to the substitution of water molecules during BEEI adsorption [58–60].

$T(K)$	K_{ads} (L mg^{-1})	ΔG°_{ads} (KJmol^{-1})	ΔH°_{ads} (KJmol^{-1})	ΔS°_{ads} (Jmol$^{-1}K^{-1}$)
293	0.9809	−25.44	−22.22	10.99
303	0.9968	−25.99		12.44
313	0.9968	−25.65		10.96

Table 4.
Thermodynamic parameters of the adsorption of BEEI on the CS in 1 M HCl at different temperatures.

6.3.2 Activation parameters of the corrosion process

The Arrhenius-type process was used to calculate the activation energies (E_a) between the corrosion rate of CS in acidic solution, from the equation given below [61]:

$$lnCR = -\frac{E_a}{RT} + lnD \tag{12}$$

Where
E_a is the apparent activation energy of the CS dissolution.
D is the Arrhenius pre-exponential factor.
The logarithm of the CR versus $1/T$ can be branded by straight lines. The activation energy values were calculated from Arrhenius plots $lnCR$ versus 1/T and registered in **Table 5**.

For the evaluation of the enthalpy ΔH_a as well as the standard adsorption entropy ΔS_a of the corrosion activation process, the alternative formulation of Arrhenius equation was employed [62, 63]:

$$ln\,\frac{CR}{T} = \left[ln\,\frac{R}{hN_a} + \frac{\Delta S_a}{R}\right] - \frac{\Delta H_a}{RT} \tag{13}$$

where
h is the Planck's constant,
N_a is the Avogadro's number.
The plot of lnCR/T vs. 1/T gave a straight line with a slope of $(-\Delta H_a/R)$ and an intercept of $\ln(R/hN_a + \Delta S_a/R)$, from which the values of ΔS_a and ΔH_a were calculated and registered in **Table 5**.

The activation energy values were higher in the presence of 500 ppm of BEEI than in its absence, which proved the adsorption of BEEI molecules on the substrate by electrostatic bonds (physisorption) [62, 64]. The positive signs of ΔH_a indicate the endothermic nature of the dissolution process [65].

The BEEI adsorption process is accompanied by a decrease in its entropy. This can be explained that before the adsorption of the extract on the steel, the disorder degree of the inhibitor molecules is high, but when the molecules are adsorbed on the surface of the substrate, there is a decrease the in the disorder (i.e. a decrease in the entropy) [66].

6.4 Mathematical regression

This investigation focused on how could the used inhibitor (BEEI) decrease the corrosion rate (CR) and offer a good inhibition efficiency (η_w).

For that, a mathematical method was employed to predict the influence of a variable (x) which is the BEEI concentration as a function of different variables (y_i) which are CR and η_w for 293 K, 303 K, and 313 K [51, 67]. The building of a

C (ppm)	E_a (kJ mol^{-1})	ΔH_a (kJ mol^{-1})	ΔS_a (J mol^{-1} K^{-1})
Blank	12.75	10.23	−3.2822
500	27.32	24.80	- 2.7202

Table 5.
Activation parameters for carbon steel in HCl with 500 ppm of BEEI.

mathematical model that can correlate between three variables (CR, η_w, and C) was also highlighted.

Based on experimental data of the CS behavior in the presence of BEEI at various temperatures, a linear regression relation between CR and η_w, was performed.

As reported by Khadom et al. [67], when the correlation coefficient of the correlation is < 0.30, the correlation is weak and when this coefficient is between 0.50 and 0.70, the correlation is important, while if it is >0.90, the correlation will be strong.

According to the plots shown in **Figure 3**, a decrease in the CR with a raise in the η_w, when adding BEEI with various concentrations at different temperatures was distinguished, and based on correlation coefficients r^2_{CR} and $r^2_{\eta_w}$ given in **Table 6**, the concordance between the experimental and the simulated results is very strong.

As we can see from the obtained results, the correlation coefficients were almost >0.90 for considered temperatures. However, as the temperature rises the correlation coefficients go up. The Eqs. (12)–(14) construe the mathematical expression obtained for the relationship between the CR and the η_w at different temperatures.

This mathematical model exhibits a decrease in the corrosion rate when inhibition efficiency rises, with reducing in temperature. These results indicate the powerful concordance between experimental and predicted results.

T = 293 K

$$\eta_w = 99.9849 - 811.470.CR \tag{14}$$

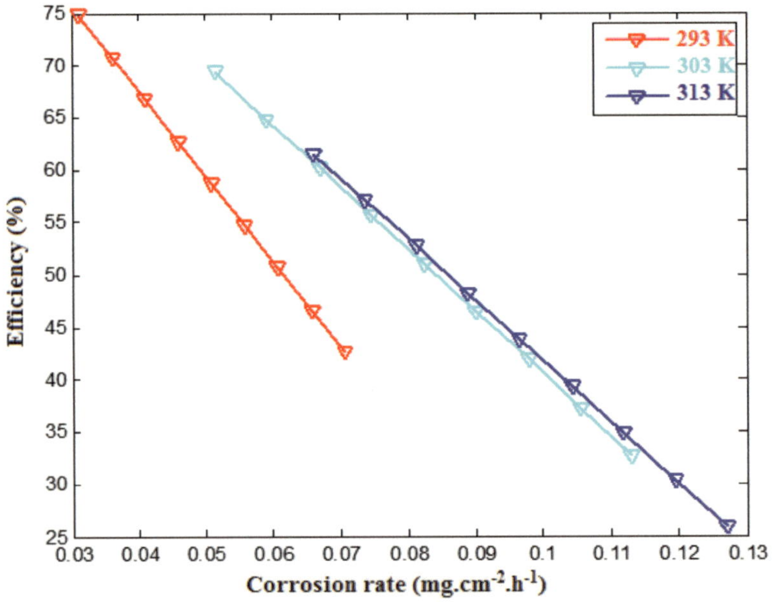

Figure 3.
Mathematical relationship between corrosion rate and inhibitory efficiency at different temperatures.

Temperature (K)	r^2_{CR}	$r^2_{\eta_w}$
293	0.9765	0.9625
303	0.9886	0.9775
313	0.9970	0.9890

Table 6.
Correlation coefficients evaluated from linear regression of CR and η_w for CS with and without BEEI solution.

T = 303 K

$$\eta_w = 99.9977 - 594.869.CR \tag{15}$$

T = 213 K

$$\eta_w = 99.9910 - 582.713.CR \tag{16}$$

7. Conclusion

In order to estimate the multiple benefits of phenolic compounds, a phenolic *Echium italicum* L. extract was evaluated as an efficient corrosion inhibitor. The subsequent conclusions can be pointed out:

- A concordance between the employed evaluating methods, suggesting that the phenolic extract could serve as an effective corrosion inhibitor for CS against corrosion.

- Depending on the gravimetric measurements, the investigated phenolic extract was spontaneously adsorbed on the CS surface following a physical model according to Freundlich isotherm.

- The prediction of the phenolic extract effect of as a corrosion inhibitor by the mathematical study was in good agreement with the experimental results, which confirms that the corrosion rate is affected by the temperature and the inhibitor concentration.

- The generated mathematical model shows a decrease in the corrosion rate with the increase of inhibition efficiency. These results indicate a strong agreement between experimental and mathematical results.

- The *Echium italicum* L. extract has prooved to be an effective corrosion inhibitor, and hence the phenolics can be useful on various existing industries to prevent losses caused by corrosion of steel without harming the environment.

For a more extensive valuation of phenolic products, many perspectives are envisaged, namely:

- Evaluate the synergy between phenolic extracts extracted from different plants looking for a better efficiency.

- The study can be extended to analyze the effect of these phenolic compounds in other corrosive media and on other types of steel.

- Perform a theoretical simulation to highlight the active compound (s) responsible for this inhibition.

Acknowledgements

The authors gratefully acknowledge the support of Tebessa University. We would like also to acknowledge the IntechOpen and the author service manager Dolores Kuzelj for their supports.

Author details

Boudiba Sameh[1*], Hanini Karima[2], Boudiba Louiza[3], Saouane Izzeddine[4] and Benahmed Merzoug[2]

1 Laboratory of Organic Materials and Heterochemistry, Tebessa University, Tebessa, Algeria

2 Laboratory of Active Molecules and Applications, Tebessa University, Tebessa, Algeria

3 Laboratory of Water and Environment, Tebessa University, Tebessa, Algeria

4 Laboratory of Physics Energy, Sciences Faculty, Brothers Mentouri University, Constantine 1, Constantine, Algeria

*Address all correspondence to: boudibasameh@gmail.com; sameh.boudiba@univ-tebessa.dz

References

[1] N. Miceli *et al.*, "Comparative study of the phenolic profile, antioxidant and antimicrobial activities of leaf extracts of five Juniperus L.(Cupressaceae) taxa growing in Turkey," vol. 34, no. 11, pp. 1636-1641, 2020.

[2] A. N. Tamfu, O. Ceylan, G. C. Fru, M. Ozturk, M. E. Duru, and F. J. M. p. Shaheen, "Antibiofilm, antiquorum sensing and antioxidant activity of secondary metabolites from seeds of *Annona senegalensis*, Persoon," vol. 144, p. 104191, 2020.

[3] A. N. Tamfu, O. Ceylan, S. Kucukaydin, and M. E. J. L. Duru, "HPLC-DAD phenolic profiles, antibiofilm, anti-quorum sensing and enzyme inhibitory potentials of *Camellia sinensis* (L.) O. Kuntze and *Curcuma longa* L," vol. 133, p. 110150, 2020.

[4] Y. Soltani, M. A. Bouzidi, F. Toumi, and A. J. P. Benyamina, "Activités anti-inflammatoire et analgésique de l'extrait hydroalcoolique des baies de Juniperus phoenicea L," vol. 17, no. 3, pp. 129-133, 2019.

[5] H. Keskes, K. Mnafgui, K. Hamden, M. Damak, A. El Feki, and N. J. A. P. j. o. t. b. Allouche, "In vitro anti-diabetic, anti-obesity and antioxidant proprieties of *Juniperus phoenicea* L. leaves from Tunisia," vol. 4, pp. S649-S655, 2014.

[6] S. Jennewein, R. J. A. m. Croteau, and biotechnology, "Taxol: biosynthesis, molecular genetics, and biotechnological applications," vol. 57, no. 1, pp. 13-19, 2001.

[7] R. Japón-Luján, P. Janeiro, M. a. D. J. J. o. A. Luque de Castro, and F. Chemistry, "Solid– Liquid Transfer of Biophenols from Olive Leaves for the Enrichment of Edible Oils by a Dynamic Ultrasound-Assisted Approach," vol. 56, no. 16, pp. 7231-7235, 2008.

[8] D. J. K. S. S. Atmanto, "Influence of the Addition of the Essential Oil of Cinnamon (Cinnamomum burmanii) in Soap Against Skin Care," pp. 587–595-587–595, 2019.

[9] K. Hanini, S. Boudiba, and M. Benahmed, "Reducing Emerging Contaminants Ensuing from Rusting of Marine Steel Installations," in *Emerging Contaminants*: IntechOpen, 2021.

[10] K. Hanini, M. Benahmed, S. Boudiba, I. Selatnia, H. Laouer, S. Akkal, "Influence of different polyphenol extracts of Taxus baccata on the corrosion process and their effect as additives in electrodeposition," vol. 14, p. 100189, 2019.

[11] B. Sameh, B. Baya, B. Louiza, H. Soraya, B. Hatem, and B. J. J. o. t. T. I. o. C. E. Merzoug, "Corrosion inhibition impact of *Pyracantha coccinea* M. Roem extracts and their use as additives in zinc electroplating: Coating morphology, electrochemical and weight loss investigations," vol. 121, pp. 337-348, 2021.

[12] S. Boudiba, K. Hanini, I. Selatnia, A. Saouane, S. Hioun, and Benahmed, "Experimental, theoretical and mathematical studies of *Echium italicum* L. extract as a corrosion inhibitor for carbon steel in acidic medium," *Materials Research Express* vol. 6, no. 8, p. 086546, 2019.

[13] M. Benahmed, I. Selatnia, N. Djeddi, S. Akkal, and H. J. C. A. Laouer, "Adsorption and corrosion inhibition properties of butanolic extract of Elaeoselinum thapsioides and its synergistic effect with Reutera lutea (Desf.) Maires (Apiaceae) on A283 carbon steel in hydrochloric acid solution," vol. 3, no. 1, pp. 251-261, 2020.

[14] K. Hanini *et al.*, "Experimental and Theoretical Studies of Taxus Baccata

Alkaloid Extract as Eco-Friendly Anticorrosion for Carbon Steel in Acidic Solution," vol. 57, no. 1, pp. 222-233, 2021.

[15] A. De Natale and A. J. J. o. E. Pollio, "Plants species in the folk medicine of Montecorvino Rovella (inland Campania, Italy)," vol. 109, no. 2, pp. 295-303, 2007.

[16] K. Morteza-Semnani, M. Saeedi, and M. J. J. o. E. O. B. P. Akbarzadeh, "Chemical composition and antimicrobial activity of essential oil of *Echium italicum* L," vol. 12, no. 5, pp. 557-561, 2009.

[17] A. E. J. I. J. o. P. Al-Snafi, "Medicinal plants possessed antioxidant and free radical scavenging effects (part 3)-A review," vol. 7, no. 4, pp. 48-62, 2017.

[18] N. Eruygur, G. Yilmaz, and O. J. F. J. o. P. S. Üstün, "Analgesic and antioxidant activity of some Echium species wild growing in Turkey," vol. 37, no. 3, p. 151, 2012.

[19] E. Retief and A. J. B. Van Wyk, "The genus Echium (Boraginaceae) in southern Africa," vol. 28, no. 2, pp. 167-177, 1998.

[20] P. Quezel and S. Santa, "Nouvelle flore de l'Algérie et des régions désertiques méridionales," 1963.

[21] I. Bošković, D. Đukić, P. Mašković, and L. J. B. C. C. Mandić, "Phytochemical composition and biological activity of *Echium italicum* L. plant extracts," vol. 49, pp. 836-845, 2017.

[22] M. Merriman, *A text book on the Method of Least Squares*. Wiley, 1909.

[23] R. Idouhli *et al.*, "Understanding the corrosion inhibition effectiveness using *Senecio anteuphorbium* L. fraction for steel in acidic media," vol. 1228, p. 129478, 2021.

[24] I. Bouali, "Étude d'inhibiteurs de corrosion métallique à base d'orthophosphates de zirconium lamellaires fonctionnalisés: synthèse, caractérisations et applications," Université de Lorraine; Université Caddi Ayyad, Marrakech (Maroc), 2018.

[25] D. Landolt, *Corrosion and surface chemistry of metals*. CRC press, 2007.

[26] P. DILLMANN and N. Delphine, "Corrosion des objets archéologiques ferreux," 2012.

[27] P. R. Roberge, *Handbook of corrosion engineering*. McGraw-Hill Education, 2019.

[28] B. Normand, *Prévention et lutte contre la corrosion: Une approche scientifique et technique*. PPUR presses polytechniques, 2004.

[29] M. J. M. Roche and Techniques, "Protection cathodique des ouvrages en milieu marin Normalisation et certification du personnel pour une protection cathodique plus efficace," vol. 101, no. 5-6, p. N18, 2013.

[30] Y.-P. Hsieh *et al.*, "Complete corrosion inhibition through graphene defect passivation," vol. 8, no. 1, pp. 443-448, 2014.

[31] C. G. Dariva and A. F. J. D. i. c. p. Galio, "Corrosion inhibitors–principles, mechanisms and applications," vol. 16, pp. 365-378, 2014.

[32] C. Verma, E. E. Ebenso, I. Bahadur, and M. J. J. o. M. L. Quraishi, "An overview on plant extracts as environmental sustainable and green corrosion inhibitors for metals and alloys in aggressive corrosive media," vol. 266, pp. 577-590, 2018.

[33] S. Akkal, S. Louaar, M. Benahmed, H. Laouer, and H. Duddeck, "A new isoflavone glycoside from the aerial

parts of Retama sphaerocarpa," Chemistry of natural compounds, vol. 46, no. 5, pp. 719-721, 2010.

[34] N. Obi-Egbedi, K. Essien, I. Obot, and E. Ebenso, "1, 2-Diaminoanthraquinone as corrosion inhibitor for mild steel in hydrochloric acid: weight loss and quantum chemical study," Int. J. Electrochem. Sci, vol. 6, pp. 913-930, 2011.

[35] M. Benahmed, I. Selatnia, A. Achouri, H. Laouer, N. Gherraf, and S. Akkal, "Steel corrosion inhibition by Bupleurum lancifolium (Apiaceae) extract in acid solution," Transactions of the Indian Institute of Metals, vol. 68, no. 3, pp. 393-401, 2015.

[36] A. Kalla, M. Benahmed, N. Djeddi, S. Akkal, and H. Laouer, "Corrosion inhibition of carbon steel in 1 MH 2 SO 4 solution by Thapsia villosa extracts," International Journal of Industrial Chemistry, vol. 7, no. 4, pp. 419-429, 2016.

[37] N. Djeddi, M. Benahmed, S. Akkal, H. Laouer, E. Makhloufi, and N. Gherraf, "Study on methylene dichloride and butanolic extracts of Reutera lutea (Desf.) Maire (Apiaceae) as effective corrosion inhibitions for carbon steel in HCl solution," Research on Chemical Intermediates, vol. 41, no. 7, pp. 4595-4616, 2015.

[38] M. Mansfield, *Text Book on the Method of Least Squares*. Hardpress Limited, 2013.

[39] M. H. Hussin, M. J. J. M. C. Kassim, and Physics, "The corrosion inhibition and adsorption behavior of Uncaria gambir extract on mild steel in 1 M HCl," vol. 125, no. 3, pp. 461-468, 2011.

[40] M. Behpour, S. Ghoreishi, M. Khayatkashani, N. J. M. C. Soltani, and Physics, "Green approach to corrosion inhibition of mild steel in two acidic solutions by the extract of *Punica granatum* peel and main constituents," vol. 131, no. 3, pp. 621-633, 2012.

[41] S. Deng and X. J. C. S. Li, "Inhibition by Ginkgo leaves extract of the corrosion of steel in HCl and H2SO4 solutions," vol. 55, pp. 407-415, 2012.

[42] S. Nofrizal *et al.*, "Elucidation of the corrosion inhibition of mild steel in 1.0 M HCl by catechin monomers from commercial green tea extracts," vol. 43, no. 4, pp. 1382-1393, 2012.

[43] I. Ahamad, R. Prasad, and M. Quraishi, "Adsorption and inhibitive properties of some new Mannich bases of Isatin derivatives on corrosion of mild steel in acidic media," Corrosion Science, vol. 52, no. 4, pp. 1472-1481, 2010.

[44] V. V. Torres *et al.*, "Inhibitory action of aqueous coffee ground extracts on the corrosion of carbon steel in HCl solution," Corrosion Science, vol. 53, no. 7, pp. 2385-2392, 2011.

[45] A. Satapathy, G. Gunasekaran, S. Sahoo, K. Amit, and P. Rodrigues, "Corrosion inhibition by Justicia gendarussa plant extract in hydrochloric acid solution," Corrosion science, vol. 51, no. 12, pp. 2848-2856, 2009.

[46] N. Patel, S. Jauhariand, G. Mehta, S. Al-Deyab, I. Warad, and B. J. I. J. E. S. Hammouti, "Mild steel corrosion inhibition by various plant extracts in 0.5 M sulphuric acid," vol. 8, no. 2635, p. 2655, 2013.

[47] X. Li, S. Deng, and H. Fu, "Inhibition of the corrosion of steel in HCl, H2SO4 solutions by bamboo leaf extract," Corrosion Science, vol. 62, pp. 163-175, 2012.

[48] M. Bobina, A. Kellenberger, J.-P. Millet, C. Muntean, and N. Vaszilcsin, "Corrosion resistance of carbon steel in weak acid solutions in the presence of l-histidine as corrosion inhibitor,"

Corrosion Science, vol. 69, pp. 389-395, 2013.

[49] M. A. Amin, "Weight loss, polarization, electrochemical impedance spectroscopy, SEM and EDX studies of the corrosion inhibition of copper in aerated NaCl solutions," Journal of applied electrochemistry, vol. 36, no. 2, pp. 215-226, 2006.

[50] A. El Bribri, M. Tabyaoui, B. Tabyaoui, H. El Attari, and F. Bentiss, "The use of Euphorbia falcata extract as eco-friendly corrosion inhibitor of carbon steel in hydrochloric acid solution," Materials Chemistry and Physics, vol. 141, no. 1, pp. 240-247, 2013.

[51] A. S. Yaro, A. A. Khadom, and R. K. Wael, "Apricot juice as green corrosion inhibitor of mild steel in phosphoric acid," Alexandria Engineering Journal, vol. 52, no. 1, pp. 129-135, 2013.

[52] S. Deng and X. Li, "Inhibition by Ginkgo leaves extract of the corrosion of steel in HCl and H2SO4 solutions," Corrosion Science, vol. 55, pp. 407-415, 2012.

[53] L. Li, X. Zhang, J. Lei, J. He, S. Zhang, and F. Pan, "Adsorption and corrosion inhibition of Osmanthus fragran leaves extract on carbon steel," Corrosion Science, vol. 63, pp. 82-90, 2012.

[54] N. Djeddi, M. Benahmed, S. Akkal, H. Laouer, E. Makhloufi, and N. J. R. o. C. I. Gherraf, "Study on methylene dichloride and butanolic extracts of Reutera lutea (Desf.) Maire (Apiaceae) as effective corrosion inhibitions for carbon steel in HCl solution," vol. 41, no. 7, pp. 4595-4616, 2015.

[55] C. M. Goulart, A. Esteves-Souza, C. A. Martinez-Huitle, C. J. F. Rodrigues, M. A. M. Maciel, and A. J. C. S. Echevarria, "Experimental and theoretical evaluation of semicarbazones and thiosemicarbazones as organic corrosion inhibitors," vol. 67, pp. 281-291, 2013.

[56] A. Ostovari, S. Hoseinieh, M. Peikari, S. Shadizadeh, and S. J. C. S. Hashemi, "Corrosion inhibition of mild steel in 1 M HCl solution by henna extract: A comparative study of the inhibition by henna and its constituents (Lawsone, Gallic acid, α-d-Glucose and Tannic acid)," vol. 51, no. 9, pp. 1935-1949, 2009.

[57] M. Deyab and S. J. J. o. t. T. I. o. C. E. Abd El-Rehim, "Effect of succinic acid on carbon steel corrosion in produced water of crude oil," vol. 45, no. 3, pp. 1065-1072, 2014.

[58] H. Zarrok, A. Zarrouk, B. Hammouti, R. Salghi, C. Jama, and F. J. C. S. Bentiss, "Corrosion control of carbon steel in phosphoric acid by purpald–weight loss, electrochemical and XPS studies," vol. 64, pp. 243-252, 2012.

[59] S. Pournazari, M. H. Moayed, and M. J. C. s. Rahimizadeh, "In situ inhibitor synthesis from admixture of benzaldehyde and benzene-1, 2-diamine along with FeCl3 catalyst as a new corrosion inhibitor for mild steel in 0.5 M sulphuric acid," vol. 71, pp. 20-31, 2013.

[60] M. Benahmed, N. Djeddi, S. Akkal, and H. J. I. J. o. I. C. Laouar, "Saccocalyx satureioides as corrosion inhibitor for carbon steel in acid solution," vol. 7, no. 2, pp. 109-120, 2016.

[61] M. Lebrini, F. Robert, A. Lecante, and C. Roos, "Corrosion inhibition of C38 steel in 1 M hydrochloric acid medium by alkaloids extract from Oxandra asbeckii plant," Corrosion Science, vol. 53, no. 2, pp. 687-695, 2011.

[62] K. Tebbji, N. Faska, A. Tounsi, H. Oudda, M. Benkaddour, and B. Hammouti, "The effect of some lactones

as inhibitors for the corrosion of mild steel in 1 M hydrochloric acid," Materials Chemistry and Physics, vol. 106, no. 2-3, pp. 260-267, 2007.

[63] M. Behpour, S. Ghoreishi, M. Khayatkashani, and N. Soltani, "Green approach to corrosion inhibition of mild steel in two acidic solutions by the extract of Punica granatum peel and main constituents," Materials Chemistry and Physics, vol. 131, no. 3, pp. 621-633, 2012.

[64] K. Tebbji *et al.*, "The effect of some lactones as inhibitors for the corrosion of mild steel in 1 M hydrochloric acid," vol. 106, no. 2-3, pp. 260-267, 2007.

[65] G. Hodaifa, J. Ochando-Pulido, S. B. D. Alami, S. Rodriguez-Vives, and A. Martinez-Ferez, "Kinetic and thermodynamic parameters of iron adsorption onto olive stones," Industrial crops and products, vol. 49, pp. 526-534, 2013.

[66] Z. Zhang, N. Tian, X. Huang, W. Shang, and L. J. R. a. Wu, "Synergistic inhibition of carbon steel corrosion in 0.5 M HCl solution by indigo carmine and some cationic organic compounds: experimental and theoretical studies," vol. 6, no. 27, pp. 22250-22268, 2016.

[67] A. A. Khadom, A. N. Abd, and N. A. Ahmed, "Xanthiumstrumarium leaves extracts as a friendly corrosion inhibitor of low carbon steel in hydrochloric acid: Kinetics and mathematical studies," *south african journal of chemical engineering,* vol. 25, pp. 13-21, 2018.

Chapter 15

Modulation of Secondary Metabolites among Mexican Medicinal Plants by Using Elicitors and Biotechnology Techniques

María Adelina Jiménez-Arellanes and Mariana Z. Pérez-González

Abstract

Medicinal plants are being utilized as raw material and the use has increased in recent decades due that these biosynthesize compounds with several pharmacological activities. Some plant species with biological potential are of interest to the industry for preparation of drugs, phytodrugs, or food supplements. This causes overexploitation and deforestation, which endangers plant species-of-interest. In recent years, alternatives have been sought to eradicate this problem. A solution that was give and is maintained is plant biotechnology, which favors the production of active Secondary Metabolites (SMt). Plant biotechnology allows us to increase the yield of a compound-of-interest, reduces its production times and costs, and allows constant and controlled production of the raw material, and while aiding in the protection of medicinal plants that are found in danger of extinction. In the scientific literature, procuring the SMt by means of biotechnological processes is described, highlighting the study of four species from Mexican traditional medicine (*Lopezia racemosa*, *Galphimia glauca, Cnidoscolus chayamansa, Sphaeralceae angustifolia* and *Buddleja cordata*), and the main biological activities are as follows: anti-inflammatory, hepatoprotector, neuroprotector, anxiolytic, antitumoral, antibacterial, and antioxidant, among others.

Keywords: Elicitors, biotechnology, Mexican medicinal plants, plant tissue culture, secondary metabolites, phenolic compounds

1. Introduction

Ever since our ancestors, humanity has been dependent on the consumption of plants as a source of food, health, and for construction/ornamental. In addition, plants have developed a complex defense system against biotic and antibiotic stress: therefore, they can produce diverse secondary metabolites (SMt). The stress to which plants are submitted under natural conditions is caused by different factors, among which stand out: attack by diverse insects and/or microorganisms (viruses, bacteria, and fungi) competition for soil, light, and nutrients, and exposure to sunlight [1].

SMt are compounds that do not play a fundamental role in the vital processes of plants, but they are important as mechanisms of defense. They are responsible for organoleptic and protective properties, such as odor, flavor, color, and consistency. These SMt also act as chemoattractants or chemorepellents. In addition, they are of great interest in industry for the preparation of food additives, agrochemicals, essences, biodiesel, narcotics, insecticides, cosmetics, and aromatics, and one of the most important of these is for the production of substances with pharmaceutical interest. Frequently, the production of SMt wild-collected plant is very low (less than 1% of the plant's Dry Weight -DW-), and this depends specifically on the plant's physiological state, the geographic location, the climate, among other factors [2].

Due to the low yield of SMt in wild plants and considering its important biological activity, alternatives or tools are currently being sought to increase its yield. One of these alternatives is the application of several biotechnology processes, a discipline that is oriented toward the development and innovation of technologies that involve the management of biological material for the production of a good or service [3].

One of the advantages in the use of biotechnological processes is to increase the production of bioactive SMt and also reduce the production time, which favors their availability [4]. The purpose of this paper is to summarize all the information that exists on the use of biotechnological processes for the production of bioactive compounds from Mexican medicinal plants.

2. Products with pharmaceutical importance

Plants constitute a huge reservoir of chemical structures, the most economically important are medicinal plants, due to their diverse biological activities; which over the years have favored human survival thanks to their use in Traditional Medicine (TM) [5–7]. TM is widely used in some developing countries, where their health system is still growing and is of great economic importance. In Africa, up to 80% of the population employs TM to help satisfy its health needs. In Asia and in Latin America, the populations continue to use TM because of historical circumstances and cultural beliefs. In China, TM is of great importance due to the large percentage of population that utilizes it, being higher than 60%. In some developed countries, the percentage of the population that uses TM is 48% in Australia, 70% in Canada, 42% in the USA, 38% in Belgium and 75% in France [8].

Currently, Medicinal Plants (MP) are employed by 80% of the world population; therefore, these are overexploited not only because are source of active ingredients, also due to the high nutritional, wood, cosmetic, agricultural, and/or medicinal value that many of these have. For example, it is estimate that China exports 120,000 tons of MP and India, some 32,000 tons while Europe imports 400,000 tons of MP. This leads to overexploitation of the species and many of them are in danger of extinction [9, 10].

The World Conservation Union and the World Nature Fund report that there are between 350,000 and 550,000 species of MP in the world, of which only approximately 20% possess documented investigation of their biological potential, and nearly 15,000 species are in danger of extinction due to the overexploitation and destruction of habitats [10, 11].

Nowadays, scientific interest in MP has increased due to the high costs and adverse effects that allopathic drugs cause, in addition to the increasing appearance of strains of microorganisms that are resistant to current treatments [12–14]. It is noteworthy that almost 25% of the active principles of allopathic drugs currently used were isolated and/or semisynthesized from plants [9]. In modern medicine, digoxin is use as a cardiotonic and was isolated from *Digitalis purpurea* (purpura, its common name); escin is use as an anti-inflammatory and venotonic and was isolated from *Aesculus hippocastanum* (its common name, horse chestnut). Another compound utilized is ajmalicin, employed for circulatory disorders, and was isolated from *Rauwolfia serpentina*; paclitaxel (an anticancer drug) has been semisynthesized by Bristol-Myers Squibb since 2002, and was obtained from the compound 10-deacetylbacatin III, it was obtained from the cell suspension of the *Taxus baccata.* While diosgenin, a steroidal sapogenin, was obtained from the tubercules of several *Dioscorea* species, which was the raw material for the semisynthesis of progesterone [15].

Guanidine is a natural product with good hypoglycemic activity that was isolated from *Galega officinalis* (L); however, this compound has been reported to be toxic for human consumption. Therefore, this compound was semi-synthesized obtaining metformin (dimethylbiguanide), which is less toxic and has a pharmacologic effect similar to the original molecule and is widely used for the treatment of type II diabetes mellitus. It is worth noting, due to the high demand for SMt on the market; several companies have seen the need to discover novel sources of raw material from MP [16].

On the other hand, at present, the use of medicinal plants and/or phytodrugs is very frequent. The phytodrugs are elaborate with plant material and some derivatives of this. The main ingredient is the aerial or subterranean plant's part; as well as extracts, tinctures, juices, resins, fatty acids, and essential oils presented in pharmaceutical form. The therapeutic effectiveness and safety have been confirmed scientifically [17]. Some examples of these include ginseng, it is obtained from Panax genus (*Panax ginseng* and *P. quinquefolium*) native from Asia and America, respectively. The main biological effect of ginseng "tonic" phytodrug is that it possesses the ability to increase the capacity to tolerate tensions, which leads to increased mental and physical yield. Another phytodrug obtained from St. John's wort (*Hypericum perforatum*) is Hiperikan, which is standardized based on its content of hypericin; its principal pharmaceutical use is against depression. *Ginkgo biloba* (Ginkgo) belongs to the Ginkgoaceae family, the active compounds in the leaf's extracts are gingolides (gingolides A-C, J, and M), along with a mix of sesquiterpene lactones and flavonoids which is used against depression. The majority of commercial preparations from Ginkgo are standardize with approximately 5–7% of terpenic lactones and 22–27% of flavonoids and they are employed mainly for the treatment of the cognitive deterioration associated with alterations in blood circulation in the brain, such as dementia. The phytodrug elaborated with *Echinacea purpurea* is commercially known as EchinaCold (Schwabe Pharma) or Immulone (ATOS Pharma). These are standardized on based of the echinacosides (caffeic acid derivative) content, whose main biological effect is as an immunostimulant [18]. In Oceania region, the extract from *Piper methysticum* (from root and rhizome) has the commercial name Kava-kava (with 30% of kava lactones), and is utilized for their neurotransmitter activity [19]. Another phytodrug is Vitango, obtained from *Rhodiola rosea* (with 3.5% rosavins and 1% salidrosides), and it is employed for reducing the stress associated with physical and mental tasks [20]. Plantival has extract mixture from *Valeriana officinalis* (160 mg) and *Melissa officinalis* (80 mg) and is use in the treatment of nervousness, restlessness and insomnia as an anxiolytic and antidepressive [21]. Another phytodrug, known as Prostasan, is the extract of *Serenoa repens*, standardized at 25% of fatty acids; the dose employed is 160 mg, and its principal effect is antiandrogenic and against benign prostatic hyperplasia [22].

Due to the acceptance and growing use of phytodrugs around the world, PM are raw materials of great attention due to high consumption. In addition, MP biosynthesize several bioactive compounds, which are classified as terpenoids, alkaloids, lactones, flavonoids, coumarin, lignans and phenols, among others; many of these have restrictive taxonomical distribution. Although the SMt functions are not directly associated with the plant's basic function, these compounds carry out some interaction roles in the plant and its environments such as: protection against pathogens, protection against abiotic tensions (ultraviolet radiation radiation), they possess the function of attracting pollinating insects, and they are signaling molecules and active ingredients for drugs [23–25].

It is estimated that around 50% of the drugs approved by the Federal Drug Administration are products derived from natural sources or analogs deriving from plants or microorganisms [26]. However, raw material can be limited, and its exploitation is one of the main ecological concerns. One of the key objectives of plant biotechnology is the development of large-scale production methods of pharmacologically active products. Additionally, the massive biosynthetic potential of plants has not been completely exploited yet and biotechnology can be employed to generate new chemical compounds that possess unknown biological activities and/or with a different mechanism of action, or a better one, than those in existence [23].

3. Production of SMt by biotechnology tools

There are distinct strategies to optimize the production and modulation SMt in medicinal plants and food. The main strategies are by uses the elicitors (molecules capable of inducing defense in the plant) [1], which are classified as biotic and abiotic. Biotics are of biological origin, while abiotics can be physical or chemical. Some examples of physical abiotics are the weather, bacteria, and plagues, among others, while chemical abiotics possess an intense variety, with those most utilized being jasmonic acid and salicylic acid [27, 28]. One of the advantages of using elicitors treatment is that they function as signaling compounds for the mechanisms of defense; thus, they increase the production of SMt in an effective and rapid manner [29]. There is great specify in the interaction of plant-elicitor species which implies that the adequate one for each culture, the time of adding it, and the concentration for obtaining best response should be selected [30].

There is other technique very used to obtain SMt *in vitro*, it focuses on obtaining the roots, which is known as "hairy roots" or transformed roots; for this, the bacterium *Agrobacterium rhizogenes* is very used. This microorganism transfers the plasmid of the Transfer-DNA (T-DNA) of the T-DNA to the plant cell, to verify whether a root transformation was obtained, this can be confirmed by Southern hybridation analysis (this technique permits the detection of a specific DNA sequence in a complex mixture). A main advantage of these is that they have the capacity of rapid growth without the external administration of Plant Growth Regulators (PGR); the majority of these do not require a light supply, and their yield of metabolites is constant due to their genetic stability [1]. Another internal factor is the culture medium added with macro- and micronutrients, as well the external factors, such as light intensity, temperature, humidity, and stirring speed [31].

In general, formulation of the culture medium begins with the base medium, being the most utilized Muashige & Skoog (MS), B5 of Gamborg and Linsmaier and Skoog (LS), and Nitsch and Nitsch (NN) [32]. These culture mediums contain minerals, vitamins, and a carbon source, normally sucrose and sometimes fructose is used. Although plant cell cultures typically are initiate in solid medium, they require liquid medium for production on a large scale. The mineral content and/or the carbon source in culture medium have a profound impact on biosynthesis of

SMt employed in the manufacturing of phytodrugs and/or compound-of--pharmaceutical-interest [33].

Other tools very used to obtain SMt by biotechnological process is through the use of BioCatalyzers; this method has been used to transform polyphenols compounds; for example, *Bouvardia ternifolia* is utilized for the production of a BioC denominated dehydrodiisoeugenol, which was obtained from the supernatant of cells suspension, demonstrating a yield of around 77%. The dehydrodiisoeugenol obtained from *B. ternifolia* allows the production of isoeugenol by biotransformation; it is known that plant peroxides transform phenols substituted for by a methyl group orto position to the corresponding O-radical, which, on establishing itself by resonance, produces a C-radical; the latter is that which leads to dimerization, producing a dimer. This biotransformation represents a clean and green alternative with respect to traditional chemical methods, in which oxidative bonding reactions are affected using catalysts such as $FeCl_3$, $K_3(FeCN)_6$, and Cu (OH)Cl [34].

Recently, interest in research and development of *in vitro* plant tissue cultures from MP has grown; however, there are scarce studies, to our knowledge, in which the biological activities of these SMt obtained by this process are described. The majority of works published only mention the conditions of the biotechnological process and the final concentrations of the different metabolites produced, but do not evaluate the pharmacological activity of these SMt, and the authors solely cite that these have been reported in previous works.

In **Table 1** and **Figure 1**, some examples are described. It is important to mention that on some occasions is difficult to establish the biotechnological process conditions to induce the biosynthesis of bioactive SMt from a MP.

Species	Products	Use	References
Alanthus altissima	Alkaloids	Antimicrobial	[35]
Ajuga reptans	Antocyanins	Antioxidant	[36, 37]
Alanthus altissima	Cantinones alkaloids	Antimicrobial	[38]
Ammi majus	Coumarins	Anticoagulant	[39]
Anchusa officinalis	Rosmarinic acid	Antioxidant	[40, 41]
Anthoceros agrestis	Rosmarinic acid and Glycosides	Antioxidant	[42]
Arachis hypogea	Piceatannol	Antioxidant	[43]
Artemisia annua	Artemisinin	Antimalaric	[30, 44]
Artemisia judaica	Flavonoids	Antioxidant	[45]
Bouvardia ternifolia	Dehydrodiisoeugenol	Biocatalyst	[34]
Beta vulgaris	Betalains	Antioxidant	[46, 47]
Buddleja cordata	Verbascoside (**1**), linarin (**2**)	Anti-inflammatory, antioxidant	[48]
Buddleja cordata	Phenylpropanoids	Antioxidant	[49]
Caesalpinia pulcherrima	Homo isoflavones	Antimicrobial, antitumoral	[50]
Calophyllum inophyllum	Dipyranocoumarins	AntiHIV	[51]
Camelia sinensis	Thiamine or theanine	Antihypertensive	[52]
Capsicum frutescens	Capsaicin (**3**)	Irritant	[53]
Cassia acutifolia	Antraquinones	Antimicrobial	[54]

Species	Products	Use	References
Castilleja tenuiflora	Verbascoside (**1**), Isoverbascoside, aucubin	Anti-inflammatory, Antispasmodics	[55]
Catharanthus roseus	Ajmalicin	Antihypertensive	[56]
Catharanthus roseus	Vinblastin	Anticancer	[57]
Cecropia obustifolia *Cecropia peltata*	Chlorogenic acid, Isoorientin	Hypoglycemic	[58]
Cephaelis ipecacuana	Emetin	Antiparasitic	[59]
Cephalo-taxus fortunei	Abietane diterpenoids	Antitumoral	[60]
Choisya ternata	Furanocoumarins	Antitumoral, Antioxidant	[61]
Choisya ternata	Furoquinolin alkaloids	Antitumoral, Antimicrobial	[62]
Cinchona robusta	Robustaquiones	Antimalarial	[63]
Cistanche deserticola	Glycosides	Antioxidant	[64]
Cistanche salsa	Glucophenyletanoids	Aphrodisiac	[65]
Colchium autumnale	Colchicine (**4**)	Antitumoral	[66]
Coleus forskolii	Forskolin (**5**)	Asthma	[67]
Comptotheca acuminate	Camptotecin (**6**)	Antitumoral	[68]
Coptis japonica	Berberin (**7**)	Intestinal infection	[69]
Cornus kousa	β-glucogallin, (+)-Catechin, (+)-gallocatechin, procyanidin B-3	Hyperglycemic and antimicrobial	[70]
Coscinium fenestratum	Berberin	Antioxidant, Antidiabetic	[71]
Crocus sativus	Crocin	Anticancer	[72]
Cynara cardunculus	Cinarin, Chlorogenic acid	Antioxidant	[73]
Daucus corata	Antocyanins	Lipoperoxidation	[74]
Digitalis lanata	Digoxin (**8**)	Cardiostimulant	[75]
Dioscorea deltoide	Diosgenin (**9**)	Steroidal stimulant	[76]
Drosophyllum lusitanicum	Plumbagin (**10**)	Anticancer, Antimicrobial	[77]
Eleutherococcus sessiliflorus	Eleuteroside	Anti-inflammatory, diuretic, analgesic, antipyretic	[78]
Eriobotrya japonica	Triterpenes	Anti-inflammatory, antidiabetic, antitumoral	[79]
Eucommia ulmoides	Chlorogenic acid (**11**)	Antimicrobial, Antioxidant	[80]
Fagopyrum esculentum	Rutin (**12**)	Antioxidant	[81]
Fragaria ananassa	Antocyanins	Antioxidant	[82]
Galphimia glauca	Galphimine B (**27**)	Central nervous system disorders	[83]
Glehnia littoralis	Antocyanins	Antioxidant	[84, 85]

Species	Products	Use	References
Gymnema sylvestre	Gymnemicanor	Antidiabetic	[86]
Helianthus tuberosus	Inulin	Antidiabetic	[87]
Hemidesmus indicus	Rutin (**12**)	Antioxidant	[88]
Hypericum perforatum	Hypericin	Antidepressive	[89]
Hyssopus officinalis	Rosmarinic acid (**13**)	Antioxidant	[90]
Hyssopus officinalis	Lithospermic acid	Antioxidant	[91]
Ipomoea batatas	Antocyanins	Antioxidant	[92–94]
Larrea divaricata	Nordihydroguayaretic acid, Quercetin	Antiarthritic, digestive, against venereal diseases	[95]
Lavandula vera	Rosmarinic acid (**13**)	Hepatoprotective	[96, 97]
Leucophyllun frutescens	Coumarins, lactones, flavonids	Antioxidants	[98]
Lithospermum erythrorhizon	Shikonin	Antibacterial	[99]
Lopezia racemosa	6-O-palmitoyl-3-O-β-D-glucopyranosylcampesterol, 6-O-palmi-toyl-3-O-β-D-gluco-pyranosyl-β-sitosterol	Anti-inflammatory	[100]
Lycium chinense	Cerebroside	Cellular Growth Regulator	[101]
Morinda elliptica	Antraquinones	Antimicrobial	[102]
Mucuna pruriens	L-dihydroxyphenylalanine	AntiParkinson	[103]
Ochrosia elliptica	Elipticin	Antitumoral	[104]
Ocimum basilium	Rosmarinic acid (**13**)	Antioxidant	[105]
Panax ginseng	Ginkgolides (**14**)	Cognitive deterioration	[106]
Panax ginseng	Ginsenosides (**15**)	Immunomodulator	[107]
Papaver somniferum	Codeine (**16**)	Sedative	[108]
Papaver somniferum	Morphine (**17**)	Sedative	[109]
Papaver somniferum	Sanguinarin (**18**)	Platelet stimulator	[110]
Passiflora quadrangularis	Orientin, Isoorientin, Vitexin, Isovitexin	Antioxidant	[111]
Petroselinum sativum	Flavonolides	Antioxidant	[112]
Picrasma quassioides	Cuasin	Antiphylogistic	[113]
Piqueria trinerva	Monoterpene	Antifungal	[114]
Podophyllum hexandrum	Podophylotoxin (**19**)	Antitumoral	[115]
Psoralea corylifolia	Genistein and Daidzein	Tonic	[116]
Rauwolfia serpentina	Reserpin (**20**)	Antihypertensive	[117]
Rubia tinctorum	Antraquin-ones	Antimicrobial	[118]
Salvia miltiorrhiza	Cryptotanshinone	Antioxidant, antimicrobial	[119]
Salvia miltiorrhiza	Tanshinone	Cardiac problems	[120]
Saussurea medusa	Hispiduline (**21**), Jaceosidine	Antitumoral	[121]
Silybum marianum	Silymarin	Hepatoprotective	[122]

Species	Products	Use	References
Solanum malacoxylon	Cholecalcipherol	Aids calcium absorption	[123]
Sphaeralcea angustifolia	Sphaeralcic acid (**22**), Tomentin (**23**)	Antiinflammatory	[124]
Swietenia humilis Zucc.	Alkaloids	Cytotoxic	[125]
Tanacetum parthenium	Parthenolide (**24**)	Anticancer	[126]
Taxus brevifolia	Taxol (**25**)	Anticancer	[127]
Vitis vinífera	Antocyanins	Antioxidant	[128]
Vitis vinífera	Resveratrol (**26**)	Antioxidant, Hepatoprotective	[129]
Withania somnifera	Withanolide A	Antioxidant, Antistress	[130]

Table 1.
Secondary metabolites obtained for cellular cultures from medicinal plant tissues in vitro *and their biological activity.*

4. Anti-inflammatory activity of SMt isolated from vegetal material obtained by biotechnological processes

Some SMt with significant anti-inflammatory activity have been obtain from MP through employment some biotechnological processes. From cell suspension cultures *Sphaeralcea angustifolia*, two compounds with important anti-inflammatory activity (evaluated in murine models) were isolated. The cell suspension was developed in MS medium with total nitrate 2.74 mM, under this condition was obtained scopoletin, sphaeralcic acid (**22**) and tomentin (**23**). From the CH_2Cl_2:CH_3OH extract sphaeralcic acid (**22**) and tomentin (**23**) were isolated; these compounds showed 58 and 66% anti-inflammatory activity, in the carrageenin model at 45 mg/kg administered by intraperitoneal (i.p.) route. On the other hand, in the topical anti-inflammatory model (TPA, 12-O-TetradecanoylPhorbol-13-Acetate), tomentin (225 mM/ear) exhibited 57% inhibition in the formation of auricular edema, while sphaeralcic acid (174 mM/ear) revealed 86% inhibition with a dose-dependent effect and one half of the Effective Dose (ED_{50}) = 93 mM. Sphaeralcic acid is the most active compound in both models (topical as well as systemic) [124, 131].

In another study, the anti-inflammatory activity of the cell suspension culture from *S. angustifolia* is described. In this case, aseptic-leaf explants and Naphthalene Acetic Acid (NAA, such as auxin) in several concentrations (0, 0.5, 1.0, and 2.0 mg/L) in combination with a constant concentration of Kinetin (KIN) were used. For the cell suspension culture, they utilized 4% initial inoculum in MS medium with 2.74 mM of the total nitrates, 1 mg/L of NAA and 0.1 mg/L of KIN and supplemented with 30 g/L of sucrose. The main SMt identified in this suspension cultures were the same compounds (scopoletin, tomentin, and sphaeralcic acid). Scopoletin was excrete in the culture medium, although it also accumulated in the biomass. For evaluation of the anti-inflammatory activity, the authors prepared the CH_2Cl_2:MeOH extract of the cell's suspension from *S. angustifolia* and this extract was administered i.p. in male ICR mice (35 g) employing the carrageenin model. This extract showed ED_{50} = 137.63 mg/kg; sphaeralcic aid and tomentin at 45 mg/kg inhibited 67 and 62%, respectively on carrageenan assay and sphaeralcic acid at 1 mg/ear was more active in TPA assay, showed ED_{50} = 93 mM and tomentin showed 48% of inhibition at 1 mg/ear [132, 133]. In addition, the same extract from

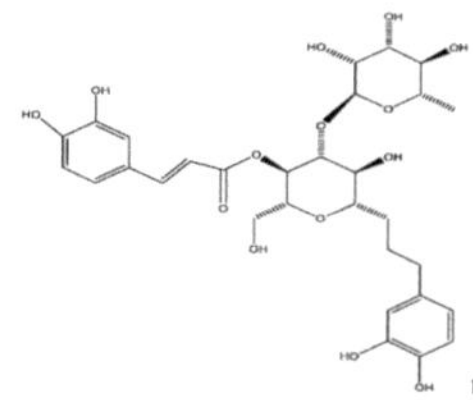
1

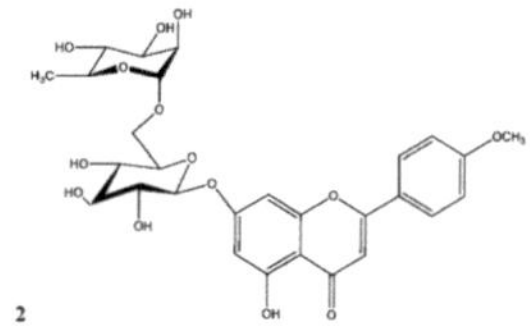
2

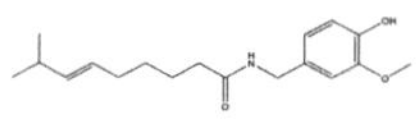
3

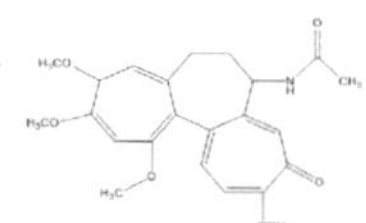
4

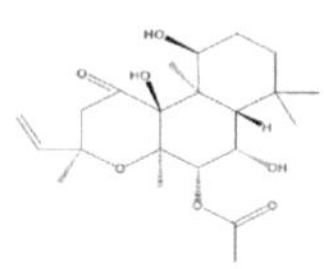
5

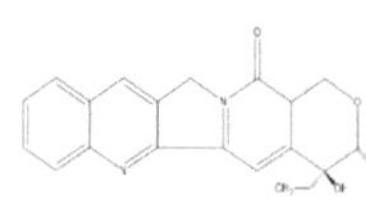
6

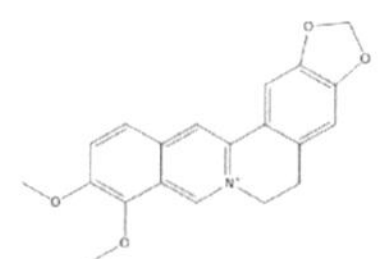
7

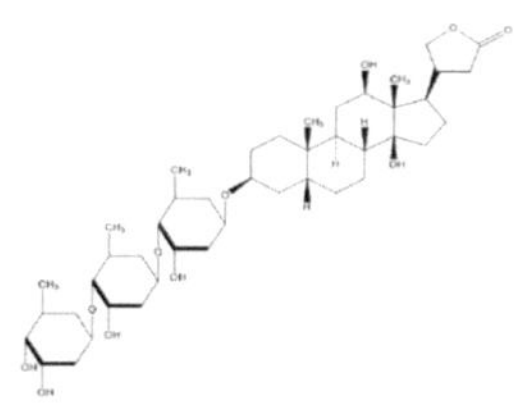
8

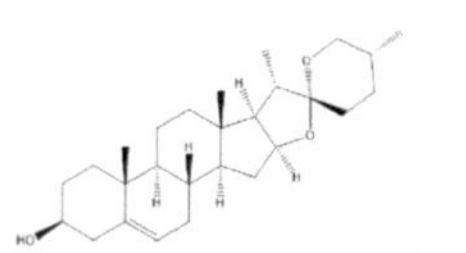
9

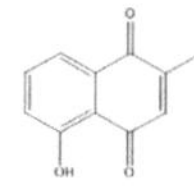
10

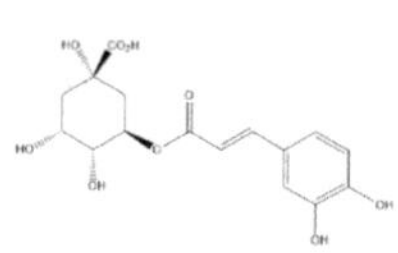
11

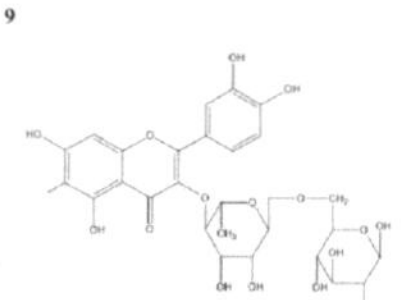
12

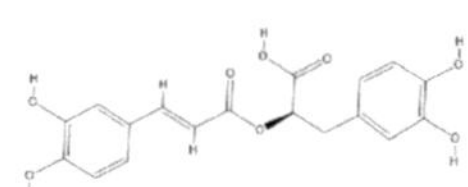
13

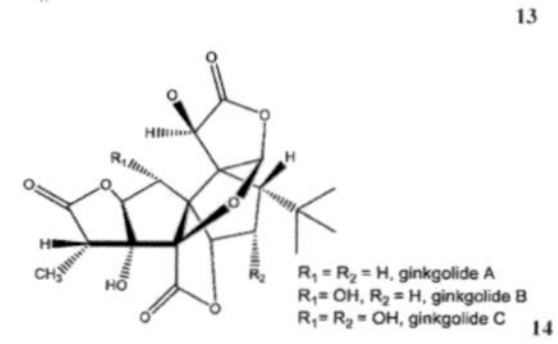

14

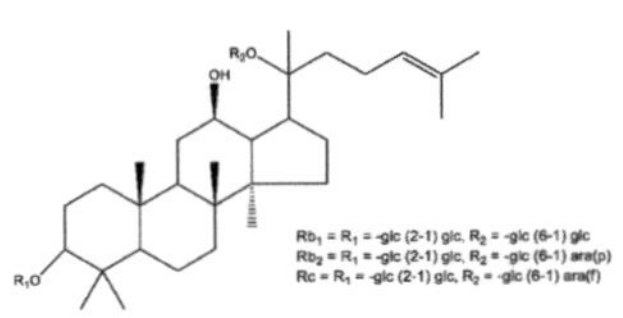

15

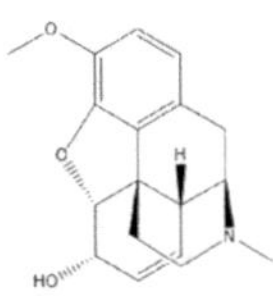

16

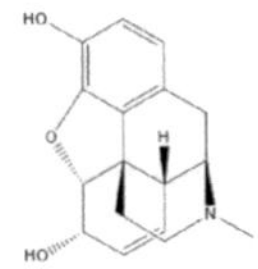

17

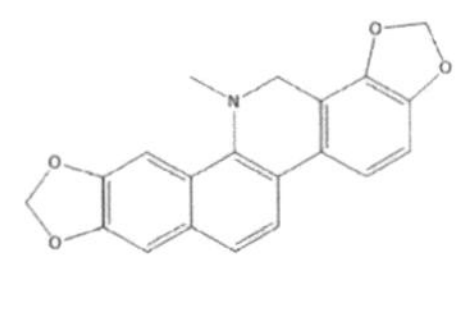

18

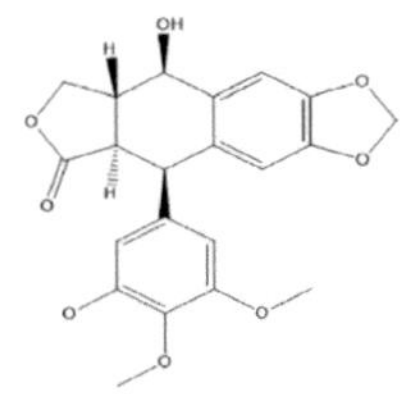

19

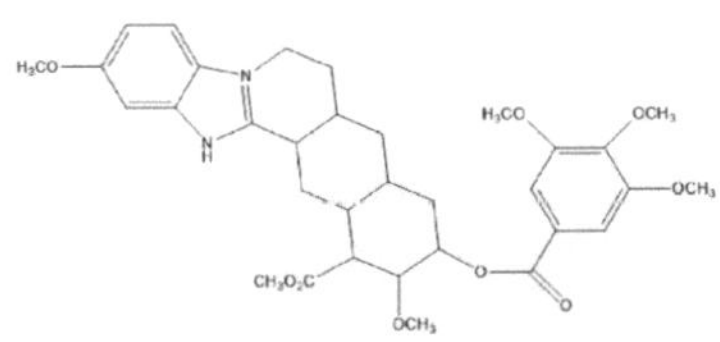

20

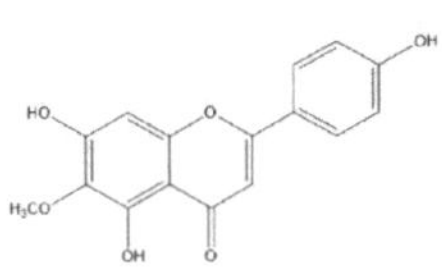

21

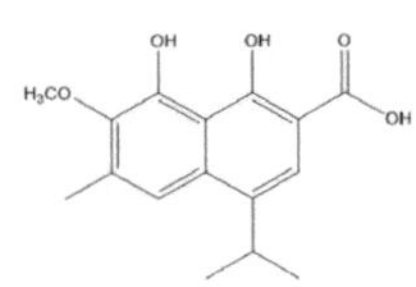

22

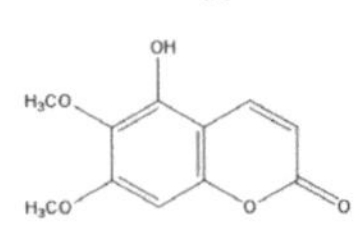

23

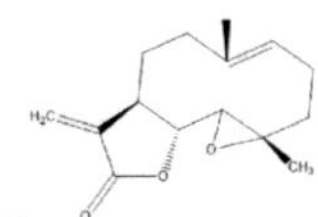

24

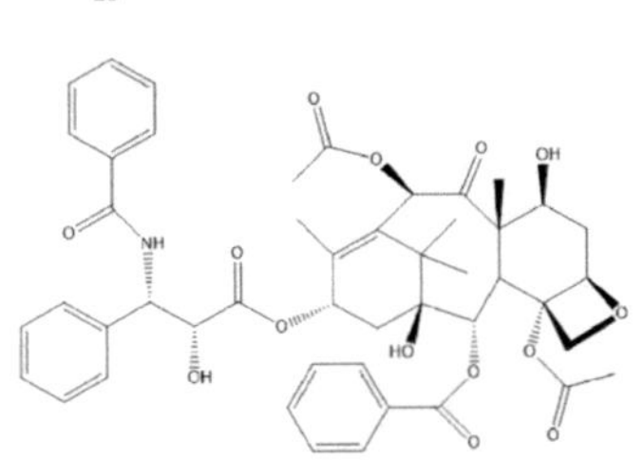

25

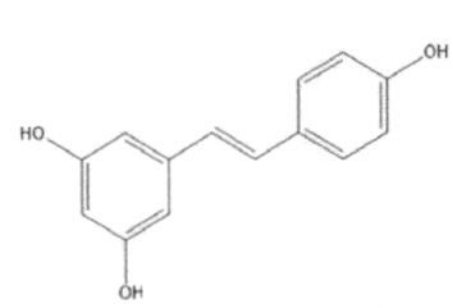

26

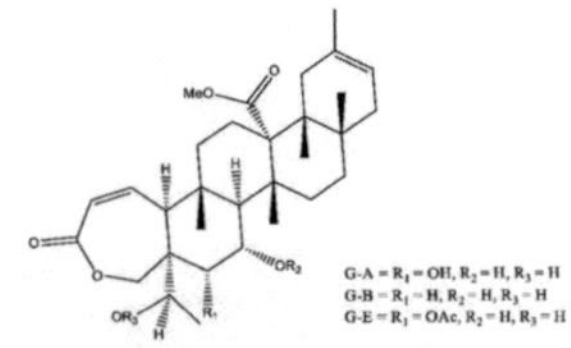

27

Figure 1.
Chemical structure of some polyphenols and other SMt with biological activity.

biomass of cells in suspension of *S. angustifolia* at 100 mg/kg (with 0.10 mg of scopoletin, 0.10 mg of tomentin and 0.19 mg of sphaerelcic acid), as well as tomentin (20 mg/kg) were active as anti-inflammatory agent and reduced the mean body weight lost in Freund adjuvant- and kaolin/carrageenan-induced arthritis, respectively. In this assay, the organic extract and tomentin reduced the levels of pro-inflammatory interleukins such as IL-1β, IL-6 and TNF-α and increased levels of IL-4 and IL-10 (anti-inflammatory cytokines) [133].

In parallel with obtaining cells *in vitro* of *S. angustifolia*, the authors performed a preclinical phase study (in rats). The CH_2Cl_2 extract of the aerial parts of *S. angustifolia* (wild material) was tested in chronic inflammation model induced with complete Freund's adjuvant (polyarthritis) The administration of the extract at 100 mg/kg/day during 8 days showed sustained and significant inhibition of edema, being of 62.6% [134]. A double-blind clinical phase study with the extract of *S. angustifolia* (wild material) standardized at 1% hydroxycoumarin content was conducted; the experiment was performed on 130 patients diagnosed with osteoarthritis. 55 of them were treated with standardized extract of *S. angustifolia* (gel) and 75 patients were treated with Diclofenac (2%). The therapeutic effectiveness of the gel administered topically for 4 weeks was 89%, while that of the control group (Diclofenac) was 91.3%; it was highlighted in the study that patients who received the treatment (gel of the standardized extract) did not exhibit adverse effects and did show an improvement in their disorder [135].

Another plant utilized in Mexican ethnomedicine is *Lopezia racemosa* Cav. Callus cultures in MS medium were obtain with variable amounts of NAA, 2,4-Dichlorophenoxiacetic acid (2,4-D) and 6-BenzylAminoPurine (BAP). The authors carried out 10 treatments with the previously mentioned PGR. In this case, they employed three types of explants (hypocotyl, stem nodule, and leaf) and several treatments. The combination of 1.0 mg/L of 2,4-D plus 0.5 mg/L of BAP was the best. From these callus material two novel compounds: 6-O-palmitoyl-3-O-β-D-glucopyranosylcampesterol (174.0 μg/g of biomass) and 6-O-palmitoyl-3-O-β-D-glucopyranosyl-β-sitosterol were isolated. When quantifying these compounds, the authors observed that the wild plant contains less quantity than the callus. The topical anti-inflammatory activity of the biomass obtained from the callus was evaluate in the TPA model on CD-1 male mice at 1 mg/ear. Three extracts (hexanic, CH_2Cl_2, and methanol), was tested and showed 48.74, 57.14, and 16. 81% of inhibition, respectively. The CH_2Cl_2 extract was the most active, with a half-maximal Inhibitory Concentration (IC_{50}) = 0.93 mg/ear. On the other hand, the pure compound (6-O-palmitoyl-3-O-β-D-glucopyranosyl-campesterol) was tested in the same model at 1 mg/ear showing a 57.14% inhibition, with IC_{50} = 0.45 mg/ear [100].

The lipophilic extract containing beta-carotene (LMBC) from plant cell cultures of *Cleome spinosa* was evaluate in two *in vivo* models to determine the anti-inflammatory and antinociceptive activities in Swiss Webster (SW) mice of both sexes. The callus culture was obtained of the MS medium supplemented with 1 mg/L of 4- amino-3,5,6-trichloropicolinic acid (picloram) and sub-cultured to culture medium with the same composition at 4-week intervals. The anti-inflammatory activity in carrageenan model at 10 mg/kg by i.p. via was evaluated. LMBC was inactive with respect to extract from whole plant, which showed more than 50% inhibition of edema at the same dose. On the other hand, the LMBC (at 50 mg/kg) showed around 68% decrease in writhes, these data were very similar to that shown in wild plant, and the effect was better than dipyrone (at 100 mg/kg) used as positive control. The authors concluded that the results of LMBC are particularly important; since this active SMt of medicinal interest can be continuously obtain from callus cultures [136].

Buddleja cordata is other medicinal species utilized to treat diseases related with inflammation. This cell suspension was obtained in MS medium supplemented with

NAA (9.05 μM) and Kin (2.32 μM). The anti-inflammatory activity of the extracts from wild plant and of the cell suspension cultures were describe. In both extracts, the verbascoside content was quantified by HPLC methods. The extract of the cells suspension has 87.48 mg verbascoside/g Dry Matter (DM), while that the same extract from wild plant only contained 47.34 mg of verbascoside/g DM. In addition, acute toxicity in Balb/C mice of the both extracts were also determined, with half of a Lethal Dose (LD_{50}) of >2 g/kg. On the other hand, the topical anti-inflammatory effect of the wild plant extract and of the cell suspension was assay. The ED_{50} values was 3.93 and 1.26 mg/ear, respectively, cell suspension extract was the most active due to its greater content of verbascoside. Evaluation of both extracts in the carrageenan model (systemic inflammation), showed ED_{50} = 251.26 and 204.62 mg/kg for wild plant and cells suspension extracts, respectively; in this case, the latter extract was more active. In the chronic inflammation model (the arthritis model induced with complete Freud's adjuvant), both extracts showed moderate anti-inflammatory activity (<35%) and favored weight increase in animals with arthritis. The authors concluded that the cell suspension culture of *Buddleja cordata* obtained through the biotechnological process contained a better anti-inflammatory activity; therefore, it represents a source for obtaining this type of secondary metabolite-of-pharmacological-interest [48].

Cnidoscolus chayamansa is medicinal plant whit anti-inflammatory, antiprotozoal, hepatoprotective, hypoglycemic and antimycobacterial activities [137–139]. Recently, a biotechnology processes was described to obtain callus using BAP (5 mg/L) and 2,4-D (2.5 or 5 mg/L), this callus was used as a biotechnological alternative for *in vitro* propagation of this plant [140]. After that, this callus was use to establish a cell suspension culture. From the cell suspension, organic extract was prepared and its antioxidant, antibacterial and anti-inflammatory activities were determined, as well as the main SMt was quantified by HPLC analysis. In cell suspension, lupeol acetate (38.1 mg/g DW) was obtained as a main constituent and scopoletin (3.6 mg/g DW) was also quantified; in wild material, both compounds were isolated in low quantity. The organic extract was active against *Staphylococcus aureus, S. coagulase* and *Listeria monocytogenes,* and a moderate antioxidant and anti-inflammatory activities (in TPA and carrageenan models) showed [28, 141].

5. Antineoplasic activity of the plant material obtained by biotechnological cultures

From the callus culture of *Eriobotrya japonica*, nine triterpenes (ursolic acid; oleanolic acid; maslinic acid; tormentic acid; 2α,19α-dihydroxy-3-oxo-urs-12-en-28-oic acid; 2α-hydroxyursolic acid; hyptadienic acid, and the mixture of 3-O-*cis*-p-coumaroyl tormentic acid and 3-O-*trans*-p-coumaroyl tormentic acid) were isolated. The main triterpenes of the callus tissues were tormentic acid (50 mg/g DW) and 2α,19α-dihydroxy-3-oxo-urs-12-en-28-oic acid (11.8 mg/g DW), the latter compound (2a,19α-dihydroxy-3-oxo-urs-12-en-28-oic acid) is known as a potent protease inhibitor of the human immunodeficiency virus. All these triterpenes were tested in two cell lines (HSC-2 and HSC); seven of the nine triterpenes were active. Showing mean cytotoxic concentration (CC_{50}) between 10 and 48 μg/ml, while the oleanolic acid and 2α,19α-dihydroxy-3-oxo-urs-12-en-28-oic acid exhibited weak cytotoxic activity. Additionally, the authors evaluated the *in vivo* antitumor activity of the 2α,19α-dihydroxy-3-oxo-urs-12-en-28-oic acid in female ICR mouse skin (n = 15) during two stages of carcinogenesis; in this assay, carcinogenesis was induced topically with (+)-(E)-4-methyl 2[(E)-hydroxyimino]-5-Nitro-6-methoxy-3-hexenamide (NOR1) at a dose of 90 μg/0.1 mL of acetone. One week after NOR1 administration started, TPA (1 μg/0.1 mL of acetone) was administered twice weekly, yielding as a result a weak inhibition of the carcinogenesis. On the other hand, the authors mention that 2α,19α-dihydroxy-3-oxo-urs-12-en-28-oic acid is an antiproliferative agent and that the number of papillomas diminished by 40% in 20 weeks, indicating that this compound possesses potential for the delay of carcinogen in mouse skin [142].

6. Biological effect of SMt isolated from cell cultures of *Galphimia glauca*

Galphimia glauca is widely used in Mexican traditional medicine. From this species, some triterpenes such as Galphimine-A, B and E (**27**) have been isolated. These compounds showed a neuroprotective effect, when these were evaluated in mice convulsions model. To induce the seizure, the authors used strychnine or pentylenetratrazole administered by i.p. or subcutaneous route. In the study's results, the depressor effects observed on motor activity directed toward an objective or an aim [143]. The pharmacological effect of galphimine B (G-B) was due to selectively inhibiting the discharge of dopaminergic neurons in the central area in *in-vivo* models [144]. Due to its therapeutic importance of G-B, the authors proceeded to induce the production of this homogeneous raw material through a biotechnological process.

A first step was to obtain callus from hypocotyl explants in MS medium for 30 days with a combination of NAA and KIN; under these conditions, only great cell growth was obtained, and with 2,4-D at 4 mg/L the G-B production was stimulated with a yield of 0.154 mg/g DW. In addition, under this condition, G-E was also obtained but at less concentration (0.057 mg/g DW). Also, friable callus from suspension culture in MS medium with NAA and KIN (2:2 mg/L) was obtain, denominating this line as ggxl. By means of a growth kinetic, galphimines were shown to be produced in the culture's stationary stage [83, 145]. The next step was to carry out the scaling of galphimine production in the 5-liter *airlift* bioreactor and in one with mechanical stirring; the growth indices were 11.66 and 1.7, respectively. However, the authors observed that neither the biomass production, nor the time exerted an influence on the yield of G-B. Because the *airlift* produced a greater biomass but with lower yield of G-B (255 mg/L), while the stirring bioreactor at day 10 shown an intracellular as well as an extracellular content of 1381 mg of G-B/L, 5.4-times higher than the *airlift* at day 25 [146]. Once the biotechnological conditions for the production of G-B were established, this allowed having raw material to carry out the pharmacological evaluation in different models.

7. Toxicologic effect of Galphimines

Aqueous extract from material obtained by bioreactor was prepared, whose galphamine content was G-A, G-B and G-E = 0.6, 1.034 mg/g, and 1.12 mg/g, respectively. Meanwhile, the content of these galphamine in the ethanolic extract was G-A = 5.35 mg/g, G-B = 18.8 mg/g, and G-E = 17.49 mg/g and the MeOH extract content G-A = 7.29 mg/g, G-B = 17.47 mg/g, and G-E = 11.6 mg/g. Afterward, each extract was administered to Balb/C male and female mice for 28 days (2.5 g/kg). During the study period, there were no deaths, and in the histopathological analysis of the different organs; the latter did not present alterations. Also, analyzed the behavioral parameters, demonstrating a reduction in spontaneous activity. Administration of these extracts for 56 days (2.5 g/kg) in mice did not cause any change in liver-function biochemical parameters. With regard to the cytotoxic evaluation in KB, UISO, and OVCAR-5 cell lines, no cytotoxic effects were found, but all of these extracts specifically inhibited growth of the colon-cancer cell line with ED_{50} of <2 μg/mL. On the genotoxicity test *in vitro*, the extracts were evaluated at three concentrations (250, 100, and 50 mg/mL) and none of the three *G. glauca* extracts showed a genotoxic effect [147].

8. Evaluation of the MeOH extract of *Galphimia glauca* in Behavioral models of anxiety

The anxiolytic and anti-depressive effects were evaluated for the *G. glauca* MeOH extract (wild material) standardized with content of G-B (8.3 mg/g), using the elevated light–dark labyrinth and forced swimming in albino (ICR) mice. The extract, administered orally, three times (24, 18, and 1 h prior to the test) at doses of 125, 250, 500, 1,000, and 2,000 mg/kg was capable of significantly increasing ($p < 0.05$) the number of entries, as well as time spent on the elevated labyrinth's open arms, which indicates an anti-anxiolytic effect. A similar effect was observed in the light–dark paradigm test: time spent in the light box increased in treated mice. However, this treatment was not able to change any parameter in the forced swimming test [148].

9. Conclusions

The MP form part of the daily life of the worldwide population. It is currently of scientific interest due to its high consumption, as an alternative treatment and/or co-administered with allopathic treatments for the improvement of chronic-degenerative diseases. On the other hand, the population has been responsible for affording a great boost to the use of MP; therefore, its consumption generates a great demand and consequently overexploitation. This overexploitation is a danger in the extinction of species of pharmaceutical interest. Another problem regarding the consumption of MP is that not all the population has access to species that are endemic and that have great biological potential. All the above led to the search for methods to achieve the production and induction of SMt biosynthesis with important biological activity in less time, with constant, controlled and standardized production. Besides helping to preserve plant species without altering the ecosystem.

In some cases, has been reported that cell suspension cultures increase by up to 300% the production of SMt with biological interest respect to wild plant material. In addition, to the increase in SMt production, these are obtained in less complex mixtures, which facilitates the purification process. In the present work, we describe several SMt obtained for biotechnological processing; however, many of these SMt have not been submitted to *in vivo* studies that prove their potential biological activity. Therefore, it is necessary to develop projects aimed at obtaining metabolites by biotechnological processes and demonstrate their biological activity in *in vivo* models.

Acknowledgements

Susan Drier for English language corrections.

Funding

This manuscript is a review and we did not have funding.

Competing interest

The author declare no competing interest.

Ethical approval

This article does not contain any studies with human participants or animals performed by any of the authors.

Nomenclature

SMt	Secondary Metabolites
SMBB	Sociedad Mexicana de Biotecnología y Bioingeniería
TM	Tradicional Medicine
MP	Medicinal Plants
DNA	Deoxyribonucleic acid
PGR	Plany Growth Regulators
BioC	BioCatalyzers
TPA	12-O-Tetradecanoyl Phorbol 13-Acetate
ED_{50}	Half of the Effective Dose
NAA	Naphthalene Acetic Acid
KIN	Kinetina
2,4-D	Dichlophenoxiacetic acid
BAP	6-BenzylAminoPurine
IC50	Hal-maximal Inhibitory concentration
DM	Dry Metter
LD_{50}	Half of a Lethal Dose
DW	Dry Metter
CC_{50}	Mean Cytotoxic Concentration
i.p.	Intraperitoneal
S.C.	Subcutaneous
G-A	Galphimine A
G-B	Galphimine B
G-E	Galphimine E
C.N	Kinetine

Author details

María Adelina Jiménez-Arellanes* and Mariana Z. Pérez-González
Unidad de Investigación Médica en Farmacología, UMAE Hospital de Especialidades, Centro Médico Nacional Siglo XXI (CMN-SXXI), Instituto Mexicano del Seguro Social (IMSS), Ciudad de México (CdMx), Mexico

*Address all correspondence to: adelinajim08@prodigy.net.mx

References

[1] Rao SR, Ravishankar GA (2002) Plant cell cultures: chemical factories of secondary metabolites. Biotechnol Adv 20:101-153. doi: 10.1016/s0734-9750(02)00007-1

[2] Akula R, Ravishankar GA (2011) Influence of abiotic stress signals on secondary metabolites in plants. Plant Signal Behav 6:1720-1731. doi: 10.4161/psb.6.11.17613

[3] SMBB, Sociedad Mexicana de Biotecnología y Bioingeniería. 2016. https://smbb.mx/ Access: january 2021.

[4] Yue W, Ming QL, Lin B, Rahman K, Zheng CJ, Han T, Qin LP (2016) Medicinal plant cell suspension cultures: pharmaceutical applications and high-yielding strategies for the desired secondary metabolites. Crit Rev Biotechnol 36:215-232. doi: 10.3109/07388551.2014.923986

[5] Vanisree M, Tsay HS (2007) Chapter 12: Plant cell cultures: production of biologically important secondary metabolites from medicinal plants of Taiwan. In: Kayser O, Quax WJ (Eds) Medical Plant Biotechnology: from Basic Research to Industrial Applications. Willey Verlag Gmb H & Co. K GaA, pp 267-285. doi: 10.1002/9783527619771.ch12

[6] Vanisree M, Lee CY, Lo SF, Nalawade SM, Lin CY, Tsay HS (2004) Studies on the production of some important secondary metabolites from medicinal plants by plant tissue cultures. Bot Bull Acad Sin 45:1-22. https://www.researchgate.net/publication/228860673

[7] Newman DJ, Cragg GM (2020) Natural Products as Sources of New Drugs over the Nearly Four Decades from 01/1981 to 09/2019. J Nat Prod 83: 770-803. https://doi.org/10.1021/acs.jnatprod.9b01285

[8] WHO. Traditional Medicine Strategy 2002-2005. World Health Organization; Geneva, Switzerland: 2002. https://apps.who.int/iris/handle/10665/67163

[9] Beyene B, Beyene B, Deribe H (2016) Review on application and management of medicinal plants for the livelihood of the local community. J Resour Develop Managem 22:33-39. https://iiste.org/Journals/index.php/JRDM/article/view/31747/32619

[10] Chen SL, Yu H, Luo HM, Wu Q, Li CF, Steinmetz A (2016) Conservation and sustainable use of medicinal plants: problems, progress, and prospects. Chin Med 11:1-37. doi:10.1186/s13020-016-0108-7

[11] Sen T, Samanta SK (2015) Medicinal plants, human health and biodiversity: a broad review. Adv Biochem Eng Biothechnol 147:59-110. doi: 10.1007/10_2014_273

[12] Bahare S, Konovalov DA, Fru P, Kapewangolo P, Peron G, Ksenija M, Pereira OR, Nigam M, Nicola S, Pignata G, Rapposelli S, Sestito S, Anil Kumar NV, Cadiz_Gurrea ML, Segura-Carretero A, Mishra AP, Sharifi-Rad M, Cho WC, Taheri Y, Setzer WN, Sharifi-Rad J (2020) Areca catechu-From farm to food and biomedical applications. Phytother Res 34: 2140-2158

[13] Bahare S, Selamoglu Z, Sener B, Kilic M, Kumar Jugran A, de Tommasi N, Sinisgalli C, Milella L, Rajkovic J, Morais-Braga MFB, Bezerra CF, Rocha JE, Coutinho HDM, Oluwaseun AA, Khan SZ, Ahnmad SJ, Erol E, Ali Z, Ostrander EA, Sharifi-Rad J, Cadiz.Gurrea ML, Taheri Y, Martorell M, Segura-Carretero A, Cho WC (2019a) Berberis Plants-Drifting from Farm to Food Applications, Phytotherapy and Phytopharmacology. Foods 8:522

[14] Bahare S, Sener B, Kilic M, Sharifi-Rad J, Naz R, Yousaf Z, Nixwell MF, Tosouh KPV, Ezzat SM, Bishbishy MH, Taheri Y, Lucariello G, Durazzo A, Lucarini M, Rasul Suleira HA, Santini A (2019) Dioscorea Plants: A genus rich in vital nutrapharmaceuticals-A review. Iran J Pharm Res 18:68-69

[15] Farag M, Mohammed MS, Foud I, Ahmed WJ, Malik S, Mohamed MS (2015) The role of natural products in drug discovery and development. World J Pharm Res 4:22-33. https://www.researchgate.net/publication/291830543

[16] Krentz AJ, Bailey CJ (2005) Oral antidiabetic agents: current role in type 2 diabetes mellitus. Drugs 65:385-411. doi: 10.2165/00003495-200565030-00005

[17] Romero-Cerecero O, Reyes-Morales H, Torija-Barrio IT, Herrera-Arellano A, Tortoriello J (2005) Conocimiento sobre fitofármacos en médicos de atención primaria del estado de Morelos. Rev Med Inst Mex Seguro Soc 43:281-286. https://www.medigraphic.com/cgi-bin/new/resumen.cgi?IDARTICULO=4919

[18] Briskin PD (2000) Medicinal plants and Phytomedicines. Linking plant biochemistry and physiology to human health. Plant Physiol 124:507-514. doi: 10.1104/pp.124.2.507

[19] Ulbricht C, Basch E, Boon H, Ernst E, Hammerness P, Sollars D, Tsourounis C, Woods J, Bent S (2005) Safety review of kava *(Piper methysticum)* by the Natural Standard Research Collaboration. Expert Opin Drug Saf 4:779-794. doi: 10.1517/14740338.4.4.779

[20] Iheozor-Ejiofor P, Dey ES (2009) Extraction of rosavin from *Rhodiola rosea* root using supercritical carbon dioxide with water. J Supercrit Fluids 50:29-32. https://doi.org/10.1016/j.supflu.2009.04.011

[21] Müller S F, Klement S (2006) A combination of valerian and lemon balm is effective in the treatment of restlessness and dyssomnia in children. Phytomedicine 13:383-387. doi: 10.1016/j.phymed.2006.01.013

[22] Vela-Navarrete R, Escribano-Burgos M, Farré AL, García-Cardoso J, Manzarbeitia F, Carrasco C (2005) *Serenoa repens* treatment modifies bax/bcl-2 index expression and caspase-3 activity in prostatic tissue from patients with benign prostatic hyperplasia. J Urol 173:507-510. doi: 10.1097/01.ju.0000150533.94952.25

[23] Gandhi SG, Mahajan V, Bedi YS (2015) Changing trends in biotechnology of secondary metabolism in medicinal and aromatic plants. Planta 241:303-317. doi: 10.1007/s00425-014-2232-x

[24] Sharifi-Rad J, Taheri Y, Abdulmajid AS, Naderi N, Anil Kumar NK, Koirala N, Khadka S, Karazhan N, Shahinozzaman Md, Sen S, Acharya K, Dey A, Martorell M, Martins N (2020) Biological activities and health-promoting effects of Pyracantha genus: A key approach to the phytochemical´s potential. Cell Mol Biol 66:20-27

[25] Sharifi-Rad J, Rayess YE, Abi Rizk A, Sadaka C, Zgheib R, Zam W, Sestito S, Rapposelli S, Neffe-Skocińska K, Zielińska D, Salehi B, Setzer WN, Dosoky NS, Taheri Y, Beyrouthy ME, Martorell M, Ostrander EA, Rasul Suleira HA, Cho WC, Mroyi A, Martins N (2020). Turmeric and Its Major Compound Curcumin on Health: Bioactive Effects and Safety Profiles for Food, Pharmaceutical, Biotechnological and Medicinal Applications. Front Pharmacol 11:01021

[26] Vuorela P, Leinonen M, Saikku P, Tammela P, Rauha JP, Wennberg T, Vuorela H (2004) Natural products in

the process of finding new drug candidates. Curr Med Chem 11: 1375-1389. doi: 10.2174/ 0929867043365116

[27] Patel H, Krishnamurthy R (2013) Elicitors in plant tissue culture. J Pharmacog Phytochem 2:60-65. https:// www.phytojournal.com/vol2Issue2/ Issue_july_2013/9.1.pdf

[28] Pérez-González MZ, Juárez-Vázquez MdelC Jiménez-Arellanes MA (2020) Establishment of a cell suspension culture with anti-inflammatory activity and the effect of salicylic acid on the production of callus from *Cnidoscolus chayamansa*. Int J Appl Microb Biotechnol Res 8:40-50. doi.org/ 10.33500/ijambr.2020.08.005

[29] DiCosmo F, Misawa M (1995) Plant cell and tissue culture: alternatives for metabolite production. Biotechnol Adv 13:425-453. doi: 10.1016/07349750(95) 02-005-n

[30] Paniego NB, Giulietti AM (1995) Artemisia annua L.: dedifferentiated and differentiated cultures. Plant Cell Tissue Organ Cul 36 (2), 163-168 doi: 10.1007/ BF00037715

[31] Schmitz C, Fritsch L, Fischer R, Schillberg S, Rasche S (2016) Statistical experimental designs for the production of secondary metabolites in plant cell suspension cultures. Biotechnol Lett 38: 2007-2014. doi: 10.1007/s10529-016-2206-0

[32] Gaurav N, Singh AP, Kumar A, Som D, Kumar D, Komal HSG (2016) *In vitro* tissue culture studies from nodal and petiole explants of wild and cultivated traits of *Withania somnifera* in MS and B5 medium. J Med Plant Stud 4:92-96. https://www.researchgate.net/ publication/328699276

[33] McCoy E, O'Connor SE (2008) Natural products from plant cell cultures. Prog Drug Res 65:329-370. https://pubmed.ncbi.nlm.nih.gov/ 18084920/

[34] Hernández-Vázquez L, Olivera-Flores M, de Jesús T, Luna H, Navarro-Ocaña A (2017) Obtención de dehidrodiisoeugenol por dimerización de isoeugenol con cultivos celulares de *Bouvardia ternifolia* (trompetilla) TIP. Rev Esp Cienc Quim-Biol 20:15-22. http://dx.doi.org/10.1016/j.recqb. 2017.04.002

[35] Crespi-Perellino N, Guicciardi A, Malyszko G, Arlandi E, Ballobio E, Ballobio, M, Minghetti A (1986) Occurrence of indole alkaloids in Ailanthus altissima cell cultures. J Nat Prod 49:1010-1014. https://doi.org/ 10.1021/np50048a007

[36] Callebaut A, Hendrickx G, Voets AM, Motte JC (1990) Anthocyanins in cell cultures of *Ajuga reptans*. Phytochemistry 29:2153-2158. doi: 10.1016/0031-9422(90)83027

[37] Terahara N, Callebaut A, Ohba R, Nagata T, Ohnishi-Kameyama M, Suzuki M (2001) Acylated anthocyanidin 3-sophoroside-5-glycosides from *Ajuga reptans* flowers and the corresponding cell cultures. Phytochemistry 58:493-500. doi: 10.1016/s0031-9422(01)00172-8

[38] Anderson LA, Harris A, Phillipson JD (1983) Production of cytotoxic canthin-6-one alkaloids by *Ailanthus altissima* plant cell cultures. J Nat Prod 46:374-378. doi: 10.1021/np50027a014

[39] Staniszewska I, Królicka A, Maliński E, Łojkowska E, Szafranek J (2003) Elicitation of secondary metabolites in *in vitro* cultures of *Ammi majus* L. Enzyme Microb Technol 33: 565-568. https://doi.org/10.1016/ S0141-0229(03)00180-7

[40] De-Eknamkul W, Ellis BE (1984) Rosmarinic acid production and growth characteristics of *Anchusa officinalis* cell

suspension cultures. Planta Med 50: 346-350. doi: 10.1055/s-2007-969728

[41] Su WW, Lei F, Su LY (1993) Perfusion strategy for rosmarinic acid production by *Anchusa officinalis*. Biotechnol Bioeng 42:884-890. doi: 10.1002/bit.260420713

[42] Vogelsang K, Schneide B, Petersen M (2006) Production of rosmarinic acid and a new rosmarinic acid 3′-O-β-D-glycoside in suspension cultures of the hornwort *Anthoceros agrestis* Paton. Planta 223:369-373. doi: 10.1007/s00425-005-0089-8

[43] Ku KL, Chang PS, Cheng Y C, Lien CY (2005) Production of stilbenoids from the callus of *Arachis hypogaea*: a novel source of the anticancer compound piceatannol. J Agric Food Chem 53:3877-3881. doi: 10.1021/jf050242o

[44] Baldi A, Dixit VK (2008) Yield enhancement strategies for artemisinin production by suspension cultures of *Artemisia annua*. Bioresour Technol 99: 4609-4614. doi: 10.1016/j.biortech.2007.06.061

[45] Liu CZ, Murch SJ, El-Demerdash M, Saxena PK. (2004) *Artemisia judaica L.*: micropropagation and antioxidant activity. J Biotechnol 110:63-71. doi: 10.1016/j.jbiotec.2004.01.011.

[46] Hunter CS, Kilby NJ (1999) Betalains. Their accumulation and release in vitro. In: Dall RD (Eds) Plant Cell Culture Protocols. Methods in Molecular Biology, vol 11. Humana Press, pp 403-410. doi: 10.1385/1-59259-583-9:403

[47] Pavlov A, Kovatcheva P, Tuneva D, Ilieva M, Bley T (2005a) Radical scavenging activity and stability of betalains from *Beta vulgaris* hairy root culture in simulated conditions of human gastrointestinal tract. Plant Foods Hum Nutr 60:43-47. https://doi.org/10.1007/s11130-005-5098-z

[48] Gutiérrez-Rebolledo GA, Estrada-Zúñiga ME, Nieto-Trujillo A, Cruz-Sosa F, Jiménez-Arellanes MA (2018) *In vivo* anti-inflammatory activity and acute toxicity of methanolic extracts from wild plant leaves and cell suspension cultures of *Buddleja cordata* Kunth (Buddlejaceae). Rev Mex Ing Quim 17: 317-330. http://rmiq.org/ojs311/index.php/rmiq/article/view/42/26

[49] Estrada-Zúñiga ME, Aarland RC, Rivera-Cabrera F, Bernabé-Antonio A, Buendía-González L, Cruz-Sosa F (2016) Micropropagation of *Buddleja cordata* and the content of verbascoside and total phenols with antioxidant activity of the regenerated plantlets. Rev Mex Ing Quim 15:333-346. http://rmiq.org/ojs311/index.php/rmiq/article/view/1003

[50] Zhao P, Iwamoto Y, Kouno I, Egami Y, Yamamoto H (2004) Stimulating the production of homoisoflavonoids in cell suspension cultures of *Caesalpinia pulcherrima* using cork tissue. Phytochemistry 65: 2455-2461. https://doi.org/10.1016/j.phytochem.2004.08.004

[51] Pawar KD, Thengane SR (2009) Influence of hormones and medium components on expression of dipyranocoumarins in cell suspension cultures of *Calophyllum inophyllum* L. Process Biochem 44: 916-922. https://doi.org/10.1016/j.procbio.2009.03.005

[52] Orihara Y, Furuya T (1990) Production of theanine and other γ-glutamyl derivatives by *Camellia sinensis* cultured cells. Plant Cell Rep 9:65-68. doi: 10.1007/BF00231550

[53] Sudha G, Ravishankar GA (2003) Influence of methyl jasmonate and salicylic acid in the enhancement of capsaicin production in cell suspension cultures of *Capsicum frutescens* Mill. Curr Sci 85:1212-1217. https://www.jstor.org/stable/24108623

[54] Nazif NM, Rady MR, El-Nasr M S (2000) Stimulation of anthraquinone production in suspension cultures of *Cassia acutifolia* by salt stress. Fitoterapia 71:34-40. https://doi.org/10.1016/S0367-326X(99)00101-X

[55] Cortes-Morales JA, Lopez-Laredo AR, Zamilpa A, Bermudez-Torres K, Trejo-Espino JL, Trejo-Tapia G (2018) Morphogenesis and secondary metabolites production in the medicinal plant *Castilleja tenuiflora* benth. Under nitrogen deficiency and starvation stress in a temporary immersion system. Rev Mex Ing Quim 17:229-242. https://doi.org/10.24275/uam/izt/dcbi/revmexingquim/2018v17n1/Cortes

[56] Ten Hoopen HJG, Vinke JL, Moreno PRH, Verpoorte R, Heijnen JJ (2002) Influence of temperature on growth and ajmalicine production by *Catharantus roseus* suspension cultures. Enzyme Microb Technol 30:56-65. doi: 10.1016/S0141-0229(01)00456-2

[57] Endo T, Goodbody A, Misawa M (1987) Alkaloid production in root and shoot cultures of *Catharanthus roseus*. Planta Med 53:479-482. doi:10.1055/s-2006-962777

[58] Nicasio-Torres M, Erazo-Gómez J C, Cruz-Sosa F (2009) *In vitro* propagation of two antidiabetic species known as guarumbo: *Cecropia obtusifolia* and *Cecropia peltata*. Acta Physiol Plant 31: 905-914. https://doi.org/10.1007/s11738-009-0304-5

[59] Jha S, Sahu NP, Sen J, Jha TB, Mahato SB (1991) Production of emetine and cephaeline from cell suspension and excised root cultures of *Cephaelis ipecacuanha*. Phytochemistry 30:3999-4003. https://doi.org/10.1016/0031-9422(91)83452-Q

[60] Xu XH, Zhang W, Cao XP, Xue S (2011) Abietane diterpenoids synthesized by suspension-cultured cells of *Cephalotaxus fortunei*. Phytochem Lett 4:52-55. https://doi.org/10.1016/j.phytol.2010.12.003

[61] Kitamura Y, Ikenaga T, Ooe Y, Hiraoka N, Mizukami H (1998) Induction of furanocoumarin biosynthesis in *Glehnia littoralis* cell suspension cultures by elicitor treatment. Phytochemistry 48:113-117. https://doi.org/10.1016/S0031-9422(97)00849-2

[62] Creche J, Trémouillaux-Guiller J, Gröger D, Chénieux J.C, Rideau M (1993) *Choisya ternate* (Mexican Orange): In Vitro Culture and the Production of Alkaloids. In: Bajaj Y.P.S. (Eds) Medicinal and Aromatic Plants V. Biotechnology in Agriculture and Forestry, vol 24. Springer, Berlin, pp 107-121. https://doi.org/10.1007/978-3-642-58062-8_8

[63] Schripsema J, Ramos-Valdivia A, Verpoorte R (1999) Robustaquinones, novel anthraquinones from an elicited *Cinchona robusta* suspension culture. Phytochemistry 51:55-60. https://doi.org/10.1016/S0031-9422(98)00470-1

[64] Cheng XY, Wei T, Guo B, Ni W, Liu CZ (2005) *Cistanche deserticola* cell suspension cultures: phenylethanoid glycosides biosynthesis and antioxidant activity. Process Biochem 40:3119-3124. doi: 10.1016/j.procbio.2005.03.012

[65] Liu JY, Guo ZG, Zeng ZL (2007) Improved accumulation of phenylethanoid glycosides by precursor feeding to suspension culture of *Cistanche salsa*. Biochem Eng J 33:88-93. https://doi.org/10.1016/j.bej.2006.09.002

[66] Yoshida K, Hayashi T, Sano K (1988) Colchicoside in *Colchicum autumnale* bulbs. Agr Biol Chem 52: 593-594. https://doi.org/10.1080/00021369.1988.10868674

[67] Mukherjee S, Ghosh B, Jha S (2000) Establishment of forskolin yielding transformed cell suspension cultures of

Coleus forskohlii as controlled by different factors. J Biotechnol 76:73-81. doi: 10.1016/s0168-1656(99)00181-9

[68] Pasqua G, Monacelli B, Mulinacci N, Rinaldi S, Giaccherini C, Innocenti M, Vinceri FF (2005) The effect of growth regulators and sucrose on anthocyanin production in *Camptotheca acuminata* cell cultures. Plant Physiol Biochem 43: 293-298. https://doi.org/10.1016/j.plaphy.2005.02.009

[69] Hara Y, Yamagata H, Morimoto T, Hiratsuka J, Yoshioka T, Fujita Y, Yamada Y (1989) Flow cytometric analysis of cellular berberine contents in high-and low-producing cell lines of *Coptis japonica* obtained by repeated selection. Planta Med 55:151-154. doi: 10.1055/s-2006-961910

[70] Ishimaru K, Arakawa H, Neera S (1993) Polyphenol production in cell cultures of *Cornus kousa*. Phytochemistry 32:1193-1197. https://doi.org/10.1016/S0031-9422(00)95090-8

[71] Kashyap S, Kapoor N, Kale RD. (2016) *Coscinium fenestratum*: callus and suspension cell culture of the endangered medicinal plant using vermicompost extract and coelomic fluid as plant tissue culture media. Am J Plant Sci 7:899-906. doi: 10.4236/ajps.2016.76085

[72] Chen SA, Wang X, Zhao B, Yuan X, Wang Y (2003) Production of crocin using *Crocus sativus* callus by two-stage culture system. Biotechnol Lett 25: 1235-1238. https://link.springer.com/article/10.1023/A:1025036729160

[73] Trajtemberg SP, Apóstolo NM, Fernández G (2006) Calluses of *Cynara cardunculus* var. *cardunculus cardoon* (Asteraceae): determination of cynarine and chlorogenic acid by automated high-performance capillary electrophoresis. In Vitro Cell Dev Biol-Plant 42:534-537. https://doi.org/10.1079/IVP2006803

[74] Ravindra PV, Narayan MS (2003) Antioxidant activity of the anthocyanin from carrot (*Daucus carota*) callus culture. Int J Food Sci Nut 54:349-355. doi: 10.1080/09637480120092134

[75] Kreis W, Reinhard E (1992) 12β-Hydroxylation of digitoxin by suspension-cultured *Digitalis lanata* cells: production of digoxin in 20-litre and 300-litre air-lift bioreactors. J Biotechnol 26:257-273. https://doi.org/10.1016/0168-1656(92)90011-W

[76] Tal B, Gressel J, Goldberg I (1982) The effect of medium constituents on growth and diosgenin production by *Dioscorea deltoidea* cells grown in batch cultures. Planta Med 44:111-115. doi: 10.1055/s-2007-971414

[77] Komaraiah P, Ramakrishna SV, Reddanna P, Kishor PK (2003) Enhanced production of plumbagin in immobilized cells of *Plumbago rosea* by elicitation and *in situ* adsorption. J Biotechnol 101:181-187. doi: 10.1016/s0168-1656(02)00338-3.

[78] Shohael AM, Ali MB, Yu KW, Hahn EJ, Islam R, Paek KY (2006) Effect of light on oxidative stress, secondary metabolites and induction of antioxidant enzymes in *Eleutherococcus senticos*us somatic embryos in bioreactor. Process Biochem 41: 1179-1185. https://doi.org/10.1016/j.procbio.2005.12.015

[79] Ho HY, Liang KY, Lin WC, Kitanaka S, Wu JB (2010) Regulation and improvement of Triterpene formation in plant cultured cells of *Eriobotrya japonica* Lindl. J Biosci Bioeng 110:588-592. doi: 10.1016/j.jbiosc.2010.06.009

[80] Wang J, Liao X, Zhang H, Du J, Chen P (2003a) Accumulation of chlorogenic acid in cell suspension cultures of *Eucommia ulmoides*. Plant Cell Tiss Organ Cult 74:193-195. https://doi.org/10.1023/A:1023957129569

[81] Lee SY, Cho SI, Park MH, Kim YK, Choi JE, Park SU (2007) Growth and rutin production in hairy root cultures of buckwheat (*Fagopyrum esculentum* M.) Prep Biochem Biotechnol 37: 239-246. doi: 10.1080/10826060701386729

[82] Zhang W, Seki M, Furusaki S (1997) Effect of temperature and its shift on growth and anthocyanin production in suspension cultures of strawberry cells. Plant Sci 127:207-214. https://doi.org/10.1016/S0168-9452(97)00124-6

[83] Osuna L, Pereda-Miranda R, Villarreal ML (2002) *In vitro* production of sedative galphimine B by cell suspension cultures of *Galphimia glauca*. Biotechnol Lett 24:257-261. https://doi.org/10.1023/A:1014028510775

[84] Miura H, Kitamura Y, Ikenaga T, Mizobe K, Shimizu T, Nakamura M, Kato Y, Yamada T, Maitani T, Goda Y (1998) Anthocyanin production of *Glehnia littoralis* callus cultures. Phytochemistry 48:279-283. https://doi.org/10.1016/S0031-9422(97)01115-1

[85] Kitamura Y, Ohta M, Ikenaga T, Watanabe M (2002) Responses of anthocyanin-producing and non-producing cells of *Glehnia littoralis* to radical generators. Phytochemistry 59: 63-68. doi: 10.1016/s0031-9422(01)00428-9

[86] Praveen N, Murthy HN, Chung IM (2011) Improvement of growth and gymnemic acid production by altering the macro elements concentration and nitrogen source supply in cell suspension cultures of *Gymnema sylvestre* R. Br. Ind Crops Prod 33: 282-286. https://doi.org/10.1016/j.indcrop.2010.12.015

[87] Taha HS, El-Kawy AA, Fathalla MA (2012) A new approach for achievement of inulin accumulation in suspension cultures of Jerusalem artichoke (*Helianthus tuberosus*) using biotic elicitors. J Genet Eng Biotechnol 10: 33-38. https://doi.org/10.1016/j.jgeb.2012.02.002

[88] Misra N, Misra P, Datta SK, Mehrotra S (2005) *In vitro* biosynthesis of antioxidants from *Hemidesmus indicus* R. Br. Cultures. In Vitro Cell Dev Biol-Plant 41:285-290. https://doi.org/10.1079/IVP2004627

[89] Walker TS, Bais HP, Vivanco JM (2002) Jasmonic acid-induced hypericin production in cell suspension cultures of *Hypericum perforatum* L. (St. John's wort). Phytochemistry 60:289-293. doi: 10.1016/s0031-9422(02)00074-2

[90] Murakami Y, Omoto T, Asai I, Shimomura K, Yoshihira K, Ishimaru K (1998) Rosmarinic acid and related phenolics in transformed root cultures of *Hyssopus officinalis.* Plant Cell Tiss Org 53:75-78. https://doi.org/10.1023/A:1006007707722

[91] Kochan E, Wysokinska H, Chmiel A, Grabias B (1999) Rosmarinic acid and other phenolic acids in hairy roots of *Hyssopus officinalis*. Z Naturforsch *C* 54c:11-16. https://doi.org/10.1515/znc-1999-1-204

[92] Konczak-Islam I, Yoshimoto M, Hou D. X, Terahara N, Yamakawa O (2003) Potential chemopreventive properties of anthocyanin-rich aqueous extracts from *in vitro* produced tissue of sweet potato (*Ipomoea batatas* L.). J Agric Food Chem 51:5916-5922. https://doi.org/10.1021/jf030066o

[93] Terahara N, Konczak I, Ono H, Yoshimoto M, Yamakawa O (2004) Characterization of acylated anthocyanins in callus induced from storage root of purple-fleshed sweet potato, Ipomoea batatas L. J Biomed Biotechnol 2004:279-286. doi: 10.1155/S1110724304406056

[94] Terahara N, Konczak-Islam I, Nakatani M, Yamakawa O, Goda Y,

Honda T (2000) Anthocyanins in callus induced from purple storage root of *Ipomoea batatas* L. Phytochemistry 54: 919-922. https://doi.org/10.1016/S0031-9422(00)00099-6

[95] Palacio L, Cantero JJ, Cusidó RM, Goleniowski ME (2012) Phenolic compound production in relation to differentiation in cell and tissue cultures of *Larrea divaricata* (Cav.). Plant Sci 193-194:1-7. doi: 10.1016/j.plantsci.2012.05.007

[96] Georgiev MI, Kuzeva SL, Pavlov AI, Kovacheva EG, Ilieva MP (2007) Elicitation of rosmarinic acid by *Lavandula vera* MM cell suspension culture with abiotic elicitors. World J Microbiol Biotechnol 23:301-304. doi: 10.1007/s11274-006-9214-5

[97] Pavlov AI, Georgiev MI, Panchev IN, Ilieva MP (2005b) Optimization of rosmarinic acid production by *Lavandula vera* MM plant cell suspension in a laboratory bioreactor. Biotechnol Prog 21:394-396. doi: 10.1021/bp049678z

[98] Espinosa-Leal C, Treviño-Neávez JF, Garza-Padrón RA, Verde-Star MJ, Rivas-Morales C, Morales-Rubio ME (2015) Contenido de fenoles totales y actividad anti-radical de extractos metanólicos de la planta silvestre y cultivada *in vitro* de *Leucophyllum frutescens.* Rev Mex Cienc Farm 46:52-56. https://www.redalyc.org/articulo.oa?id=57945705005

[99] Yamamoto H, Inoue K, Yazaki K (2000) Caffeic acid oligomers in *Lithospermum erythrorhizon* cell suspension cultures. Phytochemistry 53: 651-657. https://doi.org/10.1016/S0031-9422(99)00623-8

[100] Salinas R, Arellano-García J, Perea-Arango I, Álvarez L, Garduño-Ramírez ML, Marquina S, Zamilpa, A, Castillo-España P (2014) Production of the anti-Inflammatory compound 6-O-Palmitoyl-3-O-β-D-glucopyranosylcampesterol by callus cultures of *Lopezia racemosa* Cav. (Onagraceae) Molecules 19:8679-8690

[101] Jang YP, Lee YJ, Kim YC, Huh H (1998) Production of a hepatoprotective cerebroside from suspension cultures of *Lycium chinense*. Plant Cell Rep 18: 252-254. doi: 10.1007/s002990050566

[102] Chiang L, Abdullah MA (2007) Enhanced anthraquinones production from adsorbent-treated *Morinda elliptica* cell suspension cultures in production medium strategy. Process Biochem 42:757-763. doi: 10.1016/j.procbio.2007.01.005

[103] Pras N, Woerdenbag HJ, Batterman S, Visser JF, Van Uden W (1993) *Mucuna pruriens*: improvement of the biotechnological production of the anti-Parkinson drug L-dopa by plant cell selection. Pharm World Sci 15: 263-268. doi: 10.1007/BF01871128

[104] Kouamo K, Creche J, Chénieux JC, Rideau M, Viel C (1985) Alkaloid production by *Ochrosia elliptica* cell suspension cultures. J Plant Physiol 118: 277-283. doi: 10.1016/S0176-1617(85)80229-7

[105] Kintzios S, Kollias H, Straitouris E, Makri O (2004) Scale-up micropropagation of sweet basil (*Ocimum basilicum* L.) in an airlift bioreactor and accumulation of rosmarinic acid. Biotechnol Lett 26: 521-523. doi: 10.1023/b:bile.0000019561.89044.30

[106] Kang TH, Park HM, Kim YB, Kim H, Kim N, Do JH, Kang C, Cho Y, Kim SY (2009) Effects of red ginseng extract on UVB irradiation-induced skin aging in hairless mice. J Ethnopharmacol 123:446-451. doi: 10.1016/j.jep.2009.03.022

[107] Zhong JJ, Bai Y, Wang SJ (1996) Effects of plant growth regulators on

cell growth and ginsenoside saponin production by suspension cultures of *Panax quinquefolium*. J Biotechnol 45: 227-234. https://doi.org/10.1016/0168-1656(95)00170-0

[108] Tam WJ, Constabel F, Kurz WG (1980) Codeine from cell suspension cultures of *Papaver somniferum*. Phytochemistry 19:486-487. doi: 10.1016/0031-9422(80)83215-8

[109] Huang FC, Kutchan TM (2000) Distribution of morphinan and benzo [c] phenanthridine alkaloid gene transcript accumulation in *Papaver somniferum*. Phytochemistry 53:555-564. https://doi.org/10.1016/S0031-9422(99)00600-7

[110] Archambault J, Williams RD, Bédard C, Chavarie C (1996) Production of sanguinarine by elicited plant cell culture I. Shake flask suspension cultures. J Biotech 46:95-105. doi: 10.1016/0168-1656(95)00184-0

[111] Antognoni F, Zheng S, Pagnucco C, Baraldi R, Poli F, Biondi S (2007) Induction of flavonoid production by UV-B radiation in *Passiflora quadrangularis* callus cultures. Fitoterapia 78: 345-352. doi: 10.1016/j.fitote.2007.02.001

[112] Hempel J, Pforte H, Raab B, Engst W, Böhm H, Jacobasch G (1999) Flavonols and flavones of parsley cell suspension culture change the antioxidative capacity of plasma in rats. Mol Nutr Food Res 43:201-204 doi: 10.1002/(SICI)1521-3803(19990601)43:3<201::AID-FOOD201>3.0.CO;2-1

[113] Scragg AH, Allan EJ (1986) Production of the triterpenoid quassin in callus and cell suspension cultures of *Picrasma quassioides* Bennett. Plant Cell Rep 5:356-359. doi: 10.1007/BF00268601

[114] Saad I, Díaz E, Chávez I, Reyes-Chilpa R, Rubluo A, Jiménez-Estrada M (2000) Antifungal monoterpene production in elicited cell suspension cultures of *Piqueria trinervia*. Phytochemistry 55:51-57. https://doi.org/10.1016/S0031-9422(00)00211-9 doi: 10.3390/molecules19068679

[115] Chattopadhyay S, Srivastava AK, Bhojwani SS, Bisaria VS (2002) Production of podophyllotoxin by plant cell cultures of *Podophyllum hexandrum* in bioreactor. J Biosc Bioeng 93:215-220. doi: 10.1016/S1389-1723(02)80017-2

[116] Shinde AN, Malpathak N, Fulzele DP (2009) Studied enhancement strategies for phytoestrogens production in shake flasks by suspension culture of *Psoralea corylifolia*. Bioresour Technol 100:1833-1839. https://doi.org/10.1016/j.biortech.2008.09.028

[117] Yamamoto O, Yamada Y (1986) Production of reserpine and its optimization in cultured *Rauwolfia serpentina* Benth. Cells. Plant Cell Rep 5: 50-53. doi: 10.1007/BF00269717

[118] Perassolo M, Quevedo CV, Giulietti AM, Talou JR (2011) Stimulation of the proline cycle and anthraquinone accumulation in *Rubia tinctorum* cell suspension cultures in the presence of glutamate and two proline analogs. Plant Cell Tiss Organ Cult 106: 153-159. https://doi.org/10.1007/s11240-010-9903-5

[119] Tsutomu N, Hitoshi M, Masao N, Hideko H, Kaisuke Y (1983) Production of cryptotanshinone and ferruginol in cultured cells of *Salvia miltiorrhiza*. Phytochemistry 22:721-722. https://doi.org/10.1016/S0031-9422(00)86969-1

[120] Chen H, Yuan JP, Chen F, Zhang YL, Song JY (1997) Tanshinone production in Ti-transformed *Salvia miltiorrhiza* cell suspension cultures. J Biotechol 58:147-156. doi: 10.1016/s0168-1656(97)00144-2

[121] Zhao DX, Fu CX, Han YS, Lu DP (2005) Effects of elicitation on

jaceosidin and hispidulin production in cell suspension cultures of *Saussurea medusa*. Process Biochem 40:739-745. https://doi.org/10.1016/j.procbio.2004.01.040

[122] Sánchez-Sampedro MA, Fernández-Tárrago J, Corchete P (2005) Yeast extract and methyl jasmonate-induced silymarin production in cell cultures of *Silybum marianum* (L.) Gaertn. J Biotechnol 119:60-69. doi: 10.1016/j.jbiotec.2005.06.012

[123] Aburjai T, Bernasconi S, Manzocchi LA, Pelizzoni F (1997) Effect of calcium and cell immobilization on the production of choleocalciferol and its derivatives by *Solanum malacoxylon* cell cultures. Phytochemistry 46:1015-1018. doi: 10.1016/S0031-9422(97)00408-1

[124] Pérez-Hernández J, González-Cortázar M, Marquina S, Herrera-Ruiz M, Meckes-Fischer M, Tortoriello J, Cruz-Sosa F, Nicasio-Torres MP (2014) Sphaeralcic acid and Tomentin, Anti-inflammatory compounds produced in cell suspension cultures of *Sphaeralcea angustifolia*. Planta Med 80:209-214. doi: 10.1055/s-0033-1360302

[125] Rico-Rodríguez L, Ortiz-Butrón R, Silva-Torres R, Díaz-Cedillo F, Franco-Colín M (2015) Diseño bifactorial de reguladores de crecimiento para la obtención de un alcaloide a partir del cultivo celular de *Swietenia humilis* Zucc. Rev Mex Cienc Farm 46:57-61. https://www.redalyc.org/articulo.oa?id=57945705006

[126] Nieto-Trujillo A, Buendía-González L, García-Morales, C, Román-Guerrero, A, Cruz-Sosa F, Estrada-Zúñiga ME (2017) Phenolic compounds and parthenolide production from *in vitro* cultures of *Tanacetum parthenium*. Rev Mex Ing Qum 16: 371-387. https://www.redalyc.org/pdf/620/62052087004.pdf

[127] Yukimune Y, Tabata H, Higashi Y, Hara Y (1996) Methyl jasmonate-induced overproduction of Paclitaxel and Baccatin III in Taxus cell suspension cultures. Nat Biotechnol 14:1129-1132. https://doi.org/10.1038/nbt0996-1129

[128] Wang Q, Mawassi M, Li P, Gafny R, Sela I, Tanne E (2003b) Elimination of grapevine virus A (GVA) by cryopreservation of *in vitro*-grown shoot tips of *Vitis vinifera* L. Plant Sci 165:321-327. https://doi.org/10.1016/S0168-9452(03)00091-8

[129] Yue X, Zhang W, Deng M (2011) Hyper-production of 13 C-labeled trans-resveratrol in *Vitis vinifera* suspension cell culture by elicitation and *in situ* adsorption. Biochem Eng J 53:292-296. https://doi.org/10.1016/j.bej.2010.12.002

[130] Nagella P, Murthy HN (2010) Establishment of cell suspension cultures of *Withania somnifera* for the production of withanolide A. Bioresour Technol 101:6735-6739. https://doi.org/10.1016/j.biortech.2010.03.078

[131] Pérez-Hernández J, Nicasio-Torres MP, Sarmiento-López LG, Rodríguez-Monroy M (2019) Production of anti-inflammatory compounds in Sphaeralcea angustifolia cell suspension cultivated in stirred tank biorreactor. Eng Life Sci 19:196-205. doi: 10.1002/elsc.201800134

[132] Nicasio-Torres MP, Pérez-Hernández J, González-Cortázar M, Meckes-Fischer M, Tortoriello J, Cruz-Sosa F (2016) Production of potential anti-inflammatory compounds in cell suspension cultures of *Sphaeralcea angustifolia* (Cav.) G. Don. Acta Physiol Plant 38:209. doi:10.1007/s11738-016-2211-x

[133] Nicasio-Torres MP, Serrano-Román J, Pérez-Hernández J, Jiménez-Ferrer E, Herrera-Ruiz M (2017) Effect of dichloromethane-methanol extract and tomentin obtained from Sphaeralcea angustifolia cell suspensions in a model of

kaolin/carrageenan-induced arthritis. Planta Med Int Open 4, e35-e42. doi: 10.1055/s-0043-108760

[134] García-Rodríguez RV, Chamorro-Cevallos G, Siordia G, Jiménez-Arellanes MA, Chávez-Soto MA, Meckes-Fischer M (2012) *Sphaeralcea angustifolia* (Cav.) G. Don extract, a potential phytomedicine to treat chronic inflammation. BLACPMA 11:454-463. https://www.redalyc.org/articulo.oa?id p=1&id=85624131008&cid=47965

[135] Romero-Cerecero O, Meckes-Fischer M, Zamilpa A, Jiménez-Ferrer JE, Nicasio-Torres P, Pérez-García D, Tortoriello J (2013) Clinical trial for evaluating the effectiveness and tolerability of topical *Sphaeralcea angustifolia* treatment in hand osteoarthritis. J Ethnopharmacol 147: 467-473. doi: 10.1016/j.jep.2013.03.040

[136] Albarello N, Simotilde C, de Castro TC, Gayer CRM, Coelho MGP, de Moura R.S, Mansur E (2013) Anti-inflammatory and antinociceptive activity of field-growth plants and tissue culture of *Cleome spinosa* (Jacq.) in mice. J Med Plants Rese 7:1043-1049. doi: 10.5897/JMPR12.153

[137] Pérez-González MZ, Gutiérrez-Rebolledo GA, Yépez-Mulia L, Rojas-Tomé IS, Luna-Herrera J. Jiménez-Arellanes MA (2107) Antiprotozoal, antimycobacterial and anti-inflammatory evaluation of *Cnidoscolus chayamansa* (Mc Vaugh) extract and the isolated compounds. Biomed Pharmacother 89:89-97. doi: 10.1016/j.biopha.2017.02.021

[138] Pérez-González MZ, Siordia-Reyes AG, Damián-Nava P, Hernández-Ortega S, Macías-Rubalcava ML, Jiménez-Arellanes MA (2018) Hepatoprotective and anti-inflammatory activities of the *Cnidoscolus chayamansa* (Mc Vaugh) leaf extract in chronic models. Evid Based Complement Alternat Med 2018: 3896517. doi: 10.1155/2018/3896517

[139] Pérez-González MZ, Macias-Rubalcava ML, Hernández-Ortega S, Siordia-Reyes AG, Jiménez-Arellanes MA (2019) Additional compounds and the therapeutic potential of *Cnidoscolus chayamansa* (McVaugh) against hepatotoxicity induced by antitubercular drugs. Biomed Pharmacother 117:1091140, 1-10. doi: 10.1016/j.biopha.2019.109140

[140] Pérez-González MZ, Nieto-Trujillo A, García-Martínez I, Estrada-Zuñiga ME, Bernabe-Antonio A, Jiménez-Arellanes MA, Cruz-Sosa F (2018) Establishing a callus culture of *Cnidoscolus chayamansa* Mc Vaugh: A species with ethnopharmacological value. Adv Biochem Biotech 2018:1-5. https://gavinpublishers.com/admin/assets/articles_pdf/1544189353article_pdf1396193972.pdf

[141] Pérez-González MZ, Nieto-Trujillo A, García-Martínez I, Estrada-Zúñiga ME, Bernabé-Antonio A, Jiménez-Arellanes MA Cruz-Sosa F (2019) Lupeol acetate production and antioxidant activity of a cell suspension culture from *Cnidoscolus chayamansa* leaves. S Afr J Bot 125:30-38. https://doi.org/10.1016/j.sajb.2019.06.030

[142] Taniguchi S, Imayoshi Y, Kobayashi E, Takamatsu Y, Ito H, Hatano T, Harukuni S, Tokudac H, Nishinoc H, Sugitad D, Shimurad S, Yoshida T (2002) Production of bioactive triterpenes by *Eriobotrya japonica* calli. Phytochemistry 59: 315-323. https://doi.org/10.1016/S0031-9422(01)00455-1

[143] Tortoriello J, Ortega A (1993) Sedative effect of galphimine B, a nor-seco-triterpenoid from *Galphimia glauca.* Planta Med 59:398-400. doi: 10.1055/s-2006-959717

[144] Tortoriello J, Ortega A, Herrera-Ruíz M, Trujillo J, Reyes-Vázquez C (1998) Galphimine-B modifies electrical activity of ventral tegmental area

neurons in rats. Planta Med 64:309-313. doi: 10.1055/s-2006-957440

[145] Osuna L, Pereda-Miranda R, Tortoriello J, Villarreal ML (1999) Production of the sedative triterpene galphimine B in *Galphimia glauca* tissue culture. Planta Med 65:149-152. doi: 10.1055/s-1999-14057

[146] Osuna L, Moyano E, Mangas S, Bonfill M, Cusidó RM, Piñol MT, Zamilpa A, Tortoriello L, Palazón J (2008) Immobilization of *Galphimia glauca* plant cell suspensions for the production of enhanced amounts of Galphimine-B. Planta Med 74:94-99. doi: 10.1055/s-2007-993763

[147] Aguilar-Santamaría L, Ramírez G, Herrera-Arellano A, Zamilpa A, Jiménez JE, Alonso-Cortés D, Cortés-Gutiérrez EI, Ledesma N, Tortoriello J (2007) Toxicological and cytotoxic evaluation of standardized extracts of *Galphimia glauca*. J Ethnopharmacol 109:35-40. doi: 10.1016/j.jep.2006.06.013

[148] Herrera-Ruiz M, Jiménez-Ferrer JE, De Lima TCM, Avilés-Montes D, Pérez-García D, González-Cortázar M, Tortoriello J (2006) Anxiolytic and antidepressant-like activity of a standardized extract from *Galphimia glauca*. Phytomedicine 13:23-28. doi: 10.1016/j.phymed.2005.03.003

Section 4

Biological and Therapeutic Applications of Phenolic Compounds

Chapter 16

Phenolic Compounds – An Emerging Group of Natural Compounds against Leukaemia: *in vitro*, *in vivo* and Clinical Applications

Lucienne Gatt and Pierre Schembri Wismayer

Abstract

Leukaemia is the most common cancer in children under 15 years of age as well as the most common blood cancer in people older than 55. The use of all *trans* retinoic acid (ATRA) in combination with arsenic trioxide (ATO) for acute promyelocytic leukaemia (APL) and tyrosine kinase inhibitors for chronic myeloid leukaemia (CML) respectively, have improved survival rates. However, new, natural therapies are constantly being sought after to overcome issues with resistance, side effects and specificity. As a result of their range of health benefits, including anticancer properties, phenolic compounds have been extensively studied over the past two decades. One on hand, *in vitro* and *in vivo* studies highlight both the inhibitory as well as differentiation inducing effects of phenolics on different leukaemia types. On the other hand, clinical trials to date have shown their beneficial effects (decrease in the absolute lymphocyte count and lymphadenopathy) in CLL (Chronic lymphoblastic leukaemia) patients. Promising therapeutic candidates for future use include epigallocatechin-3-gallate, coumarin, and gallic acid, with the latter ideally used in combination with the conventional drugs daunorubicin and cytarabine.

Keywords: Leukaemia, phenols, anticancer, differentiation therapy, apoptosis

1. Introduction

Leukaemia is a malignancy which is characterised by an uncontrolled increase in immature blood cells, termed blasts, in the bone marrow [1]. As a result, these cells permeate the bone marrow and prevent haematopoiesis from occurring normally. Such blasts eventually penetrate into the bloodstream and spread into organs [2]. The earliest observations and descriptions of cases of leukaemia were recorded by Alfred Velpeau, Alfred Donné and John Hughes Bennett [3–5]. Rudolf Virchow is credited with coining the term 'leukaemia' in 1847, from the two Greek words 'leukos' and 'helma', which mean 'white blood' [6].

Broadly, leukaemia can be classified as either acute or chronic. In acute leukaemia, the proliferating cells are very immature, while in chronic leukaemia, these cells have a more mature phenotype [7]. Furthermore, both types are subdivided into myeloid, lymphoid and mixed lineages [8]. On one hand, in acute myeloid leukaemia (AML), these blasts are termed myeloblasts while they are lymphoblasts in acute lymphoblastic leukaemia (ALL). On the other hand, the mature cells are granulocytes or neutrophils in chronic myeloid leukaemia (CML) and are lymphocytes in chronic lymphatic leukaemia (CLL). In general, both chronic leukaemias and AML are more common in adults while ALL is generally prevalent in children [9–12].

Acute and chronic leukaemias differ in terms of onset time. In acute leukaemia, cell proliferation occurs rapidly in days, while in chronic leukaemia, the process is slower and takes months or years [13, 14]. As a result, in acute leukaemia, lack of treatment results in death within a time frame of weeks or months while in chronic leukaemia, this may be either months or years. The signs and symptoms of both types of leukaemia also vary. In acute leukaemia, the rapid proliferation of white blood cells causes bone discomfort, aches as well as swelling in the lymph nodes. The initial symptoms include anaemia, fatigue, fever and swelling in the liver and the spleen [15]. Patients with chronic leukaemia may also show similar symptoms but if anaemia is evident, it is milder than in acute leukaemia. Moreover, most patients diagnosed with chronic leukaemia do not show symptoms at the time of diagnosis [16].

In the following sections, current treatments for different leukaemia subtypes are discussed, as well as their drawbacks. Such disadvantages pave the way for the need for alternative therapies, whereby studies show that phenolic compounds are very promising candidates in this regard.

2. Leukaemia prevalence, current treatments and challenges

Leukaemia is the most common cancer in children under 15 years of age and accounts for 32% of cancers in children of this age. For patients under 20 years of age, leukaemia accounts for 25% of cancers. The most common childhood cancer, ALL, constitutes 23% of childhood cancers and between 75% to 80% of childhood leukaemia cases. AML follows ALL and encompasses between 15% to 20% of childhood leukaemia [17, 18]. Leukaemia is also the most common blood cancer in people older than 55.

Though the treatment offered to a patient diagnosed with leukaemia depends on the leukaemia type, the primary options for treatment of leukaemia remain chemotherapy and radiotherapy. Chemotherapy drugs for AML include cytarabine, daunorubicin, doxorubicin and idarubicin [19]. Where possible, a bone marrow or stem cell transplant is also used following remission. In the latter, though the procedure may result in complications, recovery rates are good [15]. For AML, the intensive chemotherapy treatment administered to the patient as an induction treatment and as consolidation treatment. In the former, the aim is to achieve remission, while the purpose of the latter is relapse prevention [20, 21]. The use of induction and consolidation therapy together with an autologous stem cell transplant results in both a high relapse risk and a high mortality, while the use of consolidation therapy together with allotransplation results in a lower relapse risk but a higher mortality due to risks associated with graft versus host [22].

Although chemotherapy is widely used to treat a variety of cancers, it is broadly cytotoxic to normal tissues. Chemotherapy needs to be administered in more than one cycle since both the proliferating and resting phase cells possess the genetic abnormality. As a result, one chemotherapy cycle alone is not enough to kill all the leukaemic cells [23]. Chemotherapy drugs are classified into five major classes based on their structure and mechanistic action. These are: alkylating agents, topoisomerase inhibitors, antitumour antibiotics, antimetabolites and microtubule inhibitors. Alkylating agents such as cisplatin act by damaging DNA and inhibiting transcription and protein synthesis [24]. Topoisomerase inhibitors like etoposide inhibit DNA topoisomerase from releasing supercoils during DNA replication [25]. Standard chemotherapy drugs such as daunorubicin and doxorubicin fall under the class of antitumour antibiotics which inhibit enzymes involved in DNA replication [26], while cytarabine is an antimetabolite which disrupts the S phase of the cell cycle [27]. Finally, microtubule inhibitors such as paclitaxel interfere with the M phase of the cell cycle, which results in the inhibition of mitosis [28].

For AML patients younger than sixty years of age, chemotherapy results in remission rates of between 50% to 75%, with most suffering a relapse. The incidence in AML is bimodal, with remission rates being lower for older patients and relapse rates being higher [21]. This relapse is a result of haematopoietic stem cells which survive the chemotherapeutic drug treatment and regenerate. Currently five year survival rates are estimated to be around 30% for AML [29, 30]. Moreover, standard chemotherapy for AML may result in side effects including myelosuppression, tumour lysis syndrome and hepatotoxicity [31].

While chemotherapy remains the standard treatment for AML, the use of other drugs has greatly improved survival rates for about 30% of AML cases. Such patients present with FLT3 mutations, with FLT3 being a tyrosine kinase vital for the differentiation of progenitor cells into both myeloid and lymphoid lineages. The first drug approved as an FLT3 inhibitor was Midostaurin, which since 2017 has been used a treatment for FLT3 mutant AML in combination with standard chemotherapy [32, 33]. In 2018, the second FLT3 inhibitor Gliteritinib was approved as a treatment for patients who were found to be resistant to other treatments [34]. Patients with FLT3 mutations are likely to relapse as elimination of cells harbouring the FLT3 mutation is very problematic. Moreover, some patients also become resistant to FLT3 inhibitors after treatment [35].

In AML subtype APL, treatment involves the use of all *trans* retinoic acid (ATRA) and arsenic trioxide (ATO) as induction therapy, combined with mild chemotherapy. This treatment, termed differentiation therapy, has converted the prognosis of APL from poor to favourable. Through differentiation therapy, blasts differentiate, resulting in a decline in proliferative capacity, followed by apoptosis or terminal differentiation initiation. This method contrasts highly with chemotherapy which is generally nonspecific and is often accompanied by highly toxic side effects [36]. Moreover, it is also advantageous in that while it causes terminal differentiation, it does not result in bone marrow hypoplasia, and unlike chemotherapy, the proliferating cells are not killed but their maturity is induced, leading to death [37–40].

More than 98% of APL patients possess the characteristic translocation t(15;17), which results in the fusion between two genes - the PML gene and the RARα. As a result, the fusion protein PML-RARα is formed. PML-RARα is conformationally changed by ATRA at concentrations between 10^{-7} and 10^{-6} M, resulting in co-repressor dissociation and co-activator activation, leading to a relaxation in chromatin, the activation of transcription of genes involved in differentiation, resulting in the terminal differentiation of promyelocytes to granulocytes [41, 42].

Three decades ago, APL was fatal as a result of coagulation disorders, and via anthracycline based chemotherapy, the prognosis was still poor for approximately 70% of patients. Differentiation therapy using ATRA and ATO has resulted in complete remission (CR) for around 85% of patients, and 70% of patients being cured. The use of ATRA as a differentiating agent to differentiate promyelocytes into granulocytes was first discovered by Breitman *et al* in 1980. A problem that has been encountered with the use of ATRA is ATRA resistance. This has improved through the use of ATO combined with ATRA, yet drug resistance to ATRA and ATO remains an issue [43].

Moreover, though ATRA has been pivotal in the treatment of APL, this treatment may result in another complication known as retinoic acid syndrome or differentiation syndrome (DS). It has been found to occur in around 2% to 27% of children with APL who are treated with ATRA, and in up to 50% of patients. This may result in pulmonary haemorrhage, renal failure, as well as heart failure and for this reason is termed life threatening. Differentiation syndrome typically occurs around a week or two following the start of ATRA and/or ATO therapy [44]. If DS is severe and has resulted in pulmonary or renal dysfunction, the use of ATRA is ceased [45–48]. Compared to patients who do not develop this complication, patients with DS have a lower overall free survival and event free survival [49]. Though the exact mechanism of DS is not fully known, the main key player is thought to be an excessive inflammatory response. This response stems from leukaemic cells during their differentiation process, and is due to a higher level of chemokine production and adhesion molecules on APL cells. Inflammation leads to capillary leak syndrome and blast cells infiltrating organs such as the lungs, and organ failure [50]. Treatment for DS is required early in the diagnosis, and the corticosteroid dexamethasone is administered intravenously. Corticosteroids decrease chemokine production and stop lung infiltration [51].

Reported benefits of other agents of differentiation include histone deacetylase (HDAC) and DNA methyltransferase (DNMT) inhibitors. A key DNMT inhibitor is 5-aza-cytidine while examples of HDAC inhibitors include sodium butyrate and valproic acid [52]. Moreover, for HDAC inhibitors, the combination of both valproic acid and ATRA has been found to be beneficial for older patients with AML [53, 54]. On one hand, HDAC inhibitors act by remodeling chromatin by subduing the activity of HDACs leading to histone acetylation. This results in the expression of genes involved in the processes of differentiation as well as apoptosis. On the other hand, the effect of DNMT inhibitors is DNA hypomethylation, which leads to the re-activation of tumour-suppressor genes silenced by methylation. The use of such inhibitors stems from the fact that the differentiation block of leukaemic cells may be a result of epigenetic changes including histone acetylation and DNA hypermethylation, which may be reversed through the action of these inhibitors [55].

In contrast to other leukaemias, in CML, the genetic abnormality, termed the Philadelphia chromosome is a result of the bcr-abl protein, which was identified by Nowell and Hungerford in 1960 [56]. This oncogenic protein leads to an upregulation of tyrosine kinase and inactivation of phosphoinositide-3-kinase resulting in the proliferation of myelocytes. Imatinib is a tyrosine kinase inhibitor which acts by binding to the bcr-abl protein. This inhibition allows the cells to differentiate into mature granulocytes and subsequently die by apoptosis [57–59]. Following imatinib administration, the cytogenic response to the treatment can be at one of three levels – cytogenic response, major cytogenic response and complete cytogenic response. In 80% of the patients, it is the latter that results, and following imatinib administration, most remain stable. However in some patients, mutations in the bcr-abl tyrosine kinase domain result in lack of inhibition by Imatinib. This leads patients to rely on chemotherapy and stem cell transplantation [60, 61]. A number of unfavorable effects following Imatinib treatment have been reported and include episodic bone pain, fluid retention, lethargy and weight gain. These usually occur within the first two years of treatment, and through continued treatment, they may also be reversed [62].

For ALL, 80% of cases occur in children, and like AML, its distribution is bimodal. Though the outcomes for children have greatly improved, the same cannot be said for elderly patients, with remission rates lying between 30 and 40% for this age group [63, 64]. Treatment involves the use of chemotherapy as induction treatment, consolidation therapy and also maintenance. For the former, an anthracycline, vincristine as well as corticosteroids [65] or the Hyper-CVAD chemotherapy regimen are used [66]. For ALL patients who are Ph-positive, survival rates have improved through the use of second generation tyrosine kinase inhibitors coupled to Hyper-CVAD [67]. Recently, great advancements have been made for relapsed or refractory ALL patients through CAR-T cell therapy [68]. Between 70-90% of these patients respond well to this treatment, however it is associated with challenges such as antigen escape, toxicity and tumour infiltration [69].

Contrastingly, many patients with CLL have indolent disease and are asymptomatic. For patients with active CLL, treatment involves the use of chemoimmunotherapy such as a combination of fludarabine, cyclophosphamide and rituximab (FCR) or bendamustine and rituximab (BR) [70, 71]. For patients with high risk ALL, other targeted treatment agents include venetoclax, ibutrinib and idelasilib [72–74]. Though toxicity and resistance remain challenges, these may potentially be alleviated by combination therapy.

3. Methods

This chapter discusses studies that have been published to date, that assess the anti-leukaemic effect of phenolic compounds. These studies are grouped into the following three categories: *in vitro*, *in vivo* and clinical trials, as outlined in the following flow chart (**Figure 1**). Both *in vitro* and *in vivo* fall under the term 'preclinical trials', which are vital prior to moving to clinical trials and aim to determine the usefulness of a drug as therapy, as well as whether treatment is accompanied by any toxicity effect. Coming from the Latin "in glass", *in vitro* refers to experimental work carried out in a laboratory, as opposed to "within the living" for *in vivo*, where experimental work is performed using living organisms. With regards leukaemia, *in vitro* studies include experimental work performed using leukaemia cell lines while *in vivo* studies utilize animal models such as mice injected with leukaemic cells. Lastly, clinical trials are performed using human subjects, and are used to confirm *in vitro* and *in vivo* results, as well as determine drug efficacy and safety, amongst other parameters.

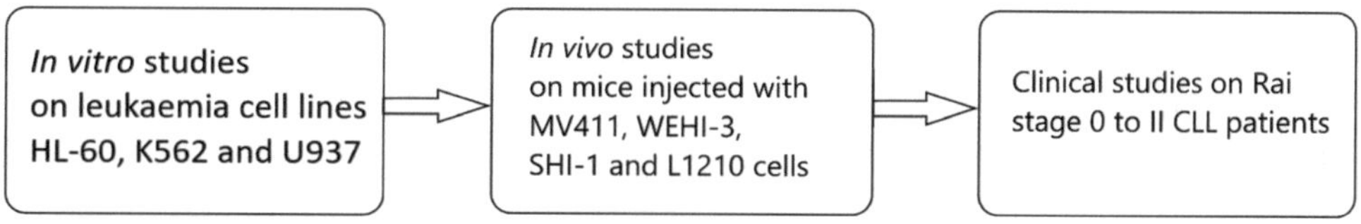

Figure 1.
A flow chart outlining the different types of studies recording the effects of phenolics on leukaemia.

4. Phenolic compounds: chemicals with a wide spectrum of bioactivity

Due to the challenges posed by the current treatments, therapies that may improve patient survival are needed. Novel treatments that are more specific and generally less toxic than conventional chemotherapy, are highly in demand. Due to their health benefits, the interest in natural products, specifically phenolic compounds, has greatly increased, making phenolics the subject of a number of research efforts over the past decade. Even more so, toxicity studies have shown that phenolics are safe and less toxic than a number of other synthetic and semi-synthetic compounds [75].

In plants, phenolic compounds are secondary metabolites consisting of an aromatic ring with one or more hydroxyl groups, which are involved in defending the plant against stress caused by drought, low or high temperatures, pathogens, restricted soil fertility and ultraviolet radiation [76, 77]. There is a wide range of such compounds and to date around 8000 of them have been identified and grouped into the following classes: phenolic acids (hydroxycinnamic and hydroxybenzoic acids), lignans, stilbenes, coumarins, xanthones and flavonoids [78–80]. Examples of each class of phenolic compounds that have been tested on leukaemia are presented in **Table 1**.

Such phenolics are distributed to varying degrees in particular parts of plants. Caffeic acid, a major phenolic acid is widely present in fruits, tannins are high in fruit pods, wood as well as bark, and flowers are rich in flavonoids [105, 106]. These compounds have been used by man for many years in the field of traditional medicine [76]. Several studies have been carried out which demonstrate the beneficial health effects of phenols. These compounds have been found to inhibit the oxidation of low density lipoprotein (LDL) *in vivo* where LDL oxidation is associated with the formation of atherosclerotic plaques, which play a role in coronary heart disease [107]. Even more so, the phenolic compound hydroxytyrosol has been found to decrease the risk of atherosclerosis and coronary heart disease [108].

Phenolic compounds such as hydroxytyrosol, hydroxytyrosol acetate and oleuropein have also been found to hinder platelet aggregation, in so doing, decreasing the synthesis of eicosanoids such as thromboxane and thus preventing thrombosis [109, 110]. Another antiatherogenic property of phenols is their ability to reduce endothelial activation by decreasing the mRNA levels of vascular adhesion molecule-1, hence resulting in a decline in its expression. Due to this, adhesion of monocytes to endothelial cells decreases, hence preventing endothelial malfunction [111].

The antioxidant capacity of phenolic compounds has also been widely investigated. Antioxidants are vital in protecting the plant from oxidative stress [112]. Compounds possessing an *ortho*-diphenolic structure are known to display antioxidant behaviour. Examples of such compounds include the phenols hydroxytyrosol and oleuropein, whose scavenging capabilities were compared to those of 2,2-diphenyl-1-picrylhydrazyl (DPPH) [113].

Additionally to antioxidant behaviour, hydroxytyrosol and oleuropein have been found to possess antimicrobial activity against a variety of American type culture

Class	Examples	Study	Mode of action	References
Flavonoids	Quercetin	*In vitro*	G_2M arrest	[81–83]
	Curcumin	*In vivo*	Tumour inhibition	[84–86]
	Epigallocatechin-3-gallate	Phases I and II	Decline in absolute lymphocyte count	[87, 88]
Phenolic acids				
- Hydroxybenzoic acids	Gallic acid	*In vitro* *In vivo*	G_0/G_1 arrest Tumour inhibition	[89, 90] [91, 92]
- Hydroxycinnamic acids	Cinnamic acid	*In vitro*	G_0/G_1 arrest and differentiation	[93]
Xanthones	α-mangostin	*In vitro*	Apoptosis	[94, 95]
Stilbenes	Resveratrol	*In vitro*	Apoptosis	[96]
Lignans	Syringaresinol	*In vitro*	G_0/G_1 arrest	[97]
Tannins	Tannic acid	*In vitro*	Apoptosis	[98]
Coumarins	Coumarin	*In vitro*	Apoptosis	[99–101]
Phenolic alcohols	Hydroxytyrosol	*In vitro*	Apoptosis and differentiation	[102]
Secoiridoids	Oleuropein	*In vitro*	Differentiation	[103, 104]

Table 1.
The major classes of phenolics and respective examples found to have an effect on leukaemia.

collection (ATCC) bacterial strains and clinical bacterial strains [114]. Moreover, such compounds are also anti-inflammatory agents. This is because they have been found to reduce both the release of arachidonic acid as well as production of arachidonic acid metabolites which play central roles in inflammation [115]. Also crucial to inflammation are the enzymes cyclooxygenase-1 (COX-1) and cyclooxygenase-2 (COX-2). It has been reported that such enzymes are inhibited by the phenolic compound oleocanthal, in a mechanism like that of ibupforen [116].

5. Phenolic compounds and leukaemia: *in vitro* studies

Various *in vitro* studies indicate that phenols possess anticancer properties [117–120]. Phenols are commonly found in foods including nuts, fruits, vegetables and oil. Studies have shown that diets rich in phenols help prevent a variety of cancers [121–124].

Structure-activity-relationship studies have shown that the anticancer properties of these compounds vary as a result of the functional groups present in the structure, where both the hydroxylic groups present as well as the aromatic ring play an important role. With regards to hydroxylic groups, the more the number of such groups, the higher the anticancer properties. Moreover, the presence of a side chain consisting of an unsaturated fatty acid makes the phenolic compound more effective (**Figure 2**) [125, 126].

In general, phenols act by inhibiting the cell cycle, leading to apoptosis (**Figure 3**) [127–129]. In addition, phenols appear to subdue the expression of chemokines as well as cytokines and angiogenesis is stopped. Both of these are vital for tumour development regulation [130–132].

Though a number of *in vitro* studies have focused on the effect of phenols on carcinomas, gliomas, melanomas, lung cancer and breast cancer, other studies have reported the inhibitory effects of phenolic compounds on leukaemia cell lines, with most studies focusing on HL-60, U937 and K562 cells.

The HL-60 cell line was isolated in 1977 and is classified as acute myeloblastic leukaemia with maturation (M2 category in the French-American-British classification) [133, 134]. In this suspension culture, a vast majority of the cells are promyelocytes which can be induced to differentiate into monocytes or granulocytes respectively by a number of compounds such as Phorbol 12-myristate 13-acetate (PMA), sodium butyrate, dimethyl sulfoxide (DMSO) as well as all-*trans* retinoic acid (ATRA) [135]. The U937 cell line was isolated in 1974 from a patient with histiocytic lymphoma and is classified under the M4 category in the

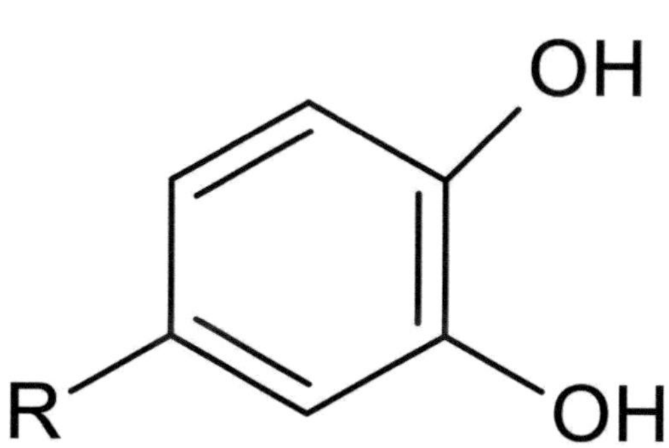

Figure 2.
The aromatic ring, the number and position of OH groups, and the presence of the unsaturated fatty acid side chain (R) influence activity.

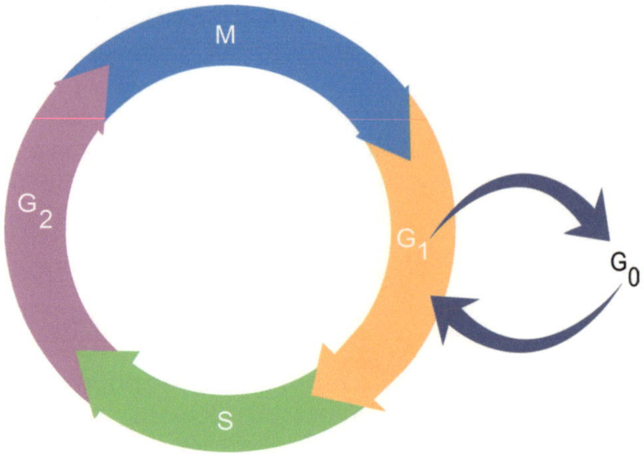

Figure 3.
The cell cycle – a process inhibited by phenolics. G_1 = Gap 1, S = Synthesis phase, G_2 = Gap 2, M = Mitosis, G_0 = resting phase.

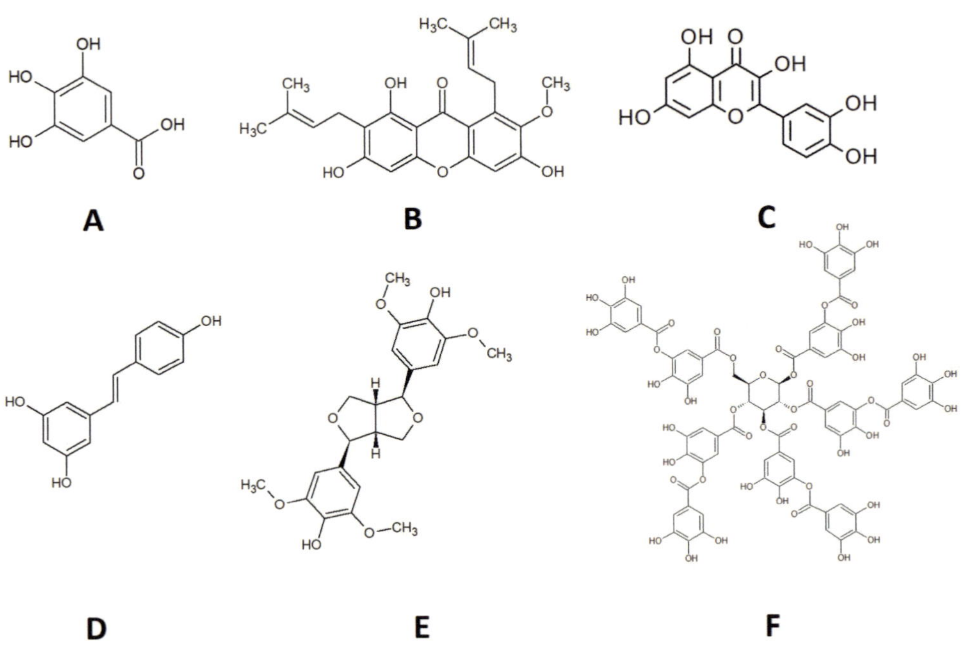

Figure 4.
Some phenolics found to possess anti-leukaemic activity. A = Gallic acid, B = Mangostin, C = Quercetin, D = Resveratrol, E = Syringaresinol and F = Tannic acid.

French-American-British classification. The cells are promonocytes and can be driven towards monocytic differentiation by PMA. For this reason, the cell line is used as a model for both monocyte and macrophage differentiation [136]. K562 is an example of erythroleukaemia [137]. It was isolated from a patient diagnosed with CML "in blast crisis", which is the final phase of the disease [138, 139]. The K562 cell line is positive for the Philadelphia chromosome, which is present in the vast majority of patients (>95%) diagnosed with CML. The cell line is termed to be proerythroblastic and studies have shown that it can be induced to monocytic, megakaryocytic and erythroid differentiation using chemicals such as proanthocyanidins for the former lineage, PMA for both the megakaryocytic and monocytic lineage and 5-azacytidine, butyric acid and hemin for the latter lineage [140, 141]. The *in vitro* effects of phenolics on the above-mentioned cell lines will be discussed below, and the structures of some of these phenolics are shown in **Figure 4**.

5.1 Flavonoids

A number of phenolic compounds belonging to the flavonoid class have been found to have an effect on leukaemia cell lines. Quercetin is a flavonol that has been reported to inhibit the proliferation of HL-60 cells and induce their apoptosis by the activation of caspase-3, the downregulation of Bcl-2 protein and the upregulation of the Bax protein. Its effects have been found to be both dose and time dependent. The action of quercetin on the mitochondrial pathway of apoptosis also involves the inhibition of COX-2 [81]. It has also been suggested that its antiproliferative effect may be due to its capacity to inhibit both cytosolic protein kinase C as well as tyrosine protein kinase [82]. Quercetin has been found to arrest the cell cycle of both HL-60 cells and U937 cells, with treatment resulting in an increase in the number of cells in G_2M phase. For U937 cells, this effect was coupled to a decrease in cyclins D, E, E2F1 and E2F2 [83]. Furthermore, the treatment of K562 cells with quercetin results in a number of morphological changes which include nuclear fragmentation as well as nuclear chromatin condensation. It has been found to inhibit the synthesis of heat shock protein 70, which is known to be involved in regulating the processes of both cell proliferation and differentiation [142, 143].

Within the same class of flavonols, both galangin and kaempferol have been found to inhibit the growth of HL-60 cells in a dose dependent manner. For kaempferol, this was attributed to both apoptotic and non-apoptotic effects but for galangin, the increased level of caspase-3 is suggestive of apoptosis [144]. These effects were also observed for two major flavones apigenin and luteolin. For the former, treatment resulted in an increase in both caspase-3 and caspase-9 proteases as well as cytochrome c [145–147]. Furthermore, the treatment of U937 cells with apigenin resulted in the cleavage of Poly (ADP-ribose) polymerase (PARP) as well as in the activation of caspase-3, caspase-7 and caspase-9. As for quercetin, down-regulation of Bcl-2 also occurs [148].

It has been shown that, similarly to quercetin, the flavone chrysin induces both U937 cell proliferation decline and DNA fragmentation. Its apoptotic effect on this cell line has been found to involve activation of caspase-3 as well as the inactivation of Akt (protein kinase B) [149, 150]. A methylated form of chrysin, termed 5,7-dimethoxyflavone was found to inhibit the growth of YCUB leukaemia cell lines in a dose and time dependent manner. Though this effect was seen on both YCUB-2 and YCUB-5 cells, for the former, an accumulation of reactive oxygen species was observed, but this was absent in the latter, suggesting a potentially different mechanism of action. Moreover, when 5,7-dimethoxyflavone was tested in combination with anticancer drugs such as cytarabine, an antagonistic effect was observed, suggesting the use of the compound as a single agent [151].

As a flavanol, epigallocatechin-3-gallate (EGCG) has been found to induce apoptosis in both acute and chronic myeloid leukaemia. For the former, a decline in death associated protein kinase 2 is observed, and an increase in neutrophil differentiation results on treatment of acute promyelocytic leukaemia with both ATRA and EGCG [152]. For the latter, the use of both EGCG and ponatinib results in a synergistic apoptotic effect which involves the downregulation of the CyclinD1 gene and the upregulation of TGF-β2 gene [153]. It has been reported that epigallocatechin-3-gallate causes the downregulation of the 67LR gene, and the induction of apoptosis is selective to cancer cells [154].

The anthocyanin delphinidin-3-sambubioside induces apoptosis in HL-60 cells through activation of three caspases which are caspase-3, caspase-8 and caspase-9, and causes DNA fragmentation [155].

Finally, the flavonoid curcumin and the metabolite tetrahydrocurcumin have both been found to induce apoptosis and autophagy respectively both in HL-60 cells as well as in HL-60 cells resistant to cytarabine [156]. This finding has very promising applications to overcome the issues with drug resistance. Furthermore, the combination of two flavonoids curcumin and quercetin induces mitochondrial apoptosis in CML. Since used in combination, any toxic effects on normal cells are unlikely since the treatment dose is lowered [157].

5.2 Phenolic acids and their derivatives

For hydroxybenzoic acids, gallic acid has been found to possess cytotoxic activity on HL-60 cells. Furthermore, gallic acid inhibits ribonucleotide reductase and arrests the cell cycle at the G_0/G_1 phase [89, 90]. The apoptosis of HL-60 cells by derivatives of gallic acid has also been investigated, and it has been concluded that apoptosis is greater in the presence of a long hydrophobic chain [158]. One of the derivatives of gallic acid, ellagic acid has been found to accumulate HL-60 cells in the S phase as well as induce their apoptosis with an increase in caspase-3 expression and PARP cleavage. Moreover, ellagic acid also enhances the differentiation effect of ATRA on HL-60 cells, and thus may be useful in overcoming ATRA resistance [159]. Ellagic acid has also been found to induce apoptosis in B-lymphocytes obtained from untreated CLL patients. This apoptotic effect involved the formation of reactive oxygen species, activation of caspase-3 and release of cytochrome c. Interestingly, this effect was selective to cancerous B-lymphocytes, and no toxic effect was seen for B-lymphocytes obtained from healthy donors [160].

With respect to hydroxycinnamic acids, caffeic acid phenethyl ester (CAPE) and cinnamic acid were found to induce apoptosis in HL-60 cells and K562 cells respectively, where for CAPE, protein, DNA and RNA synthesis in HL-60 cells were found to be inhibited [93, 161]. CAPE treatment resulted in the stimulation of Bax, downregulation of Bcl-2 as well as activation of caspase-3, signifying an apoptotic mechanism [162]. Apoptosis of U937 cells following CAPE treatment has also been recorded, with this effect being accompanied by an increase in cytochrome c [163]. For the cinnamic acid, a dose dependent arrest in the G_0/G_1 phase has been observed. Cinnamic acid has also been found to induce differentiation in K562 cells [93].

5.3 Xanthones and stilbenes

For xanthones, the effect of α-mangostin on HL-60 cells was investigated and its apoptotic effect was found to be caspase-3 dependent [94]. Apart from α-mangostin, β-mangostin also inhibits the growth of HL-60 cells, arrests them at the G_0/G_1 phase and induces intrinsic apoptosis through the activation of caspases-3, 7 and 9 and Bax, as well as the down-regulation of Bcl-2. Like quercetin, β-mangostin inhibits heat shock protein 70 [95].

With respect to stilbenes, studies have mainly focused on resveratrol, which has been found to be a differentiation inducing agent, as well as an inducer of apoptosis. This has been observed on NB4 cells, which are a type of APL. Like the xanthone α-mangostin, treatment with resveratrol results in an increase in caspase-3 activity. Nonetheless, for both α-mangostin and resveratrol, treatment on HL-60 and NB4 cells respectively does not have an effect on the Bcl-2 protein levels. Hence this is suggestive of an alternative apoptosis pathway. Differentiation of NB4 cells with resveratrol is completely effective when the cells are treated with both ATRA and resveratrol [96]. Furthermore, synthesized resveratrol analogues also arrest the cell cycle of HL-60 cells but do so at all three phases, G_0/G_1, S and G_2/M, contrasting with resveratrol which has been found to be phase specific [164]. It is relevant to highlight that though resveratrol is effective, it is limited by its poor bioavailability [165, 166]. Another two stilbenes namely piceatannol and sophorastilbene A both possess dose dependent cytotoxic activity on HL-60 cells with caspases 3, 8 and 9 being activated, with no changes in Bcl-2 protein expression being recorded [167].

5.4 Lignans

Within the class of lignans, (-)syringaresinol possesses anti-leukaemic behaviour. This is because it induces G_0/G_1 HL-60 cell cycle arrest in a manner that is both dose and time dependent. This is accompanied by the activation of both caspase-3 and caspase-9, DNA fragmentation and the release of cytochrome c [97].

5.5 Tannins

Tannins such as woodfordin C, cuphiin D1, cuphiin D2 and oenothein B have been found to possess cytotoxic behaviour on HL-60 cells [145, 168]. Tannic acid also induces apoptosis in HL-60 in both a time and dose dependent manner. The apoptotic mechanism was noted to involve the activation of caspases, PARP cleavage and cytochrome c release. Interestingly, tannic acid enhanced the cytotoxic effect of arsenic trioxide on HL-60 cells. This finding suggests the potential use of tannic acid in combination with arsenic trioxide [98].

5.6 Coumarins

Apoptotic activity on HL-60 cells was also recorded following treatment with 4-substituted coumarins, as well as furanone-coumarins, with an enhanced activity of caspases -3 and 9 also being recorded [99, 100]. Moreover, interestingly, coumarin was also found to induce cell death in drug resistant HL-60 cells when combined with doxorubicin [101]. This combination has great potential in overcoming the issue of drug resistance.

5.7 Phenolic alcohols and secoiridoids

While most of the effects reported referred to the inhibitory effect of phenolics, some studies have focused on their differentiating activity. Such studies have focused mainly on HL-60 cells, while other cell lines have been overlooked. Polyphenols from pomegranates and green tea, proanthocyanidins from barley and ellagic acid from fruits such as blackberries, pomegranates and strawberries have been found to induce differentiation HL-60 differentiation [159, 169–171]. Another three studies have focused on phenols from olive oil and the use of an olive leaf extract [102–104]. Two of these studies further confirm that phenolic compounds are capable of inhibiting cell proliferation and inducing differentiation in HL-60 cells. For the olive leaf extract study, the differentiation inducing compounds were found to be oleuropein and apigenin 7-glucoside [103]. The results from the study using olive oil on HL-60 cells show that dialdehydic compounds of elenoic acid with tyrosol and hydroxytyrosol are capable of inducing apoptosis and differentiation. It was reported that the effect of these two compounds was only a minor percentage of the total effect seen using the crude phenol extract [102]. Results from another study using an olive leaf extract with oleuropein as the major constituent show that the extract is capable of inducing both apoptosis as well as differentiation in K562 cells, along the monocyte/macrophage lineage [104].

The apoptotic effects recorded for phenolic compounds on leukaemia cell lines are potentially more similar to those of antitumour antibiotics as opposed to microtubule inhibitors and alkylating agents. This is beneficial as the latter two categories are highly unspecific as they target cells by mitotic spindle inhibition or DNA adduct formation respectively.

6. Phenolic compounds and leukaemia: *In vivo* studies

In addition to the *in vitro* effects of phenolic compounds, some *in vivo* studies have also been conducted. These studies focus on the use of gallic acid, curcumin, and resveratrol, and will be discussed in this section.

Using AML xenograft tumour NOD/SCID mice models injected with MV411 leukaemia cells, the effect of gallic acid in combination with daunorubicin and cytarabine was investigated. The results show that when gallic acid was used in combination with such drugs, tumour inhibition was observed when compared to the use of the drugs alone as single agents [91].

Interestingly, both gallic acid and curcumin were found to inhibit WEHI-3 leukaemia cells *in vivo*. Using BALB/c mice injected with WEHI-3 cells, both gallic acid and curcumin caused a reduction in the weights of the livers and spleens of such mice. For gallic acid, it has been postulated that this effect occurs through the increase in macrophage phagocytosis. This finding is particularly interesting in that it contrasts highly with the enlarged spleen associated with WEHI-3 leukaemia. Moreover, both phenolics caused a reduction in the Mac-3 marker (macrophage precursor) percentage [84, 92].

For curcumin, an inhibition of CML was recorded using CML xenograft SCID mice and mice treated with curcumin had smaller tumours. Moreover, plasma exosomes of treated mice were found to contain higher levels of miR-21 [85]. Curcumin also inhibits the growth of SHI-1 leukaemia cells in SHI-1 injected SCID mice. The mechanism involves signaling of NF-kB and ERK pathways, and an activation of JNK and p38 [86].

Using mice treated with L1210 cells, resveratrol was found to increase the life span of such mice, as well as the activity of NK cells, which is an important mechanism for eradication of a tumour. Furthermore, lymphocyte proliferation and the humoral immune response were found to be enhanced following resveratrol treatment [172].

7. Phenolic compounds and leukaemia: clinical trials

In addition to *in vitro* and *in vivo* studies, the beneficial effects of phenolic compounds have also been made evident through clinical trials. The main sources of phenolics investigated in such studies have been olive oil, pomegranate juice, *Curcuma longa* and green tea. A daily short-term consumption of olive oil has been found to affect a number of biomarkers related to oxidative stress, with an increase in high density lipoprotein cholesterol and a decrease plasma oxidized low density lipoprotein being observed dose-dependently according to the phenolic content of olive oil [173, 174]. For cancer, clinical trials have shown phenolics to be effective against prostate and colorectal cancer [175, 176].

For leukaemia, the clinical trials that have been performed to date have focused on CLL and utilized olive oil and a green tea extract as the polyphenolic sources. For one study, an olive oil rich in oleocanthal and oleacin at concentrations of 416 mg/kg and 284 mg/kg respectively, was selected. For this trial, performed in 2019, a cohort of 21 patients with CLL Rai stage 0 to II were chosen, who were not receiving any treatment. The effect of daily ingestion of 40 mL of olive oil per day for a period of six months was tested through the analysis of a number of molecular, haematological and biochemical markers at different time points. Such tests included liver function, kidney function, glucose profile, lipidemic profile and an analysis of apoptotic markers CCK18, Apo1-Fas and anti-apoptotic protein survivin. The glucose and lipidemic profiles of such patients were found to improve, the levels of the apoptotic markers CCK18 and Apo1-Fas increased and survivin decreased [177].

Similarly, on Rai stage 0 to II CLL patients, phase I (33 patients) and phase II (42 patients) clinical trials were conducted using a green tea extract (Polyphenon E), containing a standardized dose of EGCG. The results from the phase I clinical trial showed a good toleration of the extract in patients at doses ranging from 400 mg to 2000 mg twice daily, as well as a decline in both the absolute lymphocyte count as well as in lymphadenopathy. The same positive results were obtained in the phase II clinical trial, this time with a twice daily dose of 2000 mg. The side effects were reported to include nausea, transaminitis, and abdominal pain [87, 88].

8. Future perspective

The anti-leukaemic potential of phenolic compounds has been well documented through both *in vivo* and *in vitro* studies. While a number of clinical trials also show the promise of such compounds as treatments for a variety of cancers, more clinical trials on leukaemia are needed in order to ensure that the findings from *in vitro* and *in vivo* studies are confirmed, as well as determine the safety and efficacy of such treatments.

9. Conclusions

The studies presented in this chapter show the benefits of phenolic compounds, both as anti-proliferative agents as well as differentiation agents for leukaemia. These compounds have been found to arrest the cell cycle of leukaemia cells, as well as to induce apoptosis and differentiation. In a number of phenolics, such effects were noted to be selective, in contrast to chemotherapy. The promising results offer a potential alternative to the current standard treatments, in the hope that being natural products, are less toxic and are accompanied by less adverse effects. Furthermore, some phenolics show great therapeutic potential in multi-drug resistance leukaemia patients.

Conflict of interest

The authors declare no conflict of interest.

Author details

Lucienne Gatt[1,2*] and Pierre Schembri Wismayer[2,3]

1 Department of Physiology and Biochemistry, Faculty of Medicine and Surgery, University of Malta, Msida, Malta

2 Centre for Molecular Medicine and Biobanking, University of Malta, Msida, Malta

3 Department of Anatomy, Faculty of Medicine and Surgery, University of Malta, Msida, Malta

*Address all correspondence to: lucienne.gatt@um.edu.mt

References

[1] Dong Y, Shi O, Zeng Q, Lu X, Wang W, Li Y, Wang Q. Leukemia incidence trends at the global, regional, and national level between 1990 and 2017. Exp Hematol Oncol. 2020; 9:14. doi: 10.1186/s40164-020-00170-6b

[2] Estey EH. Acute myeloid leukemia: 2012 update on diagnosis, risk stratification, and management. Am J Hematol. 2012; 87(1):89-99. doi: 10.1002/ajh.22246

[3] Donné, A. Cours de microscopie complémentaire des études médicales. Anatomie microscopique et physiologie des fluides de l'Economie. Paris: Ballière, 1844, 135-136

[4] Bennett, J. Case of hypertrophy of the spleen and liver in which death took place from suppuration of the blood. Edinburgh Medical and Surgical Journal, 1845, 64, 413-423

[5] Kampen KR. The discovery and early understanding of leukemia. Leuk Res. 2012; 36(1):6-13. doi: 10.1016/j.leukres.2011.09.028

[6] Virchow, R. Weisses blut. Froriep's Notizen. 1847, 36, 151-156

[7] Szczepański T, van der Velden VH, van Dongen JJ. Classification systems for acute and chronic leukaemias. Best Pract Res Clin Haematol. 2003; 16(4):561-82. doi: 10.1016/s1521-6926(03)00086-0

[8] Rabbitts TH. Translocations, master genes, and differences between the origins of acute and chronic leukemias. Cell. 1991; 67(4):641-4. doi: 10.1016/0092-8674(91)90057-6

[9] Gale RP. Advances in the treatment of acute myelogenous leukemia. N Engl J Med. 1979; 300(21):1189-99. doi: 10.1056/NEJM197905243002105

[10] Faderl S, Talpaz M, Estrov Z, O'Brien S, Kurzrock R, Kantarjian HM. The biology of chronic myeloid leukemia. N Engl J Med. 1999; 341(3):164-72. doi: 10.1056/NEJM199907153410306

[11] Hamblin TJ, Davis Z, Gardiner A, Oscier DG, Stevenson FK. Unmutated Ig V(H) genes are associated with a more aggressive form of chronic lymphocytic leukemia. Blood. 1999; 94(6):1848-54

[12] Pui CH, Robison LL, Look AT. Acute lymphoblastic leukaemia. Lancet. 2008; 371(9617):1030-43. doi: 10.1016/S0140-6736(08)60457-2

[13] McKenzie SB. Advances in understanding the biology and genetics of acute myelocytic leukemia. Clin Lab Sci. 2005 Winter;18(1):28-37. Erratum in: Clin Lab Sci. 2005; 18(3):149

[14] Passegué E, Jamieson CH, Ailles LE, Weissman IL. Normal and leukemic hematopoiesis: are leukemias a stem cell disorder or a reacquisition of stem cell characteristics? Proc Natl Acad Sci U S A. 2003; 100 Suppl 1(Suppl 1):11842-9. doi: 10.1073/pnas.2034201100.

[15] Blesi, M., Wise, B., & Kelley-Arney, C. Workbook for Blesi/Wise/Kelly-arney's medical assisting administrative and clinical competencies. United States of America: Delmar Cengage Learning. 2011

[16] Smolej L, Turcsányi P, Kubová Z, Zuchnická J, Mihályová J, Šimkovič M, Vodárek P, Krčméryová M, Móciková H, Brejcha M, Špaček M; Czech CLL Study Group. External validation of International Prognostic Score for asymptomatic early stage chronic lymphocytic leukaemia and proposal of an alternative score. Br J Haematol. 2021; 193(1):133-137. doi: 10.1111/bjh.17074

[17] Baird, K, Wayne AS. Chapter 6 Childhood leukemias. Treleaven J, Barrett AJ. Hematopoietic Stem Cell Transplantation in Clinical Practice. 2009. 55-70 DOI: 10.1016/B978-0-443-10147-2.50010-2

[18] Hunger SP, Raetz EA. How I treat relapsed acute lymphoblastic leukemia in the pediatric population. Blood. 2020; 136(16):1803-1812. doi: 10.1182/blood.2019004043

[19] Dombret H, Gardin C. An update of current treatments for adult acute myeloid leukemia. Blood. 2016; 127(1):53-61. doi: 10.1182/blood-2015-08-604520.

[20] Rowe JM, Tallman MS. Intensifying induction therapy in acute myeloid leukemia: has a new standard of care emerged? Blood. 1997; 90(6):2121-6.

[21] Burnett A, Wetzler M, Löwenberg B. Therapeutic advances in acute myeloid leukemia. J Clin Oncol. 2011; 29(5):487-94. doi: 10.1200/JCO.2010.30.1820. Epub 2011 Jan 10. Erratum in: J Clin Oncol. 2011; 29(16):2293

[22] Löwenberg B, Downing JR, Burnett A. Acute myeloid leukemia. N Engl J Med. 1999 Sep 30;341(14):1051-62. doi: 10.1056/NEJM199909303411407

[23] Sell S. Leukemia: stem cells, maturation arrest, and differentiation therapy. Stem Cell Rev. 2005;1(3):197-205. doi: 10.1385/SCR:1:3:197

[24] Finch GL, Burns-Naas LA. Cancer Chemotherapeutic Agents. Wexler P editor. Encyclopedia of Toxicology (Third Edition). Academic Press; 2014. 630-641 p. DOI: 10.1016/B978-0-12-386454-3

[25] Liu LF. DNA topoisomerase poisons as antitumor drugs. Annu Rev Biochem. 1989;58:351-75. doi: 10.1146/annurev.bi.58.070189.002031

[26] Bharti AC, Vishnoi K, Singh SM, Aggarwal BB. Chapter 1 - Pathways Linked to Cancer Chemoresistance and Their Targeting by Nutraceuticals. Bharti AC, Aggarwal BB editors. In Cancer Sensitizing Agents for Chemotherapy, Role of Nutraceuticals in Cancer Chemosensitization. Academic Press; 2018. 1-30 p. DOI: 10.1016/B978-0-12-812373-7.00001-2

[27] Preobrazhenskaya MN, Tevyashova AN, Olsufyeva EN, Huang KF, Huang HS. Second generation drugs-derivatives of natural antitumor anthracycline antibiotics daunorubicin, doxorubicin and carminomycin. Journal of Medical Sciences (Taiwan). 2006; 26(4):119-128

[28] Quinn BA, Lee NA, Kegelman TP, Bhoopathi P, Emdad L, Das SK, Pellecchia M, Sarkar D, Fisher PB. The Quest for an Effective Treatment for an Intractable Cancer: Established and Novel Therapies for Pancreatic Adenocarcinoma. Adv Cancer Res. 2015;127:283-306. doi: 10.1016/bs.acr.2015.04.009

[29] Tallman MS, Gilliland DG, Rowe JM. Drug therapy for acute myeloid leukemia. Blood. 2005; 106(4):1154-63. doi: 10.1182/blood-2005-01-0178. Epub 2005 May 3. Erratum in: Blood. 2005; 106(7):2243

[30] Kohrt HE, Coutre SE. Optimizing therapy for acute myeloid leukemia. J Natl Compr Canc Netw. 2008; 6(10):1003-16. doi: 10.6004/jnccn.2008.0076

[31] Watts J, Nimer S. Recent advances in the understanding and treatment of acute myeloid leukemia. F1000Res. 2018; 7:F1000 Faculty Rev-1196. doi: 10.12688/f1000research.14116.1

[32] Stone RM. What FLT3 inhibitor holds the greatest promise? Best Pract Res Clin Haematol. 2018; 31(4):401-404. doi: 10.1016/j.beha.2018.09.008.

[33] Levis M. Midostaurin approved for FLT3-mutated AML. Blood. 2017; 129(26):3403-3406. doi: 10.1182/blood-2017-05-782292.

[34] Dhillon S. Gilteritinib: First Global Approval. Drugs. 2019; 79(3):331-339. doi: 10.1007/s40265-019-1062-3

[35] Ivey A, Hills RK, Simpson MA, Jovanovic JV, Gilkes A, Grech A, Patel Y, Bhudia N, Farah H, Mason J, Wall K, Akiki S, Griffiths M, Solomon E, McCaughan F, Linch DC, Gale RE, Vyas P, Freeman SD, Russell N, Burnett AK, Grimwade D; UK National Cancer Research Institute AML Working Group. Assessment of Minimal Residual Disease in Standard-Risk AML. N Engl J Med. 2016; 374(5):422-33. doi: 10.1056/NEJMoa1507471

[36] Leszczyniecka M, Roberts T, Dent P, Grant S, Fisher PB. Differentiation therapy of human cancer: basic science and clinical applications. Pharmacol Ther. 2001; 90(2-3):105-56. doi: 10.1016/s0163-7258(01)00132-2

[37] Lo Coco F, Nervi C, Avvisati G, Mandelli F. Acute promyelocytic leukemia: a curable disease. Leukemia. 1998 Dec;12(12):1866-80. doi: 10.1038/sj.leu.2401230.

[38] Kogan SC, Bishop JM. Acute promyelocytic leukemia: from treatment to genetics and back. Oncogene. 1999; 18(38):5261-7. doi: 10.1038/sj.onc.1202996

[39] Petrie K, Zelent A, Waxman S. Differentiation therapy of acute myeloid leukemia: past, present and future. Curr Opin Hematol. 2009; 16(2):84-91. doi: 10.1097/MOH.0b013e3283257aee

[40] Grimwade D, Mistry AR, Solomon E, Guidez F. Acute promyelocytic leukemia: a paradigm for differentiation therapy. Cancer Treat Res. 2010;145:219-35. doi: 10.1007/978-0-387-69259-3_13

[41] Glass CK, Rosenfeld MG. The coregulator exchange in transcriptional functions of nuclear receptors. Genes Dev. 2000;14(2):121-41

[42] Minucci S, Maccarana M, Cioce M, De Luca P, Gelmetti V, Segalla S, Di Croce L, Giavara S, Matteucci C, Gobbi A, Bianchini A, Colombo E, Schiavoni I, Badaracco G, Hu X, Lazar MA, Landsberger N, Nervi C, Pelicci PG. Oligomerization of RAR and AML1 transcription factors as a novel mechanism of oncogenic activation. Mol Cell. 2000; 5(5):811-20. doi: 10.1016/s1097-2765(00)80321-4

[43] Jimenez JJ, Chale RS, Abad AC, Schally AV. Acute promyelocytic leukemia (APL): a review of the literature. Oncotarget. 2020; 11(11):992-1003. doi: 10.18632/oncotarget.27513

[44] Ross FJ, Latham GJ. 11 - Perioperative Management of the Oncology Patient. Coté CJ, Lerman J, Anderson BJ editors. A Practice of Anesthesia for Infants and Children (Sixth Edition). Elsevier; 2019. 240-256. e5 p. DOI: 10.1016/B978-0-323-42974-0.00011-2

[45] Frankel SR, Eardley A, Lauwers G, Weiss M, Warrell RP Jr. The "retinoic acid syndrome" in acute promyelocytic leukemia. Annals of Internal Medicine. 1992; 117(4):292-296. DOI: 10.7326/0003-4819-117-4-292

[46] Tallman MS. Therapy of acute promyelocytic leukemia: all-trans retinoic acid and beyond. Leukemia. 1998; 12 Suppl 1:S37-40

[47] Sanz MA, Grimwade D, Tallman MS, Lowenberg B, Fenaux P, Estey EH, Naoe T, Lengfelder E, Büchner T, Döhner H, Burnett AK, Lo-Coco F. Management of acute promyelocytic leukemia: recommendations from an expert panel on behalf of the European

LeukemiaNet. Blood. 2009; 113(9):1875-91. doi: 10.1182/blood-2008-04-150250.

[48] Redner A, Kessel R. Chapter 19 - Acute Myeloid Leukemia. Lanzkowsky P, Lipton JM, Fish JD editors. Lanzkowsky's Manual of Pediatric Hematology and Oncology (Sixth Edition). Academic Press; 2016. 390-406 p. DOI: 10.1016/B978-0-12-801368-7.00019-3

[49] De Botton S, Dombret H, Sanz M, Miguel JS, Caillot D, Zittoun R, Gardembas M, Stamatoulas A, Condé E, Guerci A, Gardin C, Geiser K, Makhoul DC, Reman O, de la Serna J, Lefrere F, Chomienne C, Chastang C, Degos L, Fenaux P. Incidence, clinical features, and outcome of all trans-retinoic acid syndrome in 413 cases of newly diagnosed acute promyelocytic leukemia. The European APL Group. Blood. 1998; 92(8):2712-8

[50] Rego EM, De Santis GC. Differentiation syndrome in promyelocytic leukemia: clinical presentation, pathogenesis and treatment. Mediterr J Hematol Infect Dis. 2011; 3(1):e2011048. doi: 10.4084/MJHID.2011.048

[51] Tsai WH, Shih CH, Lin CC, Ho CK, Hsu FC, Hsu HC. Monocyte chemotactic protein-1 in the migration of differentiated leukaemic cells toward alveolar epithelial cells. Eur Respir J. 2008; 31(5):957-62. doi: 10.1183/09031936.00135707

[52] Zhu WG, Otterson GA. The interaction of histone deacetylase inhibitors and DNA methyltransferase inhibitors in the treatment of human cancer cells. Curr Med Chem Anticancer Agents. 2003; 3(3):187-99. doi: 10.2174/1568011033482440

[53] Pilatrino C, Cilloni D, Messa E, Morotti A, Giugliano E, Pautasso M, Familiari U, Cappia S, Pelicci PG, Lo Coco F, Saglio G, Guerrasio A. Increase in platelet count in older, poor-risk patients with acute myeloid leukemia or myelodysplastic syndrome treated with valproic acid and all-trans retinoic acid. Cancer. 2005; 104(1):101-9. doi: 10.1002/cncr.21132

[54] Fredly H, Gjertsen BT, Bruserud O. Histone deacetylase inhibition in the treatment of acute myeloid leukemia: the effects of valproic acid on leukemic cells, and the clinical and experimental evidence for combining valproic acid with other antileukemic agents. Clin Epigenetics. 2013; 5(1):12. doi: 10.1186/1868-7083-5-12

[55] Zelent A, Petrie K, Chen Z, Lotan R, Lübbert M, Tallman MS, Ohno R, Degos L, Waxman S. Molecular target-based treatment of human cancer: summary of the 10th international conference on differentiation therapy. Cancer Res. 2005; 65(4):1117-23. doi: 10.1158/0008-5472.CAN-04-3603

[56] Nowell, PC., & Hungerford, D. A.. A minute chromosome in human chronic granulocytic leukemia. Science, 1960; 132, 1488-1501.

[57] Randolph TR. Chronic myelocytic leukemia--Part I: History, clinical presentation, and molecular biology. Clin Lab Sci. 2005 Winter;18(1):38-48. Erratum in: Clin Lab Sci. 2005; 18(3):149

[58] Tefferi A, Dewald GW, Litzow ML, Cortes J, Mauro MJ, Talpaz M, Kantarjian HM. Chronic myeloid leukemia: current application of cytogenetics and molecular testing for diagnosis and treatment. Mayo Clin Proc. 2005; 80(3):390-402. doi: 10.4065/80.3.390

[59] Kharas, MG, Fruman, DA. ABL oncogenes and phosphoinositide 3-kinase: Mechanism of activation and downstream effectors. Cancer Research, 2005; 65(6), 2047-2053.

[60] Pulsipher MA. Treatment of CML in pediatric patients: should imatinib mesylate (STI-571, Gleevec) or allogeneic hematopoietic cell transplant be front-line therapy? Pediatr Blood Cancer. 2004; 43(5):523-33. doi: 10.1002/pbc.20062

[61] McKenzie SB. Advances in understanding the biology and genetics of acute myelocytic leukemia. Clin Lab Sci. 2005 Winter;18(1):28-37. Erratum in: Clin Lab Sci. 2005; 18(3):149

[62] Hochhaus A, O'Brien SG, Guilhot F, Druker BJ, Branford S, Foroni L, Goldman JM, Müller MC, Radich JP, Rudoltz M, Mone M, Gathmann I, Hughes TP, Larson RA; IRIS Investigators. Six-year follow-up of patients receiving imatinib for the first-line treatment of chronic myeloid leukemia. Leukemia. 2009; 23(6):1054-61. doi: 10.1038/leu.2009.38

[63] Paul S, Kantarjian H, Jabbour EJ. Adult Acute Lymphoblastic Leukemia. Mayo Clin Proc. 2016; 91(11):1645-1666. doi: 10.1016/j.mayocp.2016.09.010

[64] Jabbour E, O'Brien S, Konopleva M, Kantarjian H. New insights into the pathophysiology and therapy of adult acute lymphoblastic leukemia. Cancer. 2015; 121(15):2517-28. doi: 10.1002/cncr.29383.

[65] Scavino HF, George JN, Sears DA. Remission induction in adult acute lymphocytic leukemia. Use of vincristine and prednisone alone. Cancer. 1976; 38(2):672-7. doi: 10.1002/1097-0142(197608)38:2<672::aid-cncr2820380208>3.0.co;2-c

[66] Kantarjian HM, O'Brien S, Smith TL, Cortes J, Giles FJ, Beran M, Pierce S, Huh Y, Andreeff M, Koller C, Ha CS, Keating MJ, Murphy S, Freireich EJ. Results of treatment with hyper-CVAD, a dose-intensive regimen, in adult acute lymphocytic leukemia. J Clin Oncol. 2000; 18(3):547-61. doi: 10.1200/JCO.2000.18.3.547

[67] Ravandi F, O'Brien SM, Cortes JE, Thomas DM, Garris R, Faderl S, Burger JA, Rytting ME, Ferrajoli A, Wierda WG, Verstovsek S, Champlin R, Kebriaei P, McCue DA, Huang X, Jabbour E, Garcia-Manero G, Estrov Z, Kantarjian HM. Long-term follow-up of a phase 2 study of chemotherapy plus dasatinib for the initial treatment of patients with Philadelphia chromosome-positive acute lymphoblastic leukemia. Cancer. 2015; 121(23):4158-64. doi: 10.1002/cncr.29646

[68] Pehlivan KC, Duncan BB, Lee DW. CAR-T Cell Therapy for Acute Lymphoblastic Leukemia: Transforming the Treatment of Relapsed and Refractory Disease. Curr Hematol Malig Rep. 2018;13(5):396-406. doi: 10.1007/s11899-018-0470-x

[69] Sterner RC, Sterner RM. CAR-T cell therapy: current limitations and potential strategies. Blood Cancer J. 2021; 11(4):69. doi: 10.1038/s41408-021-00459-7

[70] Hallek M, Pflug N. Chronic lymphocytic leukemia. Ann Oncol. 2010 Oct;21 Suppl 7:vii154-64. doi: 10.1093/annonc/mdq373. Erratum in: Ann Oncol. 2011; 22(2):492

[71] Eichhorst B, Fink AM, Bahlo J, Busch R, Kovacs G, Maurer C, Lange E, Köppler H, Kiehl M, Sökler M, Schlag R, Vehling-Kaiser U, Köchling G, Plöger C, Gregor M, Plesner T, Trneny M, Fischer K, Döhner H, Kneba M, Wendtner CM, Klapper W, Kreuzer KA, Stilgenbauer S, Böttcher S, Hallek M; international group of investigators; German CLL Study Group (GCLLSG). First-line chemoimmunotherapy with bendamustine and rituximab versus fludarabine, cyclophosphamide, and rituximab in patients with advanced chronic lymphocytic leukaemia (CLL10): an international, open-label, randomised, phase 3, non-inferiority trial. Lancet Oncol. 2016; 17(7):928-942. doi: 10.1016/S1470-2045(16)30051-1

[72] Burger JA, Tedeschi A, Barr PM, Robak T, Owen C, Ghia P, Bairey O, Hillmen P, Bartlett NL, Li J, Simpson D, Grosicki S, Devereux S, McCarthy H, Coutre S, Quach H, Gaidano G, Maslyak Z, Stevens DA, Janssens A, Offner F, Mayer J, O'Dwyer M, Hellmann A, Schuh A, Siddiqi T, Polliack A, Tam CS, Suri D, Cheng M, Clow F, Styles L, James DF, Kipps TJ; RESONATE-2 Investigators. Ibrutinib as Initial Therapy for Patients with Chronic Lymphocytic Leukemia. N Engl J Med. 2015; 373(25):2425-37. doi: 10.1056/NEJMoa1509388

[73] Roberts AW, Davids MS, Pagel JM, Kahl BS, Puvvada SD, Gerecitano JF, Kipps TJ, Anderson MA, Brown JR, Gressick L, Wong S, Dunbar M, Zhu M, Desai MB, Cerri E, Heitner Enschede S, Humerickhouse RA, Wierda WG, Seymour JF. Targeting BCL2 with Venetoclax in Relapsed Chronic Lymphocytic Leukemia. N Engl J Med. 2016 ; 374(4):311-22. doi: 10.1056/NEJMoa1513257

[74] Furman RR, Sharman JP, Coutre SE, Cheson BD, Pagel JM, Hillmen P, Barrientos JC, Zelenetz AD, Kipps TJ, Flinn I, Ghia P, Eradat H, Ervin T, Lamanna N, Coiffier B, Pettitt AR, Ma S, Stilgenbauer S, Cramer P, Aiello M, Johnson DM, Miller LL, Li D, Jahn TM, Dansey RD, Hallek M, O'Brien SM. Idelalisib and rituximab in relapsed chronic lymphocytic leukemia. N Engl J Med. 2014; 370(11):997-1007. doi: 10.1056/NEJMoa1315226.

[75] Habauzit V, Morand C. Evidence for a protective effect of polyphenols-containing foods on cardiovascular health: an update for clinicians. Ther Adv Chronic Dis. 2012; 3(2):87-106. doi: 10.1177/2040622311430006

[76] Asensi M, Ortega A, Mena S, Feddi F, Estrela JM. Natural polyphenols in cancer therapy. Crit Rev Clin Lab Sci. 2011; 48(5-6):197-216. doi: 10.3109/10408363.2011.631268.

[77] Russell W, Duthie G. Plant secondary metabolites and gut health: the case for phenolic acids. Proc Nutr Soc. 2011; 70(3):389-96. doi: 10.1017/S0029665111000152

[78] Huang WY, Cai YZ, Zhang Y. Natural phenolic compounds from medicinal herbs and dietary plants: potential use for cancer prevention. Nutr Cancer. 2010; 62(1):1-20. doi: 10.1080/01635580903191585

[79] Tsao R. Chemistry and biochemistry of dietary polyphenols. Nutrients. 2010; 2(12):1231-46. doi: 10.3390/nu2121231. Epub 2010 Dec 10

[80] Anantharaju PG, Gowda PC, Vimalambike MG, Madhunapantula SV. An overview on the role of dietary phenolics for the treatment of cancers. Nutr J. 2016; 15(1):99. doi: 10.1186/s12937-016-0217-2

[81] Niu G, Yin S, Xie S, Li Y, Nie D, Ma L, Wang X, Wu Y. Quercetin induces apoptosis by activating caspase-3 and regulating bcl-2 and cyclooxygenase-2 pathways in human HL-60 cells. Acta Biochim Biophys Sin. 2011; 43(1), 30-37

[82] Kang TB, Liang NC. Studies on the inhibitory effects of quercetin on the growth of HL-60 leukemia cells. Biochem Pharmacol. 1997; 54(9):1013-8. doi: 10.1016/s0006-2952(97)00260-8

[83] Lee TJ, Kim OH, Kim YH, Lim JH, Kim S, Park JW, Kwon TK. Quercetin arrests G2/M phase and induces caspase-dependent cell death in U937 cells. Cancer Lett. 2006; 240(2):234-42. doi: 10.1016/j.canlet.2005.09.013

[84] Ho CC, Lin SY, Yang JS, Liu KC, Tang YJ, Yang MD, Chiang JH, Lu CC, Wu CL, Chiu TH, Chung JG. Gallic acid inhibits murine leukemia WEHI-3 cells in vivo and promotes macrophage phagocytosis. In Vivo. 2009; 23(3):409-13

[85] Taverna S, Giallombardo M, Pucci M, Flugy A, Manno M, Raccosta S, Rolfo C, De Leo G, Alessandro R. Curcumin inhibits in vitro and in vivo chronic myelogenous leukemia cells growth: a possible role for exosomal disposal of miR-21. Oncotarget. 2015; 6(26):21918-33. doi: 10.18632/oncotarget.4204

[86] Zhu G, Shen Q, Jiang H, Ji O, Zhu L, Zhang L. Curcumin inhibited the growth and invasion of human monocytic leukaemia SHI-1 cells *in vivo* by altering MAPK and MMP signalling. Pharm Biol. 2020; 58(1):25-34. doi: 10.1080/13880209.2019.1701042

[87] Shanafelt TD, Call TG, Zent CS, LaPlant B, Bowen DA, Roos M, Secreto CR, Ghosh AK, Kabat BF, Lee MJ, Yang CS, Jelinek DF, Erlichman C, Kay NE. Phase I trial of daily oral Polyphenon E in patients with asymptomatic Rai stage 0 to II chronic lymphocytic leukemia. J Clin Oncol. 2009; 27(23):3808-14. doi: 10.1200/JCO.2008.21.1284

[88] Shanafelt TD, Call T, Zent CS, LaPlant B, Leis JF, Bowen D, et al. Phase II trial of daily, oral green tea extract in patients with asymptomatic, Rai stage 0-II chronic lymphocytic leukemia (CLL). Journal of Clinical Oncology. 2010; doi: 10.1200/jco.2010.28.15_suppl.6522

[89] Madlener S, Illmer C, Horvath Z, Saiko P, Losert A, Herbacek I, Grusch M, Elford HL, Krupitza G, Bernhaus A, Fritzer-Szekeres M, Szekeres T. Gallic acid inhibits ribonucleotide reductase and cyclooxygenases in human HL-60 promyelocytic leukemia cells. Cancer Lett. 2007; 245(1-2):156-62. doi: 10.1016/j.canlet.2006.01.001

[90] Yeh RD, Chen JC, Lai TY, Yang JS, Yu CS, Chiang JH, Lu CC, Yang ST, Yu CC, Chang SJ, Lin HY, Chung JG. Gallic acid induces G_0/G_1 phase arrest and apoptosis in human leukemia HL-60 cells through inhibiting cyclin D and E, and activating mitochondria-dependent pathway. Anticancer Res. 2011; 31(9):2821-32

[91] Gu R, Zhang M, Meng H, Xu D, Xie Y. Gallic acid targets acute myeloid leukemia via Akt/mTOR-dependent mitochondrial respiration inhibition. Biomed Pharmacother. 2018; 105:491-497. doi: 10.1016/j.biopha.2018.05.158

[92] Su CC, Yang JS, Lin SY, Lu HF, Lin SS, Chang YH, Huang WW, Li YC, Chang SJ, Chung JG. Curcumin inhibits WEHI-3 leukemia cells in BALB/c mice in vivo. In Vivo. 2008; 22(1):63-8

[93] Zhang J, Xiao A, Wang T, Liang X, Gao J, Li P, Shi T. Effect and mechanism of action of cinnamic acid on the proliferation and apoptosis of Leukaemia cells. Biomedical Research (India). 2014; 25. 405-408

[94] Matsumoto K, Akao Y, Yi H, Ohguchi K, Ito T, Tanaka T, Kobayashi E, Iinuma M, Nozawa Y. Preferential target is mitochondria in alpha-mangostin-induced apoptosis in human leukemia HL60 cells. Bioorg Med Chem. 2004; 12(22):5799-806. doi: 10.1016/j.bmc.2004.08.034

[95] Omer FAA, Hashim NBM, Ibrahim MY, Dehghan F, Yahayu M, Karimian H, Salim LZA, Mohan S. Beta-mangostin from Cratoxylum arborescens activates the intrinsic apoptosis pathway through reactive oxygen species with downregulation of the HSP70 gene in the HL60 cells associated with a G0/G1 cell-cycle arrest. Tumour Biol. 2017; 39(11):1010428317731451. doi: 10.1177/1010428317731451

[96] Cao Y, Wang F, Liu HY, Fu ZD, Han R. Resveratrol induces apoptosis and differentiation in acute promyelocytic leukemia (NB4) cells. J Asian Nat Prod Res. 2005; 7(4):633-41. doi: 10.1080/1028602032000169523

[97] Park BY, Oh SR, Ahn KS, Kwon OK, Lee HK. (-)-Syringaresinol inhibits proliferation of human promyelocytic HL-60 leukemia cells via G1 arrest and apoptosis. International Immunopharmacology. 2008; 8(7):967-973. DOI: 10.1016/j.intimp.2008.02.012

[98] Chen KS, Hsiao YC, Kuo DY, Chou MC, Chu SC, Hsieh YS, Lin TH. Tannic acid-induced apoptosis and -enhanced sensitivity to arsenic trioxide in human leukemia HL-60 cells. Leuk Res. 2009; 33(2):297-307. doi: 10.1016/j.leukres.2008.08.006

[99] Ito C, Murata T, Itoigawa M, Nakao K, Kaneda N, Furukawa H. Apoptosis inducing activity of 4-substituted coumarins from Calophyllum brasiliense in human leukaemia HL-60 cells. J Pharm Pharmacol. 2006; 58(7):975-80. doi: 10.1211/jpp.58.7.0013

[100] Murata T, Itoigawa M, Ito C, Nakao K, Tsuboi M, Kaneda N, Furukawa H. Induction of apoptosis in human leukaemia HL-60 cells by furanone-coumarins from Murraya siamensis. J Pharm Pharmacol. 2008; 60(3):385-9. doi: 10.1211/jpp.60.3.0015

[101] Al-Abbas NS, Shaer NA. Combination of coumarin and doxorubicin induces drug-resistant acute myeloid leukemia cell death. Heliyon. 2021; 7(3):e06255. doi: 10.1016/j.heliyon.2021.e06255

[102] Fabiani R, De Bartolomeo A, Rosignoli P, Servili M, Selvaggini R, Montedoro GF, Di Saverio C, Morozzi G. Virgin olive oil phenols inhibit proliferation of human promyelocytic leukemia cells (HL60) by inducing apoptosis and differentiation. J Nutr. 2006; 136(3):614-9. doi: 10.1093/jn/136.3.614

[103] Abaza L, Talorete TP, Yamada P, Kurita Y, Zarrouk M, Isoda H. Induction of growth inhibition and differentiation of human leukemia HL-60 cells by a Tunisian gerboui olive leaf extract. Biosci Biotechnol Biochem. 2007; 71(5):1306-12. doi: 10.1271/bbb.60716

[104] Samet I, Han J, Jlaiel L, Sayadi S, Isoda H. Olive (Olea europaea) leaf extract induces apoptosis and monocyte/macrophage differentiation in human chronic myelogenous leukemia K562 cells: insight into the underlying mechanism. Oxid Med Cell Longev. 2014;2014:927619. doi: 10.1155/2014/927619

[105] Achakzai AK, Achakzai P, Masood A, Kayani SA, Tareen RB. Response of plant parts and age on the distribution of secondary metabolites on plants found in Quetta. Pak. J. Bot. 2009; 41(5):2129-35.

[106] Ververidis F, Trantas E, Douglas C, Vollmer G, Kretzschmar G, Panopoulos N. Biotechnology of flavonoids and other phenylpropanoid-derived natural products. Part II: Reconstruction of multienzyme pathways in plants and microbes. Biotechnol J. 2007; 2(10):1235-49. doi: 10.1002/biot.200700184

[107] Visioli F, Bellomo G, Montedoro G, Galli C. Low density lipoprotein oxidation is inhibited in vitro by olive oil constituents. Atherosclerosis. 1995; 117(1):25-32. doi: 10.1016/0021-9150(95)05546-9

[108] Grignaffini P, Roma P, Galli C, Catapano AL. Protection of low-density lipoprotein from oxidation by 3,4-dihydroxyphenylethanol. Lancet. 1994; 343(8908):1296-7. doi: 10.1016/s0140-6736(94)92186-5

[109] Petroni A, Blasevich M, Salami M, Papini N, Montedoro GF, Galli C. Inhibition of platelet aggregation and eicosanoid production by phenolic components of olive oil. Thromb Res. 1995; 78(2):151-60. doi: 10.1016/0049-3848(95)00043-7

[110] González-Correa JA, Navas MD, Lopez-Villodres JA, Trujillo M, Espartero JL, De La Cruz JP. Neuroprotective effect of hydroxytyrosol and hydroxytyrosol acetate in rat brain slices subjected to hypoxia-reoxygenation. Neurosci Lett. 2008; 446(2-3):143-6. doi: 10.1016/j.neulet.2008.09.022

[111] Carluccio MA, Siculella L, Ancora MA, Massaro M, Scoditti E, Storelli C, Visioli F, Distante A, De Caterina R. Olive oil and red wine antioxidant polyphenols inhibit endothelial activation: antiatherogenic properties of Mediterranean diet phytochemicals. Arterioscler Thromb Vasc Biol. 2003; 23(4):622-9. doi: 10.1161/01.ATV.0000062884.69432.A0

[112] Mittler R. Oxidative stress, antioxidants and stress tolerance. Trends Plant Sci. 2002; 7(9):405-10. doi: 10.1016/s1360-1385(02)02312-9

[113] Tuck KL, Hayball PJ. Major phenolic compounds in olive oil: metabolism and health effects. J Nutr Biochem. 2002; 13(11):636-644. doi: 10.1016/s0955-2863(02)00229-2

[114] Bisignano G, Tomaino A, Lo Cascio R, Crisafi G, Uccella N, Saija A. On the in-vitro antimicrobial activity of oleuropein and hydroxytyrosol. J Pharm Pharmacol. 1999; 51(8):971-4. doi: 10.1211/0022357991773258

[115] Moreno JJ. Effect of olive oil minor components on oxidative stress and arachidonic acid mobilization and metabolism by macrophages RAW 264.7. Free Radic Biol Med. 2003; 35(9):1073-81. doi: 10.1016/s0891-5849(03)00465-9

[116] Beauchamp GK, Keast RS, Morel D, Lin J, Pika J, Han Q, Lee CH, Smith AB, Breslin PA. Phytochemistry: ibuprofen-like activity in extra-virgin olive oil. Nature. 2005; 437(7055):45-6. doi: 10.1038/437045a

[117] Gomes, CA, da Cruz, TG, Andrade, JL, Milhazes, N, Borges, F, & Marques, MP. Anticancer activity of phenolic acids of natural or synthetic origin: A structure-activity study. *Journal of Medicinal Chemistry, 2003; 46*(25), 5395-5401. Doi: 10.1021/jm030956v

[118] McCann MJ, Gill CI, O' Brien G, Rao JR, McRoberts WC, Hughes P, McEntee R, Rowland IR. Anti-cancer properties of phenolics from apple waste on colon carcinogenesis in vitro. Food Chem Toxicol. 2007; 45(7):1224-30. doi: 10.1016/j.fct.2007.01.003

[119] Nandi S, Vracko M, Bagchi MC. Anticancer activity of selected phenolic compounds: QSAR studies using ridge regression and neural networks. Chem Biol Drug Des. 2007; 70(5):424-36. doi: 10.1111/j.1747-0285.2007.00575.x.

[120] Spilioti E, Jaakkola M, Tolonen T, Lipponen M, Virtanen V, Chinou I, Kassi E, Karabournioti S, Moutsatsou P. Phenolic acid composition, antiatherogenic and anticancer potential of honeys derived from various regions in Greece. PLoS One. 2014; 9(4):e94860. doi: 10.1371/journal.pone.0094860

[121] Block G, Patterson B, Subar A. Fruit, vegetables, and cancer prevention: a review of the epidemiological evidence. Nutr Cancer. 1992;18(1):1-29. doi: 10.1080/01635 589209514201

[122] Mathew A, Peters U, Chatterjee N, Kulldorff M, Sinha R. Fat, fiber, fruits, vegetables, and risk of colorectal adenomas. Int J Cancer. 2004; 108(2):287-92. doi: 10.1002/ijc.10984

[123] Gandini S, Merzenich H, Robertson C, Boyle P. Meta-analysis of studies on breast cancer risk and diet: the role of fruit and vegetable consumption and the intake of associated micronutrients. Eur J Cancer. 2000; 36(5):636-46. doi: 10.1016/s0959-8049(00)00022-8

[124] Temple NJ, Gladwin KK. Fruit, vegetables, and the prevention of cancer: research challenges. Nutrition. 2003; 19(5):467-70. doi: 10.1016/s0899-9007(02)01037-7

[125] Lee YJ, Liao PH, Chen WK, Yang CY. Preferential cytotoxicity of caffeic acid phenethyl ester analogues on oral cancer cells. Cancer Lett. 2000; 153(1-2):51-6. doi: 10.1016/s0304-3835(00)00389-x

[126] Chen M, Meng H, Zhao Y, Chen F, Yu S. Antioxidant and in vitro anticancer activities of phenolics isolated from sugar beet molasses. BMC Complement Altern Med. 2015; 15:313. doi: 10.1186/s12906-015-0847-5

[127] Duthie SJ. Berry phytochemicals, genomic stability and cancer: evidence for chemoprotection at several stages in the carcinogenic process. Mol Nutr Food Res. 2007; 51(6):665-74. doi: 10.1002/mnfr.200600257

[128] Fresco P, Borges F, Diniz C, Marques MP. New insights on the anticancer properties of dietary polyphenols. Med Res Rev. 2006; 26(6):747-66. doi: 10.1002/med.20060

[129] Meeran SM, Katiyar SK. Proanthocyanidins inhibit mitogenic and survival-signaling in vitro and tumor growth in vivo. Front Biosci. 2008; 13:887-97. doi: 10.2741/2729

[130] Aneja R, Odoms K, Denenberg AG, Wong HR. Theaflavin, a black tea extract, is a novel anti-inflammatory compound. Crit Care Med. 2004; 32(10):2097-103. doi: 10.1097/01.ccm.0000142661.73633.15

[131] Aggarwal BB, Shishodia S. Molecular targets of dietary agents for prevention and therapy of cancer. Biochem Pharmacol. 2006; 71(10):1397-421. doi: 10.1016/j.bcp.2006.02.009

[132] Porath D, Riegger C, Drewe J, Schwager J. Epigallocatechin-3-gallate impairs chemokine production in human colon epithelial cell lines. The Journal of Pharmacology and Experimental Therapeutics. 2005; 315(3):1172-1180. DOI: 10.1124/jpet.105.090167

[133] Collins SJ, Gallo RC, Gallagher RE. Continuous growth and differentiation of human myeloid leukaemic cells in suspension culture. Nature. 1977; 270, 347-349.

[134] Dalton WT, Ahearn MJ Jr, McCredie KB, Freireich EJ, Stas, SA, Trujillo JM. HL-60 cell line was derived from a patient with FAB-M2 and not FAB-M3. Blood. 1988; 71(1), 242-247

[135] Breitman TR, Selonick SE, Collins SJ. Induction of differentiation of the human promyelocytic leukemia cell line (HL-60) by retinoic acid. Proceedings of the National Academy of Sciences. 1980; 77(5), 2936-2940

[136] Chun EM, Park YJ, Kang HS, Cho HM, Jun DY, Kim YH. Expression of the apolipoprotein C-II gene during myelomonocytic differentiation of human leukemic cells. J Leukoc Biol. 2001; 69(4):645-50

[137] Andersson LC, Nilsson K, Gahmberg CG. K562-a human erythroleukemic cell line. International Journal of Cancer. 1979; 23(2), 143-147

[138] Klein E, VÃ¡nky F, Ben-Bassat H, Neumann H, Ralph P, Zeuthen J, Polliack A. Properties of the K562 cell line, derived from a patient with chronic myeloid leukemia. International Journal of Cancer. 1976; 18(4), 421-431

[139] Calabretta B, Perrotti D. The biology of CML blast crisis. Blood. 2004; 103(11), 4010-4022

[140] Bianchi N, Ongaro F, Chiarabelli C, Gualandi L, Mischiati C, Bergamini P, Gambari R. Induction of erythroid differentiation of human K562 cells by

cisplatin analogs. Biochemical Pharmacology. 2000; 60(1), 31-40

[141] Tetteroo PA, Massaro F, Mulder A, Schreuder-van Gelder R, von dem Borne AE. Megakaryoblastic differentiation of proerythroblastic K562 cell-line cells. Leukemia Research. 1984; 8(2), 197-206

[142] Elia G, Santoro MG. Regulation of heat shock protein synthesis by quercetin in human erythroleukaemia cells. Biochem J. 1994; 300 (Pt 1)(Pt 1): 201-9. doi: 10.1042/bj3000201

[143] Wei YQ, Zhao X, Kariya Y, Fukata H, Teshigawara K, Uchida A. Induction of apoptosis by quercetin: involvement of heat shock protein. Cancer Res. 1994; 54(18):4952-7

[144] Bestwick CS, Milne L. Influence of galangin on HL-60 cell proliferation and survival. Cancer Lett. 2006; 243(1):80-9. doi: 10.1016/j.canlet.2005.11.025

[145] Wang CC, Chen LG, Yang LL. Antitumor activity of four macrocyclic ellagitannins from Cuphea hyssopifolia. Cancer Lett. 1999; 140(1-2):195-200. doi: 10.1016/s0304-3835(99)00071-3

[146] Ko WG, Kang TH, Lee SJ, Kim YC, Lee BH. Effects of luteolin on the inhibition of proliferation and induction of apoptosis in human myeloid leukaemia cells. Phytother Res. 2002; 16(3):295-8. doi: 10.1002/ptr.871

[147] Cheng AC, Huang TC, Lai CS, Pan MH. Induction of apoptosis by luteolin through cleavage of Bcl-2 family in human leukemia HL-60 cells. Eur J Pharmacol. 2005; 509(1):1-10. doi: 10.1016/j.ejphar.2004.12.026

[148] Budhraja A, Gao N, Zhang Z, Son YO, Cheng S, Wang X, Ding S, Hitron A, Chen G, Luo J, Shi X. Apigenin induces apoptosis in human leukemia cells and exhibits anti-leukemic activity in vivo. Mol Cancer Ther. 2012; 11(1):132-42. doi: 10.1158/1535-7163.MCT-11-0343

[149] Woo KJ, Jeong YJ, Park JW, Kwon TK. Chrysin-induced apoptosis is mediated through caspase activation and Akt inactivation in U937 leukemia cells. Biochemical and Biophysical Research Communications. 2004; 325(4):1215-1222. DOI: 10.1016/j.bbrc.2004.09.225

[150] Monasterio A, Urdaci MC, Pinchuk IV, López-Moratalla N, Martínez-Irujo JJ. Flavonoids induce apoptosis in human leukemia U937 cells through caspase- and caspase-calpain-dependent pathways. Nutr Cancer. 2004;50(1):90-100. doi: 10.1207/s15327914nc5001_12

[151] Goto H, Yanagimachi M, Goto S, Takeuchi M, Kato H, Yokosuka T, Kajiwara R, Yokota S. Methylated chrysin reduced cell proliferation, but antagonized cytotoxicity of other anticancer drugs in acute lymphoblastic leukemia. Anticancer Drugs. 2012; 23(4):417-25. doi: 10.1097/CAD.0b013e32834fb731

[152] Britschi A, Simon HU, Tobler A, Fey MF, Tschan MP. Epigallocatechin-3-gallate induces cell death in acute myeloid leukaemia cells and supports all-trans retinoic acid-induced neutrophil differentiation via death-associated protein kinase 2. Br J Haematol. 2010; 149(1):55-64. doi: 10.1111/j.1365-2141.2009.08040.x

[153] Goker B, Caliskan C, Onur Caglar H, Kayabasi C, Balci T, Erbaykent Tepedelen B, Aygunes D, Yilmaz Susluer S, Mutlu Z, Selvi Gunel N, Korkmaz M, Saydam G, Gunduz C, Biray Avci C. Synergistic effect of ponatinib and epigallocatechin-3-gallate induces apoptosis in chronic myeloid leukemia cells through altering expressions of cell cycle regulatory genes. J BUON. 2014; 19(4):992-8

[154] Okada N, Tanabe H, Tazoe H, Ishigami Y, Fukutomi R, Yasui K,

Isemura M. Differentiation-associated alteration in sensitivity to apoptosis induced by (-)-epigallocatechin-3-O-gallate in HL-60 cells. Biomed Res. 2009; 30(4):201-6. doi: 10.2220/biomedres.30.201

[155] Hou DX, Tong X, Terahara N, Luo D, Fujii M. Delphinidin 3-sambubioside, a Hibiscus anthocyanin, induces apoptosis in human leukemia cells through reactive oxygen species-mediated mitochondrial pathway. Arch Biochem Biophys. 2005; 440(1):101-9. doi: 10.1016/j.abb.2005.06.002

[156] Tseng YH, Chiou SS, Weng JP, Lin PC. Curcumin and tetrahydrocurcumin induce cell death in Ara-C-resistant acute myeloid leukemia. Phytother Res. 2019; 33(4):1199-1207. doi: 10.1002/ptr.6316

[157] Mutlu Altundağ E, Yılmaz AM, Koçtürk S, Taga Y, Yalçın AS. Synergistic Induction of Apoptosis by Quercetin and Curcumin in Chronic Myeloid Leukemia (K562) Cells. Nutr Cancer. 2018; 70(1):97-108. doi: 10.1080/01635581.2018.1380208

[158] Htay, H. H., Tsubouchi, R., Haneda, M., Murakami, K., & Yoshino, M. Induction of apoptosis of HL60 cells by gallic acid derivatives. *Biomedical Research, 2002; 23*(3), 127-134

[159] Hagiwara Y, Kasukabe T, Kaneko Y, Niitsu N, Okabe-Kado J. Ellagic acid, a natural polyphenolic compound, induces apoptosis and potentiates retinoic acid-induced differentiation of human leukemia HL-60 cells. Int J Hematol. 2010; 92(1):136-43. doi: 10.1007/s12185-010-0627-4

[160] Salimi A, Roudkenar MH, Sadeghi L, Mohseni A, Seydi E, Pirahmadi N, Pourahmad J. Ellagic acid, a polyphenolic compound, selectively induces ROS-mediated apoptosis in cancerous B-lymphocytes of CLL patients by directly targeting mitochondria. Redox Biol. 2015; 6:461-471. doi: 10.1016/j.redox.2015.08.021

[161] Chen JH, Shao Y, Huang MT, Chin CK, Ho CT. Inhibitory effect of caffeic acid phenethyl ester on human leukemia HL-60 cells. Cancer Lett. 1996; 108(2):211-4. doi: 10.1016/s0304-3835(96)04425-4

[162] Chen YJ, Shiao MS, Wang SY. The antioxidant caffeic acid phenethyl ester induces apoptosis associated with selective scavenging of hydrogen peroxide in human leukemic HL-60 cells. Anticancer Drugs. 2001; 12(2):143-9. doi: 10.1097/00001813-200102000-00008

[163] Jin UH, Song KH, Motomura M, Suzuki I, Gu YH, Kang YJ, Moon TC, Kim CH. Caffeic acid phenethyl ester induces mitochondria-mediated apoptosis in human myeloid leukemia U937 cells. Mol Cell Biochem. 2008; 310(1-2):43-8. doi: 10.1007/s11010-007-9663-7

[164] Simoni D, Roberti M, Invidiata FP, Aiello E, Aiello S, Marchetti P, Baruchello R, Eleopra M, Di Cristina A, Grimaudo S, Gebbia N, Crosta L, Dieli F, Tolomeo M. Stilbene-based anticancer agents: resveratrol analogues active toward HL60 leukemic cells with a non-specific phase mechanism. Bioorg Med Chem Lett. 2006; 16(12):3245-8. doi: 10.1016/j.bmcl.2006.03.028

[165] Baur JA, Sinclair DA. Therapeutic potential of resveratrol: the in vivo evidence. Nat Rev Drug Discov. 2006; 5(6):493-506. doi: 10.1038/nrd2060

[166] Gambini J, Inglés M, Olaso G, Lopez-Grueso R, Bonet-Costa V, Gimeno-Mallench L, Mas-Bargues C, Abdelaziz KM, Gomez-Cabrera MC, Vina J, Borras C. Properties of Resveratrol: In Vitro and In Vivo Studies about Metabolism, Bioavailability, and Biological Effects in Animal Models and

Humans. Oxid Med Cell Longev. 2015; 2015:837042. doi: 10.1155/2015/837042

[167] Chowdhury SA, Kishino K, Satoh R, Hashimoto K, Kikuchi H, Nishikawa H, Shirataki Y, Sakagami H. Tumor-specificity and apoptosis-inducing activity of stilbenes and flavonoids. Anticancer Res. 2005; 25(3B):2055-63

[168] Wang CC, Chen LG, Yang LL. Cuphiin D1, the macrocyclic hydrolyzable tannin induced apoptosis in HL-60 cell line. Cancer Lett. 2000 Feb 28;149(1-2):77-83. doi: 10.1016/s0304-3835(99)00344-4

[169] Kawaii S, Lansky EP. Differentiation-promoting activity of pomegranate (Punica granatum) fruit extracts in HL-60 human promyelocytic leukemia cells. J Med Food. 2004; 7(1):13-8. doi: 10.1089/109662004322984644

[170] Tamagawa K, Fukushima S, Kobori M, Shinmoto H, Tsushida T. Proanthocyanidins from barley bran potentiate retinoic acid-induced granulocytic and sodium butyrate-induced monocytic differentiation of HL60 cells. Biosci Biotechnol Biochem. 1998; 62(8):1483-7. doi: 10.1271/bbb.62.1483

[171] Britschgi A, Simon HU, Tobler A, Fey MF, Tschan MP. Epigallocatechin-3-gallate induces cell death in acute myeloid leukaemia cells and supports all-trans retinoic acid-induced neutrophil differentiation via death-associated protein kinase 2. Br J Haematol. 2010; 149(1):55-64. doi: 10.1111/j.1365-2141.2009.08040.x

[172] Li T, Fan GX, Wang W, Li T, Yuan YK. Resveratrol induces apoptosis, influences IL-6 and exerts immunomodulatory effect on mouse lymphocytic leukemia both in vitro and in vivo. Int Immunopharmacol. 2007; 7(9):1221-31. doi: 10.1016/j.intimp.2007.05.008

[173] Weinbrenner T, Fitó M, De la Torre R, Saez GT, Rijken P, Tormos C, Coolen S, Albaladejo MF, Abanades S, et al. Olive oils high in phenolic compounds modulate oxidative/antioxidative status in men. J Nutr. 2004;134:2314-21

[174] Marrugat J, Covas MI, Fitó M, Schroder H, Miro-Casas E, Gimeno E, Lopez-Sabater MC, De la Torre R, Farrè M, SOLOS Investigators. Effect of differing phenolic content in dietary olive oils on lipids and LDL oxidation. Eur J Nutr. 2004;43:140-7

[175] Paller CJ, Ye X, Wozniak PJ, Gillespie BK, Sieber PR, Greengold RH, Stockton BR, Hertzman BL, Efros MD, Roper RP, Liker HR, Carducci MA. A randomized phase II study of pomegranate extract for men with rising PSA following initial therapy for localized prostate cancer. Prostate Cancer Prostatic Dis. 2013; 16(1):50-5. doi: 10.1038/pcan.2012.20

[176] Sharma RA, McLelland HR, Hill KA, Ireson CR, Euden SA, Manson MM, Pirmohamed M, Marnett LJ, Gescher AJ, Steward WP. Pharmacodynamic and pharmacokinetic study of oral Curcuma extract in patients with colorectal cancer. Clin Cancer Res. 2001; 7(7):1894-900

[177] ClinicalTrials.gov [Internet]. National and Kapodistrian University of Athens and Harokopio University. 2020 Jan 2. Identifier NCT04215367, Dietary Intervention With High Phenolic EVOO in CLL; 2020 Jan 2 [cited 2021 Jun 6]; Available from: https://clinicaltrials.gov/ct2/show/NCT04215367

Chapter 17

Can Polyphenols be Used as Anti-Inflammatory Agents against Covid-19 (SARS-CoV-2)-Induced Inflammation?

Volkan Gelen, Abdulsamed Kükürt, Emin Şengül, Ömer Faruk Başer and Mahmut Karapehlivan

Abstract

Covid-19 is the causative agent of a beta coronavirus that causes severe inflammatory pneumonia, so excessive inflammation is considered a risk factor for the disease. In Covid-19 disease, an inflammatory response develops in the body. It has been reported as a result of various studies that this response causes damage to various organs and tissues, especially the lungs. According to reports, cytokine storms are largely responsible for death in such patients. Some of the consequences of severe inflammation and cytokine storms include acute respiratory distress syndrome, acute lung injury, and multiple organ dysfunction syndromes. Many studies are showing that there may be various agents to prevent or treat these effects of Covid-19 disease. Some of these agents are phenolic compounds. Phenolic compounds are the most abundant substances in vegetables and fruits. Inflammasomes, their function. It has been stated that phenolic compounds inhibit inflammation by inhibiting cytosolic multiprotein complexes that assemble in response to cytosolic pathogen-associated molecular patterns (PAMPs), and damage-associated molecular patterns (DAMPs) to form active forms of IL-1β and IL-18. It suggested that Apigenin, Resveratrol, Morin, and Silymarin an anti-inflammatory, antioxidant, anti-viral, and anti-microbial compound could be a potential therapeutic agent for severe inflammation from Covid-19.

Keywords: anti-inflammatory, apigenin, covid-19, resveratrol, morin, silymarin

1. Introduction

Treatment of Covid-19 (SARS-CoV-2) disease which is characterized by acute respiratory syndrome and continues widely in the world and causes a serious number of deaths, is among the discussed topics [1]. The clinical symptoms of this disease, such as fatigue, headache, diarrhea, cough, fever, and dyspnea, occur after an incubation period (about 5–7 days) [2]. In some patients, respiratory failure, acute respiratory distress syndrome (ARDS), or multiple organ failure may take shape. In most patients, it can be asymptomatic or mild [1–3]. However, some conditions such as old age cardiovascular diseases, chronic kidney disease, diabetes, hypertension, and chronic obstructive pulmonary disease (COPD) predispose to severe Covid-19 disease. The covid-19 disease can cause several complications such as COPD, coagulation dysfunction, septic shock, metabolic acidosis, cardiac arrhythmia, heart failure, liver dysfunction, kidney damage, or secondary infections [2]. Many studies have noted that inflammation is a natural defense mechanism against various pathogens and its association with oxidative stress in various pathological conditions [4–12]. There is a great deal of evidence that systemic hyper-inflammation plays a role in the occurrence of lung and multi-organ failure in Covid-19 patients [1]. High levels of ferritin, fibrinogen, D-dimer, interleukin-6 (IL-6), C-reactive protein, and procalcitonin were found in the sera of Covid-19 patients. It has been determined that these laboratory and clinical signs are associated with macrophage activation syndrome and hyper inflammation [3]. Macrophages and monocytes play an important role in the inflammatory reactions that accompany severe Covid-19 infection [13]. These immune cells secrete large amounts of proinflammatory cytokines (Tumor necrosis factor-alpha (TNF-α), interleukin-1β (IL-1β), interleukin-6 (IL-6), interleukin-8 (IL-8)) typical for critically ill patients with Covid-19 [14–17]. Cytokine excessive release in Covid-19 disease causes acute heart damage, acute respiratory failure, or the development of multi-organ failure and worsening of the situation [2]. For this reason, the use of anti-inflammatory agents in the treatment of Covid-19 disease plays a very important role in preventing the severity of the disease. Identifying new agents in addition to existing agents will contribute to developing new strategies to overcome the pandemic [1].

Apigenin is a yellow-colored flavone with a closed formula of C15H10O5 and a molecular weight of 270.24 g/mol. It is chemically known as 4′,5,7-trihydroxyflavone or 5,7-dihydroxy-2-(4-hydroxyphenyl)-4H-1-benzopyren-4-one (**Figure 1A**). Apigenin is mostly found in the flowers of *Matricaria chamomilla* (German chamomile) from the Asteraceae family, but it is also abundant in *Apium graveolens* (celery) leaves, *Allium sativum L.* (garlic) and *Petroselinum crispum L.* (parsley) species [18–20]. It was determined that it was found at a higher rate in the leaf part of the plants [21]. Resveratrol is in the structure of 3,4′,5 trihydroxystilbene and has two isomers as trans and cis isomers (**Figure 1B**). Trans isomers have higher biological activities than cis isomers. The chemical structure of resveratrol is similar to the synthetic estrogen, diethylstilbestrol. It is also the main component of a molecular family that includes glucosides and polymers, and has

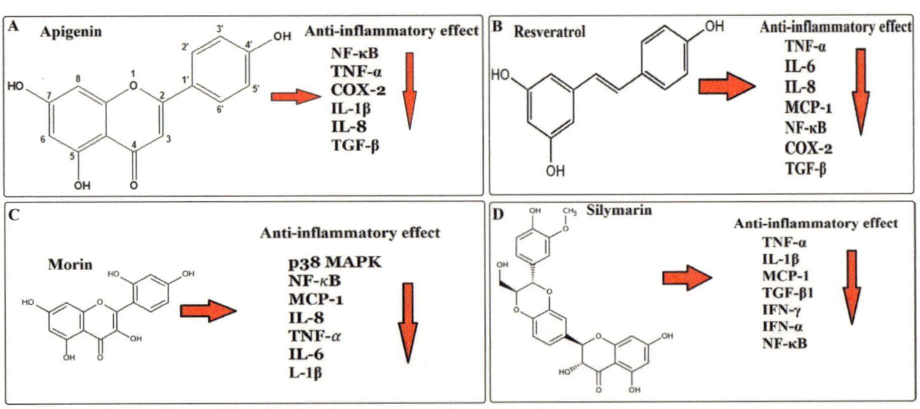

Figure 1.
Chemical structures and anti-inflammatory effects of related phenolic compounds. A (Apigenin), B (Resveratrol), C (Morin), and D (Silymarin).

been shown to be found in grapevines, peanuts, and mulberries [22, 23]. Morine has been named a natural polyphenol (3, 5, 7, 20, 40-pentahydroxyflavone). The hydroxyl groups at the 3 and 4′ positions in morin can be electrochemically oxidized to form the corresponding quinones (**Figure 1C**) [24, 25]. The chemical formula of Silymarin is C25H22O10 (**Figure 1D**). The main ingredient of silymarin is silybin. Flavolignans constitute 70–80% of silymarin. 20–30% consists of polyphenolic components. Silydiadin, silychristin and isosilybin make up the remaining 40% of the compound [26, 27].

Polyphenols are plant-derived phenolic compounds. Polyphenols have been characterized by extensive biological activities in a variety of mammalian systems. These compounds act as free radical scavengers and exhibiting anti-mutagenic, anti-inflammatory, antioxidants, and antiviral effects [28]. In various studies conducted recently, the use of phenolic compounds as anti-inflammatory and antioxidant has become widespread [29–37]. Some factors such as the cheapness of flavonoids and the absence of side effects also increase their usability [38]. As such, the use of flavonoids as an anti-inflammatory will be effective in suppressing hyper-inflammation caused by Covid-19 disease, which is quite common and quite deadly worldwide and thus decreases the mortality rates by reducing the severity of the disease. Therefore, in this study, it will be emphasized that Apigenin, Resveratrol, Morin, and Silymarin, which are natural flavonoids, can be potential agents that can suppress hyper-inflammation in Covid-19 patients.

2. Virus morphology and way of attachment to the cell

When you look at the morphological structure of the Coronavirus, the Virus is a member of a single-stranded (+) RNA enveloped virus family. This virus was identified by scientists in the United States and the United Kingdom in the sixties as a causative agent of the common cold in humans [39]. Coronaviruses are pleomorphic or spherical and are 80–120 nm in diameter. As a result of research conducted in 1968, electron microscope images determined that this family has virus crown-like structures resembling "solar corona", whose name is derived from the Latin word "coronavirus" [40]. It has been determined that there are four main structural proteins in the structure of the coronavirus. These proteins: The first is the trimeric Spike glycoprotein, localized on the surface of the virus envelope and required for virus entry into cells, and this

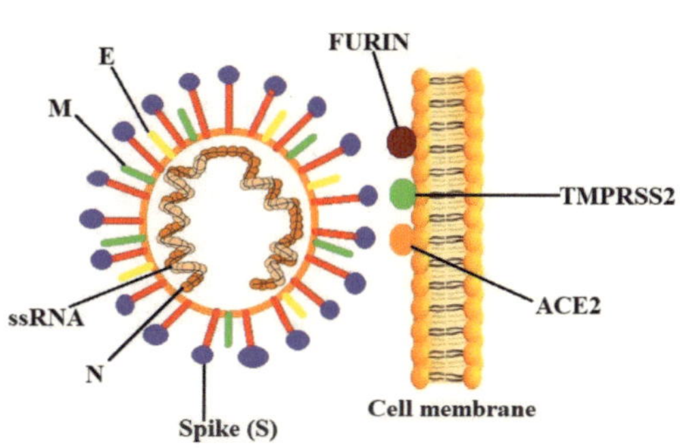

Figure 2.
The structure of the coronavirus and its entryway into the cell. ssRNA: Single-stranded RNA, N: Nucleocapsid proteins. S: The trimeric spike glycoprotein. It recognizes the ACE2 receptor on the cell membrane after cleavage and activation by two serine proteases: FURIN and TMPRSS2. M: Membrane or matrix protein, E: Small envelope protein.

protein is named S. The second is called matrix or membrane protein, and is named M. The third is the small envelope protein required for the collection and release of virions and is named E. The fourth is called the nucleocapsid protein and is named N, which helically binds to the RNA genome forming the symmetrical nucleocapsid (**Figure 2**) [41]. However, homology modeling revealed that the new virus has a similar receptor binding domain structure (RBD) to that of SARS-CoV, despite amino acid variation at several key residues. It was hypothesized that the virus entered cells using the Angiotensin Receptor Enzyme-2 (ACE2) protein, which is widely expressed in the kidney, heart, lung, testis, and gastrointestinal tract [42]. ACE2 is a membrane-bound protein responsible for the reduction of Ang II to Ang 1–7 [43]. Several steps are required to initiate and complete the Covid-19 infection cycle: These steps 1. Recognize and bind to the cellular receptor (s). The second is that changes occur in the structure and proteolysis of the S protein. The third is fusion to the cellular membrane. The fourth is the entry of the virus into host cells by endocytosis [44]. In host cells, the virus uses an endogenous cellular mechanism to replicate viral RNA. It is well known that the spiky glycoprotein S located on the surface of the viral phospholipidic membrane is very important for coronavirus pathogenesis and infection. The life cycle of SARS-CoV-2 begins with the RBD of the S protein in contact with the ACE2 receptor in cells [45, 46]. It was determined that two host serine proteases, TMPRSS2 and the endo-protease Furin, were involved in this event (**Figure 2**).

3. Cytokine storm and inflammatory pathways associated with Covid-19

In Covid-19, clinical deterioration and a high risk of death may be associated with the cytokine storm that develops as a result of the inflammatory response stimulated [14]. Blood levels of various cytokines such as monocyte chemoattractant protein 1 (MCP1), and interferon-alpha (IFN-α), IL-1ß, interferon-gamma (IFN-γ), induced protein 10 (IP10) increased in Covid-19 patients. Also, it has been determined that IL-10, IL-7, IL-2, macrophage inflammatory protein 1-α, IP10, granulocyte colony-stimulating factor (G-CSF), MCP1, and TNF-α levels are quite high in severe Covid-19 patients [47]. It was determined that those who had the severe Covid-19 disease and died had very high IL-6 levels [48]. This shows the importance of cytokines in the severe course of Covid-19. In a study, cytokine storm was divided into two stages [49]. The first stage is an immunodeficiency state. The secondary stage is an overactive immune state that appears to be a clinical manifestation of a cytokine storm [50]. Experimental studies have determined that the effect of coronavirus on cytokines stimulates the delayed secretion of type I and III IFNs including IFN-α/ß in the early stage and the excessive secretion of pro-inflammatory cytokines from mononuclear macrophages in the next stage [51]. It has been shown that impaired type 1 IFN responses and hyperinflammatory responses involving IL-6 and TNF-α occur with the low level of IFN activity and down-regulation of IFN-induced genes [52]. Based on this information, it is understood why COPD accompanies severe Covid-19. Failure of the immune response in the initial period of infection causes general hyper-inflammation of the lung leading to acute lung injury and COPD. In some studies, it has been determined that there is a genetic predisposition that makes some patients more sensitive to cytokine storms in Covid-19 disease [53–57].

4. Flavonoids and phenolic compounds in COVID-19

Various studies have shown that the use of some natural substances with anti-inflammatory properties can prevent inflammation-induced tissue damage [58–65]. Flavonoids are one of these natural ingredients. Flavonoids and phenolic compounds have significant anti-oxidant, anti-bacterial, anti-cancer, immunomodulatory, and anti-inflammatory abilities [66–71]. Additionally, flavonoids and phenolic compounds exhibit a strong anti-viral capability in multiple pathologies [72–75]. More importantly, flavonoids and phenolic compounds have been determined to exhibit immunomodulatory and anti-viral activities against coronaviruses [76, 77]. Therefore, the anti-viral abilities of flavonoids and phenolic compounds may also apply in the current Covid-19 pandemic. The potentially beneficial role of polyphenols in the Covid-19 pandemic is currently a widely debated topic [78–80]. One of the recommended targets of SARS-CoV-2 treatments is the ACE-2 receptor [81]. Moreover, the biological activity of flavonoids and phenolic compounds predetermines their efficacy in the modulation of the immune and inflammatory pathways of the pathology associated with SARS-CoV-2.

4.1 Anti-inflammatory effects of Apigenin

Among the flavonoids, Apigenin is one of the most widely found and most studied phenolics in the plant kingdom. Apigenin is commonly found in many fruits, vegetables, and plants, mainly in parsley, celery, artichoke, onion, spinach, chamomile, thyme, basil, wheat sprouts, and oranges [82, 83]. Apigenin has been found to have an anti-inflammatory effect by suppressing lipopolysaccharide (LPS)-induced *Cyclooxygenase-2* (COX-2) and nitric oxide synthetase-2 activities and expressions in mouse macrophages [84]. It has been reported that Apigenin regulates different anti-inflammatory pathways including PI3K/Akt and p38/Mitogen-activated protein kinase (MAPK), also prevents inhibitory kB (IKB) degradation and nuclear translocation of *nuclear factor kappa B* (NF-κB), and reduced COX-2 activity [85–87]. Inhibition of NF-κB activation occurs by preventing the inhibitory kB (IkB) degradation [88]. Nitric oxide (NO) is an important intra and intercellular signal molecule that plays a role in the regulation of physiological and pathophysiological mechanisms. It relaxes vascular smooth muscles, inhibits platelet aggregation, stimulates angiogenesis, lowers blood pressure, transmits neuronal signals, activates macrophages, and can act as a cytotoxic agent in inflammation [89, 90]. The anti-inflammatory properties of apigenin are formed by the dose-dependent suppression of the inflammatory mediator's prostaglandin and NO by inhibition of *inducible nitric oxide synthase* (iNOS), and COX-2 in BV-2 murine microglial cells [91]. It has been reported that Apigenin exerts most of its effects in both human and murine cell culture models through interactions with signaling molecules in the 3 major MAPK pathways (p38, JNK, and ERK) [92, 93]. Apigenin suppresses TNF-α-induced NF-κB transcriptional activation [94]. Apigenin suppresses LPS -induced NF-κB activity in lung tissue, reduces the infiltration of inflammatory cells, and reduces the accumulation of chemotactic factors [95]. Apigenin inhibits the production of proinflammatory cytokines IL-1β, IL-8, and TNF-α by suppressing NF-κB activity in mouse macrophages stimulated by LPS, and that apigenin suppresses inflammation and modulates immune responses [96]. It has been determined that dietary apigenin administration to ovalbumin-sensitized BALB/c mice inhibits the release of interleukin-4 (IL-4) from Th2 cells [97].

Apigenin has been reported to have anti-inflammatory potential by suppressing T helper cell-1 and -2 (Th1-Th2) related chemokine production by human monocyte cells by modulating mitogen-activated protein kinase pathways [86]. Prophylactic administration of apigenin in mice with intratracheal acute lung injury caused increased levels of IL-6, IL-1β, and TNF-α, leukocyte count, and percentage of neutrophils in bronchoalveolar lavage fluid by suppressing COX-2 and NF-κB pathways. It has an anti-inflammatory effect by reducing it [98]. In a study investigating the effects and molecular mechanisms of apigenin on cisplatin-induced kidney damage in mice; It has been shown that apigenin improves the pathological changes induced by cisplatin in a dose-dependent manner and decreases the increases in TNF-α, IL-1β, and transforming growth factor-beta (TGF-β) mRNA expressions in a dose-dependent manner [99]. Apigenin also strongly suppressed CD40, TFN-α, and IL-6 production levels in murine microglia through inhibition of IFN- γ induced phosphorylation of signal transducer and activator of transcription 1 (STAT1) [100]. Apigenin has demonstrated neuroprotective properties against apoptosis induced by endoplasmic reticulum stress in HT22 murine hippocampal neuronal cells through reduction of ROS, mitochondrial damage, and endoplasmic reticulum-stress-related proteins [101].

4.2 Anti-inflammatory effects of Resveratrol

Resveratrol is a polyphenolic compound found in peanuts, carob molasses, blueberries, grapes, and red wine [102, 103]. It has been reported in various studies that it stimulates nitric oxide synthesis while suppressing oxidative stress [104–109]. Besides, studies have reported that resveratrol plays a protective role in major respiratory diseases such as ARDS, COPD, and allergic inflammation [110]. These diseases increase the susceptibility to Covid-19 disease and the probability of death increases [22]. In vitro studies have reported that resveratrol has anti-inflammatory and antioxidant properties in COPD patients. It has been reported that resveratrol reduces glutathione (GSH) consumption by activating the nuclear factor (erythroid derivative 2) derivative (Nrf2) pathway, which is a redox-sensitive transcription factor [111]. In other studies, resveratrol has also been reported to inhibit COPD-associated inflammatory mediators such as TNF-α, IL-6, IL-8, MCP-1, and granulocyte-macrophage colony-stimulating factor (GM-CSF) and reduced nuclear NF-κB expression [112–114]. In another study conducted using cigarette smoke, resveratrol reduced the histological damage of the lung, lowered pro-inflammatory protein levels TNF-α, IL-17, IL-6, and transforming growth factor TGF-beta, and prevented airway remodeling, and It has been reported to reduce excessive mucus secretion [115]. Resveratrol SIRT1 and PGC-1 have also been reported to reduce inflammation and restructuring of small airways in lung tissue by increasing α expression [116]. Consistent with in vitro data, resveratrol treatment has been reported to increase superoxide dismutase (SOD) and catalase (CAT) activities and glutathione (GSH) levels, and in addition to preventing NF-κB translocation and binding activity to the nucleus [111]. In-vivo studies conducted over the past few years have shown that resveratrol can effectively control asthma in mouse models [110]. Resveratrol exerts its anti-inflammatory effect by suppressing the passage of inflammatory cells, especially eosinophils, to bronchoalveolar lavage fluid (BALF) and lung tissue by suppressing AHR [101]. Total immunoglobulin E (IgE) and ovalbumin (OVA) specific IgE levels were reported to be decreased in the OVA-induced asthma model and decreased levels of TNF-*α*, IL-4 and IL-5 cytokines [110]. In another study, it was reported that TGF and TGF-B1/phosphorylated Smad 2/3 receptor expression levels decreased significantly as a result of treatment with resveratrol [117, 118]. Currently, there is still no effective treatment for COPD, but resveratrol has been added to existing treatment protocols for its beneficial effect against lung damage and its beneficial effect in reducing inflammation through several possible molecular mechanisms. Resveratrol reduces myeloperoxidase protein expression and activity in the treatment of structural changes in the lung, reducing pulmonary edema, improving lung functions, decreasing neutrophil infiltration. Regarding cytokines, resveratrol IL-1ß, IL-18, IL-6; It has been reported that COX-2 and macrophage inflammatory protein-1 (MIP-1) significantly modulate BALF and systemic TNF-α. Considering the findings obtained in these studies, it is thought that resveratrol can prevent inflammation caused by Covid-19 as in other respiratory system diseases.

4.3 Anti-inflammatory effects of Morin

Morin, a natural bioflavonoid belonging to the family Moraceae, is found in the structure of many plants commonly used in alternative medicine [119, 120]. Morin has antihyperglycemic and hepatoprotective effects. Morin's anti-inflammatory effects have been reported in many studies [121–124]. MAPK signaling pathway plays an important role in the transcription of some proinflammatory cytokines as eotaxin-1, MCP-1, and IL-8, which leads to a worsened airway inflammation [125]. Morin attenuates inflammation by regulating MAPK signaling pathway in ovalbumin-induced airway inflammation [126]. Eotaxin-1 provides the delivery of eosinophils to airways and could cause tissue injury and heavy inflammation. It is known that eotaxin-1 expression is regulated by TNF-*α* via the p38 MAPK/NF-*κ*B signaling pathways [127]. MCP-1 stimulates histamine release from basophils and TNF-*α* stimulates MCP-1 secretion from airway smooth muscle cells [128]. IL-8 has proinflammatory effects on immune cells and stimulates the infiltration of neutrophils into the airways in asthma [129]. In the study has been determined that Morin significantly reduced the increases in eotaxin-1, MCP1, and IL-8 in human and Morin inhibits lung inflammation with these effects [123]. NF-κB pathway activation is considered to respond to oxidative stress [130] and leads to an increase in the expression of inflammatory cytokines and consequently, inflammation develops. It has been reported that Morin administration caused NF-κB inhibition in the Parkinson model which was experimentally created in mice [124]. It has been determined that Morin prevents inflammatory damage by regulating the NF-κB pathway in indomethacin-induced gastric ulcer [131]. In another study was determined that Morin attenuates the expression of inflammatory cytokine with downregulation of MAPK and NF-κB signaling pathways in LPS-induced primary bovine mammary epithelial cells [132]. Tian et al. has been determined that Morin has hepatoprotective effects by inhibiting to NF-κB/TLR4 signaling pathway in LPS/D-galactosamine-induced acute liver injury [127]. Also, Morin prevents inflammation by inhibiting PI3K/AKT/NF-κB signaling pathway the cigarette smoke-induced lung inflammation in mice. Morin significantly inhibits the levels of proinflammatory cytokines as TNF-α, and IL-1β and reduces the inflammatory cells, including neutrophils and macrophages [133]. NF-κB-signaling pathway is a crucial regulator of proinflammatory cytokines such as TNF-α, IL-6, L-1β, and levels of proinflammatory cytokines increase inflammation. It was observed that Morin has protective effects by inhibiting proinflammatory cytokines in LPS-induced mastitis [134]. TNF-α, IL-6, and IL-1β promote the development of lung fibrosis and proinflammatory cytokines expression has increased in bleomycin-induced pulmonary fibrosis [135]. Morin inhibited the increase of inflammatory cells such as eosinophils, macrophages, and lymphocytes and reduces total IL-4, IL-13, and IgE levels in OVA-induced mice. Overexpression of Th2 and IgE cytokines causes eosinophil-rich inflammation, mucus hypersecretion, and increased collagen deposition in the lungs. Therefore, Morin prevents mucus hypersecretion, inflammatory cell infiltration, and collagen deposition/fibrosis. In another study reported that TNF-α, IL-6, IL-18, and IL-1β levels importantly increased in bronchoalveolar lavage fluid after LPS-induced Acute Lung Injury, and Morin treatment markedly decreased to these raises due to its anti-inflammatory effects [136].

4.4 Anti-inflammatory effects of Silymarin

The main content of Silybin, which is a complex compound obtained from the seeds of *Silybum marianum* is composed of slybin, and it contains isosilybin, silychristin, silydianin and taxifolin, which is a flavonoid, in its structure [137]. Milk thistle extract is noted to be anti-carcinogenic in human prostate cancer. It is stated that silibinin can be anti-carcinogenic through insulin-like growth factor receptor type I (IGF-I), epidermal growth factor receptor, and NF-κB signaling [138]. Silymarin regulates inflammatory mediators such as interleukins, TNF-α, and inhibits NF-κB activation [139–142]. Silymarin inhibits the inflammatory cytokines (IFN-γ, IFN-α, and IL-1β) [27]. It is well known that silymarin generally has antioxidative and chemo-protective properties in the liver. It is thought that the hepatoprotective activity of silymarin is due to its antioxidant and membrane stabilizing properties. Silymarin shows hepatoprotective activity by inhibiting the function of Kupfer cells and the formation of leukotriene. Silymarin shows strong antioxidant, cytoprotective, anti-inflammatory, and anti-carcinogenic activities [143, 144]. In a rat sepsis model, Silymarin has been shown to suppress transcription of the transporter gene that binds NF-κB. It was also shown in the same study that silymarin showed anti-inflammatory activity by inhibiting prostaglandin-E2 and cyclooxygenase-2 in macrophages stimulated with LPS [145]. Silymarin reduces the increase in TNF-α, IL-1β, MCP-1, TGF-β1, and CRP levels with oxidative stress caused by sodium nitrite, and also, DNA fragmentation due to decrease in cytochrome C oxidase and increase in caspase-3 activity significantly. It has been reported to improve [146]. In the Methotrexate-induced nephrotoxicity model, it was noted that the increase in NF-KB, TNF-α, IL-6, and IL - 1β levels caused by Methotrexate decreased silymarin and prevented inflammatory responses by suppressing the activation of COX-2 and iNOS. Also, silymarin has been reported to play a protective role against apoptosis and autophagy by reducing caspase-3 and light chain 3D activities.

5. Conclusion

As a result, a more effective treatment method has not yet been found against the highly contagious and deadly coronavirus epidemic. This situation encourages

DOI: http://dx.doi.org/10.5772/intechopen.98684

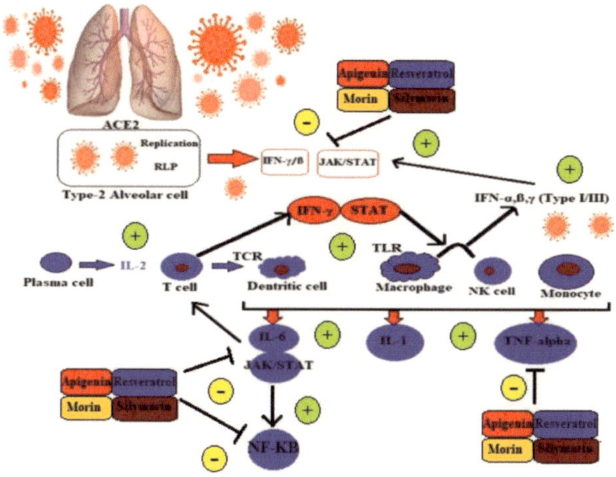

Figure 3.

Possible anti-inflammatory role of Apigenin, resveratrol, Morin, and Silymarin in the treatment of Covid-19. IFN: Interferon; IL: Interleukin; JAK/STAT; Janus kinase-signal transducer and activator of transcription; NK: Natural killer; RLR: Retinoic acid-inducible gene-1-like receptor; TCR: T cell receptor; TLR: Toll-like receptor; TNF-α: Tumor necrosis factor-alpha.

scientists to look for alternatives to human coronavirus infections. Looking at various studies, it is known that Apigenin, Resveratrol, Morin, and Silymarin play an important role in relieving inflammation in various tissues. It is seen that coronavirus causes severe inflammation in various tissues and death after tissue damage. In this context, we believe that the flavonoids and phenolic compounds mentioned can be an alternative to the agents currently used in preventing/treating these adverse effects caused by coronavirus (**Figure 3**).

Author details

Volkan Gelen[1*], Abdulsamed Kükürt[2], Emin Şengül[3], Ömer Faruk Başer[4] and Mahmut Karapehlivan[4]

1 Department of Physiology, Faculty of Veterinary Medicine, Kafkas University, Kars, Turkey

2 Department of Biochemistry, Faculty of Veterinary Medicine, Kafkas University, Kars, Turkey

3 Department of Physiology, Faculty of Veterinary Medicine, Atatürk University, Erzurum, Turkey

4 Department of Biochemistry, Faculty of Medicine, Kafkas University, Kars, Turkey

*Address all correspondence to: gelen_volkan@hotmail.com

References

[1] Huang Q, Wu X, Zheng X, Luo S, Xu S, Weng J. Targeting inflammation and cytokine storm in COVID-19. Pharmacol Res 2020;159:105051. https://doi.org/10.1016/j.phrs.2020.105051.

[2] Costela-Ruiz VJ, Illescas-Montes R, Puerta-Puerta JM, Ruiz C, Melguizo-Rodríguez L. SARS-CoV-2 infection: The role of cytokines in COVID-19 disease. Cytokine Growth Factor Rev 2020;54:62-75. https://doi.org/10.1016/j.cytogfr.2020.06.001.

[3] Soy M, Keser G, Atagündüz P, Tabak F, Atagündüz I, Kayhan S. Cytokine storm in COVID-19: pathogenesis and overview of anti-inflammatory agents used in treatment. Clin Rheumatol 2020;39:2085-94. https://doi.org/10.1007/s10067-020-05190-5.

[4] Gelen V, Şengül E, Yıldırım S, Çelebi F, Çınar A. Ratlarda Cyclophosphamide ile İndüklenen Hemorajik Sistitte Mesane Kontraktilitesi ve Histopatolojisi Üzerine Rutin'in Etkileri. Atatürk Üniversitesi Vet Bilim Derg 2018;13:337-46. https://doi.org/10.17094/ataunivbd.370609.

[5] Gelen V, Şengül E. Antioxidant, anti-inflammatory and antiapoptotic effects of naringin on cardiac damage induced by cisplatin. Indian J Tradit Knowl 2020;19:459-65.

[6] Gelen V, Şengül E, Yıldırım S, Atila G. The protective effects of naringin against 5-fluorouracil-induced hepatotoxicity and nephrotoxicity in rats. Iran J Basic Med Sci 2018;21:404-10. https://doi.org/10.22038/ijbms.2018.27510.6714.

[7] Gelen V, Şengül E. Hematoprotective Effect of Naringin on 5-FU Toxicity in Rats. Chem Reseach 2018;3:127-30.

[8] Kükürt A, Gelen V, Başer ÖF, Deveci HA, Karapehlivan M. Thiols: Role in Oxidative Stress-Related Disorders. Lipid Peroxidation [Working Title], IntechOpen; 2021. https://doi.org/10.5772/intechopen.96682.

[9] Kara A, Gedikli S, Sengul E, Gelen V, Ozkanlar S. Oxidative Stress and Autophagy. Free Radicals Dis., InTech; 2016. https://doi.org/10.5772/64569.

[10] Volkan Gelen, Emin Şengül, Dursun Ali Çınar. The effects of rutin and quercetin on ECG parameters in 5-FU-induced cardiotoxicity rat model. World J Adv Res Rev 2021;9:253-7. https://doi.org/10.30574/wjarr.2021.9.3.0104.

[11] Sengul E, Gelen V. Protective effects of naringin in indomethacin-induced gastric ulcer in rats. GSC Biol Pharm Sci 2019;8:006-14. https://doi.org/10.30574/gscbps.2019.8.2.0132.

[12] Kaya İ, Deveci HA, Karapehlivan M, Kükürt A. Investigation of oxidative stress index in pyridine and ellagic acid treated mice. Eurasian J Vet Sci 2015;31:148-148. https://doi.org/10.15312/EurasianJVetSci.2015310971.

[13] Merad M, Martin JC. Pathological inflammation in patients with COVID-19: a key role for monocytes and macrophages. Nat Rev Immunol 2020;20:355-62. https://doi.org/10.1038/s41577-020-0331-4.

[14] Mehta P, McAuley DF, Brown M, Sanchez E, Tattersall RS, Manson JJ. COVID-19: consider cytokine storm syndromes and immunosuppression. Lancet 2020;395:1033-4. https://doi.org/10.1016/S0140-6736(20)30628-0.

[15] Chen G, Wu D, Guo W, Cao Y, Huang D, Wang H, et al. Clinical and

immunological features of severe and moderate coronavirus disease 2019. J Clin Invest 2020;130. https://doi.org/10.1172/JCI137244.

[16] Zhang W, Zhao Y, Zhang F, Wang Q, Li T, Liu Z, et al. The use of anti-inflammatory drugs in the treatment of people with severe coronavirus disease 2019 (COVID-19): The Perspectives of clinical immunologists from China. Clin Immunol 2020;214:108393. https://doi.org/10.1016/j.clim.2020.108393.

[17] Gelen V, Kükürt A, Şengül E. Role of the Renin-Angiotensin-Aldosterone System in Various Disease Processes: An Overview. Renin-Angiotensin Aldosterone Syst. [Working Title], IntechOpen; 2021. https://doi.org/10.5772/intechopen.97354.

[18] Li B, Robinson DH, Birt DF. Evaluation of Properties of Apigenin and [G-3H]Apigenin and Analytic Method Development. J Pharm Sci 1997;86:721-5. https://doi.org/10.1021/js960383s.

[19] Venigalla M, Gyengesi E, Münch G. Curcumin and Apigenin - novel and promising therapeutics against chronic neuroinflammation in Alzheimer's disease. Neural Regen Res 2015;10:1181. https://doi.org/10.4103/1673-5374.162686.

[20] Miean KH, Mohamed S. Flavonoid (myricetin, quercetin, kaempferol, luteolin, and apigenin) content of edible tropical plants. J Agric Food Chem 2001;49. https://doi.org/10.1021/jf000892m.

[21] Justesen U, Knuthsen P, Leth T. Quantitative analysis of flavonols, flavones, and flavanones in fruits, vegetables and beverages by high-performance liquid chromatography with photo-diode array and mass spectrometric detection. J Chromatogr A 1998;799:101-10. https://doi.org/10.1016/S0021-9673(97)01061-3.

[22] Kaldas MI, Walle UK, Walle T. Resveratrol transport and metabolism by human intestinal Caco-2 cells. J Pharm Pharmacol 2010;55:307-12. https://doi.org/10.1211/002235702612.

[23] Bradamante S, Barenghi L, Villa A. Cardiovascular protective effects of resveratrol. Cardiovasc Drug Rev 2004;22. https://doi.org/10.1111/j.1527-3466.2004.tb00139.x.

[24] Khokhar TS, Memon S, Memon AA, Bhatti AA, Bhatti AA. Improved Solubility of Morin Using p -sulphonatocalix[4]arene as Encapsulating Agent: HPLC Analysis and their Molecular Modelling. Polycycl Aromat Compd 2020;40:609-21. https://doi.org/10.1080/10406638.2018.1464037.

[25] Jacob V, Hagai T, Soliman K. Structure-Activity Relationships of Flavonoids. Curr Org Chem 2011;15:2641-57. https://doi.org/10.2174/138527211796367309.

[26] Fraschini F, Demartini G, Esposti D. Pharmacology of silymarin. Clin Drug Investig 2002;22. https://doi.org/10.2165/00044011-200222010-00007.

[27] Kren V, Walterova D. Silybin and silymarin - new effects and applications. Biomed Pap 2005;149:29-41. https://doi.org/10.5507/bp.2005.002.

[28] Levy E, Delvin E, Marcil V, Spahis S. Can phytotherapy with polyphenols serve as a powerful approach for the prevention and therapy tool of novel coronavirus disease 2019 (COVID-19)? Am J Physiol Metab 2020;319:E689-708. https://doi.org/10.1152/ajpendo.00298.2020.

[29] Karamese M, Guvendi B, Karamese SA, Cinar I, Can S, Erol HS, et al. The protective effects of epigallocatechin gallate on lipopolysaccharide-induced hepatotoxicity: An in vitro study on Hep3B cells. Iran J Basic

Med Sci 2016;19:483-9. https://doi.org/10.22038/ijbms.2016.6932.

[30] Şengül E, Gelen V, Gedikli S, Özkanlar S, Gür C, Çelebi F, et al. The protective effect of quercetin on cyclophosphamide-Induced lung toxicity in rats. Biomed Pharmacother 2017;92:303-7. https://doi.org/10.1016/j.biopha.2017.05.047.

[31] Gelen V, Şengül E, Gedikli S, Atila G, Uslu H, Makav M. The protective effect of rutin and quercetin on 5-FU-induced hepatotoxicity in rats. Asian Pac J Trop Biomed 2017;7: 647-53. https://doi.org/10.1016/j.apjtb.2017.06.013.

[32] Gelen V, Şengül E, Gedikli S, Gür C, Özkanlar S. Therapeutic effect of quercetin on renal function and tissue damage in the obesity induced rats. Biomed Pharmacother 2017;89:524-8. https://doi.org/10.1016/j.biopha.2017.02.057.

[33] Sengul E, Gelen V, Gedikli S. Cardioprotective Activities of Quercetin and Rutin in Sprague Dawley Rats Treated with 5-Fluorouracil. J Anim Plant Sci 2020;31:423-31. https://doi.org/10.36899/JAPS.2021.2.0231.

[34] Ogun M, Ozcan A, Karaman M, Merhan O, Ozen H, Kukurt A, et al. Oleuropein ameliorates arsenic induced oxidative stress in mice. J Trace Elem Med Biol 2016;36:1-6. https://doi.org/10.1016/j.jtemb.2016.03.006.

[35] Kaya İ, Kaya MM, Kükürt A, Özcan A, Karaman M, Deveci HA, et al. Effect of Ellagic Acid on Some Oxidative Stress Parameters and Cyclooxygenase-2 Reactivity in Mice with Experimental Gastric Injury. Japanese J Gastroenterol Hepatol 2019;2:1-9.

[36] Deveci HA, Nur G, Kırpık MA, Kükürt A, Karapehlivan M. Akut klorprifos-etil toksikasyonunda kafeik asit fenetil esterin total sialik asit, nitrik oksit ve aminotransferaz düzeylerine etkisi. Kafkas Üniversitesi Fen Bilim Enstitüsü Derg 2016;9:39-45.

[37] Deveci HA, Karapehlivan M, Kaya İ, Kükürt A, Alpay M. Protective role of caffeic acid phenethyl ester against to chlorpyrifos-ethyl acute poisoning. Ankara Üniversitesi Vet Fakültesi Derg 2015;62:255-60. https://doi.org/10.1501/Vetfak_0000002689.

[38] Ginwala R, Bhavsar R, Chigbu DI, Jain P, Khan ZK. Potential Role of Flavonoids in Treating Chronic Inflammatory Diseases with a Special Focus on the Anti-Inflammatory Activity of Apigenin. Antioxidants 2019;8:35. https://doi.org/10.3390/antiox8020035.

[39] Kapikian AZ, James HD, Kelly SJ, Dees JH, Turner HC, McIntosh K, et al. Isolation from Man of 'Avian Infectious Bronchitis Virus-like' Viruses (Coronaviruses) similar to 229E Virus, with Some Epidemiological Observations. J Infect Dis 1969;119: 282-90. https://doi.org/10.1093/infdis/119.3.282.

[40] Zhong N, Zheng B, Li Y, Poon L, Xie Z, Chan K, et al. Epidemiology and cause of severe acute respiratory syndrome (SARS) in Guangdong, People's Republic of China, in February, 2003. Lancet 2003;362:1353-8. https://doi.org/10.1016/S0140-6736(03)14630-2.

[41] Fung TS, Liu DX. Human Coronavirus: Host-Pathogen Interaction. Annu Rev Microbiol 2019;73:529-57. https://doi.org/10.1146/annurev-micro-020518-115759.

[42] Lu R, Zhao X, Li J, Niu P, Yang B, Wu H, et al. Genomic characterisation and epidemiology of 2019 novel coronavirus: implications for virus origins and receptor binding. Lancet

2020;395:565-74. https://doi.org/10.1016/S0140-6736(20)30251-8.

[43] Başer ÖF, Kükürt A, Karapehlivan M. Oksidatif stresin azaltılmasında anjiyotensin dönüştürücü enzimin rolü. In: Evereklioğlu C, editor. Sağlık Bilim. Teor. ve Araştırmalar II, Gece Kitaplığı; 2020, p. 243-53.

[44] Pillay TS. Gene of the month: the 2019-nCoV/SARS-CoV-2 novel coronavirus spike protein. J Clin Pathol 2020;73:366-9. https://doi.org/10.1136/jclinpath-2020-206658.

[45] Hoffmann M, Kleine-Weber H, Pöhlmann S. A Multibasic Cleavage Site in the Spike Protein of SARS-CoV-2 Is Essential for Infection of Human Lung Cells. Mol Cell 2020;78:779-784.e5. https://doi.org/10.1016/j.molcel.2020.04.022.

[46] Hoffmann M, Kleine-Weber H, Schroeder S, Krüger N, Herrler T, Erichsen S, et al. SARS-CoV-2 Cell Entry Depends on ACE2 and TMPRSS2 and Is Blocked by a Clinically Proven Protease Inhibitor. Cell 2020;181:271-280.e8. https://doi.org/10.1016/j.cell.2020.02.052.

[47] Huang C, Wang Y, Li X, Ren L, Zhao J, Hu Y, et al. Clinical features of patients infected with 2019 novel coronavirus in Wuhan, China. Lancet 2020;395:497-506. https://doi.org/10.1016/S0140-6736(20)30183-5.

[48] Ruan Q, Yang K, Wang W, Jiang L, Song J. Clinical predictors of mortality due to COVID-19 based on an analysis of data of 150 patients from Wuhan, China. Intensive Care Med 2020;46:846-8. https://doi.org/10.1007/s00134-020-05991-x.

[49] McGonagle D, Sharif K, O'Regan A, Bridgewood C. The Role of Cytokines including Interleukin-6 in COVID-19 induced Pneumonia and Macrophage Activation Syndrome-Like Disease. Autoimmun Rev 2020;19:102537. https://doi.org/10.1016/j.autrev.2020.102537.

[50] Ye Q, Wang B, Mao J. The pathogenesis and treatment of the `Cytokine Storm' in COVID-19. J Infect 2020;80:607-13. https://doi.org/10.1016/j.jinf.2020.03.037.

[51] Blanco-Melo D, Nilsson-Payant BE, Liu W-C, Uhl S, Hoagland D, Møller R, et al. Imbalanced Host Response to SARS-CoV-2 Drives Development of COVID-19. Cell 2020;181:1036-1045.e9. https://doi.org/10.1016/j.cell.2020.04.026.

[52] Hadjadj J, Yatim N, Barnabei L, Corneau A, Boussier J, Smith N, et al. Impaired type I interferon activity and inflammatory responses in severe COVID-19 patients. Science (80-) 2020;369:718-24. https://doi.org/10.1126/science.abc6027.

[53] Jenkins MR, Rudd-Schmidt JA, Lopez JA, Ramsbottom KM, Mannering SI, Andrews DM, et al. Failed CTL/NK cell killing and cytokine hypersecretion are directly linked through prolonged synapse time. J Exp Med 2015;212:307-17. https://doi.org/10.1084/jem.20140964.

[54] Vastert SJ, van Wijk R, D'Urbano LE, de Vooght KMK, de Jager W, Ravelli A, et al. Mutations in the perforin gene can be linked to macrophage activation syndrome in patients with systemic onset juvenile idiopathic arthritis. Rheumatology 2010;49:441-9. https://doi.org/10.1093/rheumatology/kep418.

[55] Wulffraat NM. Reduced perforin expression in systemic juvenile idiopathic arthritis is restored by autologous stem-cell transplantation. Rheumatology 2003;42:375-9. https://doi.org/10.1093/rheumatology/keg074.

[56] Trouillet-Assant S, Viel S, Gaymard A, Pons S, Richard J-C,

Perret M, et al. Type I IFN immunoprofiling in COVID-19 patients. J Allergy Clin Immunol 2020;146:206-208.e2. https://doi.org/10.1016/j.jaci.2020.04.029.

[57] The COVID-19 Host Genetics Initiative. The COVID-19 Host Genetics Initiative, a global initiative to elucidate the role of host genetic factors in susceptibility and severity of the SARS-CoV-2 virus pandemic. Eur J Hum Genet 2020;28:715-8. https://doi.org/10.1038/s41431-020-0636-6.

[58] Volkan Gelen, Emin Şengül. Protective effects of resveratrol on kidney function tests and renal histopathology in carbon tetrachloride-induced renal toxicity in rats. World J Adv Res Rev 2021;10:156-61. https://doi.org/10.30574/wjarr.2021.10.1.0155.

[59] Sengul E, Gelen V, Yildirim S, Tekin S, Dag Y. The Effects of Selenium in Acrylamide-Induced Nephrotoxicity in Rats: Roles of Oxidative Stress, Inflammation, Apoptosis, and DNA Damage. Biol Trace Elem Res 2021;199:173-84. https://doi.org/10.1007/s12011-020-02111-0.

[60] Gedikli S, Gelen V, Sengul E, Ozkanlar S, Gur C, Agırbas O, et al. Therapeutic Effects of Melatonin On Liver And Kidney Damages In Intensive Exercise Model of Rats. Endocrine, Metab Immune Disord Targets 2015;15:308-14. https://doi.org/10.2174/1871530315666150827103043.

[61] Karamese M, Aydin H, Gelen V, Sengul E, Karamese SA. The anti-inflammatory, anti-oxidant and protective effects of a probiotic mixture on organ toxicity in a rat model. Future Microbiol 2020;15:401-12. https://doi.org/10.2217/fmb-2020-0005.

[62] Gelen V, Gelen SU, Celebi F, Cinar A, Yildirim S, Eser G. The protective effect of *Lactobacillus rhamnosus*, *Lactobacillus fermentum* and lactobacillus brevis against cisplatin-induced hepatic damage in rats. Fresenius Environ Bull 2019;28.

[63] Karamese M, Aydin H, Sengul E, Gelen V, Sevim C, Ustek D, et al. The Immunostimulatory Effect of Lactic Acid Bacteria in a Rat Model. Iran J Immunol 2016;13:220-8. https://doi.org/IJIv13i3A7.

[64] Gedikli S, Ozkanlar S, Gur C, Sengul E, Gelen V. Preventive effects of quercetin on liver damages in high-fat diet-induced obesity. J Histol Histopathol 2017;4:7. https://doi.org/10.7243/2055-091X-4-7.

[65] Sengul E, Gelen S, Yıldırım S, Çelebi F, Çınar A. Probiotic bacteria attenuates cisplatin-induced nephrotoxicity through modulation of oxidative stress, inflammation and apoptosis in rats. Asian Pac J Trop Biomed 2019;9:116. https://doi.org/10.4103/2221-1691.254605.

[66] Liskova A, Koklesova L, Samec M, Smejkal K, Samuel SM, Varghese E, et al. Flavonoids in Cancer Metastasis. Cancers (Basel) 2020;12:1498. https://doi.org/10.3390/cancers12061498.

[67] Abotaleb M, Samuel S, Varghese E, Varghese S, Kubatka P, Liskova A, et al. Flavonoids in Cancer and Apoptosis. Cancers (Basel) 2018;11:28. https://doi.org/10.3390/cancers11010028.

[68] Senthilvel P, Lavanya P, Kumar KM, Swetha R, Anitha P, Bag S, et al. Flavonoid from *Carica papaya* inhibits NS2B-NS3 protease and prevents Dengue 2 viral assembly. Bioinformation 2013;9:889-95. https://doi.org/10.6026/97320630009889.

[69] Andreu L, Nuncio-Jáuregui N, Carbonell-Barrachina ÁA, Legua P, Hernández F. Antioxidant properties and chemical characterization of Spanish *Opuntia ficus-indica* Mill. cladodes and fruits. J Sci Food Agric

2018;98. https://doi.org/10.1002/jsfa.8628.

[70] Meng XH, Liu C, Fan R, Zhu LF, Yang SX, Zhu HT, et al. Antioxidative Flavan-3-ol Dimers from the Leaves of Camellia fangchengensis. J Agric Food Chem 2018;66. https://doi.org/10.1021/acs.jafc.7b04572.

[71] Gelen V, Şengül E, Yıldırım S, Senturk E, Tekin S, Kükürt A. The protective effects of hesperidin and curcumin on 5-fluorouracil–induced nephrotoxicity in mice. Environ Sci Pollut Res 2021. https://doi.org/10.1007/s11356-021-13969-5.

[72] Kang SY, Kang J-Y, Oh M-J. Antiviral activities of flavonoids isolated from the bark of *Rhus verniciflua* stokes against fish pathogenic viruses In Vitro. J Microbiol 2012;50:293-300. https://doi.org/10.1007/s12275-012-2068-7.

[73] Fukuchi K, Okudaira N, Adachi K, Odai-Ide R, Watanabe S, Ohno H, et al. Antiviral and Antitumor Activity of Licorice Root Extracts. In Vivo (Brooklyn) 2016;30:777-86. https://doi.org/10.21873/invivo.10994.

[74] Roschek B, Fink RC, McMichael MD, Li D, Alberte RS. Elderberry flavonoids bind to and prevent H1N1 infection in vitro. Phytochemistry 2009;70:1255-61. https://doi.org/10.1016/j.phytochem.2009.06.003.

[75] Ngwa W, Kumar R, Thompson D, Lyerly W, Moore R, Reid T-E, et al. Potential of Flavonoid-Inspired Phytomedicines against COVID-19. Molecules 2020;25:2707. https://doi.org/10.3390/molecules25112707.

[76] Song J-W, Long J-Y, Xie L, Zhang L-L, Xie Q-X, Chen H-J, et al. Applications, phytochemistry, pharmacological effects, pharmacokinetics, toxicity of Scutellaria baicalensis Georgi. and its probably potential therapeutic effects on COVID-19: a review. Chin Med 2020;15:102. https://doi.org/10.1186/s13020-020-00384-0.

[77] Li T, Li X, Dai T, Hu P, Niu X, Liu C, et al. Binding mechanism and antioxidant capacity of selected phenolic acid - β-casein complexes. Food Res Int 2020;129. https://doi.org/10.1016/j.foodres.2019.108802.

[78] Korkmaz H. Could Sumac Be Effective on COVID-19 Treatment? J Med Food 2020:jmf.2020.0104. https://doi.org/10.1089/jmf.2020.0104.

[79] Hamza M, Ali A, Khan S, Ahmed S, Attique Z, Ur Rehman S, et al. nCOV-19 peptides mass fingerprinting identification, binding, and blocking of inhibitors flavonoids and anthraquinone of *Moringa oleifera* and hydroxychloroquine. J Biomol Struct Dyn 2020:1-11. https://doi.org/10.1080/07391102.2020.1778534.

[80] Solnier J, Fladerer J-P. Flavonoids: A complementary approach to conventional therapy of COVID-19? Phytochem Rev 2020. https://doi.org/10.1007/s11101-020-09720-6.

[81] Muchtaridi M, Fauzi M, Khairul Ikram NK, Mohd Gazzali A, Wahab HA. Natural Flavonoids as Potential Angiotensin-Converting Enzyme 2 Inhibitors for Anti-SARS-CoV-2. Molecules 2020;25:3980. https://doi.org/10.3390/molecules25173980.

[82] Sung B, Chung HY, Kim ND. Role of Apigenin in Cancer Prevention via the Induction of Apoptosis and Autophagy. J Cancer Prev 2016;21:216-26. https://doi.org/10.15430/JCP.2016.21.4.216.

[83] Hostetler GL, Ralston RA, Schwartz SJ. Flavones: Food Sources, Bioavailability, Metabolism, and Bioactivity. Adv Nutr An Int Rev J 2017;8:423-35. https://doi.org/10.3945/an.116.012948.

[84] Liang Y-C, Huang Y-T, Tsai S-H, Lin-Shiau S-Y, Chen C-F, Lin J-K. Suppression of inducible cyclooxygenase and inducible nitric oxide synthase by apigenin and related flavonoids in mouse macrophages. Carcinogenesis 1999;20:1945-52. https://doi.org/10.1093/carcin/20.10.1945.

[85] Lee J-H, Zhou HY, Cho SY, Kim YS, Lee YS, Jeong CS. Anti-inflammatory mechanisms of apigenin: inhibition of cyclooxygenase-2 expression, adhesion of monocytes to human umbilical vein endothelial cells, and expression of cellular adhesion molecules. Arch Pharm Res 2007;30:1318-27. https://doi.org/10.1007/BF02980273.

[86] Huang C-H, Kuo P-L, Hsu Y-L, Chang T-T, Tseng H-I, Chu Y-T, et al. The Natural Flavonoid Apigenin Suppresses Th1- and Th2-Related Chemokine Production by Human Monocyte THP-1 Cells Through Mitogen-Activated Protein Kinase Pathways. J Med Food 2010;13:391-8. https://doi.org/10.1089/jmf.2009.1229.

[87] Lapchak PA, Boitano PD. Effect of the Pleiotropic Drug CNB-001 on Tissue Plasminogen Activator (tPA) Protease Activity in vitro: Support for Combination Therapy to Treat Acute Ischemic Stroke. J Neurol Neurophysiol 2014;5:214.

[88] Madunić J, Madunić IV, Gajski G, Popić J, Garaj-Vrhovac V. Apigenin: A dietary flavonoid with diverse anticancer properties. Cancer Lett 2018;413:11-22. https://doi.org/10.1016/j.canlet.2017.10.041.

[89] Kükürt A. Doğal bir antioksidan olarak propolis tedavisinin koruyucu etkileri. In: Evereklioğlu C, editor. Sağlık Bilim. Teor. ve Araştırmalar II, Gece Kitaplığı; 2020, p. 501-15.

[90] Kükürt A, Kuru M, Karapehlivan M. Nitrik oksit, nitrik oksit sentaz ve dişi üreme sistemindeki rolleri. In: Evereklioğlu C, editor. Sağlık Bilim. Alanında Akad. Çalışmalar - II, Gece Kitaplığı; 2020, p. 113-23.

[91] Ha SK, Lee P, Park JA, Oh HR, Lee SY, Park J-H, et al. Apigenin inhibits the production of NO and PGE2 in microglia and inhibits neuronal cell death in a middle cerebral artery occlusion-induced focal ischemia mice model. Neurochem Int 2008;52:878-86. https://doi.org/10.1016/j.neuint.2007.10.005.

[92] Patel D, Shukla S, Gupta S. Apigenin and cancer chemoprevention: Progress, potential and promise (Review). Int J Oncol 2007;30. https://doi.org/10.3892/ijo.30.1.233.

[93] Shukla S, Gupta S. Apigenin: A Promising Molecule for Cancer Prevention. Pharm Res 2010;27:962-78. https://doi.org/10.1007/s11095-010-0089-7.

[94] Funakoshi-Tago M, Nakamura K, Tago K, Mashino T, Kasahara T. Anti-inflammatory activity of structurally related flavonoids, Apigenin, Luteolin and Fisetin. Int Immunopharmacol 2011;11:1150-9. https://doi.org/10.1016/j.intimp.2011.03.012.

[95] Cardenas H, Arango D, Nicholas C, Duarte S, Nuovo G, He W, et al. Dietary Apigenin Exerts Immune-Regulatory Activity in Vivo by Reducing NF-κB Activity, Halting Leukocyte Infiltration and Restoring Normal Metabolic Function. Int J Mol Sci 2016;17:323. https://doi.org/10.3390/ijms17030323.

[96] Nicholas C, Batra S, Vargo MA, Voss OH, Gavrilin MA, Wewers MD, et al. Apigenin Blocks Lipopolysaccharide-Induced Lethality In Vivo and Proinflammatory Cytokines Expression by Inactivating NF-κB through the Suppression of p65 Phosphorylation. J Immunol 2007;179:7121-7. https://doi.org/10.4049/jimmunol.179.10.7121.

[97] Yano S, Umeda D, Yamashita T, Ninomiya Y, Sumida M, Fujimura Y, et al. Dietary flavones suppresses IgE and Th2 cytokines in OVA-immunized BALB/c mice. Eur J Nutr 2007;46:257-63. https://doi.org/10.1007/s00394-007-0658-7.

[98] Wang J, Liu Y-T, Xiao L, Zhu L, Wang Q, Yan T. Anti-Inflammatory Effects of Apigenin in Lipopolysaccharide-Induced Inflammatory in Acute Lung Injury by Suppressing COX-2 and NF-kB Pathway. Inflammation 2014;37:2085-90. https://doi.org/10.1007/s10753-014-9942-x.

[99] He X, Li C, Wei Z, Wang J, Kou J, Liu W, et al. Protective role of apigenin in cisplatin-induced renal injury. Eur J Pharmacol 2016;789:215-21. https://doi.org/10.1016/j.ejphar.2016.07.003.

[100] Rezai-Zadeh K, Ehrhart J, Bai Y, Sanberg PR, Bickford P, Tan J, et al. Apigenin and luteolin modulate microglial activation via inhibition of STAT1-induced CD40 expression. J Neuroinflammation 2008;5:41. https://doi.org/10.1186/1742-2094-5-41.

[101] Choi AY, Choi JH, Lee JY, Yoon K-S, Choe W, Ha J, et al. Apigenin protects HT22 murine hippocampal neuronal cells against endoplasmic reticulum stress-induced apoptosis. Neurochem Int 2010;57:143-52. https://doi.org/10.1016/j.neuint.2010.05.006.

[102] Ergin K, Yaylalı A. Resveratrol ve Etkileri Üzerine Bir Gözden Geçirme. SDÜ Tip Fak Derg 2013;20:115-20. https://doi.org/10.17343/sdutfd.94607.

[103] Signorelli P, Ghidoni R. Resveratrol as an anticancer nutrient: molecular basis, open questions and promises. J Nutr Biochem 2005;16:449-66. https://doi.org/10.1016/j.jnutbio.2005.01.017.

[104] Barreiro-Hurlé J, Colombo S, Cantos-Villar E. Is there a market for functional wines? Consumer preferences and willingness to pay for resveratrol-enriched red wine. Food Qual Prefer 2008;19:360-71. https://doi.org/10.1016/j.foodqual.2007.11.004.

[105] Fernández-Mar MI, Mateos R, García-Parrilla MC, Puertas B, Cantos-Villar E. Bioactive compounds in wine: Resveratrol, hydroxytyrosol and melatonin: A review. Food Chem 2012;130:797-813. https://doi.org/10.1016/j.foodchem.2011.08.023.

[106] Çetin A, Sağdiç O. A concise review: Antioxidant effects and bioactive constituents of grape. Erciyes Tip Derg 2009;31:369-75.

[107] Cenesiz S, Ciftci G, Gulbahar MY, Yarım GF, Nisbet HO, Kabak Y, et al. Investigation of the Effects of N-Acetyl Cysteine and Resveratrol to Acute Phase Response in Rats with Experimentally Induced Arthritis. Kafkas Univ Vet Fak Derg 2012;18:991-6. https://doi.org/10.9775/kvfd.2012.6900.

[108] Cenesiz S, Yarim GF, Karabulut AB, Ara C. Changing of antioxidant enzyme activity on the biliary obstructed rats treated with resveratrol. Dtsch Tierarztl Wochenschr 2007;114:345-8. https://doi.org/10.2377/0341-6593-114-345.

[109] Işık Bircan C, Merhan O. Kadmiyum Uygulanan Farelerde Oluşturulan Oksidatif Strese Karşı Resveratrolün Koruyucu Etkisinin Araştırılması*. Erciyes Üniversitesi Vet Fakültesi Derg 2020:215-20. https://doi.org/10.32707/ercivet.828324.

[110] de Sá Coutinho D, Pacheco M, Frozza R, Bernardi A. Anti-Inflammatory Effects of Resveratrol: Mechanistic Insights. Int J Mol Sci 2018;19:1812. https://doi.org/10.3390/ijms19061812.

[111] Liu H, Ren J, Chen H, Huang Y, Li H, Zhang Z, et al. Resveratrol Protects against Cigarette Smoke-Induced Oxidative Damage and

Pulmonary Inflammation. J Biochem Mol Toxicol 2014;28:465-71. https://doi.org/10.1002/jbt.21586.

[112] Liu X-J, Bao H-R, Zeng X-L, Wei J-M. Effects of resveratrol and genistein on nuclear factor-κB, tumor necrosis factor-α and matrix metalloproteinase-9 in patients with chronic obstructive pulmonary disease. Mol Med Rep 2016;13:4266-72. https://doi.org/10.3892/mmr.2016.5057.

[113] Knobloch J, Sibbing B, Jungck D, Lin Y, Urban K, Stoelben E, et al. Resveratrol Impairs the Release of Steroid-Resistant Inflammatory Cytokines from Human Airway Smooth Muscle Cells in Chronic Obstructive Pulmonary Disease. J Pharmacol Exp Ther 2010;335:788-98. https://doi.org/10.1124/jpet.110.166843.

[114] Knobloch J, Hag H, Jungck D, Urban K, Koch A. Resveratrol Impairs the Release of Steroid-resistant Cytokines from Bacterial Endotoxin-Exposed Alveolar Macrophages in Chronic Obstructive Pulmonary Disease. Basic Clin Pharmacol Toxicol 2011;109:138-43. https://doi.org/10.1111/j.1742-7843.2011.00707.x.

[115] Chen J, Yang X, Zhang W, Peng D, Xia Y, Lu Y, et al. Therapeutic Effects of Resveratrol in a Mouse Model of LPS and Cigarette Smoke-Induced COPD. Inflammation 2016;39:1949-59. https://doi.org/10.1007/s10753-016-0430-3.

[116] Wang X-L, Li T, Li J-H, Miao S-Y, Xiao X-Z. The Effects of Resveratrol on Inflammation and Oxidative Stress in a Rat Model of Chronic Obstructive Pulmonary Disease. Molecules 2017;22:1529. https://doi.org/10.3390/molecules22091529.

[117] Lee HY, Kim IK, Yoon HK, Kwon SS, Rhee CK, Lee SY. Inhibitory Effects of Resveratrol on Airway Remodeling by Transforming Growth Factor-β/Smad Signaling Pathway in Chronic Asthma Model. Allergy Asthma Immunol Res 2017;9:25. https://doi.org/10.4168/aair.2017.9.1.25.

[118] Royce SG, Dang W, Yuan G, Tran J, El Osta A, Karagiannis TC, et al. Resveratrol has protective effects against airway remodeling and airway hyperreactivity in a murine model of allergic airways disease. Pathobiol Aging Age-Related Dis 2011;1:7134. https://doi.org/10.3402/PBA.v1i0.7134.

[119] Caselli A, Cirri P, Santi A, Paoli P. Morin: A Promising Natural Drug. Curr Med Chem 2016;23:774-91. https://doi.org/10.2174/0929867323666160106150821.

[120] Wu L, Wang Y, Chi G, Shen B, Tian Y, Li Z, et al. Morin reduces inflammatory responses and alleviates lipid accumulation in hepatocytes. J Cell Physiol 2019;234:19785-98. https://doi.org/10.1002/jcp.28578.

[121] Kapoor R, Kakkar P. Protective Role of Morin, a Flavonoid, against High Glucose Induced Oxidative Stress Mediated Apoptosis in Primary Rat Hepatocytes. PLoS One 2012;7:e41663. https://doi.org/10.1371/journal.pone.0041663.

[122] El Sayed NF, Abdallah DM, Awad AS, Ahmed KA, El-Abhar HS. Novel peripheral role of Nurr-1/GDNF/AKT trajectory in carvedilol and/or morin hydrate hepatoprotective effect in a model of hepatic ischemia/reperfusion. Life Sci 2021;273:119235. https://doi.org/10.1016/j.lfs.2021.119235.

[123] Jung H-J, Kim S-J, Song Y-S, Park E-H, Lim C-J. Evaluation of the Antiangiogenic, Anti-inflammatory, and Antinociceptive Activities of Morin. Planta Med 2010;76:273-5. https://doi.org/10.1055/s-0029-1186079.

[124] Lee KM, Lee Y, Chun HJ, Kim AH, Kim JY, Lee JY, et al. Neuroprotective and anti-inflammatory effects of morin

in a murine model of Parkinson's disease. J Neurosci Res 2016;94:865-78. https://doi.org/10.1002/jnr.23764.

[125] Zhang Y, Li X. Lipopolysaccharide-regulated production of bone sialoprotein and interleukin-8 in human periodontal ligament fibroblasts: the role of toll-like receptors 2 and 4 and the MAPK pathway. J Periodontal Res 2015;50:141-51. https://doi.org/10.1111/jre.12193.

[126] Ma Y, Ge A, Zhu W, Liu Y-N, Ji N-F, Zha W-J, et al. Morin Attenuates Ovalbumin-Induced Airway Inflammation by Modulating Oxidative Stress-Responsive MAPK Signaling. Oxid Med Cell Longev 2016;2016:1-13. https://doi.org/10.1155/2016/5843672.

[127] Roh K-B, Jung E, Park D, Lee J. Fumaric acid attenuates the eotaxin-1 expression in TNF-α-stimulated fibroblasts by suppressing p38 MAPK-dependent NF-κB signaling. Food Chem Toxicol 2013;58:423-31. https://doi.org/10.1016/j.fct.2013.05.020.

[128] Patel JK, Clifford RL, Deacon K, Knox AJ. Ciclesonide inhibits TNFα- and IL-1β-induced monocyte chemotactic protein-1 (MCP-1/CCL2) secretion from human airway smooth muscle cells. Am J Physiol Cell Mol Physiol 2012;302:L785-92. https://doi.org/10.1152/ajplung.00257.2011.

[129] Hollander C, Sitkauskiene B, Sakalauskas R, Westin U, Janciauskiene SM. Serum and bronchial lavage fluid concentrations of IL-8, SLPI, sCD14 and sICAM-1 in patients with COPD and asthma. Respir Med 2007;101:1947-53. https://doi.org/10.1016/j.rmed.2007.04.010.

[130] Tian Y, Li Z, Shen B, Zhang Q, Feng H. Protective effects of morin on lipopolysaccharide/ d -galactosamine-induced acute liver injury by inhibiting TLR4/NF-κB and activating Nrf2/HO-1 signaling pathways. Int Immunopharmacol 2017;45:148-55. https://doi.org/10.1016/j.intimp.2017.02.010.

[131] Sinha K, Sadhukhan P, Saha S, Pal PB, Sil PC. Morin protects gastric mucosa from nonsteroidal anti-inflammatory drug, indomethacin induced inflammatory damage and apoptosis by modulating NF-κB pathway. Biochim Biophys Acta - Gen Subj 2015;1850:769-83. https://doi.org/10.1016/j.bbagen.2015.01.008.

[132] Wang J, Guo C, Wei Z, He X, Kou J, Zhou E, et al. Morin suppresses inflammatory cytokine expression by downregulation of nuclear factor-κB and mitogen-activated protein kinase (MAPK) signaling pathways in lipopolysaccharide-stimulated primary bovine mammary epithelial cells. J Dairy Sci 2016;99:3016-22. https://doi.org/10.3168/jds.2015-10330.

[133] Cai B, Gan X, He J, He W, Qiao Z, Ma B, et al. Morin attenuates cigarette smoke-induced lung inflammation through inhibition of PI3K/AKT/NF-κB signaling pathway. Int Immunopharmacol 2018;63:198-203. https://doi.org/10.1016/j.intimp.2018.07.035.

[134] Yu S, Liu X, Yu D, Changyong E, Yang J. Morin Protects LPS-Induced Mastitis via Inhibiting NLRP3 Inflammasome and NF-κB Signaling Pathways. Inflammation 2020;43:1293-303. https://doi.org/10.1007/s10753-020-01208-x.

[135] Shi J, Zhou L, Wang X, Du J, Jiang M, Song Z, et al. KLF2 attenuates bleomycin-induced pulmonary fibrosis and inflammation with regulation of AP-1. Biochem Biophys Res Commun 2018;495:20-6. https://doi.org/10.1016/j.bbrc.2017.10.114.

[136] Tianzhu Z, Shihai Y, Juan D. The Effects of Morin on Lipopolysaccharide-Induced Acute Lung Injury by

Suppressing the Lung NLRP3 Inflammasome. Inflammation 2014;37. https://doi.org/10.1007/s10753-014-9930-1.

[137] Ramasamy K, Agarwal R. Multitargeted therapy of cancer by silymarin. Cancer Lett 2008;269: 352-62. https://doi.org/10.1016/j.canlet.2008.03.053.

[138] Singh RP, Agarwal R. A cancer chemopreventive agent silibinin, targets mitogenic and survival signaling in prostate cancer. Mutat Res Mol Mech Mutagen 2004;555:21-32. https://doi.org/10.1016/j.mrfmmm.2004.05.017.

[139] Flaig TW, Gustafson DL, Su L-J, Zirrolli JA, Crighton F, Harrison GS, et al. A phase I and pharmacokinetic study of silybin-phytosome in prostate cancer patients. Invest New Drugs 2006;25:139-46. https://doi.org/10.1007/s10637-006-9019-2.

[140] Au AY, Hasenwinkel JM, Frondoza CG. Silybin inhibits interleukin-1β-induced production of pro-inflammatory mediators in canine hepatocyte cultures. J Vet Pharmacol Ther 2011;34:120-9. https://doi.org/10.1111/j.1365-2885.2010.01200.x.

[141] de Souza CO, Peraçoli MTS, Weel IC, Bannwart CF, Romão M, Nakaira-Takahagi É, et al. Hepatoprotective and anti-inflammatory effects of silibinin on experimental preeclampsia induced by l-NAME in rats. Life Sci 2012;91:159-65. https://doi.org/10.1016/j.lfs.2012.06.036.

[142] Tyagi A, Agarwal C, Dwyer-Nield LD, Singh RP, Malkinson AM, Agarwal R. Silibinin modulates TNF-α and IFN-γ mediated signaling to regulate COX2 and iNOS expression in tumorigenic mouse lung epithelial LM2 cells. Mol Carcinog 2012;51:832-42. https://doi.org/10.1002/mc.20851.

[143] Wu J-W, Lin L-C, Tsai T-H. Drug–drug interactions of silymarin on the perspective of pharmacokinetics. J Ethnopharmacol 2009;121:185-93. https://doi.org/10.1016/j.jep.2008.10.036.

[144] Karapehlivan M, Kükürt A, Özen H, Kaya İ, Kasacı T, Deveci HA, et al. The beneficial effect of silymarin administration against nicotine-Induced oxidative stress in mice. J Cell Neurosci Oxidative Stress 2016;8:561.

[145] Kang JS, Jeon YJ, Park S-K, Yang K-H, Kim HM. Protection against lipopolysaccharide-induced sepsis and inhibition of interleukin-1β and prostaglandin E2 synthesis by silymarin. Biochem Pharmacol 2004;67:175-81. https://doi.org/10.1016/j.bcp.2003.08.032.

[146] Sherif IO, Al-Gayyar MMH. Antioxidant, anti-inflammatory and hepatoprotective effects of silymarin on hepatic dysfunction induced by sodium nitrite. Eur Cytokine Netw 2013;24:114-21. https://doi.org/10.1684/ecn.2013.0341.

Chapter 18

Polyphenols of *Salvia miltiorrhiza* in Aging-Associated Cardiovascular Diseases and Cancer

Yu-Chen Cheng, Yu-Chiang Hung and Wen-Long Hu

Abstract

With the increasing lifespan of human, cardiovascular diseases (CVDs) and cancer are the main diseases leading to the death in the world. Aging is related to a progressive decline in cardiovascular function and structure. While human body suffer from oxidative stress, reactive oxygen species (ROS) are generated as metabolic by-products, which lead to inactivate proteins, damage nucleic acids, and alter the fatty acids of lipids. The accumulation of this oxidative damage contributes to the development of heart disease, diabetes, chronic inflammatory diseases, and cancer. Polyphenols have been widely studied as an anti-oxidant agent in the world. Danshen, the dried root or rhizome of *Salvia miltiorrhiza* Bunge. is a common Traditional Chinese medicine used in cardiovascular disease and cancer. The main polyphenols in Danshen are phenolic acids (including Salvianolic acids A and B, rosmarinic acid, and their derivatives) and flavonoids. Salvianolic acids have potent anti-oxidative capabilities due to their polyphenolic structure and exhibit cardiovascular protection through mechanisms of ROS scavengers, reduction of leukocyte-endothelial adherence, inhibition of inflammation and indirect regulation of immune function. Salvianolic acids A and B have been reported to owe anti-cancer, anti-inflammatory activities not only through inducing apoptosis, halting cell cycle and adjourning metastasis by targeting multiple deregulated signaling networks of cancer but also sensitizing cancer cells to chemotherapeutic agents.

Keywords: *Salvia miltiorrhiza*, polyphenol, Traditional Chinese medicine, cardiovascular disease, cancer

1. Introduction

With the increasing lifespan of human, cardiovascular diseases (CVDs) and cancer are the main diseases leading to the death in the world [1]. Aging is related to a progressive decline in cardiovascular function and structure. The major CVDs include ischemic heart disease, cardiomyopathy, hypertensive heart disease, atrial fibrillation, stroke, aortic aneurysm, rheumatic heart disease, endocarditis, and peripheral arterial disease [2].

There are many oxidants surrounding our environment even persisted inside the human body. While human body suffer from oxidative stress, reactive oxygen species (ROS) are produced from the respiratory chain and leading the electron transfer. Superoxide radical ($O_2^{\bullet -}$) which dismutates from hydrogen peroxide (H_2O_2) and molecular oxygen (O_2) is a toxic compound after the ROS stimulation [3, 4]. ROS are related to inactivate proteins, damage nucleic acids, and alter the fatty acids of lipids. When those oxidative intracellular components in turn to perturbations in membrane structure and function, those reaction might lead to cell damage. The accumulation of this oxidative damage for a long period of time will leading the development of heart disease, diabetes, chronic inflammatory diseases, cancer, and several neurodegenerative diseases in the aging process.

Polyphenols have been widely studied as an anti-oxidant agent in the world. They are common nutrient antioxidants, mainly derived from fruits, vegetables, tea, coffee, cocoa, mushrooms, beverages, and Traditional Chinese medicine [5, 6]. Traditional Chinese medicine (TCM) are widely used for a long time in Asia countries. Most TCM source come from plants, including leaf, stem, roots or whole plants. Polyphenols are content rich in plants, and so are TCM. Danshen, the dried root or rhizome of *Salvia miltiorrhiza* Bunge. is a common TCM used in cardiovascular disease and cancer [7–9]. Following, we will make a discussion of aging-associated CVDs, cancer and *Salvia miltiorrhiza* (Danshen).

2. The monographs of aging-associated cardiovascular disease, cancer and *Salvia miltiorrhiza*

2.1 Aging-associated cardiovascular disease

The epidemic of CVDs has taken on a global dimension. CVDs now represent more than 30% of all deaths worldwide. According to the World Health Report, CVDs were responsible for 15 million annual deaths worldwide. Especially in developing countries, 9 million deaths every year while 2 million deaths in economies in transition [10].

CVD is positive related to human's age. By 2030, approximately 20% of the population will be aged 65 or older. At that time, the prevalence of CVD will exponential increase due to the fact that additional 27 million people will have hypertension, 8 million coronary heart disease, 4 million stroke and 3 million heart failure [11]. In this age group, CVDs will result in 40% of all deaths and rank as the leading cause and cost triple payment for treatment [12, 13].

Consistently, researchers have found that many of the factors underlying age-related changes in the arteries are also implicated in the development of CVD [14]. The incidence and prevalence of common CVDs such as hypertension, atherosclerosis, coronary and cerebral artery disease are increasing at about age 45 in men and age 55 in women [15]. These diseases may develop to increase in the prevalence of congestive heart failure and stroke during aging.

Aging is accompanied by changes in vascular structure and function, especially in the large arteries [16]. The aging cardiovascular tissues are exemplified by pathological alterations including hypertrophy, altered left ventricular (LV) diastolic function, and diminished LV systolic reverse capacity [17], increased arterial stiffness, and impaired endothelial function.

Endothelial dysfunction [18] is one of the major pathologic change of CVDs, besides, increasing intima media thickness, vascular stiffness [19], vesicular smooth muscle cells hypertrophy and proliferation and increasing vessel diameter are related to aging vessels. Impaired endothelial vasodilation is an early sign of arterial aging before the clinical manifestations of vascular dysfunction [20]. As endothelial cells age, they exhibit a reduction in endothelial nitric oxide synthetase (eNOS) activity, reducing the abundance of nitric oxide (NO) [21]. NO is a vasodilator produced by endothelial cells, and related to regulate vascular tone, inhibiting vascular inflammation, thrombotic events, and aberrant cellular proliferation [22].

Aging has also a remarkable effect on the heart [23]. The number of cardiac myocytes lessen while heart weight gains with age. The functional cardiac cell continued loss come with the lower regenerative activity from 1% to 0.4% per year of age 20 to 75 years [24]. Most of researches found no obvious difference between male and female in increasing atrial volume [25] and cardiac fibrosis [26]. Although one study of cardiac extracellular matrix proteins found that senior women had a greater amount of collagen and other extracellular matrix proteins in the LV than senior men [27]. A recent work has clearly demonstrated that age-dependent mitochondrial DNA damage is an important substrate underpinning the pathophysiology of cardiac arrhythmias [28]. Another important pathological feature associated with aging is the calcification of aortic and mitral valves which triggers stenosis/insufficiency resulting in cardiac pressure/volume overload [29].

2.2 Cancer

Cancer is the second leading cause of death globally after ischemic heart disease, accounting for an estimated 9.6 million deaths, or one in six deaths, in 2018 and accounting for nearly 10 million deaths in 2020, but will likely become the first for nearly 18.63 million deaths in 2060 [30, 31]. Lung, prostate, colorectal, stomach and liver cancer are the most common types of cancer in men, while breast, colorectal, lung, cervical and thyroid cancer are the most common among women. It might prevent about one-third to half of cancer death after modifying or avoiding key risk factors and reduce the cancer burden through early detection of cancer. Prevention is the most important and effective long-term strategy for cancer control [32].

Cancer is a multistage process that involves mutational changes and uncontrolled cell proliferation. The etiology of cancer is linked to environmental and genetic inheritance causes. The physical (such as ultraviolet and ionizing radiation), chemical (such as asbestos, components of tobacco smoke, aflatoxin, and arsenic) and biological carcinogens (infections from certain viruses, bacteria, or parasites) may play a role in tumor genesis. The accumulation of molecular damage in DNA, proteins and lipids during the aging progress is also characterized by an increase in intracellular oxidative stress due to the progressive decrease of the intracellular ROS scavenging [33]. Therefore, oxidative stress and the resulting oxidative damage are important contributors to the formation and progression of cancer [34].

2.3 Bioactive components of *Salvia miltiorrhiza* (Danshen)

Salvia miltiorrhiza (Danshen) belongs to the *Lamiaceae* family. There are at least 49 diterpenoid quinones, more than 36 hydrophilic phenolic acids, and 23 essential oil constituents have been isolated and identified from Danshen [35]. Our previous population-based studies demonstrated that Danshen is the most common herbal drug used to treat ischemic heart disease [36] and ischemic stroke [37].

The predominant bioactive compounds in Danshen contains two major groups of chemicals [8, 38]. The first group includes lipophilic compounds (Terpenoids) such as tanshinone I, tanshinone IIA, acetyltanshinone IIA, cryptotanshinone, isocryptotanshinone, dihydrotanshinone, 15,16-dihydrotanshinone I, and miltirone (**Figure 1b**). These terpenoids possess a wide range of biological activities including antioxidant [39], antibacterial [40], anti-inflammatory [41], antiatherogenic, neuroprotective [42], antitumor [43, 44], and antidiabetic [39] effects.

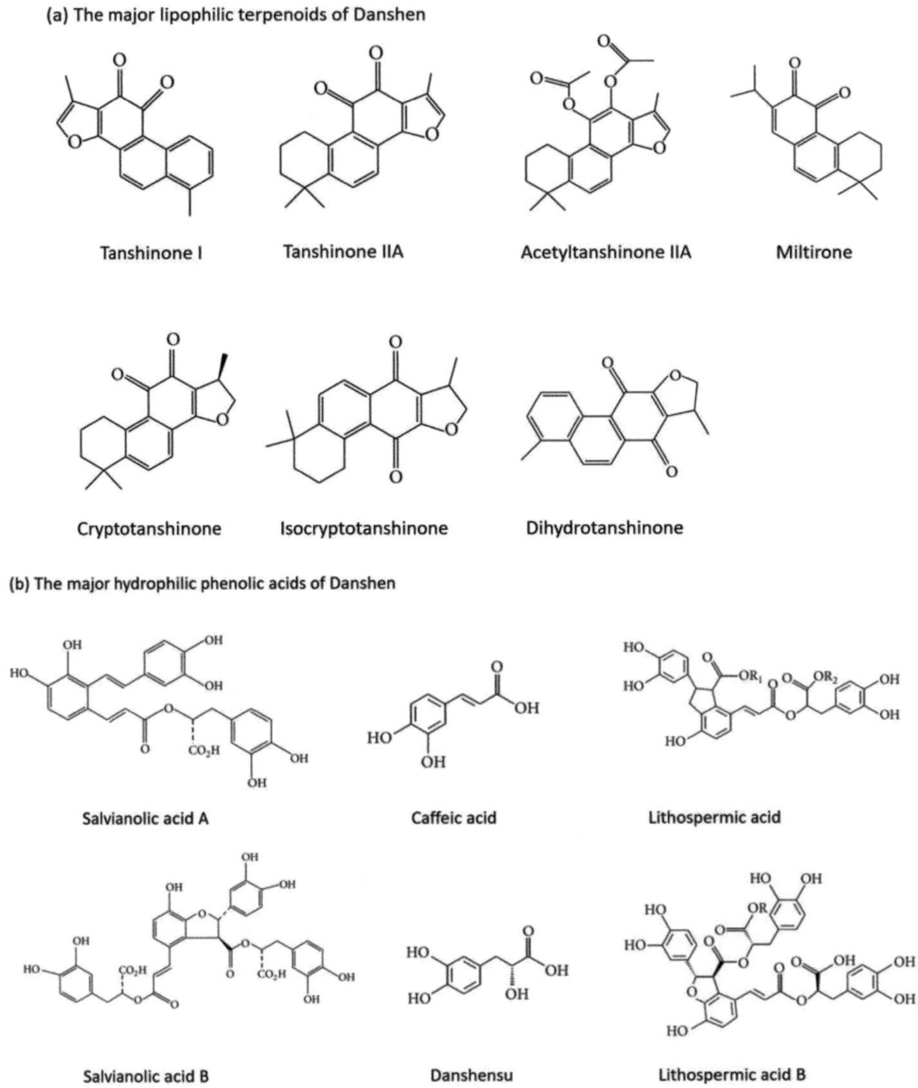

Figure 1.
The chemical structures of major (a) lipophilic terpenoids and (b) hydrophilic phenolic acids of Danshen.

The second group includes the hydrophilic phenolic acids such as caffeic acid, danshensu, salvianolic acid A(SalA), salvianolic acid B(SalB), lithospermic acid and lithospermic acid B (**Figure 1b**). Tanshinones show antibacterial, antioxidant, and antineoplastic activities, whereas phenolic acids possess more antioxidant and anticoagulant activities [45]. The classification of polyphenols mainly includes flavonoids (60%), phenolic acids (30%), and other polyphenols (including stilbenes and lignans) [46]. The main polyphenols in Danshen are phenolic acids (including SalA, SalB, rosmarinic acid, and their derivatives) and flavonoids, which exhibit anti-oxygenation, anti-ischemia–reperfusion injury, anti-thrombosis, anti-tumor, and other therapeutic effects [47]. The main polyphenolic compounds are based on caffeic acid (3,4-dihydroxycinnamic acid), one of the most common phenolic acids, formed from two to four or more caffeic acid units, is one of the most common phenolic acids, frequently exist in fruits, grains, as well as TCM [48].

3. Oxidative stress in aging-associated cardiovascular disease and cancer

3.1 Oxidative stress and aging-associated cardiovascular disease

Decreasing in absolute number of cardiomyocytes due to increased apoptosis and necrosis and decreasing in repopulation of cardiomyocytes from cardiac stem cell reserves were occurred in aging heart [49, 50]. The increase in oxidative stress due to the increase in ROS production with age results in an overall enhancement in the rate of cardiomyocyte death with age. With advancing age, we accumulate mutations in our somatic cells. The expression of such factors as p53, p21, p16, senescence-associated β-galactosidase activity and phosphorylation status of γ-H2Ax are widely used to detect the DNA damage. These biomarkers of aging can be used in cardiac tissue to assess how modulation of longevity genes influences the rate and degree of cardiovascular aging at the cellular level [51, 52].

Many aging-associated CVDs including ischemia/reperfusion, hypertensive heart disease and diabetes are related to oxidative stress and that will exhibit cytokines. In addition, increased ROS-responsive signaling pathways are objective by inflammatory oxidative stress and ROS generative system like unfolded protein response of the endoplasmic reticulum or NADPH oxidase activation [53].

The Apoptosis signal-regulating kinase 1(ASK1)-signalosome regulates p38 MAPK and SAPK/JNK and NFκB signaling networks promote senescence (in vitro) and aging (in vivo, animal models and human cohorts) in response to oxidative stress and inflammation leading to age-associated CVDs. Furthermore, their inhibition delays the onset of these CVDs as well as senescence and aging [53, 54].

The Energy generation from mitochondria is through oxidative phosphorylation and will also increase in ROS production which leads to free radical–imposed damage to macromolecules and cellular component. $p66^{Shc}$, a mitochondrial adaptor, plays an important role in the generation of ROS and as a molecular effector which may explain how aging is connected with CVD and metabolic disease [55]. Several studies show that increased $p66^{Shc}$ expression with time may promote ROS accumulation with subsequent deregulation of pathways implicated in mitochondrial dysfunction, fat accumulation, insulin resistance and diabetes [56–58].

The AMPK-SIRT1 pathway is involved in energy metabolism in cell. The functional AMP-activated protein kinase (AMPK) is a heterotrimer consisting of a catalytic alpha (α), a regulatory gamma (γ) and a scaffolding beta (β) subunit and is activated by low cellular energy status [59]. AMPK activates eNOS, and facilitates autophagy and mitophagy, thus preventing mitochondrial insufficiency, inflammation and cellular death [60]. Sirtuin 1 (SIRT1) is a NAD^+-dependent class III histone deacetylase (HDAC) that mediates the effects of caloric restriction on lifespan and metabolic pathways in various organisms. SIRT1 prevents cardiovascular aging by activating of eNOS [61].

3.2 Oxidative stress in cancer

Cancer is a multistage process defined by at least three stages: initiation, promotion, and progression [62]. ROS from both endogenous and exogenous sources result in increased oxidative stress in the cell. Oxidative stress modulates gene expression of downstream targets involved in DNA repair, cell proliferation and in part through activation or inhibition of transcription factors and second messengers. The role of single nuclear polymorphism for oxidative DNA repair and enzymatic antioxidants is important in determining the potential human cancer risk [34].

ROS regulates tumor development including following steps: transformation [63], survival [64], proliferation [65], invasion [66], metastasis [67], and angiogenesis [68]. One study showed the oxidative stress may be positive correlation with lung cancer staging [69]. In breast carcinomas, 8-OHdG (a most widely used fingerprint of radical attack towards DNA) might be increased 8- to 17-fold in breast primary tumors compared with non-malignant breast tissue [70].

H_2O_2 plays an important role in carcinogenesis because it is capable of diffusing throughout the mitochondria and across cell membranes and producing many types of cellular injury [71]. ROS may down-regulate the expression of the DNA mismatch repair genes (mutS homolog 2 and 6) and inhibit its enzymatic activity. ROS also induce the expression of DNA methyltransferases, leading to a total hypermethylation of the genome [72]. DNA methylation silence several tumor suppressor genes promoter, such as adenomatous polyposis coli (APC), cyclin-dependent kinase inhibitor-2 (CDKN-2), breast cancer susceptibility gene 1 (BRCA1), retinoblastoma protein (Rb), and the DNA mismatch repair gene, human mutL homolog 1 (hMLH1) [73, 74].

However, it is interesting that oxidative stress induces cancer, but also exists opposite condition. When ROS produced in large excess, they endanger the viability of the cancer cells, through the sustained activation of the cell cycle inhibitors [75]. To protect themselves from ROS-mediated toxicity, many types of cancers enhance the intrinsic antioxidant defenses, which make them dependent on the efficacy of a given ROS-detoxifying system. This poses an attractive target for anticancer therapy by using prooxidants or inhibiting of a chosen antioxidant system [76]. Whether ROS promote tumor cell survival or act as anti-tumorigenic agents depends on the cell and tissues, the location of ROS production, and the concentration of individual ROS.

4. Mechanisms of *Salvia miltiorrhiza* in aging-associated CVD and cancer

4.1 Therapeutic properties of Danshen in aging-associated CVD

Salvianolic acids, especially SalA and SalB, have potent anti-oxidative capabilities due to their polyphenolic structure. The cardiovascular protection of salvianolic acids include the following mechanisms: ROS scavengers, reduction of leukocyte-endothelial adherence, inhibition of inflammation and metalloproteinases expression from aortic smooth muscle cells, and indirect regulation of immune function, and also competitive binding to target proteins to interrupt protein–protein interactions [77].

SalA inhibits oxidative stress directly by scavenging the free radicals to improve the endothelial dysfunction [78], vascular smooth muscle cell proliferation [79], pulmonary arterial hypertension [80], and cardiac fibrosis. SalA can chelate Cu^{2+} and inhibit Cu^{2+}-promoted oxidation of low-density lipoprotein to reduce the production of malondialdehyde which is the final product of polyunsaturated fatty acids peroxidation in a cell-free system [81]. Interesting, there is a study showed both Salvianolic acid and tanshinone contribute to the cardioprotective effect of Danshen. Tanshinone mainly inhibits intracellular calcium and cell adhesion pathways at an early stage after ischemic injury whereas Salvianolic acid acts mainly by decreasing apoptosis [82].

SalB protects human endothelial progenitor cells against oxidative stress-mediated dysfunction by modulating Akt/mTOR/4EBP1, p38 MAPK/ATF2, and ERK1/2 signaling pathways and prevents oxidative-induced endothelial dysfunction via down-regulated NADPH oxidase 4 and eNOS expression [18].

Cardiac fibrosis is a chronic harmful result of hypertension which may further advance to heart failure and increased matrix metalloproteinase-9 (MMP-9) contributes to the underlying mechanism. In neonatal cardiac fibroblast, SalA inhibited fibroblast migration, blocked myofibroblast transformation, inhibited secretion of intercellular adhesion molecule (ICAM), interleukin-6 (IL-6) and soluble vascular cell adhesion molecule-1 (sVCAM-1) as well as collagen induced by MMP-9. The inhibition on MMP-9 by SalA was further confirmed in cultured cardiac H9c2 cell overexpressing MMP-9 in vitro and in heart of spontaneously hypertensive rats (SHR) in vivo [83]. SalA targeted transgelin and had a protective effect on myocardium by stabilizing the transgelin-actin complex, modulating the reorganization of the actin cytoskeleton, facilitating F-actin bundling, further enhancing the contractility and blood flows of coronary arteries, and improving outcomes of myocardial ischemia [84]. SalB facilitates angiogenesis and alleviated cardiac fibrosis and cardiac remodeling in diabetic cardiomyopathy by suppressing insulin-like growth factor-binding protein 3 (IGFBP3) [85]. SalB can alleviate Ang II-induced cardiac fibrosis via suppressing the NF-κB pathway in vitro [86]. It is reported that treatment with 5% water-soluble extract of Danshen which contained SalB for 12 weeks lowers blood cholesterol and reduces atherosclerotic plaque formation in diet-induced hypercholesterolemic rabbits, which is associated with its ROS scavenging capacity (**Table 1**) [87].

Homocysteine (Hcy), a by-product of methionine metabolism, may lead to hyperhomocysteinemia which is the risk factors responsible for the development of several vascular diseases (thromboembolism, atherosclerosis, stroke, vascular diseases and dementia). The aqueous extracts of Danshen against vascular atherosclerotic lesions though inhibiting Hcy-induced rat smooth muscle cell line(A10) growth via the PKC/MAPK-dependent pathway, attenuated carbonyl-modification of specific cytoskeleton and chaperone proteins leading to cell type transformation, also, scavenging of ROS and subsequent modulation of protein carbonylation to inhibit cell proliferation [88]. Another study demonstrated the protective effect of Danshen extract against the Hcy-induced adverse effect on human umbilical vein endothelial cell and showed different effectiveness in protection according to the following descending order: Danshen aqueous extract, 3-(3,4-dihydroxyphenyl)-2-hydroxy-propionic acid (Danshensu), protocatechuic acid, catechin and protocatechualdehyde [89]. Danshensu decreases foam cell formation by reducing the expression of TNFα, ICAM-1, and ET-1 while increasing NO production, thus protecting the vascular endothelium from injury [90]. SalA markedly attenuated induction of MKP-3(mitogen-activated protein kinase phosphatases 3) and inhibition of eNOS expression and NO formation under endothelial ischemia/reperfusion condition [91].

Some clinical studies reported that the Danshen preparations in combination with Western medicine were more effective for treatment of various CVDs including angina pectoris, myocardial infarction, hypertension, hyperlipidemia, and pulmonary heart diseases [92]. Our previous series studies showed the most common used single Chinese herbal products which prescribed by TCM Doctors during 2000–2010 in Taiwan is Danshen (16.50% in ischemic stroke; 29.30% in ischemic heart disease; 3.95% in atrill fibrillation; 5.13% in heart failure) [36, 37, 93, 94]. There was nearly one-third lower stroke risk in ischemic heart disease patients with combination TCM than patients with non-TCM treatment (95% CI = 0.11–0.84, P = .02). The higher survival rate (P < .001) and the lower incidence of hemorrhagic stroke (P = .04) in ischemic heart disease patients with TCM treatment was reported [95]. Compared to non-TCM users, the stroke risk was significantly lower in TCM users with atrial fibrillation who were female or younger than 65 years, but not in males, people more than 65 years old, or people with comorbidities [93].

Component	Pathology of CVD	Mechanism	References
Salvianolic acid A	Endothelial dysfunction	⊕ microvascular remodeling	[78]
	Vascular smooth muscle cell proliferation	⊕ p21 expression via cAMP/PKA/CREB signaling cascade	[79]
	Pulmonary arterial hypertension	↓ right ventricular systolic pressure ↓ hypertrophic damage of myocardium, parenchymal injury and collagen deposition in the lungs	[80]
	Lipid oxidation	chelate Cu^{2+} and ⊖ Cu^{2+}-mediated oxidation of LDL ↓ reducing MDA	[81]
	hypertension	⊖ MMP-9	[83]
	myocardial ischemia	stabilize the transgelin-actin complex modulate the reorganization of the actin cytoskeleton ⊕ F-actin bundling, ↑ contractility and blood flows of coronary arteries	[84]
Salvianolic acid b	Endothelial dysfunction	modulating Akt/mTOR/4EBP1, p38 MAPK/ATF2 ↑ ERK1/2 signaling pathways ↓ Nox4 and eNOS	[18]
	Atherosclerotic plaque formation	↓ LDL ⊖ atherosclerotic plaque formation ⊕ scavenging ROS	[87]
	Cardiac fibrosis	⊖ fibroblast migration and myofibroblast transformation ↓ ICAM, IL-6 and sVCAM-1 ⊖ MMP-9 ⊖ NF-κB pathway	[86]
	Diabetic cardiomyopathy	⊕ angiogenesis and cardiac remodeling ↓ cardiac fibrosis ⊖ IGFBP3	[85]

↑: increase; ↓: decrease; ↔: no change; ⊖: inhibit; ⊕: promote. cAMP, cyclic adenosine monophosphate; PKA, protein kinase A; CREB, cAMP-response element binding protein; LDL, low-density lipoprotein; MDA, malondialdehyde; MMP-9, Matrix metallopeptidase 9; Akt, protein kinase B; mTOR, mechanistic target of rapamycin; 4EBP1, Eukaryotic translation initiation factor 4E-binding protein 1; p38 MAPK, mitogen-activated protein kinases; ATF2, Activating Transcription Factor 2; ERK1, extracellular signal-regulated kinase 1; Nox4, NADPH oxidase 4; eNOS, Endothelial Nitric Oxide Synthase; ROS, reactive oxygen species; ICAM, intercellular adhesion molecule; IL-6, interleukin-6; sVCAM-1, soluble vascular cell adhesion molecule-1; NF-κB, nuclear factor kappa-light-chain-enhancer of activated B cells; IGFBP3, insulin-like growth factor-binding protein 3.

Table 1.
The main antioxidative mechanisms of Salvia miltiorrhiza *(Danshen) in CVD.*

One randomized controlled trial showed *Salvia Miltiorrhiza* Depside Salt combined with aspirin is a clinically effective and safe intervention to treat adults aged 35 and older with stable angina pectoris without adverse drug reactions such as bleeding tendency occurred [96].

4.2 Therapeutic properties of Danshen in cancer

SalA and SalB have been reported to owe anti-cancer, anti-inflammatory and cardioprotective activities not only through inducing apoptosis, halting cell cycle and adjourning metastasis by targeting multiple deregulated signaling networks of cancer but also sensitizing cancer cells to chemo-drugs [97].

Acting to protect the organism against these harmful pro-oxidants is a complex system of enzymatic antioxidants (e.g., superoxide dismutase (SOD), glutathione peroxidase, glutathione reductase, catalase) and nonenzymatic antioxidants (e.g., glutathione, vitamins C and D) [98].

SalA elevated ROS levels, downregulated P-glycoprotein, and triggered apoptosis by increasing caspase-3 activity and upregulating Bax expression, while downregulating Bcl-2 expression and disrupting the mitochondrial membrane potential in multidrug resistance MCF-7 human breast cancer cells [99]. In lung cancer, SalA could increase the chemotherapeutic efficacy of cisplatin by enhanced sensitivity to cisplatin in A549/DDP cells mainly through suppression of the c-met/AKT/mTOR signaling pathway [100]. In addition, SalA considerably suppressed the migrative and invasive activity of human NPC cells but not presented cytotoxicity. In SalA-treated NPC cells, the activity and expression of matrix metalloproteinase-2 (MMP-2), a key regulator of cancer cell invasion, were reduced. Additionally, the presence of high concentrations of SalA dramatically abolished the activation of focal adhesion kinase (FAK) and moderately inhibited the phosphorylation of Src and ERK in NPC cells [101].

The anti-tumor effect of SalB is via inhibiting the expression of glucosylceramide and GM3 synthases, and then increases the ceramide accumulation and ceramide-mediated Triple-negative breast cancer cell apoptosis [102]. One study indicated SalB induced cell death and triggered autophagy in HCT116 and HT29 cells in a dose-dependent manner, and it is as a novel autophagy inducer in colorectal cancer cells through the suppression of AKT/mTOR pathway [103]. Besides,

Component	Cancer	Mechanism	References
Salvianolic acid A	Non-small cell lung cancer	↑ efficacy of DDP ⊖c-met/AKT/mTOR signaling pathway	[100]
	Breast cancer	↑ ROS in resistant cells ↑ apoptosis via caspase-3 activity, disrupted mitochondrial membrane potential, ↓ Bcl-2 and ↑ Bax in the resistant cells ↓ P-glycoprotein	[99]
	Nasopharyngeal carcinoma	↓ MMP-2 ⊖ FAK, Src, and ERK pathways	[101]
Salvianolic acid B	Colorectal cancer	⊕ cancer cell death and autophagy ⊖ AKT/mTOR pathway	[103]
	Head and neck carcinoma	⊖ COX-2/PGE-2 pathway ⊕ the promotion of apoptosis ⊕ angiogenesis.	[107]
	Hepatocellular cancer	↓ cytotoxicity of doxorubicin ↓ ROS by enhancing the expression of SOD and decreasing NADPH oxidase	[104]
	Gastric cancer	↓ the resistance to DDP via AKT/mTOR pathway	[105]

↑: increase; ↓: decrease; ↔: no change; ⊖: inhibit; ⊕: promote. ROS, reactive oxygen species; DDP, cisplatin; AKT, protein kinase B; mTOR, mechanistic target of rapamycin; Bcl-2, B-cell lymphoma 2; Bax, Bcl-2-associated X protein; FAK, focal Adhesion Kinase; ERK, extracellular signal regulated kinase; COX-2/PGE-2; SOD; NADPH.

Table 2.
The therapeutic effect mechanism of polyphenols of Salvia miltiorrhiza (Danshen) in common cancers.

SalB reduced the cytotoxicity of doxorubicin through scavenging ROS generated by doxorubicin in HepG2 cells and enhance the expression of SOD and decrease that of NADPH oxidase, which resulted in the elimination of ROS [104]. Sal-B regulated proliferation, epithelial-mesenchymal transition (EMT) and apoptosis to reduce the resistance to cisplatin via AKT/mTOR pathway in cisplatin-resistant gastric cancer cells [105].

Rosmarinic acid (RA) inhibited non-small cell lung cancer (NSCLC) by inducing G1 phase cell cycle arrest, apoptosis and the sensitivity of cisplatin-resistant cell via activating MAPK, enhancing p21 and p53 expression, and inhibiting the expression of P-gp and MDR1 [106]. RA reverses cisplatin resistance of NSCLS by activating the MAPK signaling pathway.

Most of the currently available chemotherapeutic and radiotherapeutic agents kill cancer cells by increasing ROS stress. Thus, both ROS-elevating and ROS-eliminating strategies have been developed for cancer therapy. As we know either chemotherapy or radiotherapy was usually associate with uncomfortable side-effects which are burdens to clinical physicians. Our previous researches find the aqueous extract of Danshen has shown anticancer as well as antioxidant effects, besides, it could prevent or mitigate the causative cardiomyopathy through controlling multiple targets without compromising the efficacy of chemotherapy (**Table 2**) [108, 109].

5. Conclusion

The current epidemiologic data show the incremental trend of CVD and cancer prevalence, mortality as well as disease burden expected in the next 40 years. The prevention of disease becomes the main lesson from now on to the future. Danshen plays a role as anti-oxidative agent and its therapeutic effects in diseases including age-associated CVDs and cancer are confirmed in many studies. Traditional Chinese medicine might be an option for treatment.

Conflict of interest

The authors declare no conflict of interest.

Abbreviations

CVD	Cardiovascular disease
ROS	Reactive oxygen species
TCM	Traditional Chinese medicine
LV	Left ventricular
eNOS	Endothelial nitric oxide synthetase
NO	Nitric oxide
SalA	Salvianolic acid A
SalB	Salvianolic acid B
AMPK	AMP-activated protein kinase
SIRT1	Sirtuin 1
MMP-9	Matrix metalloproteinase-9
Hcy	Homocysteine
SOD	Superoxide dismutase
RA	Rosmarinic acid
NSCLC	Non-small cell lung cancer

Author details

Yu-Chen Cheng[1], Yu-Chiang Hung[1*] and Wen-Long Hu[1,2,3]

1 Department of Chinese Medicine, Kaohsiung Chang Gung Memorial Hospital and Chang Gung University College of Medicine, Kaohsiung, Taiwan

2 Kaohsiung Medical University College of Medicine, Kaohsiung, Taiwan

3 Fooyin University College of Nursing, Kaohsiung, Taiwan

*Address all correspondence to: hungyuchiang@gmail.com

References

[1] World Health Organization. The top 10 causes of death [Internet]. World Health Organization; [cited 2020 Dec 9]. Available from: https://www.who.int/news-room/fact-sheets/detail/the-top-10-causes-of-death.

[2] Moran AE, Roth GA, Narula J, Mensah GA. 1990-2010 global cardiovascular disease atlas. Global heart. 2014;9(1):3-16.

[3] Imlay JA. Cellular defenses against superoxide and hydrogen peroxide. Annual review of biochemistry. 2008;77:755-776.

[4] Kiley PJ, Storz G. Exploiting thiol modifications. PLoS biology. 2004;2(11):e400.

[5] Song FL, Gan RY, Zhang Y, Xiao Q, Kuang L, Li HB. Total phenolic contents and antioxidant capacities of selected chinese medicinal plants. International journal of molecular sciences. 2010;11(6):2362-2372.

[6] Cai Y, Luo Q, Sun M, Corke H. Antioxidant activity and phenolic compounds of 112 traditional Chinese medicinal plants associated with anticancer. Life sciences. 2004;74(17):2157-2184.

[7] Wu CF, Bohnert S, Thines E, Efferth T. Cytotoxicity of Salvia miltiorrhiza Against Multidrug-Resistant Cancer Cells. The American journal of Chinese medicine. 2016;44(4):871-894.

[8] Hung YC, Pan TL, Hu WL. Roles of Reactive Oxygen Species in Anticancer Therapy with Salvia miltiorrhiza Bunge. Oxidative medicine and cellular longevity. 2016;2016:5293284.

[9] Cheng YC, Sheen JM, Hu WL, Hung YC. Polyphenols and Oxidative Stress in Atherosclerosis-Related Ischemic Heart Disease and Stroke. Oxidative medicine and cellular longevity. 2017;2017:8526438.

[10] The World Health Report 1997--conquering suffering, enriching humanity. World health forum. 1997;18(3-4):248-260.

[11] Heidenreich PA, Trogdon JG, Khavjou OA, Butler J, Dracup K, Ezekowitz MD, et al. Forecasting the future of cardiovascular disease in the United States: a policy statement from the American Heart Association. Circulation. 2011;123(8):933-944.

[12] North BJ, Sinclair DA. The intersection between aging and cardiovascular disease. Circulation research. 2012;110(8):1097-1108.

[13] Fleg JL, Aronow WS, Frishman WH. Cardiovascular drug therapy in the elderly: benefits and challenges. Nature reviews Cardiology. 2011;8(1):13-28.

[14] Kovacic JC, Moreno P, Hachinski V, Nabel EG, Fuster V. Cellular senescence, vascular disease, and aging: Part 1 of a 2-part review. Circulation. 2011;123(15):1650-1660.

[15] Lakatta EG, Levy D. Arterial and cardiac aging: major shareholders in cardiovascular disease enterprises: Part I: aging arteries: a "set up" for vascular disease. Circulation. 2003;107(1):139-146.

[16] Safar ME. Arterial aging--hemodynamic changes and therapeutic options. Nature reviews Cardiology. 2010;7(8):442-449.

[17] Chen C, Sung KT, Shih SC, Liu CC, Kuo JY, Hou CJ, et al. Age, Gender and Load-Related Influences on Left Ventricular Geometric Remodeling, Systolic Mid-Wall Function, and NT-ProBNP in Asymptomatic Asian

Population. PloS one. 2016;11(6):e0156467.

[18] Tang Y, Jacobi A, Vater C, Zou X, Stiehler M. Salvianolic acid B protects human endothelial progenitor cells against oxidative stress-mediated dysfunction by modulating Akt/mTOR/4EBP1, p38 MAPK/ATF2, and ERK1/2 signaling pathways. Biochemical pharmacology. 2014;90(1):34-49.

[19] Collins JA, Munoz JV, Patel TR, Loukas M, Tubbs RS. The anatomy of the aging aorta. Clinical anatomy (New York, NY). 2014;27(3):463-466.

[20] El Assar M, Angulo J, Rodríguez-Mañas L. Oxidative stress and vascular inflammation in aging. Free radical biology & medicine. 2013;65:380-401.

[21] Collins C, Tzima E. Hemodynamic forces in endothelial dysfunction and vascular aging. Experimental gerontology. 2011;46(2-3):185-188.

[22] Heffernan KS, Fahs CA, Ranadive SM, Patvardhan EA. L-arginine as a nutritional prophylaxis against vascular endothelial dysfunction with aging. Journal of cardiovascular pharmacology and therapeutics. 2010;15(1):17-23.

[23] Stern S, Behar S, Gottlieb S. Cardiology patient pages. Aging and diseases of the heart. Circulation. 2003;108(14):e99-101.

[24] Dai DF, Chen T, Johnson SC, Szeto H, Rabinovitch PS. Cardiac aging: from molecular mechanisms to significance in human health and disease. Antioxidants & redox signaling. 2012;16(12):1492-1526.

[25] Medrano G, Hermosillo-Rodriguez J, Pham T, Granillo A, Hartley CJ, Reddy A, et al. Left Atrial Volume and Pulmonary Artery Diameter Are Noninvasive Measures of Age-Related Diastolic Dysfunction in Mice. The journals of gerontology Series A, Biological sciences and medical sciences. 2016;71(9):1141-1150.

[26] Dzeshka MS, Lip GY, Snezhitskiy V, Shantsila E. Cardiac Fibrosis in Patients With Atrial Fibrillation: Mechanisms and Clinical Implications. Journal of the American College of Cardiology. 2015;66(8):943-959.

[27] Dworatzek E, Baczko I, Kararigas G. Effects of aging on cardiac extracellular matrix in men and women. Proteomics Clinical applications. 2016;10(1):84-91.

[28] Baris OR, Ederer S, Neuhaus JF, von Kleist-Retzow JC, Wunderlich CM, Pal M, et al. Mosaic Deficiency in Mitochondrial Oxidative Metabolism Promotes Cardiac Arrhythmia during Aging. Cell metabolism. 2015;21(5): 667-677.

[29] Freeman RV, Otto CM. Spectrum of calcific aortic valve disease: pathogenesis, disease progression, and treatment strategies. Circulation. 2005;111(24):3316-3326.

[30] Mattiuzzi C, Lippi G. Current Cancer Epidemiology. Journal of epidemiology and global health. 2019;9(4):217-222.

[31] World Health Organization. Projections of mortality and causes of death, 2016 to 2060 [internet]. World Health Organization; [cited 2019 Jul 9]. Available from: https://www.who.int/healthinfo/global_burden_disease/projections/en/.

[32] World Health Organization. Cancer [Internet]. Lyon: World Health Organization; 2020 [cited 2021 Mar 3]. Available from: https://www.who.int/news-room/fact-sheets/detail/cancer.

[33] Minelli A, Bellezza I, Conte C, Culig Z. Oxidative stress-related aging:

A role for prostate cancer? Biochimica et biophysica acta. 2009;1795(2):83-91.

[34] Klaunig JE. Oxidative Stress and Cancer. Current pharmaceutical design. 2018;24(40):4771-4778.

[35] Pang H, Wu L, Tang Y, Zhou G, Qu C, Duan JA. Chemical Analysis of the Herbal Medicine Salviae miltiorrhizae Radix et Rhizoma (Danshen). Molecules (Basel, Switzerland). 2016;21(1):51.

[36] Hung YC, Tseng YJ, Hu WL, Chen HJ, Li TC, Tsai PY, et al. Demographic and Prescribing Patterns of Chinese Herbal Products for Individualized Therapy for Ischemic Heart Disease in Taiwan: Population-Based Study. PloS one. 2015;10(8): e0137058.

[37] Hung IL, Hung YC, Wang LY, Hsu SF, Chen HJ, Tseng YJ, et al. Chinese Herbal Products for Ischemic Stroke. The American journal of Chinese medicine. 2015;43(7):1365-1379.

[38] Chen X, Guo J, Bao J, Lu J, Wang Y. The anticancer properties of Salvia miltiorrhiza Bunge (Danshen): a systematic review. Medicinal research reviews. 2014;34(4):768-794.

[39] Orgah JO, He S, Wang Y, Jiang M, Wang Y, Orgah EA, et al. Pharmacological potential of the combination of Salvia miltiorrhiza (Danshen) and Carthamus tinctorius (Honghua) for diabetes mellitus and its cardiovascular complications. Pharmacological research. 2020;153: 104654.

[40] Chen BC, Ding ZS, Dai JS, Chen NP, Gong XW, Ma LF, et al. New Insights Into the Antibacterial Mechanism of Cryptotanshinone, a Representative Diterpenoid Quinone From Salvia miltiorrhiza Bunge. Frontiers in microbiology. 2021;12:647289.

[41] Cui S, Chen S, Wu Q, Chen T, Li S. A network pharmacology approach to investigate the anti-inflammatory mechanism of effective ingredients from Salvia miltiorrhiza. International immunopharmacology. 2020;81:106040.

[42] Zhang XZ, Qian SS, Zhang YJ, Wang RQ. Salvia miltiorrhiza: A source for anti-Alzheimer's disease drugs. Pharmaceutical biology. 2016;54(1): 18-24.

[43] Zhang Y, Jiang P, Ye M, Kim SH, Jiang C, Lü J. Tanshinones: sources, pharmacokinetics and anti-cancer activities. International journal of molecular sciences. 2012;13(10): 13621-13666.

[44] Pan TL, Hung YC, Wang PW, Chen ST, Hsu TK, Sintupisut N, et al. Functional proteomic and structural insights into molecular targets related to the growth inhibitory effect of tanshinone IIA on HeLa cells. Proteomics. 2010;10(5):914-929.

[45] Wang X, Morris-Natschke SL, Lee KH. New developments in the chemistry and biology of the bioactive constituents of Tanshen. Medicinal research reviews. 2007;27(1):133-148.

[46] Neveu V, Perez-Jiménez J, Vos F, Crespy V, du Chaffaut L, Mennen L, et al. Phenol-Explorer: an online comprehensive database on polyphenol contents in foods. Database: the journal of biological databases and curation. 2010;2010:bap024.

[47] Wang J, Xu J, Gong X, Yang M, Zhang C, Li M. Biosynthesis, Chemistry, and Pharmacology of Polyphenols from Chinese Salvia Species: A Review. Molecules (Basel, Switzerland). 2019;24(1).

[48] Jiang RW, Lau KM, Hon PM, Mak TC, Woo KS, Fung KP. Chemistry and biological activities of caffeic acid derivatives from Salvia miltiorrhiza.

Current medicinal chemistry. 2005;12(2):237-246.

[49] Olivetti G, Melissari M, Capasso JM, Anversa P. Cardiomyopathy of the aging human heart. Myocyte loss and reactive cellular hypertrophy. Circulation research. 1991;68(6):1560-1568.

[50] Goldspink DF, Burniston JG, Tan LB. Cardiomyocyte death and the ageing and failing heart. Experimental physiology. 2003;88(3):447-458.

[51] Szczesny B, Bhakat KK, Mitra S, Boldogh I. Age-dependent modulation of DNA repair enzymes by covalent modification and subcellular distribution. Mechanisms of ageing and development. 2004;125(10-11):755-765.

[52] Mah LJ, El-Osta A, Karagiannis TC. GammaH2AX as a molecular marker of aging and disease. Epigenetics. 2010;5(2):129-136.

[53] Papaconstantinou J. The Role of Signaling Pathways of Inflammation and Oxidative Stress in Development of Senescence and Aging Phenotypes in Cardiovascular Disease. Cells. 2019;8(11).

[54] Costantino S, Paneni F, Cosentino F. Ageing, metabolism and cardiovascular disease. The Journal of physiology. 2016;594(8):2061-2073.

[55] Paneni F, Cosentino F. p66 Shc as the engine of vascular aging. Current vascular pharmacology. 2012;10(6): 697-699.

[56] Paneni F, Costantino S, Cosentino F. p66(Shc)-induced redox changes drive endothelial insulin resistance. Atherosclerosis. 2014;236(2):426-429.

[57] Rota M, LeCapitaine N, Hosoda T, Boni A, De Angelis A, Padin-Iruegas ME, et al. Diabetes promotes cardiac stem cell aging and heart failure, which are prevented by deletion of the p66shc gene. Circulation research. 2006;99(1):42-52.

[58] Franzeck FC, Hof D, Spescha RD, Hasun M, Akhmedov A, Steffel J, et al. Expression of the aging gene p66Shc is increased in peripheral blood monocytes of patients with acute coronary syndrome but not with stable coronary artery disease. Atherosclerosis. 2012;220(1):282-286.

[59] Salminen A, Kaarniranta K. AMP-activated protein kinase (AMPK) controls the aging process via an integrated signaling network. Ageing research reviews. 2012;11(2):230-241.

[60] Lesniewski LA, Zigler MC, Durrant JR, Donato AJ, Seals DR. Sustained activation of AMPK ameliorates age-associated vascular endothelial dysfunction via a nitric oxide-independent mechanism. Mechanisms of ageing and development. 2012;133(5):368-371.

[61] Stein S, Matter CM. Protective roles of SIRT1 in atherosclerosis. Cell cycle (Georgetown, Tex). 2011;10(4):640-647.

[62] Reuter S, Gupta SC, Chaturvedi MM, Aggarwal BB. Oxidative stress, inflammation, and cancer: how are they linked? Free radical biology & medicine. 2010;49(11): 1603-1616.

[63] Jackson AL, Loeb LA. The contribution of endogenous sources of DNA damage to the multiple mutations in cancer. Mutation research. 2001;477 (1-2):7-21.

[64] Manning BD, Cantley LC. AKT/ PKB signaling: navigating downstream. Cell. 2007;129(7):1261-1274.

[65] Shaulian E, Karin M. AP-1 as a regulator of cell life and death. Nature cell biology. 2002;4(5):E131-E136.

[66] Roebuck KA. Oxidant stress regulation of IL-8 and ICAM-1 gene

expression: differential activation and binding of the transcription factors AP-1 and NF-kappaB (Review). International journal of molecular medicine. 1999;4(3):223-230.

[67] Westermarck J, Kähäri VM. Regulation of matrix metalloproteinase expression in tumor invasion. FASEB journal: official publication of the Federation of American Societies for Experimental Biology. 1999;13(8): 781-792.

[68] Folkman J. Tumor angiogenesis. Advances in cancer research. 1985;43:175-203.

[69] Xiang M, Feng J, Geng L, Yang Y, Dai C, Li J, et al. Sera total oxidant/ antioxidant status in lung cancer patients. Medicine. 2019;98(37):e17179.

[70] Marnett LJ. Oxyradicals and DNA damage. Carcinogenesis. 2000;21(3):361-370.

[71] Matés JM, Sánchez-Jiménez FM. Role of reactive oxygen species in apoptosis: implications for cancer therapy. The international journal of biochemistry & cell biology. 2000;32(2):157-170.

[72] Schetter AJ, Heegaard NH, Harris CC. Inflammation and cancer: interweaving microRNA, free radical, cytokine and p53 pathways. Carcinogenesis. 2010;31(1):37-49.

[73] Fleisher AS, Esteller M, Harpaz N, Leytin A, Rashid A, Xu Y, et al. Microsatellite instability in inflammatory bowel disease-associated neoplastic lesions is associated with hypermethylation and diminished expression of the DNA mismatch repair gene, hMLH1. Cancer research. 2000;60(17):4864-4868.

[74] Das PM, Singal R. DNA methylation and cancer. Journal of clinical oncology: official journal of the American Society of Clinical Oncology. 2004;22(22):4632-4642.

[75] Prasad S, Gupta SC, Pandey MK, Tyagi AK, Deb L. Oxidative Stress and Cancer: Advances and Challenges. Oxidative medicine and cellular longevity. 2016;2016:5010423.

[76] Firczuk M, Bajor M, Graczyk-Jarzynka A, Fidyt K, Goral A, Zagozdzon R. Harnessing altered oxidative metabolism in cancer by augmented prooxidant therapy. Cancer letters. 2020;471:1-11.

[77] Ho JH, Hong CY. Salvianolic acids: small compounds with multiple mechanisms for cardiovascular protection. Journal of biomedical science. 2011;18(1):30.

[78] Teng F, Yin Y, Cui Y, Deng Y, Li D, Cho K, et al. Salvianolic acid A inhibits endothelial dysfunction and vascular remodeling in spontaneously hypertensive rats. Life sciences. 2016;144:86-93.

[79] Sun L, Zhao R, Zhang L, Zhang W, He G, Yang S, et al. Prevention of vascular smooth muscle cell proliferation and injury-induced neointimal hyperplasia by CREB-mediated p21 induction: An insight from a plant polyphenol. Biochemical pharmacology. 2016;103:40-52.

[80] Chen YC, Yuan TY, Zhang HF, Wang DS, Yan Y, Niu ZR, et al. Salvianolic acid A attenuates vascular remodeling in a pulmonary arterial hypertension rat model. Acta pharmacologica Sinica. 2016;37(6): 772-782.

[81] Liu YL, Liu GT. [Inhibition of human low-density lipoprotein oxidation by salvianolic acid-A]. Yao xue xue bao = Acta pharmaceutica Sinica. 2002;37(2):81-85.

[82] Wang X, Wang Y, Jiang M, Zhu Y, Hu L, Fan G, et al. Differential

cardioprotective effects of salvianolic acid and tanshinone on acute myocardial infarction are mediated by unique signaling pathways. Journal of ethnopharmacology. 2011;135(3):662-671.

[83] Jiang B, Li D, Deng Y, Teng F, Chen J, Xue S, et al. Salvianolic acid A, a novel matrix metalloproteinase-9 inhibitor, prevents cardiac remodeling in spontaneously hypertensive rats. PloS one. 2013;8(3):e59621.

[84] Zhong W, Sun B, Gao W, Qin Y, Zhang H, Huai L, et al. Salvianolic acid A targeting the transgelin-actin complex to enhance vasoconstriction. EBioMedicine. 2018;37:246-258.

[85] Li CL, Liu B, Wang ZY, Xie F, Qiao W, Cheng J, et al. Salvianolic acid B improves myocardial function in diabetic cardiomyopathy by suppressing IGFBP3. Journal of molecular and cellular cardiology. 2020;139:98-112.

[86] Wang C, Luo H, Xu Y, Tao L, Chang C, Shen X. Salvianolic Acid B-Alleviated Angiotensin II Induces Cardiac Fibrosis by Suppressing NF-κB Pathway In Vitro. Medical science monitor: international medical journal of experimental and clinical research. 2018;24:7654-7664.

[87] Wu YJ, Hong CY, Lin SJ, Wu P, Shiao MS. Increase of vitamin E content in LDL and reduction of atherosclerosis in cholesterol-fed rabbits by a water-soluble antioxidant-rich fraction of Salvia miltiorrhiza. Arteriosclerosis, thrombosis, and vascular biology. 1998;18(3):481-486.

[88] Hung YC, Wang PW, Pan TL. Functional proteomics reveal the effect of Salvia miltiorrhiza aqueous extract against vascular atherosclerotic lesions. Biochimica et biophysica acta. 2010;1804(6):1310-1321.

[89] Chan K, Chui SH, Wong DY, Ha WY, Chan CL, Wong RN. Protective effects of Danshensu from the aqueous extract of Salvia miltiorrhiza (Danshen) against homocysteine-induced endothelial dysfunction. Life sciences. 2004;75(26):3157-3171.

[90] Yang RX, Huang SY, Yan FF, Lu XT, Xing YF, Liu Y, et al. Danshensu protects vascular endothelia in a rat model of hyperhomocysteinemia. Acta pharmacologica Sinica. 2010;31(10): 1395-1400.

[91] Yang D, Xie P, Liu Z. Ischemia/reperfusion-induced MKP-3 impairs endothelial NO formation via inactivation of ERK1/2 pathway. PloS one. 2012;7(7):e42076.

[92] Ren J, Fu L, Nile SH, Zhang J, Kai G. Salvia miltiorrhiza in Treating Cardiovascular Diseases: A Review on Its Pharmacological and Clinical Applications. Frontiers in pharmacology. 2019;10:753.

[93] Hung YC, Cheng YC, Muo CH, Chiu HE, Liu CT, Hu WL. Adjuvant Chinese Herbal Products for Preventing Ischemic Stroke in Patients with Atrial Fibrillation. PloS one. 2016;11(7): e0159333.

[94] Tsai MY, Hu WL, Lin CC, Lee YC, Chen SY, Hung YC, et al. Prescription pattern of Chinese herbal products for heart failure in Taiwan: A population-based study. International journal of cardiology. 2017;228:90-96.

[95] Cheng YC, Lu CN, Hu WL, Hsu CY, Su YC, Hung YC. Decreased stroke risk with combined traditional Chinese and western medicine in patients with ischemic heart disease: A real-world evidence. Medicine. 2020;99(42): e22654.

[96] Lyu J, Xue M, Li J, Lyu W, Wen Z, Yao P, et al. Clinical effectiveness and safety of salvia miltiorrhiza depside salt combined with aspirin in patients with stable angina pectoris: A multicenter,

pragmatic, randomized controlled trial. Phytomedicine: international journal of phytotherapy and phytopharmacology. 2021;81:153419.

[97] Qin T, Rasul A, Sarfraz A, Sarfraz I, Hussain G, Anwar H, et al. Salvianolic acid A & B: potential cytotoxic polyphenols in battle against cancer via targeting multiple signaling pathways. International journal of biological sciences. 2019;15(10):2256-2264.

[98] Sies H. Oxidative stress: from basic research to clinical application. The American journal of medicine. 1991;91(3c):31s-38s.

[99] Wang X, Wang C, Zhang L, Li Y, Wang S, Wang J, et al. Salvianolic acid A shows selective cytotoxicity against multidrug-resistant MCF-7 breast cancer cells. Anti-cancer drugs. 2015;26(2):210-223.

[100] Tang XL, Yan L, Zhu L, Jiao DM, Chen J, Chen QY. Salvianolic acid A reverses cisplatin resistance in lung cancer A549 cells by targeting c-met and attenuating Akt/mTOR pathway. Journal of pharmacological sciences. 2017;135(1):1-7.

[101] Chuang CY, Ho YC, Lin CW, Yang WE, Yu YL, Tsai MC, et al. Salvianolic acid A suppresses MMP-2 expression and restrains cancer cell invasion through ERK signaling in human nasopharyngeal carcinoma. Journal of ethnopharmacology. 2020;252:112601.

[102] Sha W, Zhou Y, Ling ZQ, Xie G, Pang X, Wang P, et al. Antitumor properties of Salvianolic acid B against triple-negative and hormone receptor-positive breast cancer cells via ceramide-mediated apoptosis. Oncotarget. 2018;9(91):36331-36343.

[103] Jing Z, Fei W, Zhou J, Zhang L, Chen L, Zhang X, et al. Salvianolic acid B, a novel autophagy inducer, exerts antitumor activity as a single agent in colorectal cancer cells. Oncotarget. 2016;7(38):61509-61519.

[104] Kan S, Cheung WM, Zhou Y, Ho WS. Enhancement of doxorubicin cytotoxicity by tanshinone IIA in HepG2 human hepatoma cells. Planta medica. 2014;80(1):70-76.

[105] Wang J, Ma Y, Guo M, Yang H, Guan X. Salvianolic acid B suppresses EMT and apoptosis to lessen drug resistance through AKT/mTOR in gastric cancer cells. Cytotechnology. 2021;73(1):49-61.

[106] Liao XZ, Gao Y, Sun LL, Liu JH, Chen HR, Yu L, et al. Rosmarinic acid reverses non-small cell lung cancer cisplatin resistance by activating the MAPK signaling pathway. Phytotherapy research: PTR. 2020;34(5):1142-1153.

[107] Zhao Y, Guo Y, Gu X. Salvianolic Acid B, a potential chemopreventive agent, for head and neck squamous cell cancer. Journal of oncology. 2011;2011: 534548.

[108] Hung YC, Wang PW, Lin TY, Yang PM, You JS, Pan TL. Functional Redox Proteomics Reveal That Salvia miltiorrhiza Aqueous Extract Alleviates Adriamycin-Induced Cardiomyopathy via Inhibiting ROS-Dependent Apoptosis. Oxidative medicine and cellular longevity. 2020;2020:5136934.

[109] You JS, Pan TL, Lee YS. Protective effects of Danshen (Salvia miltiorrhiza) on adriamycin-induced cardiac and hepatic toxicity in rats. Phytotherapy research: PTR. 2007;21(12):1146-1152.

Chapter 19

Seabuckthorn Polyphenols: Characterization, Bioactivities and Associated Health Benefits

Traynard Veronique, Yuen Muk Wing and Drapeau Christian

Abstract

Sea Buckthorn (*Hippophae rhamnoides*) has a long history of use as food and medicine in Tibet and Northern Asia, where the plant has been associated with a wide range of health benefits. Sea buckthorn (SB) berry, seed and leaf have been reported to contain more than 190 bioactive compounds, including polyphenols (epicatechin, epigallocatechin, gallic acid, proanthocyanidins, chloregenic acid) and flavonoids (quercetin, isorhamnetin, kampferol glycosides, lutoelin, myricetin). SB represents a good source of phenolic compounds and flavonoids acting in synergy with PUFA such as omegas 3, 6, 7 and 9, vitamins (vitamin C), and organic acids. SB exerts antioxidant, anti-inflammatory, cytoprotective, anti-cancer, hepatoprotective properties, associated with improvement in various metabolic markers such as glycemic control and lipid profile. SB polyphenol fraction also demonstrated significant cardioprotective, antihypertensive and neuroprotective actions. SB acts as a natural stem cell mobilizer associated with significant regenerative properties. As a consequence, SB polyphenol consumption stimulates pancreatic regeneration in animal model of insulin-dependent diabetes. In conclusion, SB polyphenols exert a wide range of health benefits in metabolic health including obesity, diabetes and hypertension, as well as liver, kidney and brain health, positioning sea buckthorn berry extract as an interesting and valuable dietary supplement for natural complementary therapy and for antiaging.

Keywords: Polyphenols, flavonoids, proanthocyanidins, sea buckthorn, anti-inflammatory, antioxidant, stem cell enhancer, metabolic health, cardiovascular health

1. Introduction

Although it has been used for centuries in many parts of the world, especially in Northern Asia, sea buckthorn berry and its derivatives are relatively novel ingredients in the field of dietary supplements and functional foods. Sea buckthorn (SB) berry and leaf contain a variety of polyphenols, some of them being especially abundant in SB, that have been documented to bring a wide range of health benefits. This review is aimed at describing the composition of SB derivatives and the health benefits associated with their consumption.

2. Geographical origin

Sea Buckthorn is a deciduous, branched, spiny shrub belonging to genus *Hippophae* and family *Elaeagnaceae*. *Hippophae rhamnoides L.* is synonymous with *Elaeagnus rhamnoides* (L.) *A. Nelson*. SB usually forms a shrub or a small tree of 3–4 m in height, though it can reach up to 7 m when growing at low altitude and in moderate climate. The plant originally comes from the Northern Himalayan region where it naturally grows at altitudes ranging between 1600 and 5200 m and can resist to winter temperatures down to −40°C. Sea buckthorn's natural distribution area includes Northern China, Mongolia, India, Nepal, Northern Pakistan, and Russia, though over the centuries it has spread to Europe and North America. Seven species and 11 subspecies have been identified worldwide [1].

3. Botanical identity

Sea buckthorn is dioecious, with separate male and female plants. The male plants produce brownish flowers, which produce wind-distributed pollen and female plants produce an orange berry-like fruit. The leaves are narrow, alternate, lanceolate-linear and obtuse with peltate and stellate scales on the lower surface. The fruits are sub-globose, spherical or oblate, succulent, and orange colored with a mean diameter of 5-8 mm. The seeds are solitary, uniquely lobed, light black, and stony. The seed kernel is white and oily, sour and astringent. The pulp of the fruit is oily and soft. The surface of the peel epidermal cells is polygonal with a slightly thicker vertical wall. The parenchyma cells of the pulp contain many orange-red or orange-yellow particles, along with bright yellow oil drops. The content in actives is respectively not less than 1.5% flavonoids, 0.1% isorhamnetin (Identification criteria by Chinese pharmacopeia).

Sea buckthorn has been reported to contain more than 190 bioactive compounds in the seeds, pulp, fruit, and juice. These compounds include fat-soluble vitamins (A, K, E), 22 fatty acids, 42 lipids, organic acids (malic acid, oxalic acid), amino acids, carbohydrates, vitamins C, B1, B2, B6, B12, folic acid, flavonoids (quercetin, isorhamentin, kaempferol glycosides, luteolin, myricetin), polyphenols (epicatechin, epigallocatechin, gallic acid, proanthocyanidins, chlorogenic acid), terpenes, carotenoids (zeaxanthin, beta carotene, lycopene) and tannins [2]. Sea buckthorn berry also contains twenty mineral elements, including Ca, Cd, Cr, Cu, Fe, K, Mg, Mn, Na, Ni, Pb, Zn. It is a rich source of omega 3, 6, 7, and 9. It is the only plant that offers a wide variety of fatty acids and includes a beneficial amount Omega 7 (palmitoleic acid) [3].

4. Traditional use of the plant

The genus name of Sea Buckthorn "Hippophae" originates from the Greek words "Hippo" (horse) and "Phaos" (to shine), meaning essentially "shining horse". It is said that when Alexander the Great headed back from his Asian conquest, he travelled with his troops through a desert region near today's Northern Pakistan, where he had previously abandoned a group of horses wounded in battle. The area was abundant in sea buckthorn trees and the horses had the opportunity to feed on the berries and leaves. At one point they spotted the horses looking vibrant and with a shiny coat, which is a sign of health for horses. From there sea buckthorn was brought to Greece where it was used ever since of as horse feed to keep horses healthy and strong.

It was used as a medicinal plant in Tibet as early as 900 AD. The references to the medicinal use of SB were found in the ancient Tibetan medicinal texts, including "the

RGyud Bzi" (The Four Books of Pharmacopoeia) dated to the times of Tang Dynasty (618–907) AD. In Tibetan and Mongolian traditional medicines, SB berries were used in the treatment of cough, wound healing and burns, blood circulation and digestive system support (constipation, stomach burn). Sea buckthorn berries are listed in the *Chinese Pharmacopeia* as an ingredient for the treatment of cough and for improving blood circulation and digestion. In Russia and the Indian Himalayan region, SB is used in a wide variety of therapeutical applications such as the treatment of skin diseases, jaundice, asthma, gastro-intestinal treatment, as laxative and for the treatment of rheumatism. In Central Asia, local people use SB berries for treatment of hypertension, gastric ulcers and skin diseases [4]. Sea buckthorn berries are normally not consumed as fresh fruits. However, they have become popular in jams, beverages, candies, and juices. Juice from sea buckthorn berries is a common drink in many parts of Asia and Europe. The juice is rich in protein, vitamins C and E, as well as organic acids. The leaves, either fresh or dried, can be steeped to yield a nutritional tea. The leaves were used in ancient Greece as a fodder for horses to promote weight gain and a shiny coat. Sea buckthorn has been used for centuries in both Europe and Asia as food (tea, beverages, jams...) and for its pharmaceutical properties [5, 6]. Chinese Pharmacopeia recommends for officinal use in humans a dose of 3-10 g of SB berries per day (Editorial Committee of Chinese Pharmacopeia, 2010, p184–185, 2015 (Vol1)).

5. Polyphenol composition in SB and phytochemistry

Polyphenols from SB have antioxidant [1], anti-inflammatory [7, 8], cardioprotective [9, 10] and anticancer [2] properties, associated with metabolic-health enhancement including weight management, improvement of lipid and glucose profiles, pancreatic regeneration, and reduction of hypertension (**Figure 1**) [11–16]. Particularly, kaempferol, quercetin, and their derivatives, proanthocyanidins (PAC), catechins, phenolic acid and tannins demonstrated significant health-promoting benefits. Isorhamnetin, kampferol, quercetin, catechins and procyanidins represent some active molecules with well-known health benefits.

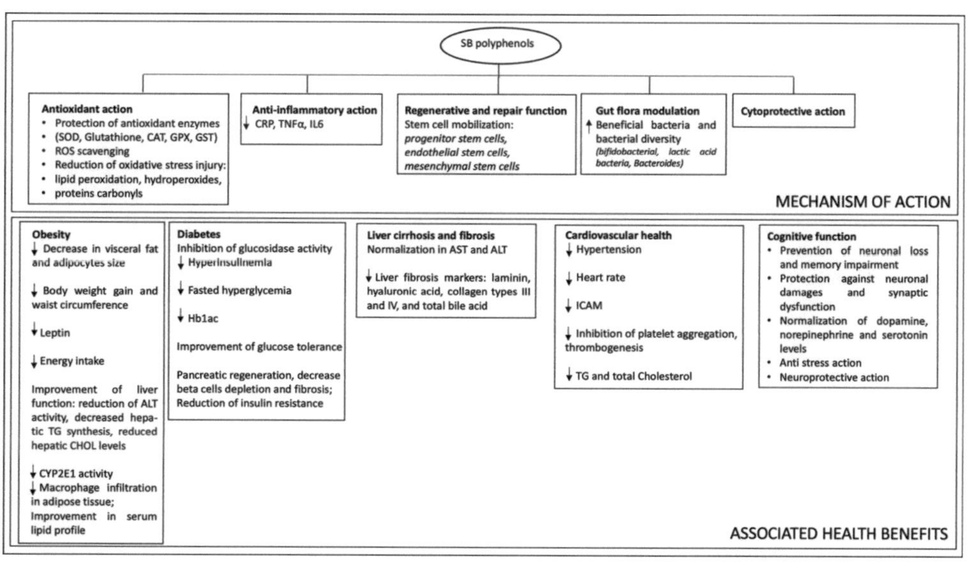

Figure 1.
Main mechanism of action and associated health benefits of Sea Buckthorn polyphenols.

	SB Berries Tibetan plateau	SB Berries*	SB berries defatted pomace ethanolic fraction
Method	**UPLC-Q-Orbitrap MS (μg/g)**	**RP-HPLC (μg/g)**	**UPLC-Q/TOF MS** (μg/g)**
Total phenolic acids		629	
Total phenolics		4730	
Total Flavones		309	
Total Flav-monoglycosides		1470	
Total Flav-diglycosides		2330	
Phloretin	310		
Gallic acid	80.9	198	
EGC			238.8
Protocatechuate	112.9	393	
Catechin	208.5	89.9	369.6
Epicatechin		21.4	123.2
Chlorogenic acid	14.04		
PAC-B2	99.12		
vanillic acid	31.7		
O-hydroxybenzene acetic acid	8.56		
coffeic acid	33.84		
P-coumaric acid	13.08		
Ferulic acid	11.28	37.6	
Salicylic acid	3.22		
Rutin	1,121.4		162.9
Ellagic acid	1.23		
Myricetin	39.96		
Naringenin	1.23		
Quercetin	36.6	55.1	
Kaempferol	117.48	12.3	
Isorhamnetin		131	*195.1*
Quercetin –3-rutinoside		329	
Quercetin-3-glucoside		397	
Isorhamnetin-3-rutinoside		586	
Isorhamnetin-3-glucoside		155	139.8
Quercetin-3-Sophoroside-7-Rhamnoside			1220
Kaempferol-3-Sophoroside-7-rhamnoside		450	1739
Isorhamnetin-3-Sophoroside-7-rhamnoside		397	1166
Isorhamnetin-3-glucoside-7-rhamnoside		1480	

	SB Berries Tibetan plateau	SB Berries*	SB berries defatted pomace ethanolic fraction
Method	**UPLC-Q-Orbitrap MS (µg/g)**	**RP-HPLC (µg/g)**	**UPLC-Q/TOF MS** (µg/g)**
Kaempferol-3-glucoside-7-rhamnoside			203.5
Reference	Jia et al. [17]	Guo et al. [18]	Dienaite et al. [19]

**Average of 4 sub-species Sinensis, Yunnanensis, Mongolica and Turkestanica.*
***Dry Weight extract.*

Table 1.
Identification and quantification of the main polyphenolic compounds present in SB; taken from [17–19].

SB represents a good source of phenolic compounds acting in synergy with PUFA such as omegas 3, 6, 7 and 9, vitamins (vitamin C), organic acids, making SB a suitable candidate for dietary supplement and food fortification. SB polyphenols are mainly phenolic acids and flavonoids. Polyphenolic content ranges from 29 to 38.8 mg/g (GAE), with more than 100 polyphenolic compounds identified (**Table 1**). The estimated content in polyphenols is higher than in mulberry, blueberry, raspberry or pomegranate [1]. Polyphenol content varies from species to species, geographical origin, the degree of maturity at the harvest, and the production process, such as drying temperature, method of extraction and storage. Comparing species, flavonoid content was the highest in *H. rhamnoides L susbp sinensis* and *yunnanensis*. Phenolic acids are divided into hydroxybenzoic acid, hydroxycinnamic acid, and their derivatives. The main phenolic acids naturally occurring in SB fruit are gallic acid, protocatechuic acid, salicylic acid, vanillic acid, caffeic acid, ferulic acid, P-coumaric acid and chlorogenic acid (**Table 1**). Flavonoids commonly found in SB include isorhamnetin, quercetin, kaempferol, myricetin, catechin, epicatechin and rutin. Condensed tannins or PAC constitutes the third category of polyphenols in SB. There is also a significant amount of carotenoids (including ß-carotene, zeaxanthin and lycopene) in SB. **Table 1** summarizes the main polyphenols identified and quantified in SB berry preparations in recent publications.

6. Effects of SB polyphenols on cellular function

Ethanolic extract of SB berry (SBB) exerts significant cytoprotective properties against sodium nitroprusside induced oxidative stress in lymphocytes [20]. SBB extract also attenuated nicotine-induced oxidative stress in rat liver and heart [21]. Moreover, the total flavones of SB provided protection against H2O2-induced apoptosis on vascular endothelial cells with the lowering the caspase-3 expression [22]. SBB also showed immunomodulating effect against T-2 toxin-induced immunodepression in 15-day-old chicks [23]. The SBB extract also had a protective effect on antioxidant enzyme levels and contributed to the reduction of lipid peroxidation, leading to reduced levels of cellular oxidation processes. Furthermore, Yasukawa et al. reported that an ethanolic fraction of SB containing (+)-catechin, (+)-gallocatechin, (−)-epigallocatechin and ursolic acid exhibited anti-tumor activity [24]. When tested on cell proliferation in the Caco-2 and HepG2 cancer cell lines, SBB extracts induced apoptotic activity and apoptotic morphological changes of the nucleus. This included chromatin condensation in HL-60 cells treated with flavonols isolated from SB such as quercetin, kaempferol and myricetin [25–27].

A flavonoid extract of SB containing isorhamnetin and quercetin exerted protective effects on myocardial ischemia and reperfusion, on microcirculation and on the regulation of thyroid function [2]. Isorhamnetin isolated from SB has also been investigated for its cytotoxicity and its influence on human hepatocellular carcinoma cells. The cytotoxic effect of isorhamnetin was showed to be dose and time-dependent against hepatocellular carcinoma cells after treatment with isorhamnetin for 72 h [28].

Polyphenolic compounds in SBB juice at different phases of digestion exerts beneficial effects on colonic microbial diversity, with an increase in total phenolic content and in total antioxidant activity during gastric and small intestine digestion, and the release of quercetin from the food matrix in the colon. Colonic fermentation resulted in an increase in quercetin and caffeic acid, along with a decrease in rutin and chlorogenic acid after 36 h of fermentation. The Shannon diversity index of beneficial groups including Lactic acid bacteria, Bacteroides/Prevotella and Bifidobacteria was increased by 35%, 71% and 17%, respectively. As a consequence, SB juice seems to represent a good source of prebiotic substrate for the proliferation of beneficial gut microbiota [29].

7. Safety of SB extracts

The safety of SB leaf and berry extracts was assessed in several studies [30–32]. In a sub-acute study, the absence of any sign of toxicity at the highest dose used established the LD50 at >10 g/kg bw for SB leaf extract. In a chronic 90-day repeated gavage administration study, no changes were observed at any of the doses used with regard to body weight and organ weight for animal of both sexes, when compared to control rats [31]. Moreover, no significant changes in biochemical parameters were noticed relative to lipid metabolism as well as renal or hepatic function. The absence of histopathological lesions in the main organs at any dose suggests a NOAEL superior to 500 mg/kg bw. In addition, the safety of herbal antioxidants composed of SB pulp and extract thereof was studied [32]. There were no significant alterations in hematological and biochemical parameters at any dose. Histopathological analysis of vital organs showed normal architecture and absence of lesions in all treated groups, which was associated with no difference in weight gain and relative organ weight in treated groups compared to controls. Even at high dose of 2,000 to 8,000 mg/kg bw [32], an absence of toxicity and side effects was reported, confirming that SBB extract is a safe product.

8. Health benefits of SB leaf and seed polyphenols

8.1 Lead intoxication model

The efficacy of SB leaf aqueous extract (SLE) was assessed in a model of lead toxicity in Wistar rat model, at a daily dose of 100 mg/kg bw for 60 days [33]. Administration of SLE to lead intoxicated Wistar rats resulted in normalization of almost all the safety parameters studied - albumin, creatinine, blood urea, total proteins. Significant improvement in total protein levels after SLE treatment in lead intoxicated animals may be due to its antioxidant properties and its hepatoprotective effect, normalizing protein synthesis. SLE treatment of lead intoxicated rats resulted in normalization of serum urea and creatinine levels, suggesting a normalization of glomerular filtration rate in kidney. Supplementation of SLE in lead intoxicated rats resulted in normalization of elevated cholesterol levels, that may due to the presence of flavonoids, terpenoids, carotenoids.

8.2 Cardiometabolic risk improvement, anti-obesity and hepatoprotective effects

SB leaf tea (SBLT) included at levels of 1 and 5% of total diet, in a high fat diet (HFD) for 6 weeks, suppressed body weight gain in a dose-dependent manner and significantly reduced visceral fat, plasma levels of leptin, triglyceride, total cholesterol and ALT activity compared with high-fat-fed control mice [34]. SBLT also decreased hepatic triglyceride, serum cholesterol and lipid accumulation. Moreover, its consumption normalized the expression of several hepatic lipid metabolic markers such as glucose-6-phosphate dehydrogenase, phosphatidate phosphohydrolase, beta-oxidation, and carnitine palmitoyltransferase. Intra-abdominal deposition of visceral adipose tissue is a major risk factor for the development of hypertension, insulin resistance, metabolic syndrome, diabetes mellitus and hyperlipidemia. SBLT supplementation seemed to have a direct effect on lipid metabolism, and it exhibited significant anti-visceral obesity property, while also reducing hepatic lipid accumulation when compared with the high-fat-fed control animals. The hypolipidemic effect of SBLT supplementation seemed likely to be due to a decrease in hepatic triglycerides synthesis through a modulation of the fatty acid esterification pathway. Compared to high-fat-fed control mice, SBLT lowered CYP2E1 activity which participates to the production of reactive oxygen species and overall oxidative stress. Both 1% and 5% SBLT supplementation effectively improved ALT activity. SBLT supplementation may prevent hepatic damage of HFD by enhancing the antioxidant defense system and the attenuation of microsomal CYP2E1 induction. Therefore, SBLT exerts antioxidant, anti-obesity and hepatoprotective effects by modulating hepatic lipid metabolism.

Total flavones from SB fruit seed residues were administered at the daily concentrations of 50, 100 or 150 mg/kg/day for 8 weeks in sucrose-fed rats model [35]. Sucrose-fed rats displayed increases of 25.6% in systolic blood pressure, 114% in plasma insulin, 85% in triglycerides (TG), as well as an increase in activated angiotensin in both heart and kidney. SB flavones significantly suppressed the elevated hypertension, hyperinsulinemia and dyslipidemia. It also led to the normalization of systolic blood pressure by at least improving insulin sensitivity and the increase in plasma angiotensin II after 8 weeks of SB consumption, especially at the daily dose of 150 mg/kg. The antihyperinsulinemia abilities of SB total flavones from fruit seed residues and irbesartan were comparable. SB flavones reversed the abnormalities in plasma triglyceride, cholesterol and FFA levels and low content of HDL. Administration of SB seed residues at a daily dose of 400 mg/kg bw for 4 weeks significantly decreased serum glucose, TG and nitric oxide levels in diabetic rats and increased serum superoxide dismutase activity and glutathione level [36]. Therefore, SB seed extract has hypoglycemic, hypolipidemic and antioxidant effects in diabetic rats.

These findings were confirmed in another study where an ethanolic extract of SB leaves (SBLE) at daily doses of 500 and 1,000 mg/kg was administered for 13 weeks to mice fed high-fat diet (HFD) [37]. Oral administration of SBLE significantly reduced energy intake, body weight gain, epididymal fat pad weight, hepatic triglyceride, hepatic and serum total cholesterol levels, as well as serum leptin levels when compared to control HFD mice. Glucose tolerance assessed by OGTT was significantly improved at both daily doses of SBLE. Lipid droplet infiltration in the liver was significantly reduced at the lower dose of SBLE and absent at the higher dose of

SBLE, confirming hepatoprotective action against triglycerides accumulation in the liver, as well as steatosis. SBLE modulates liver lipid metabolism by the upregulation of PPARa, PPARy and CPT1 and downregulation of acetylCoA carboxylase.

8.3 Stress reduction and cytoprotective effect

The anxiolytic properties of water and ethanolic extract of SB leaves (SBLE) at daily doses ranging from 50 to 300 mg/kg bw for 17 days were compared to L-theanine as positive control and saline in a model of electric foot shock stress in mice [38]. Corticosterone-induced impairment model was also studied in SH-SY5Y neuroblastoma cell line. Corticosterone-induced decrease in cellular viability was restored by different SB extracts at the concentration of 30 ug/ml. The group receiving SBLE exhibited a significantly reduced stress-induced increase in immobility times compared with the mice in the EFS group. Moreover, SBLE consumption increased climbing times in forced swim tests induced by electric foot shocks in the stressed mice. The levels of CORT, dopamine, and norepinephrine were increased, and the level of serotonin in the hippocampus was decreased in the electric foot shock stress model. The standardized SB extract effectively restored abnormal CORT and monoamine levels in the hippocampus to normal levels. These findings suggest anti-stress and neuroprotective properties of SB leaf extract in vivo.

8.4 Hepatoprotective effect

Phenolic-rich fraction of SB leaves (319.33 mg gallic acid equivalent; SBLE) was administered at doses of 25, 50 and 75 mg/kg bw for 7 days in a model of carbon tetrachloride (CCl4)-induced oxidative stress and liver injury in Sprague Dawley rats [39]. SBLE significantly protected against CCl4-induced increase in serum aspartate aminotransferase (AST), alanine aminotransferase (ALT), c-glutamyl transpeptidase (GGT), bilirubin, hepatic lipid peroxidation, hydroperoxides, protein carbonyls, as well as depletion of hepatic reduced glutathione (GSH) and decrease in the activity of hepatic antioxidant enzymes such as superoxide dismutase (SOD), catalase (CAT), glutathione peroxidase (GPX), glutathione reductase (GR) and glutathione-S-transferase (GST). SBLE protected against histopathological alterations induced by CCl4 such as liver necrosis, fatty changes, and vacuolation. SBLE demonstrated antioxidant and hepatoprotective effects against CCL4 liver injury. These observations were confirmed in another model of CCl4-induced liver injury in male albino rats fed SBLE at doses of 50, 100 and 200 mg/kg-bw for 5 days [40]. SBLE at doses of 100 and 200 mg/kg significantly restricted the CCl4-induced increase of glutamate oxaloacetate transferase, glutamate pyruvate transferase, alkaline phosphatase and bilirubin. SBLE also enhanced GSH and decreased MDA levels. SBLE (100 mg/kg) protected against CCl4-induced hepatotoxicity, as hepatic cells showed well-preserved cytoplasm and the liver showed a marked decrease in inflammatory cells. These results confirm the antioxidant and hepatoprotective effect of SBLE against CCL4 liver injury model.

9. Health benefits of SB berry polyphenols (in powders, purees, and extracts)

9.1 Anti-inflammatory effect

Two grams of frozen SBB puree containing 16.7 mg of flavanol glycosides or a placebo was consumed for 3 months in a study including 254 healthy volunteers [41].

The objective was to assess the efficacy and safety of SBB for common cold (CC), digestive tract infections (DTI), and urinary tract infections (UTI). While no difference was reported in CC and DTI frequency or duration, consumption of SBB reduced both the number and duration of UTI. A small but significant decrease in CRP was also observed in the SBB group. The decrease in the inflammatory marker CRP was confirmed in another study in which a dose of 28 g of SBB or placebo was given to 220 healthy volunteers for 90 days. SBB did not affect serum total HDL and LDL cholesterol, nor serum triacylglycerol concentrations. However, compared with placebo, there was a significant reduction in blood concentrations of CRP in the SBB group [9].

9.2 Regenerative effect for improvement of pancreatic function

SBB pulp at a daily dose of 1 or 2 ml/kg bw for 3 weeks was administered to streptozotocin-nicotinamide (STZ) induced diabetes in rats [42]. A decrease of more than 50% of fasting hyperglycemia was observed in diabetic rats, at both 1 and 2 ml/kg. Pancreatic glutathione content increased significantly in SBB treated diabetic rats Moreover, a decrease in HbA1c was reported at the highest dose. SBB decreased all histopathological changes induced by STZ, such as degenerative and lytic changes were reduced, beta cells depletion was decreased, as well as fibrosis. SBB pulp had a regenerative and protective effect on pancreatic beta cells.

9.3 Metabolic health improvement

The beneficial effect of flavonoid-rich extract of SBB was assessed in high-fat diet–induced obesity (HFDO) at daily doses of 100 and 300 mg/kg bw for 9 weeks, and compared to placebo [43]. SBB administration significantly reduced body weight gain, inhibited macrophage infiltration into adipose tissues, and downregulated TNFα mRNA expression in adipose tissue. A decrease in TG was observed but not in total cholesterol. At the highest dose of 300 mg/kg, hepatic TG was decreased by 49.56% when compared to HFDO control mice. Blood glucose concentration was 14.55% lower in the SBB treatment group (300 mg/kg), compared to the HFDO control. SBB alleviated the glucose intolerance induced by HFD, as determined by Oral Glucose Tolerance Test (OGTT). The sizes of adipocytes were considerably lower at both doses of SBB, compared to the HFDO control. Therefore, the anti-obesity activity of SBB may be attributed partly to a decrease in the volume of fat cells. Decrease in adipose tissue inflammation, anti-obesity properties, improvement of glucose tolerance and glycemia, and the decrease in hepatic TG accumulation all points to an improvement of cardiometabolic profile.

The efficacy of SBB was also evaluated in a model of spontaneously hypertensive stroke-prone rat, using a daily dose of 0.7 g/kg-bw for 2 months [44]. Mean and diastolic blood pressure, heart rate, total plasma cholesterol, triglycerides, and glycated hemoglobin were significantly decreased by the SBB treatment when compared to hypertensive control group. The number of AP-containing capillary portions fell while the number of those containing DPPIV increased. The expression of these 2 enzymes is modulated by inflammation enhanced by hypertension. Antihypertensive and cardioprotective properties measured by heart rate, blood pressure, total plasma cholesterol, TG levels improvement were thus confirmed.

The efficacy of SBB, and SBB phenolic extract on metabolic health was compared to bilberry berries (BB) in 110 overweight women [45]. The daily doses were all equivalent to 100 g of fresh berries. Each product was consumed for 35 days in a cross-over study. Decrease in waist circumference and TNFα, and a small decrease in fasting plasma glucose was observed after SBB consumption. A decrease in ICAM and TNFα was observed after consumption of SB extract. No significant difference in BP, percentage of fat mass, fasting plasma cholesterol, TG, and IL6 levels was observed. Therefore, SB products brought mild but significant improvement in metabolic, inflammatory and endothelial markers in overweight volunteers. Another clinical study demonstrated the efficacy of SB juice (SBJ) on platelet aggregation. Placebo or 300 ml of SBJ was consumed by 20 healthy volunteers for 8 weeks [46]. No difference in platelet aggregation and LDLox levels was seen between placebo and SBJ group. A non-significant increase of 20% was observed in plasma HDL-C. In another study, a crude flavone extract from SBB prevented thrombogenesis in an *in vivo* model of thrombosis in mouse femoral artery, probably by inhibition of platelet aggregation. It prolonged occlusion time at a dose of 300 μg/kg had a similar effect to aspirin at 10 mg/kg [47].

This cardioprotective effect and improvement in metabolic profile has been confirmed in clinical studies. Eighty overweight women were given either 20 g of dried SBB, or 14.6 g of sea buckthorn phenolic extract for 35 days, in a randomized cross-over study. All these daily doses represent 100 g of fresh SBB [16]. All groups using the various SB products showed significant improvement in metabolic profile, especially in individuals with higher baseline cardiometabolic risk. SB-induced effects were mainly on serum TG and very-low-density lipoprotein (VLDL) and its subclasses, which decreased in participants with higher baseline cardiometabolic risk. During SB consumption, a significant decrease in TG and creatinine was observed. To conclude, a meta-analysis including 11 RCT which enrolled 514 patients confirmed that supplementation with SBB and extracts significantly reduced total cholesterol, LDL-cholesterol and significantly increased HDL-cholesterol in subjects with cardiovascular risks [6].

9.4 Hepatoprotective effect

Sea buckthorn extract (SBE) was administered at the daily dose of 15 g, 3 times a day for 6 months, to 50 cirrhotic Child-Pugh grade A and B patients [48]. The rate of normalization in AST and ALT, was significantly higher in the groups treated with SBE: 80% in the treated group and 56% in the control group. Parameters of liver fibrosis such as serum laminin, hyaluronic acid, total bile acid (TBA), collagen types III and IV were decreased after treatment, when compared with control group. SBE decreased markers of liver fibrosis and improved the rate of AST/ALT normalization, suggesting hepatoprotective properties.

9.5 Regenerative effect

Stemberry®, a SBB aqueous extract standardized in 30% of proanthocyanidins was consumed by 12 healthy participants at a daily dose of 500 mg, compared to a placebo in one single dose. Rapid and highly selective stem cell mobilization was observed, as quantified by an increase in the number of circulating CD45dim C $D34^+$ $CD309^-$ progenitor stem cells by 24%, $CD45^-$ CD31+ $CD309^+$ endothelial stem cells by 33%, and $CD45^-$ $CD90^+$ mesenchymal stem cells by 20%. All these types of stem cells are involved in regenerative and reparative functions. Moreover, a mild significant increase was observed in the number of CD45dim $CD34^+$ $CD309^+$ pluripotential stem cells [49]. SB PAC-rich extract supports the natural ability of the body to repair and renew, suggesting regenerative properties.

9.6 Acute lung injury protection

The efficacy of SBB paste (SB total polyphenols 191.5 mg/g and SB total flavonoids 130.9 mg/g) was studied in a mouse model of LPS-induced acute lung injury [50]. SBB paste was consumed for 7 days at daily doses of 200, 400 and 800 mg/kg bw, and at day 8 LPS challenge was carried out. The loss of body weight, microstructure lesions in the lung tissue, increase in MDA, and reduction of SOD and glutathione peroxidase levels caused by LPS challenge were all significantly reduced by SB treatment in a dose dependent manner. As a consequence, SBB paste improved lung lesions such as alveolar thickness caused by edema, hemorrhage alveolus collapse, inflammatory cell inflammation were greatly reduced in the SB-treated group compared with the group acute lung injury. SB treatment provided significant protection against protein transvascular leakage. As lung lesions, oxidative stress markers are decreased, SB provides protection against acute lung injury, via partly the activation of Nrf2 pathway and redox homeostasis due its high content in polyphenols.

9.7 Cytoprotective and antioxidant effect

SBB flavones at the concentration of 100 ug/mL exert cytoprotective and antioxidant properties in a tert-Butyl hydroperoxide-induced cytotoxicity (BOOH) model in lymphocytes [51]. SBB flavones significantly inhibited tert-BOOH-induced cytotoxicity and free radical production, restored the antioxidant status, significantly maintained ATP levels comparable to control and protected the cells from tert-BOOH-induced lipid peroxidation. Treatment with SBB flavones reduced tert-BOOH-induced apoptosis and a decreased tert-BOOH-induced formation of DNA breaks by 30%. Cytoprotective and antioxidant effects suggest safety of SB berries extracts.

9.8 Anti-inflammatory and neuroprotective effect

The efficacy of a SBB extract rich in flavonoids was demonstrated in a model of high-fat high-fructose diet (HFFD) induced cognitive impairment [11]. The extract was consumed for 14 days at 2 daily doses of 100 and 500 mg/kg-bw. Compared to HFFD placebo mice, SBB consumption resulted in a reduction in body weight gain by 8.8% and a decrease in glucose intolerance. It also improves insulin sensitivity. More specifically, SBB consumption resulted in a 45–48% decrease in HOMA-IR value, a 20–30% decrease in fasting hyperglycemia, a 12–20% decrease in fasting insulinemia, a reduction in TG and total cholesterol levels, a prevention of neuronal loss and working memory impairment in behavioral tests, and a suppression of HFFD-induced synaptic dysfunction and neuronal damages. Dietary supplementation SF significantly improved the length by 37.91% and width by 10.07% of post-synaptic density in the hippocampus when compared with the HFFD group mice. SBB flavonoids also increased the levels of BDNF, NT-3, NT-4 and NGF involved in the growth, survival, and synaptic plasticity of brain neurons. SBB flavonoids also reversed HFD-induced overexpression of iNOS, and the up-regulation of p38 phosphorylation and NFkB expression, which are markers of neuro-inflammation.

As a consequence, SBB flavonoids displayed neuroprotective effects against chronic neuro and systemic inflammation observed in diabetes-induced obesity and is associated with an improvement of metabolic parameters (namely lipid and glucose profiles).

9.9 Beneficial effect in idiopathic nephrotic syndrome

Hydroalcoholic extract of leaves and fruits of SB at a daily dose of 350 mg twice daily for 12 weeks, in addition to other standard drugs, was administered to 52 patients with Idiopathic Nephrotic Syndrome [52]. Beneficial effects were reported in symptoms like anorexia and oedema. There was no statistically significant change in creatinine, phosphorous and blood urea after 12 weeks of treatment when compared to control subjects. Improvement in cholesterol, triglyceride, albumin and 24-hour urinary protein excretion in the SB group was observed. Decreased CRP and IL6 levels were also noticed in the group treated with SB, confirming a nephroprotective role of SB.

10. Conclusion

Sea buckthorn can be considered as a functional ingredient for use in cosmetics, dietary supplements, general foods and fortified foods due to its richness in antioxidant molecules, in vitamins C, A and E, in omega 3, 6, 7 and 9, and a diversity of bioactive molecules. Its polyphenolic compounds include phenolic acids, flavonoids, carotenoids associated with antioxidant, anti-inflammatory, antibacterial and anticancer properties. SB exerts cardioprotective effect including antiatherogenic properties, hepatoprotective and neuroprotective effect, improves metabolic profile (lipid profile, glycemic control, blood pressure, fat mass and waist circumference), protects against acute lung injury, and supports tissue regeneration (**Table 2**).

As a consequence, sea buckthorn offers an excellent source of bioactive molecules [3] that could enter in the formulation of nutritional beverages, yogurts, muesli, healthy snacks and bars, in dietary supplements or instant powder mixes as a superfood ingredient. SB polyphenolic extracts such as Stemberry® standardized for polyphenols and more specifically proanthocyanidins, demonstrated

Health benefits of sea buckthorn	Mechanism of action and main outcomes
Cardioprotective effect	Anti-hypertensive effect [35, 44] Improvement of lipid profile [6, 16, 43] Inhibition of atherosclerotic plaque formation [47] Preservation of cardiac function, decrease in ischemic zone, reduction of progression of infarction [2] Preservation of structural integrity of myocardium [2]
Improvement of metabolic profile	Decrease in hyperglycemia and Hb1ac [36, 42, 43] Decrease in hyperinsulinemia [11, 35] Improvement of insulin resistance [11] Anti-obesity effect [34, 37, 43] Anti-inflammatory effect (CRP, TNFa) [9, 45] Antioxidant effect (SOD, GSH, GPx) [39, 40, 50, 51]
Hepatoprotective effect	Decrease in ALAT/ASAT/GGT [34, 37, 39, 40] Decrease in histopathological lesions and markers of fibrosis [34, 37, 39, 40] Decrease in hepatic lipid accumulation [34, 37, 39, 40]
Antiatherogenic effect	Improvement in lipid profile [6] Decrease in VCAM, ICAM endothelial markers [45] Decrease in platelet aggregation [46, 47]
Tissue regeneration	Selective mobilization of several stem cell types participating to tissue renewal and repair [49]
Neuro and cytoprotective effect	Prevention of neuronal loss and memory impairment in behavioral tests Suppression of synaptic dysfunction and neuronal damages [11] Decrease in neuroinflammation [11]
Protection against acute lung injury	Preservation of lung tissue microstructure, body weight loss reduction, transvascular leakage increase reduction, MDA decrease, and increase in SOD and glutathione peroxidase levels [50]

Health benefits of sea buckthorn	Mechanism of action and main outcomes
Protection against acute intestinal injury	Decrease in injury/ulcer area size [4] Decreased in apoptotic cells [4]
Prebiotic effect and gut health support	Colonic microbial diversity increase of beneficial groups of bacteria [29]
Kidney function support	Improvement of creatinine, phosphorous, blood urea, 24 h urinary protein excretion and albumin in idiopathic nephrotic syndrome patients [52] Reduction of oedema [52]

Table 2.
Summary of health benefits of sea buckthorn berry, seed and leaf.

significant regenerative properties through a stimulation of endogenous stem cell mobilization, which has demonstrated therapeutical benefits in neurodegenerative disease, heart diseases, diabetes, and chronic inflammatory diseases by supporting the body's natural repair system. Additional studies on the regenerative effect of SB polyphenol fraction in several diseases such as diabetes, neurodegenerative, cardiovascular diseases could open new complementary therapeutical strategies in order to improve patients' quality of life [53].

Author details

Traynard Veronique[1], Yuen Muk Wing[2] and Drapeau Christian[3]*

1 VT Consulting, Ostwald, France

2 Puredia, East Kowloon, Hong Kong

3 Kalyagen, Austin, TX, USA

*Address all correspondence to: christian@kalyagen.com

References

[1] Ji M., Gong X., Li X. et al., Advanced Research on the Antioxidant Activity and Mechanism of Polyphenols from Hippophae Species—A Review, Molecules 2020, 25, 917; doi:10.3390/molecules25040917

[2] Patil S., Chaudhary A., Unexplored therapeutic treasure of Himalayan sea buckthorn berry: An opportunity for rejuvenation applications in Ayurveda, International Journal of Green Pharmacy, 2016, S10(4) :S164

[3] Wani T., Wani S., Ahmad M., et al., Bioactive profile, health benefits and safety evaluation of sea buckthorn (*Hippophae rhamnoides* L.): A review, Cogent Food & Agriculture, 2016, 2: 1128519, http://dx.doi.org/10.1080/23311932.2015.1128519

[4] Suryakumar G. & Gupta A., Medicinal and therapeutic potential of Sea buckthorn (*Hippophae rhamnoides* L.), Journal of Ethnopharmacology, 2011, 138: 268– 278

[5] Khan B. Akhtar N., Mahmood T., A Comprehensive Review of a Magic Plant, *Hippophae rhamnoides*, Pharmacognosy Journal, 2010, 2(16): 65-68

[6] Guo X., Yang B., Cai W., et al., Effect of sea buckthorn (*Hippophae rhamnoides* L.) on blood lipid profiles: a systematic review and meta-analysis from 11 independent randomized controlled trials, Trends in Food Science & Technology, 2016, 61 : 1-10

[7] Li J., Wu R., Qin X., et al., Isorhamnetin inhibits IL-1β-induced expression of inflammatory mediators in human chondrocytes, MOLECULAR MEDICINE REPORTS, 2017, 16: 4253-4258, DOI: 10.3892/mmr.2017.7041

[8] Ren Q., Li X., Li Q., et al., Total flavonoids from sea buckthorn ameliorates lipopolysaccharide/cigarette smoke-induced airway inflammation, Phytotherapy Research. 2019;1-16, DOI: 10.1002/ptr.6404

[9] Larmo P., Yang B., Hurme S., et al., Effect of a low dose of sea buckthorn berries on circulating concentrations of cholesterol, triacylglycerols, and flavonols in healthy adults, Eur J Nutr, 2009, 48:277-282

[10] Olas B., Kontek B., Szczesna M., et al., INHIBITION OF BLOOD PLATELET ADHESION BY PHENOLICS' RICH FRACTION OF *HIPPOPHAE RHAMNOIDES* L. FRUITS, JOURNAL OF PHYSIOLOGY AND PHARMACOLOGY 2017, 68, 2, 223-229

[11] Mulati A., Ma S., Zhang H., et al., Sea-buckthorn flavonoids alleviate high-fat and high-fructose diet-induced cognitive impairment by inhibiting insulin resistance and neuroinflammation, Journal of Agricultural Food Chemistry, 2020, 68(21):5835-5846, DOI: 10.1021/acs.jafc.0c00876

[12] Ambarasu S., Radhakrishnan M, Suresh A., PHYTOCHEMICAL, ETHNOMEDICINAL AND PHARMACOLOGICAL POTENTIALS OF SEABUCKTHORN- A MINI REVIEW, Int J Pharm Bio Sci, 2015, 6(3): (B) 263 – 272

[13] Singh I., Ahmad F., Gore D., et al., Therapeutic potential of seabuckthorn: a patent review (2000-2018), Expert Opinion on Therapeutic Patents, 2019, DOI:10.1080/13543776.2019.1648434

[14] Yang X., Wang Q., Pang Z., et al., Flavonoid-enriched extract from *Hippophae rhamnoides* seed reduces high fat diet induced obesity, hypertriglyceridemia, and hepatic triglyceride accumulation in C57BL/6

mice, PHARMACEUTICAL BIOLOGY, 2017, 55(1) : 1207-1214

[15] Linderborg K., Lehtonen H., Jarvinen R., et al., The fibres and polyphenols in sea buckthorn (Hippophae¨ rhamnoides) extraction residues delay postprandial lipemia, International Journal of Food Sciences and Nutrition, 2012; 63(4): 483-490; DOI: 10.3109/09637486.2011.636346

[16] Larmo P., Kangas A., Soininen P., et al., Effects of sea buckthorn and bilberry on serum metabolites differ according to baseline metabolic profiles in overweight women: a randomized crossover trial, Am J Clin Nutr, 2013, 98:941-51

[17] Jia Q., Zhang S., Zhang H., et al., A comparative study on polyphenolic composition of berries from the Tibetan plateau by UPLC-Q-Orbitrap MS system, 2020, Chem. Biodiversity 10.1002/cbdv.202000033

[18] Guo R., Guo X., Li T., et al., Comparative Assessment of Phytochemical Profiles, Antioxidant and Antiproliferative Activities in Sea buckthorn (Hippophaë rhamnoides L.) Berries, Food Chemistry, 2016, http://dx.doi.org/10.1016/j.foodchem.2016.11.063

[19] Dienaite L., Pukalskas A., Pukalskiene M., et al., Phytochemical Composition, Antioxidant and Antiproliferative Activities of Defatted Sea Buckthorn (Hippophaë rhamnoides L.) Berry Pomace Fractions Consecutively Recovered by Pressurized Ethanol and Water, Antioxidants 2020, 9, 274; doi:10.3390/antiox9040274

[20] Geetha, S., Sai Ram, M., Singh, V., Ilavazhagan, G., Sawhney, R.C., Effect of Sea buckthorn against sodium nitroprusside induced oxidative stress in murine macrophages. Biomedicine and Pharmacotherapy, 2002, 56, 463-467

[21] Taysi, S., Gumustekin, K., Demircan, B., Aktas, O., Oztasan, N., Akcay, F., Suleyman, H., Akar, S., Dane, S., Gul, M., Hippophae rhamnoides attenuates nicotineinduced oxidative stress in rat liver. Pharmaceutical Biology, 2010, 48, 488-493

[22] Cheng JY, Teng D, Li W. Protection and mechanism of total flavone of *Hippophae rhamnoides* on vascular endothelial cells. Zhongguo Zhong Xi Yi Jie He Za Zhi 2011; 31:355-8.

[23] Ramasamy T, Varshneya C, Katoch VC. Immunoprotective effect of seabuckthorn (*Hippophae rhamnoides*) and glucomannan on T-2 toxin-induced immunodepression in poultry. Vet Med Int 2010; 2010:149373.

[24] Yasukawa, K., Kitanaka, S., Kawata, K., Goto, K., Anti-tumor promoters phenolics and triterpenoid from *Hippophae rhamnoides*. Fitoterepia, 2009, 80, 164-167.

[25] Hibasami, H., Mitani, A., Katsuzaki, H., Imai, K., Yoshioka, K., Komiya, T., Isolation of five types of flavonol from seabuckthorn (*Hippophae rhamnoides*) and induction of apoptosis by some of the flavonols in human promyelotic leukemia HL-60 cells. International Journal of Molecular Medicine, 2005, 15, 805-809.

[26] Grey, C., Widen, C., Adlercreutz, P., Rumpunen, K., Duan, R.D., Antiproliferative effects of sea buckthorn (*Hippophae rhamnoides* L.) extracts on human colon and liver cancer cell lines. Food Chemistry, 2010, 120, 1004-1010.

[27] Olsson, M., Gustavsson, K., Andersson, S., Nilsson, A., and Duan, R., Inhibition of cancer cell proliferation in vitro by fruit and berry extracts and correlations with antioxidant levels. J. Agric. Food Chem., 2004, 52, 7264-7271, doi: 10.1021/jf030479p

[28] Kumar, M. S. Y., Dutta, R., Prasad, D., & Misra, K., Subcritical water extraction of antioxidant compounds from Seabuckthorn (*Hippophae rhamnoides*) leaves for the comparative evaluation of antioxidant activity. *Food Chemistry,* 2011, *127*, 1309-1316.

[29] Attri S., Sharma K., Raigond P., Goel G., Colonic fermentation of polyphenolics from Sea buckthorn (*Hippophae rhamnoides*) berries: Assessment of effects on microbial diversity by Principal Component Analysis, Food Research International, 2018, 105:324-332, https://doi.org/10.1016/j.foodres.2017.11.032

[30] Saggu S., Divekar H., Gupta V., et al., Adaptogenic and safety evaluation of seabuckthorn (*Hippophae rhamnoides*) leaf extract: A dose dependent study, Food and Chemical Toxicology, 2007, 45 :609-617

[31] Tulsawani R., Ninety day repeated gavage administration of Hipphophae rhamnoides extract in rats, Food and Chemical Toxicology, 2010, 48 ; 2483-2489

[32] Ali R., Ali R., Jaimini A., et al., Acute and sub acute toxicity and efficacy studies of *Hippophae rhamnoides* based herbal antioxidant supplement, Indian J Pharmacol., 2012, 44(4): 504-508

[33] Zarger R., Raghuwanshi P., Rastogi A., et al., Protective and ameliorative effect of sea buckthorn leaf extract supplementation on lead induced hemato-biochemical alterations in Wistar rats, Veterinary World, 2016, 9(9): 929-934

[34] Lee H., Kim M., Lee K., et al., Anti-visceral obesity and antioxidant effects of powdered sea buckthorn (*Hippophae rhamnoides* L.) leaf tea in diet-induced obese mice, Food and Chemical Toxicology, 2011, 49: 2370-2376

[35] Pang X., Zhao J., Zhang W., et al., Antihypertensive effect of total flavones extracted from seed residues of *Hippophae rhamnoides* L. in sucrose-fed rats, Journal of Ethnopharmacology, 2008, 117 :325-331

[36] Zhang W., Zhao J., ang J., et al., Hypoglycemic Effect of Aqueous Extract of Seabuckthorn (*Hippophae rhamnoides* L.) Seed Residues in Streptozotocin-induced Diabetic Rats, Phytother. Res., 2010, 24: 228-232

[37] Pichiah P., Moon H., Park J., et al., Ethanolic extract of seabuckthorn (*Hippophae rhamnoides* L) prevents high-fat diet–induced obesity in mice through down-regulation of adipogenic and lipogenic gene expression, Nutrition Research, 2012, 32 :856-864

[38] Bae D., Kim J., Oh D., et al., Multifunctional antistress effects of standardized aqueous extracts from *Hippophae rhamnoides* L., 2016, 20(6): 369-383

[39] Maeshwari D., Kumar M., Verma S., et al., Antioxidant and hepatoprotective activities of phenolic rich fraction of Seabuckthorn (*Hippophae rhamnoides* L.) leaves, Food and Chemical Toxicology, 2011, 49: 2422-2428

[40] Geetha S., Jayamurthy P., PAL K., et al., Hepatoprotective effects of sea buckthorn (*Hippophae rhamnoides* L.) against carbon tetrachloride induced liver injury in rats, J Sci Food Agric, 2008, 88:1592-1597

[41] Larmo P., Alin J., Salminen E., et al., Effects of sea buckthorn berries on infections and inflammation: a double-blind, randomized, placebo-controlled trial, European Journal of Clinical Nutrition, 2008, 62:1123-1130

[42] Sharma M., Siddique M., Shamim A., et al., Evaluation of Antidiabetic and Antioxidant Effects of Seabuckthorn (*Hippophae rhamnoides* L.)

in Streptozotocin-Nicotinamide Induced Diabetic Rats, The Open Conference Proceedings Journal, 2011, 2 :53-58

[43] Yang X., Wang Q., Pang Z., et al., Flavonoid-enriched extract from *Hippophae rhamnoides* seed reduces high fat diet induced obesity, hypertriglyceridemia, and hepatic triglyceride accumulation in C57BL/6 mice, PHARMACEUTICAL BIOLOGY, 2017, 55(1) : 1207-1214

[44] Koyama T., Taka A., Togashi H., Effects of a herbal medicine, *Hippophae rhamnoides*, on cardiovascular functions and coronary microvessels in the spontaneously hypertensive stroke-prone rat, Clinical Hemorheology and Microcirculation, 2009, 41 : 17-26

[45] Lehtonen H., Suomela J., Tahvonen R., et al., Different berries and berry fractions have various but slightly positive effects on the associated variables of metabolic diseases on overweight and obese women, European Journal of Clinical Nutrition, 2011, 65 : 394-401

[46] Eccleston C., Baoru Y., Tahvonen R., et al., Effects of an antioxidant-rich juice (sea buckthorn) on risk factors for coronary heart disease in humans, Journal of Nutritional Biochemistry, 2002, 13: 346-354

[47] Cheng J., Kondo K., Suzuki Y., et al., Inhibitory effects of total flavones of *Hippophae rhamnoides* L on thrombosis in mouse femoral artery and in vitro platelet aggregation, Life Sciences, 2003, 72 : 2263-2271

[48] Gao Z., Gu X., Cheng F., et al., Effect of Sea buckthorn on liver fibrosis: A clinical study, World J Gastroenterol, 2003, 9(7):1615-7

[49] Drapeau C., Benson K., Jensen G., Rapid and selective mobilization of specific stem cell types after consumption of a polyphenol-rich extract from sea buckthorn berries (Hippophae) in healthy human, subjects, Clinical Interventions in Aging, 2019, 14:253-263

[50] Du L., Hu X., Chen C., et al., Seabuckthorn Paste Protects Lipopolysaccharide-Induced Acute Lung Injury in Mice through Attenuation of Oxidative Stress, Oxidative Medicine and Cellular Longevity Volume 2017, Article ID 4130967, 9 pages

[51] Geetha S., Ram M., Sharma S., et al., Cytoprotective and Antioxidant Activity of Seabuckthorn (*Hippophae rhamnoides* L.), Flavones Against tert-Butyl Hydroperoxide-Induced Cytotoxicity in Lymphocytes, J Med Food, 2009, 12 (1): 151-158

[52] Singh R., Singh P., Singh P., et al., Immunomodulating and Antiproteinuric Effect of *Hippophae rhamnoides* (Badriphal) in Idiopathic Nephrotic Syndrome, J Assoc Physicians India, 2012, 61(6):397-9.

[53] Drapeau C., Eufemio, G., Mazzoni P., et al., Therapeutical potential of stimulating endogenous stem cell mobilization, chapter 8, Tissue Regeneration – From Basic Biology to Clinical Application, 2012, 167-202, Edited by Prof. Jamie Davies, Intech Publisher

Chapter 20

Phenolic Compounds in Tea: Phytochemical, Biological, and Therapeutic Applications

Jyoti V. Vastrad, Pratikhya Badanayak and Giridhar Goudar

Abstract

Phenolic compounds are one of the major and most complex groups of phytochemicals found among plant kingdom. Structurally they comprise of aromatic ring along with one or more hydroxyl groups. Based on the structure they are divided into subgroups such as phenolic acids, flavonoids, tannins, coumarins, lignans, quinones, stilbenes and carotenoids. Plant polyphenols are gaining popularity as a result of their potent antioxidant properties and notable effects in the prevention of oxidative stress-related diseases. Extraction, identification and characterisation of phenolic compounds from various plant sources has become a major area of health and medical research in the recent years. The major bioactive compounds responsible for tea's health benefits are thought to be phenolics. Catechin derivatives make up the majority of the phenolic compounds found in tea, and though flavonols and phenolic acids are also present in smaller amounts. The bioactivity of the compounds has been linked to a lower risk of serious illnesses like cancer, cardiovascular disease, and neurodegenerative disease. This chapter covers phenolic extraction, purification, analysis and quantification, as well as their antioxidant properties in different varieties of tea leaves.

Keywords: Tea leaf, phenolic compound, tannin, flavonoids, antioxidant, health benefits

1. Introduction

Tea, which is prepared from the leaves of the *C. sinensis*, and is the second most popular beverage in the world, after water. Green tea, oolong tea, and black tea are made primarily from the young green shoots of the tea plant [1]. Only the young, top leaves and the unopened leaf bud are used in fine teas, which is one of their distinguishing characteristics. As a result, immature, light-green leaves are preferred for tea production, whereas mature and old leaves, which have a darker green colour, are unsuitable due to their unpleasant flavour. Tea beverages are classified as green (unfermented), white (lightly fermented), oolong (semi-fermented), or black (fermented) based on their manufacturing method (fully fermented) [2, 3]. It is cultivated in more than 50 countries of the world and the main producers of tea are China (2,473,443 t) and India (1,325,950 t) as per 2017 statistics (**Table 1**) [7].

Compounds	Green tea	Black tea	Black tea infusion
Protein	15	15	Trace
Amino acids	4	4	3.5
Fibre	26	26	0
Carbohydrates	7	7	4
Lipids	7	7	Trace
Pigments	2	2	Trace
Minerals	5	5	4.5
Phenolic compounds	30	5	4.5
Oxidised phenolic compounds	0	25	4.5

Table 1.
Composition (%) of green, black and infusion [4–6].

Phenolic acids, terpenoids, flavonoids, alkaloids, tannins, and phytosterols are the major bioactive constituents in the plant leaves [8]. Tea chemical composition depends on the cultivars, environmental factors and different manufacturing process [9]. Flavonoids like flavanols (flavan-3-ols), flavonols, flavones, flavanones, and anthocyanidins are important components of tea leaves, accounting for up to 30% of the dry weight of the leaves [10, 11]. Catechins which are group of flavan-3-ols are the major bioactive compounds in fresh tea leaves, among them epicatechin (EC), epigallocatechin (EGC), epicatechin gallate (ECG),

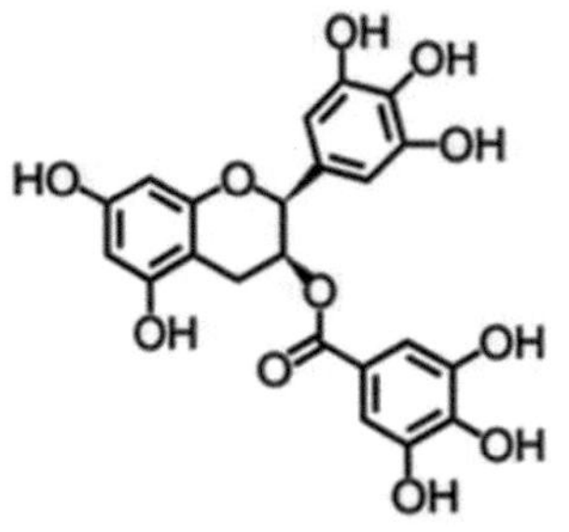

Epigallocatechin-3-gallate (EGCG)

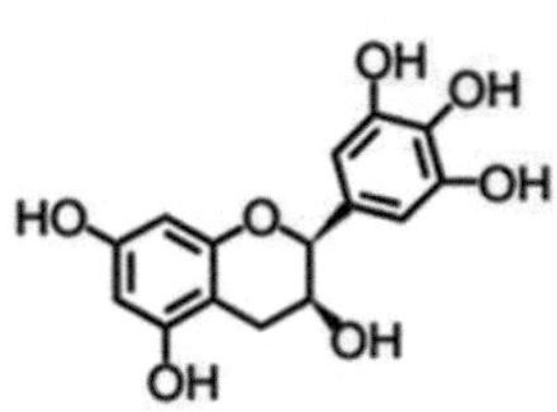

Epigallocatechin (EGC)

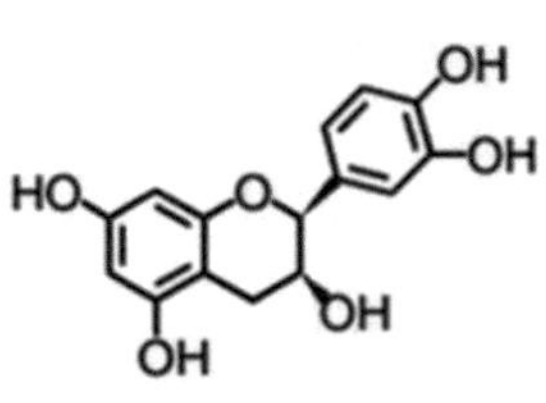

Epicatechin (EC)

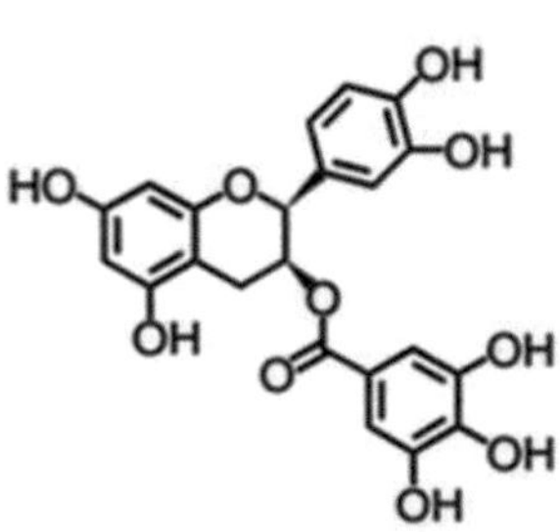

Epicatechin-3-gallate (ECG)

Figure 1.
The main phenolic compounds of tea polyphenols [12].

epigallocatechin gallate (EGCG), catechin and gallocatechin are majorily found (**Figure 1**) [13, 14].

Tea is widely acknowledged to have numerous health benefits, including antioxidant activity, anti-inflammatory activity, anti-microbial effects, and anti-carcinogenic effects when consumed regularly. The phenolic compounds in tea are thought to be responsible for these effects. As a result, tea phenolics are thought to be have valuable phytochemicals and received a great deal of attention [15].

2. Different types of tea

Type of tea produced depends on the various fermentation processes, white tea (sundried fresh leaves), green tea (heated or steamed fresh leaves), and black tea (fermented leaves) [16]. Among different tea types, black tea is produced highest about 76–78%, followed by green tea (20–22%) and oolong tea (2%) accounting for worldwide production [17].

2.1 Black tea

Black tea which is prepared from the young tender shoots of *C. sinensis* is consumed widely as a non-alcoholic beverage. It is most popular in India, Europe and North America [18]. The phenolic composition of black tea differs significantly from green tea due to the fermentation process resulting in the formation of condensation and oxidation products such as thearubigins and theaflavins, which might be due to the action of polyphenol oxidases (PPO) [19, 20]. In comparison to green tea, 85% of catechin in black tea is transformed into theaflavin-3-3′-digallate and thearubigin [21]. Astringency and brightness of tea is due to theaflavins, whereas colour and mouthfeel is because of thearubigins [18]. Theaflavins and thearubigins (2 to 6%) account for catechin content of 3 to 10% (w/w) of 10 to 20% (w/w) of the dry weight of black leaves (**Figure 2**) [18].

Rechner et al. [22] has reported numerous *in vitro* and *in vivo* effects of tea polyphenols, including antioxidant, anticarcinogenic, and hypolipidemic properties. The bioavailability and metabolism of individual polyphenolic constituents of black tea (flavan-3-ols, flavonols, hydroxybenzoic acids, hydroxycinnamates) in humans have been reported by Liebert et al. [23]. Rababah et al. [24] show that a cup of black tea, as it is traditionally brewed in the UK, is an excellent source of polyphenols, containing up to six different classes of polyphenols and having higher antioxidant activity than other dietary sources. Treatment of the black tea brew with simulated gastric juice resulted in a significant increase in the identified theaflavins, implying partial cleavage of thearubigins in the gastric lumen environment. As a result, black tea can be considered a good source of polyphenols and/or antioxidants [25].

2.2 Green tea

Green tea is prepared by leaves of *C. sinensis* which after harvest are heated with rolling immediately for inactivating the polyphenol oxidase (PPO) enzyme which is responsible for oxidation of tea catechins into theaflavins and thearubigins. Green tea is prepared by steaming of fresh leaves and drying at higher temperature to avoid the polymerisation and oxidation of polyphenols [26].

Green tea is a popular tea that is usually consumed as an infusion with a pleasant taste and is thought to have a positive effect on general health even at high doses of 8 to 16 cups per day [27]. Green tea leaves are high in bioactive compounds, especially

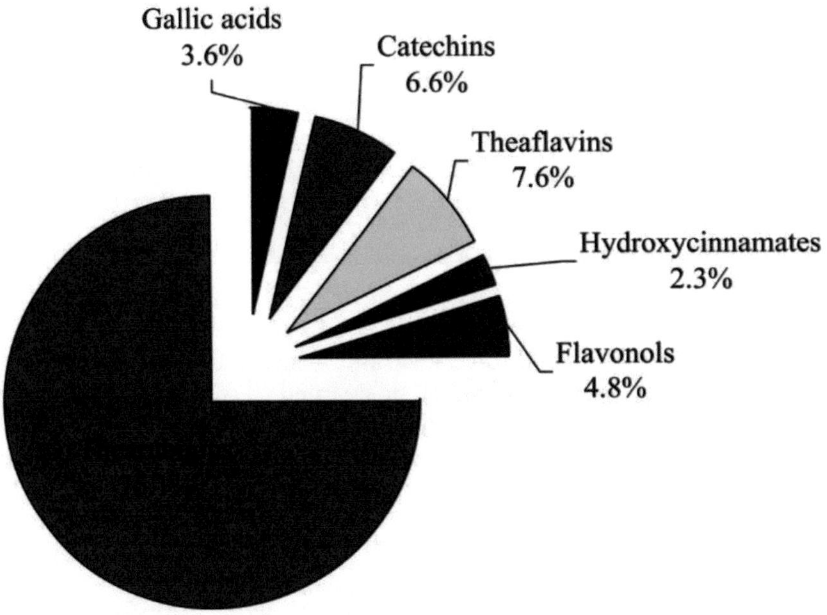

Figure 2.
Mean percentile composition of different classes of polyphenols in consumer brews of different black teas [22].

phenolic compounds with antioxidant activity (**Figure 3**). Although recent studies have identified several other phenolic compounds at lower concentrations, particularly flavonols and phenolic acids, the increased proportion of catechins is related to biological functionality [29]. The stability of green tea flavanols depend on the intactness of the plant cell [16]. Green tea consumption has been shown in scientific studies to improve general health and reduce the risk of severe diseases. This is a trend with promising and positive results to help with body weight control [30], UV radiation protection, physical functional performance [27], oral health, bone health, and other physiological effects [23]. Specific diseases, including those with severe consequences, such as neurodegenerative and cardiovascular diseases, have received special attention.

Green tea health benefits are linked to polyphenolic compounds, which have piqued the interest of the food industry and researchers [31]. Green tea can be used in the formulation of some products to boost antioxidant activity for nutritional or technological purposes. Several mechanisms, similar to those seen in biological structures, can be used to prevent lipid oxidation in food (e.g., free radical scavenging and metal-chelating activity). Lipid oxidation can change physical–chemical and sensory properties like colour, flavour, and taste. Among the many foods that require the use of antioxidants, meat and muscle products are particularly vulnerable to lipid oxidation, necessitating the addition of antioxidants to extend shelf life [23, 29, 32].

2.3 Oolong tea

Oolong tea is a semifermented tea, which is less fermented than black tea. Young green shoots (usually the top three leaves of each branch) are freshly harvested in the early morning and allowed to wither in the sunlight for a few hours before undergoing the semifermentation process, in which tea leaves are oxidised, pan fired at 200°C, rolled into a ball shape, and then dried in a specialised oven at various desired temperatures [33].

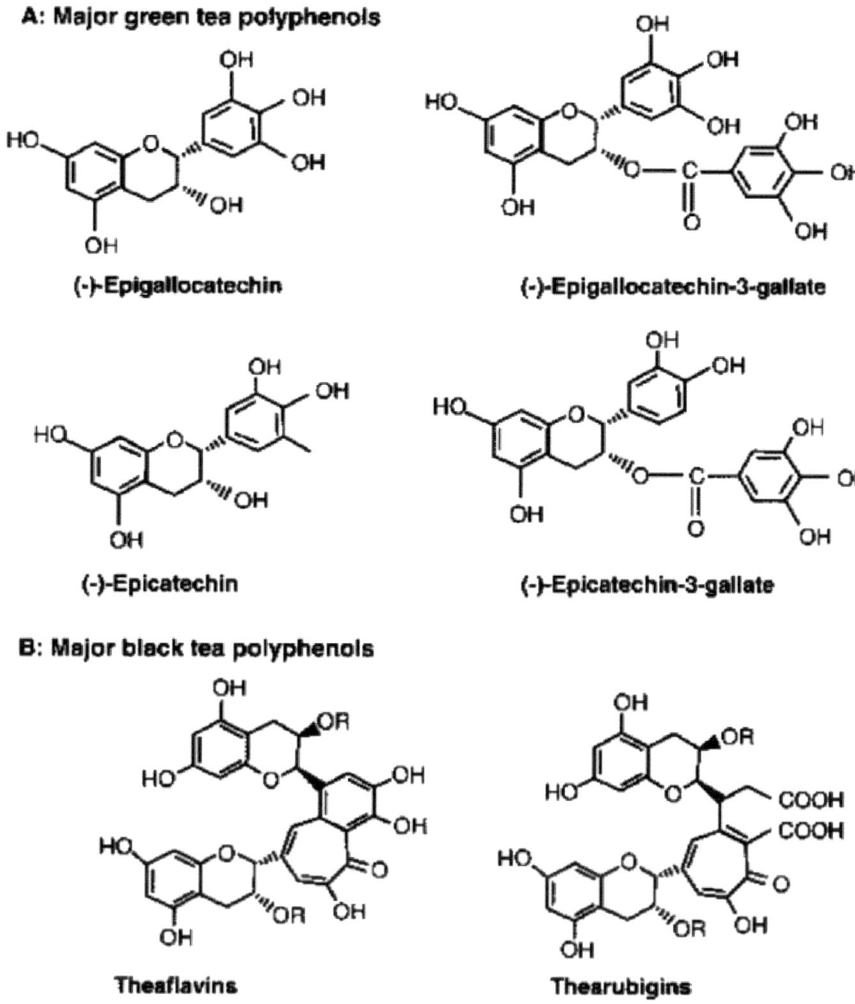

Figure 3.
Major green tea polyphenols [28].

The partial fermentation of oolong tea produces polymerised polyphenols such as procyanidins (condensed tannins) which is unique to the limited time of oxidation process. Oolong tea contain both properties of green tea and black tea with catechins and theaflavins, however it contains half the content epigallocatechin-3-gallate in comparison to green tea [19, 34]. The components of oolong tea are classified in **Table 2**.

Oolong tea, which has a taste and colour between green and black tea, is primarily produced in China's Fujian [33] and Guangdong provinces [37], as well as in Taiwan [38]. Oolong tea absorbs a lot of moisture from the air after a long time in storage, so it needs to be refined by drying on a regular basis. In general, old oolong tea refers to oolong teas that have been stored for more than five years and refined annually through a professional drying process. Experientially, the longer oolong tea is stored and gradually oxidised, the better it tastes and has beneficial effects on human health. As a result, fermentation (oxidation) and drying are two critical steps in the production of oolong tea [39].

Wang et al. [37] discovered that the main bioactive compounds in oolong tea, such as phenolics and flavonoids, have remarkable antioxidant activity and inhibitory potential on the growth of 4 T1 murine breast cancer cells in vitro. Thus, after

Compounds	Contents (mg/g)	
	Oolong tea	Green tea
Flavon-3-ol without galloyl moiety		
Catechin	10	5
Gallocatechin	30	43
Epigallocatechin	6	25
Epicatechin	2	8
Flavon-3-ol without galloyl moiety		
Catechin gallates	7	5
Epicatechin gallates	3	6
Epigallocatechin gallates	14	29
Gallocatechin gallates	16	19
Allocatechin gallates	1.85	—
Gallic acid	2.19	—
Caffeine	64	53
Polymerised	33.65	—
Total polyphenols	99.32	—

Table 2.
Components of oolong and green tea [35, 36].

further research, the phenolic enriched extracts of Tieguanyin tea are expected to have a potential application in the food and pharmaceutical industries. More alike, it has similar constituent of green tea.

3. Phenolic composition in different tea types

The tea leaves are good source of several polyphenols and oxidative enzymes, hence they are selected for preparation of different types of the tea. These tea polyphenols are basically tea flavonoids, earlier known as tannins. The catechins of flavonoid group are the predominant polyphenols of fresh tea, which account for 12–24% of dry weight. Other than catechins, the tea leaves also contain other polyphenols such as phenolic acids, anthocyanidin and flavonols along with their glycosides [40, 41]. Depending on the harvesting season, cultivars, cultivation conditions and manufacturing process the polyphenolic content varies in different types of tea. The major catechins in tea leaves are (−)-epigallocatechin gallate (EGCG; 9170–14900 mmol/100 g leaf), (−)-epi-gallocatechin (ECC; 8060–17900 mmol/100 g leaf), (−)-epicatechin gallate (ECG; 1400–2350 mmol/100 g leaf [42]. The catechin content in different tea types vary depending on fermentation process, green tea produced without fermentation contain highest amount of catechins among which EGCG is the major catechin found. Considering EGCG as an abundant catechin in all tea types, as the fermentation process is increased in different tea types the EGCG content decreased in different tea types; green tea (70.2 mg/100 g), oolong tea (34.48 mg/100 g), and black tea (9.36 mg/100 g) [43].

Flavonoids are phenolic compounds that are divided into several sub-classes: anthocyanidins, flavanones, flavanols, flavones, flavonols, and isoflavones. These sub-classes share a basic structure of 15 carbons with a three-carbon bridge connecting two aromatic rings in the C6–C3–C6 configuration. Along with flavonoids, phenolic acids, which are divided into hydroxybenzoic acids and hydroxycinnamic acids, are an important group. Gallic acid, also known as 3,4,5-trihydroxybenzoic acid, has a relatively simple structure. This compound serves as the foundation for hydroxybenzoic acids and other derivatives with antioxidant activity, such as ellagic acid [44–46]. The hydroxycinnamic acid derivatives, on the other hand, have p-coumaric acid as the basic structure, which is formed by an aromatic ring with one hydroxy substitution and one propenoic acid.

In tea leaves almost 20 different flavonols and their glycosides have been detected which include quercetin, kaempferol, and myricetin which account for 2–3% of the water-soluble extractive in green tea [47, 48]. The main flavonol glycosides found in tea are rutin, quercetin glycoside and kaempferol glycoside with 0.05–0.15%, 0.2–0.5% and 0.16–0.35%, respectively of dry weight [47]. The other group of phenolic compounds found in tea are phenolic acids which account for 5% of tea leaf dry weight. The major phenolic acids found in tea are gallic acid, chlorogenic acid and theogallin which account for 0.5–1.4%, 0.3% and 1–2%, respectively of dry weight content. Whereas ellagic acid and m-digallic acid are found in trace amounts. These phenolic acids act as precursors of catechin gallate and in association with other polyphenols have an effect on the astringency of tea beverage [40, 41]. Anthocyanidins and leucoanthocyanidins are another group of phenolic compounds in fresh tea leaves with 2–3% of dry weight content. Anthocyanidins such as cyanidin, pelargonidin, delphinidin and tricetinidin are approximately 0.01% of dry weight in tea leaves, however they may reach up to 1.0% in processed tea which gives a purple colour to the tea preparation with some bitterness [40, 41].

Infusions of black tea contain relatively high levels of catechins, ranging from 102 to 418 mg of total catechins/L [49]. The four major tea catechins are enzymatically oxidised and converted to various oxidation products containing black tea polyphenols during tea fermentation. Characteristic pigments of these oxidation products are typically classified into two major groups: theaflavins and thearubigins [48, 50]. Theaflavin content of black tea leaves ranges between 0.8 and 2.8 percent, depending on fermentation conditions. Thearubigins, on the other hand, can account for up to 60% of the solids in black tea infusions [50]. In black tea, theaflavin, theaflavin 3-gallate, theaflavin 3′-gallate, and theaflavin 3,3′-digallate are the main theaflavins formed by the reaction of quinones derived from a simple catechin and a gallocatechin [48]. With relative molecular masses ranging from 700 to 40,000 Da, thearubigins remain ambiguous, and little is known about their chemical structures [48, 50]. Aside, from these unidentified colourless oxidation products, tea fermentation produces a plethora of other unidentified colourless oxidation products.

The polyphenols in green tea accounting for 40% of the dry leaf weight include different polyphenolic groups such as flavonols (quercetin, kaempferol and rutin), phenolic acids, leucoanthocyanins, caffeine and theanine [4]. Theanine is a non-protein amino acid which account for 1.5–3.0% of dry weight of tea leaf and is almost 50% of the total amino acids content in tea. The natural theanine in tea is usually in L-form which significantly contributes to the sweet taste in green tea infusions [41].

4. Analysis of polyphenols in tea

The high level of interest in green tea composition has been linked to antioxidant activity and, as a result, elevated phenolic content. Recently, a wide range of compounds have been identified, and several methods for identifying and quantifying these compounds have been developed. Some phenolic compound

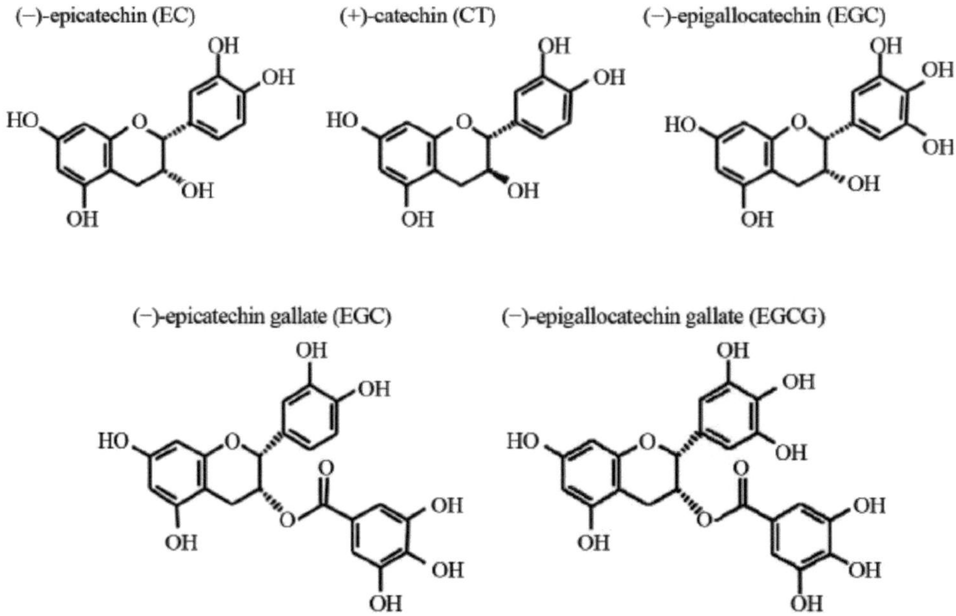

Figure 4.
Chemical structures of the green tea polyphenols (−) -epicatechin (EC), (+) catechin (CT), (−) -epigallocatechin (EGC), (−) -epicatechin gallate (ECG), and (−) epigallocatechin gallate (EGCG) [51].

properties have been considered for identifying each class of phenolic compounds in several matrices. Because degradation of important phenolic compounds in green tea can reach 70% at temperatures lower than those used in gas chromatography, thermal sensitivity necessitates the use of liquid chromatography rather than gas chromatography [43]. The double bonds in the aromatic ring of phenolic allow for UV–visible spectrophotometric measurements. The maximum absorption evaluation indicates, at the very least, the subclass (e.g., flavanol, flavonol, and flavones) or supports the identification with a standard.

The distinct fragmentation pattern of each phenolic compound enables identification in mass analyzers or provisional identification for compounds lacking a standard, even for complex and high molecular weight compounds (**Figures 1**, **3** and **4** depicts the phenolic structures present in different teas) [31, 44]. Given the aforementioned characteristics, liquid chromatography separation followed by spectrophotometry and/or mass spectrometry analysis can provide valuable information for the investigation of the phenolic profile in tea extracts. Other analyses were also carried out using nuclear magnetic resonance (NMR) to provide solid information on the phenolic profile of tea [44].

5. Health beneficial effects of tea

Epidemiological studies on the benefits of green tea consumption against major diseases, supported by *in vitro* and *in vivo* experiments have revealed promising protective effects. Catechins, the major phenolic constituents of green tea, are also the compounds linked to health benefits through modulation of relevant mechanisms altered by important diseases, as discussed in this section. Tea is high in antioxidants. Caffeine content of tea is lower than that of coffee. Tea may lower your chances of having a heart attack or having a stroke. It may aid in weight loss and beneficial to your bones. Tea may help you keep your grin bright. It helps to enhance the immune system and aid in the fight against cancer. Herbal tea may be beneficial to the digestive system. The benefits of tea are depicted in **Figure 5** and explained in detail below.

5.1 Antioxidant properties of tea polyphenols

Antioxidant activity is defined as a molecule's or ion's ability to avoid oxidative reactions with other molecules. Phenolic compounds found in green tea leaves have antioxidant potential through a variety of mechanisms, providing additional protection against oxidants as well as additional protection against oxidative reactions and reactive species [44]. The oxidative series of events proposed by Miguel et al. [39] provides an overview of the major antioxidant effects (preventive and primary antioxidants), which may also be presented by polyphenol-rich green tea extracts. Preventive antioxidants can protect against oxidative reactions by lowering local oxygen concentrations, preventing the initiation of chain reactions by scavenging radicals (e.g., HO•, O2•-), preventing radical generation, and breaking down lipid peroxides to peroxyl and alkoxyl radicals.

Primary antioxidants play a role in subsequent events by causing peroxides to decompose into nonradical products and inhibiting hydrogen removal from oxidable by intermediate radicals such as peroxyl and alkoxyl radicals. These radicals are reactive oxygen species that cause oxidative damage to biological and food systems. The major effects are related to lipid and protein oxidation, membrane damage, mutagenesis, and carcinogenesis, and it is critical to assess how natural extracts impact and mitigate these effects.

Many studies have found a strong and positive correlation ($p > 0.05$) between the phenolic compounds and their antioxidant potential in several plant species [52, 53]. This antioxidant mechanism found in plants plays an important role in reducing lipid oxidation in (plant and animal) tissues, because when it is included in the human diet, it not only preserves the food's quality, but it also lowers the risk of developing certain diseases.

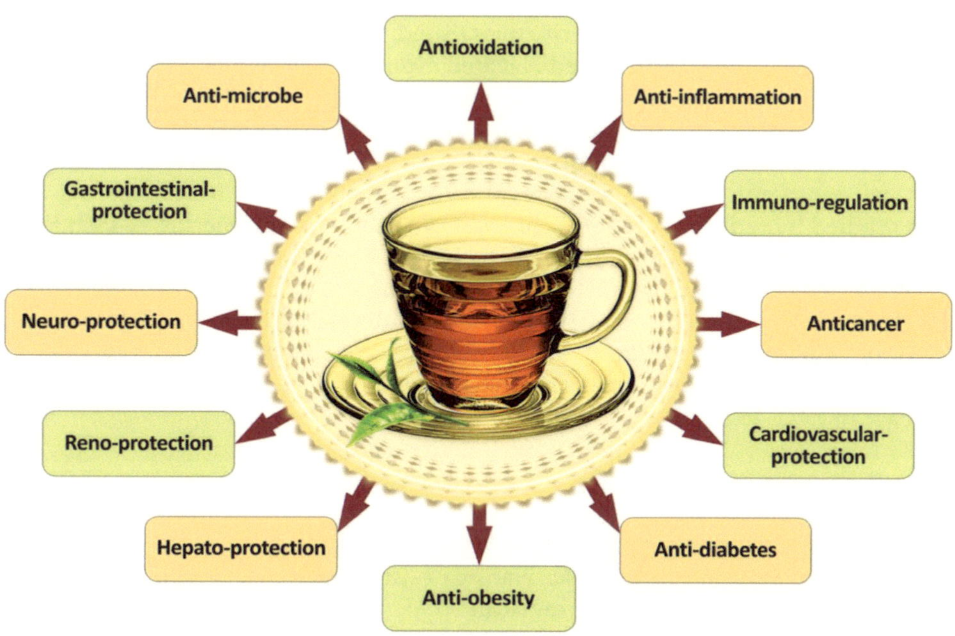

Figure 5.
Health benefits of tea.

The correlation between phenolic content and antioxidant activity as measured by multiple methods is an important finding in studies on phenolic compounds in green tea extracts. Green tea's structural differences among phenolic compounds also play an important role in antioxidant activity. Individual flavanol content of green tea was found to be inversely related to radical content of green tea leaves in a study conducted by Socha et al. [54]. The correlation coefficient for epigallocatechin gallate was higher than that of the other tested flavanols. This result was linked to the presence of hydroxyl groups in the aromatic rings of gallyl and galloyl substituents, because flavanols lacking this substituent had lower antioxidant activity.

Induction of antioxidant enzymes by EGCG with detailed molecular mechanism have been studied [9]. EGCG has been studied for antiradical activity which was proved to have stronger activity than the antioxidants such as vitamin E and vitamin C [55]. By binding to different molecules, catechins in particular EGCG modulate the compounds' activity and inturn regulates the cell-signalling pathway [56].

5.2 Cardiovascular disease

Cardiovascular diseases are one of the leading causes of death, accounting for nearly one-third of all deaths globally. Sano et al. [57] investigated the relationship between green tea consumption and the incidence of cardiovascular disease and discovered that patients without cardiovascular disease consumed more green tea than those with cardiovascular disease (5.9 and 3.5 cups, respectively). Green tea consumption was linked to a lower risk of coronary artery disease in Chinese patients, according to Wang et al. [58]. In this study, the risk was found to be inversely related to green tea consumption, with a dose-dependent effect as the frequency, period, and amount of green tea consumed increased. Fung et al. [59] reported that chronic green tea consumption results in a different pattern of behaviour. Plasma levels of selected catechin derivatives were measured after 1 and 2 hours of green tea consumption, as well as after 7 days of daily consumption. After 1 hour of tea consumption, the plasma level of epigallocatechin gallate was the highest among the catechin derivatives tested, followed by epigallocatechin and epicatechin gallate, which remained elevated even after 2 hours of green tea consumption. An unexpected result was observed in the chronic consumption evaluation because only epicatechin gallate had a higher level in plasma.

The effectiveness of tea polyphenols against CVD by regulating lipid metabolism, cell proliferation, platelet aggregation and antithrombotic activity has been studied by reduction of total cholesterol, LDL and triglycerides which are helpful in development of atherosclerosis [60]. The study on EGCG has been shown effective in reduction of lipid metabolising enzyme activities in the serum and cardiac tissue thereby resulting into less lipidemic pathologies [61]. In an in vivo study on mice, 40 mg/kg/day of EGCG was administered which resulted in decrease of LDL and the size of atherosclerotic plaques in the aortas, and increase in HDL [62].

5.3 Anti-cancer properties

Increased levels of reactive oxygen species (ROS) and oxidative stress modulation have an important role in the activation of carcinogenesis, and polyphenols act against these mechanisms by preventing or controlling the tumorigenesis [63]. Tea polyphenols have been effective in inhibiting enzymes related to cell proliferation and tumour progression [34]. Theaflavins in tea can act as anti-cancerous compounds by controlling the DNA damage which is the main cause of induction of cancer. They act by scavenging the free radicals which inhibits the mutagenicity and cleavage of single strand DNA [41].

Suppression of elevated cytochrome P450 1A1 (CYP1A1) in cells is inhibited by theaflavins which inturn prevents the cellular DNA damage, carcinogen related DNA damage and oxidative stress induced cytotoxicity [59]. EGCG has been investigated for proliferation of epidermal cell line A431 in humans. *In vitro* inhibition of protein tyrosine kinase activities of EGF-R, PDGF-R and FGF-R are strongly inhibited the DNA synthesis of A431 cells [64]. Tea polyphenols have been shown inhibition for PKC, MAPK, and AP-1 activities in NIH 3 T3 cells [65]. In mouse epidermal JB6 C1 41 cells inhibition of UVB-induced phosphatidylinositol-3-kinase (PI3K) activation was studied for tea polyphenols [66]. Some tea catechins of green and black tea have been potent inhibitor of Bcl-2 antiapoptotic family proteins, which shows a strong link of tea polyphenols and their anticancerous properties [34].

Given the growing interest in the relationship between dietary flavonoids and cancer initiation and progression, this important field is likely to see increased effort and attract and stimulate additional vigorous research [67]. In liver carcinoma cells, effect of tea catechins have studied and showed that due to the activity of EGCG, H2P2 mediated cytotoxicity was supressed with increase in cellular glutathione levels [63]. The effect of catechin, epicatechin, ECG, EGC and EGCG in A549 cells have been studied for apoptosis and cell profileration [68]. Tea polyphenols have been shown for inhibition activity for the enzymes involved in oestrogen biosynthesis, which might play role in the development of breast cancer [69].

5.4 Obesity and lipogenesis

Tea catechins have been proved to be very effective for obesity by the acting on the adipose tissue. These tea catechins have been effective for suppression of enzymes involved in fatty acid, triacylglycerols and cholesterol metabolism [70]. Rocha et al. [71] showed in rat model study that daily consumption of green tea extract decreased adipose tissue, adiposity index, cholesterol, triacylglycerols and reduction in hypertrophy of adipocytes. Green tea catechins were showed for inhibition of enzymes metabolising noradrenaline, this mechanism have been effective in lipid metabolism [72]. A study conducted in the United States on men and women for consumption of black, oolong and green tea was showed inverse association for body mass index and metabolic syndrome markers [73].

5.5 Other health beneficial effects

Significant in vitro and animal model research support the beneficial effects of polyphenols in a variety of gastrointestinal diseases [74]. Recent human studies suggest that green tea may help to promote oral health as well as other physiological functions like anti-hypertensive effect, body weight control, anti-inflammatory, anti-antibacterial, and antiviral activity, solar ultraviolet protection, bone mineral density increase, anti-fibrotic properties, and neuroprotective power [20, 75]. Tea catechins have been studied for beneficial activity on bone, wherein the cell lines and animal model studies revealed that they are effective for osteoporosis [76]. Green tea catechins have been investigated as dietary polyphenols for their neurodegenerative diseases due to their anti-amyloidogenic properties [77]. Also EGCG has been studied for neuroprotective properties by evaluation of its brain accessibility [78]. Tea catechins also have been shown effective against hyperglycemia and its related type 2 diabetes mellitus complications [79]. Green tea consumption has increased bone formation and improved bone strength; however, it decreased the process for deterioration of bone microstructure which was studied in postmenopausal women [80].

Manach et al. [81] estimated the daily intake of catechin and proanthocyanidin dimers and trimers to be 18–50 mg/d. In Caco-2 cells, efflux transport was greatest in the following order: EC > EGC > ECG = EGCG [82]. Pgp, MRP1 and MRP2 efflux transporters have also been found to play roles in the absorption and excretion of green tea catechins [83]. Recent research has shown that green tea catechins undergo methylation, glucuronidation, and sulfation in in vitro, animal, and human systems [81, 83, 84].

6. Effect of tea phenolic on iron absorption

Iron is stored in the body as ferritin and hemosiderin, which are found throughout the body, with a largest amount typically found in the liver, spleen, and bone marrow. Tea flavonoids are responsible for tea's inhibitory action on non-heme iron absorption [85]. Tea flavonoids are polyphenols with two aromatic rings and two or more hydroxyl groups as a functional group [86]. The development of a complex compound of tea flavonoids with iron is the process through which tea inhibits iron absorption. Iron is selectively bound by the galloyl group primarily present in these phenolic compounds [87]. Merhav et al. [88] revealed the iron status of Israeli infants in their investigation. They discovered an overall frequency of anaemia of 48.4% and a tenfold greater incidence of microcytic anaemia in tea-drinking neonates compared to the non-tea-drinking control group. Razagui et al. [89] investigated the iron status of 15 mentally challenged menstruation women, a population with limited food intake. They examined the link between tea drinking and iron status. It was discovered that participants with depleted iron levels consumed much more tea during meals (563 ml/meal/d) than ladies with adequate iron reserves (184 ml/meal/d). According to Zijp et al. [90], simultaneous consumption of tea reduces iron absorption from a test meal by 60 to 70 percent.

7. Application of tea phenolics in textile and allied sectors

Polyphenols can be grafted onto fibres and fabrics using both enzyme-mediated and non-enzyme-mediated techniques, and their bioactivities vary depending on the type of phenolic compound used. In the development of environmentally friendly coloration and functionalization of textiles, polyphenol grafting onto textile fibres is a promising alternative to conventional synthetic dyestuffs [91]. Cheng et al. [92] reported in the literature on the use of tea as a natural dye and flame retardant finish on silk. They discovered that the oxidative polymerisation of polyphenols during alkaline extraction resulted in the formation of macromolecular polyphenols, which could give silk flame retardancy. Postmordanting with metal salts clearly improved the poor fastness characteristics. Because sufficient tea stem extract was used, dyed silk demonstrated good flame retardant, antibacterial, and antioxidant properties. According to Bonet-Aracil et al. [93], tea extracts behave differently depending on the type of tea used (green, red, or black). Green tea has the highest total antioxidant content when it comes to antioxidant effect. While dyeing, red tea had the highest colour strength value, whereas green tea had the lowest UPF value and red and black had higher values.

8. Conclusion

Tea, which is consumed all over the world, is thought to be not only a popular beverage but also can be an antioxidative agent that is readily available in everyday life. Polyphenols such as theaflavins and thearubigins, as well as catechins, which are major constituents of tea, are thought to be primarily responsible for several beneficial effects. Tea's antioxidant properties include its ability to inhibit free radical generation, scavenge free radicals, and chelate transition metal ions. It is clear that, despite extensive research, the composition of tea is still unknown. Only 20% of the approximately 2.5 million metric tonnes of dried tea produced is green tea, and less than 2% is oolong tea. Green tea is primarily consumed in China, Japan, and a few North African and Middle Eastern countries. Fresh tea leaf is unusually high in catechins, a flavanol group of polyphenols that can account for up to 30% of the dry leaf weight. Other polyphenols include flavonols and their glycosides, as well as depsides like chlorogenic acid, coumarylquinic acid, and theogallin, which is unique to tea (3-galloylquinic acid). Caffeine is present in an average concentration of 3%, with very small amounts of the other common methylxanthines, theobromine and theophylline also present. Teas used in pharmacological studies should be classified according to their type, source, and method of production. It would be preferable to include analytical information such as caffeine and catechin content. Methods of preparation should be specified when using tea extracts or fractions. Tea polyphenols can also reduce the risk of certain types of cancer by inducing phase I and phase II metabolic enzymes that increase the formation and excretion of carcinogen-detoxified metabolites. The research interest in tea components may provide a method to reduce the incidence and mortality from a variety of diseases. In general, tea is a more affordable natural beverage than modern beverages such as soft drinks.

Author details

Jyoti V. Vastrad[1*], Pratikhya Badanayak[1] and Giridhar Goudar[2]

1 Department of Textile and Apparel Designing, College of Rural Home Science, University of Agricultural Sciences, Dharwad, Karnataka, India

2 Biochem Research and Testing Laboratory, Dharwad, Karnataka, India

*Address all correspondence to: vastradjv@uasd.in

References

[1] Zandi, P., & Gordon, M. H. (1999). Antioxidant activity of extracts from old tea leaves. Food Chemistry, *64*(3), 285-288.

[2] Chang, K. (2015). World tea production and trade: Current and future development. Food and Agriculture Organization of The United Nations, Rome.

[3] Tenore GC, Daglia M, Ciampiglia R, Novellino E (2015) Exploring the nutraceutical potential of polyphenols from black, green, and white tea infusions – an overview. Curr Pharm Biotechnol 16(3):1-7.

[4] Graham, H. N. (1992). Green tea composition, consumption, and polyphenol chemistry. Preventive medicine, *21*(3), 334-350.

[5] Liang, Y., Lu, J., Zhang, L., Wu, S., & Wu, Y. (2003). Estimation of black tea quality by analysis of chemical composition and colour difference of tea infusions. Food chemistry, *80*(2), 283-290.

[6] Reto, M., Figueira, M. E., Filipe, H. M., & Almeida, C. M. (2007). Chemical composition of green tea (Camellia sinensis) infusions commercialized in Portugal. Plant foods for human nutrition, *62*(4), 139.

[7] FAOSTAT. FAO database. Food and Agriculture Organization; United Nations. http://faostat3.fao.org/download/Q/QC/E. Accessed Jan 2019.

[8] Vastrad, J. V., Byadgi, S. A., Goudar, G., & Kotur, R. (2014). Characterization of phytoconstituents in leaf extracts of forest species for textile applications. Forest Products Journal, *64*(7-8), 259-264.

[9] Baldi, A., Abramovič, H., Poklar Ulrih, N., & Daglia, M. (2019). Tea catechins. Handbook of Dietary Phytochemicals, 1-46.

[10] Drynan, J. W., Clifford, M. N., Obuchowicz, J., & Kuhnert, N. (2010). The chemistry of low molecular weight black tea polyphenols. Natural Product Reports, *27*(3), 417-462.

[11] Lucci, P., Saurina, J., & Núñez, O. (2017). Trends in LC-MS and LC-HRMS analysis and characterization of polyphenols in food. TrAC Trends in Analytical Chemistry, *88*, 1-24.

[12] Mao, X., Gu, C., Chen, D., Yu, B., & He, J. (2017). Oxidative stress-induced diseases and tea polyphenols. Oncotarget, *8*(46), 81649.

[13] Lucena, A. P. S., Nascimento, R. J. B., Maciel, J. A. C., Tavares, J. X., Barbosa-Filho, J. M., & Oliveira, E. J. (2010). Antioxidant activity and phenolics content of selected Brazilian wines. Journal of Food Composition and Analysis, *23*(1), 30-36.

[14] Matanjun, P., Mohamed, S., Mustapha, N. M., Muhammad, K., & Ming, C. H. (2008). Antioxidant activities and phenolics content of eight species of seaweeds from north Borneo. Journal of Applied Phycology, *20*(4), 367-373.

[15] Yang, J., & Liu, R. H. (2013). The phenolic profiles and antioxidant activity in different types of tea. International Journal of Food Science & Technology, *48*(1), 163-171.

[16] Persson, I. A. L. (2013). Tea flavanols: An overview. Tea in Health and Disease Prevention, 73-78.

[17] McKay DL, Blumberg JB. 2002. The role of tea in human health: an update. J Am Coll Nutr 21(1):1-13.

[18] Sajilata, M. G., Bajaj, P. R., & Singhal, R. S. (2008). Tea polyphenols as nutraceuticals. Comprehensive reviews in food science and food safety, 7(3), 229-254.

[19] Balentine, D.A., Wiseman, S.A., Bouwens, L.C.M., 1997. The chemistry of tea flavonoids. Critical Reviews in Food Science and Nutrition 37, 693e704.

[20] Gardner, E. J., Ruxton, C. H. S., & Leeds, A. R. (2007). Black tea–helpful or harmful? A review of the evidence. European journal of clinical nutrition, *61*(1), 3-18.

[21] Evans, W.C., 2009. Trease and Evans Pharmacognosy, 16th international edition. Saunders/Elsevier Ltd, Edinburgh.

[22] Rechner, A. R., Wagner, E., Van Buren, L., Van De Put, F., Wiseman, S., & Rice-Evans, C. A. (2002). Black tea represents a major source of dietary phenolics among regular tea drinkers. Free Radical Research, *36*(10), 1127-1135.

[23] Liebert, M., Licht, U., Böhm, V., & Bitsch, R. (1999). Antioxidant properties and total phenolics content of green and black tea under different brewing conditions. Zeitschrift für Lebensmitteluntersuchung und-forschung A, *208*(3), 217-220.

[24] Rababah, T. M., Hettiarachchy, N. S., & Horax, R. (2004). Total phenolics and antioxidant activities of fenugreek, green tea, black tea, grape seed, ginger, rosemary, gotu kola, and ginkgo extracts, vitamin E, and tert-butylhydroquinone. Journal of Agricultural and Food Chemistry, *52*(16), 5183-5186.

[25] Stewart, A. J., Mullen, W., & Crozier, A. (2005). On-line high-performance liquid chromatography analysis of the antioxidant activity of phenolic compounds in green and black tea. Molecular Nutrition & Food Research, *49*(1), 52-60.

[26] Katiyar SK, Bergamo BM, Vyalil PK, Elmets CA. 2001. Green tea polyphenols: DNA photodamage and photoimmunology. J Photochem Photobiol B Biol 65:109-114.

[27] Šilarová, P., Česlová, L., & Meloun, M. (2017). Fast gradient HPLC/MS separation of phenolics in green tea to monitor their degradation. Food chemistry, *237*, 471-480.

[28] Anesini, C., Ferraro, G. E., & Filip, R. (2008). Total polyphenol content and antioxidant capacity of commercially available tea (Camellia sinensis) in Argentina. Journal of agricultural and food chemistry, *56*(19), 9225-9229.

[29] Kodama, D. H., Gonçalves, A. E. D. S. S., Lajolo, F. M., & Genovese, M. I. (2010). Flavonoids, total phenolics and antioxidant capacity: comparison between commercial green tea preparations. Food Science and Technology, *30*(4), 1077-1082.

[30] Clark, J., & You, M. (2006). Chemoprevention of lung cancer by tea. Molecular nutrition & food research, *50*(2), 144-151.

[31] Ryu, J., Kim, M. J., & Lee, J. (2019). Extraction of green tea phenolics using water bubbled with gases. Journal of food science, *84*(6), 1308-1314.

[32] Minatel, I. O., Borges, C. V., Ferreira, M. I., Gomez, H. A. G., Chen, C. Y. O., & Lima, G. P. P. (2017). Phenolic compounds: Functional properties, impact of processing and bioavailability. Phenolic Compd. Biol. Act, 1-24.

[33] Zhu, Q. Y., Hackman, R. M., Ensunsa, J. L., Holt, R. R., & Keen, C. L. (2002). Antioxidative activities of oolong tea. Journal of Agricultural and Food Chemistry, *50*(23), 6929-6934.

[34] Lin, J. K. (2009). Mechanisms of cancer chemoprevention by tea and tea polyphenols. Tea and Tea Products Chemistry and Health-Promoting Properties; Ho, CT, Lin, JK, Shahidi, F., Eds, 161-176.

[35] Toyoda-Ono, Y., Yoshimura, M., Nakai, M., Fukui, Y., Asami, S., Shibata,

H., Kiso, Y. & Ikeda, I. (2007). Suppression of postprandial hypertriglyceridemia in rats and mice by oolong tea polymerized polyphenols. Bioscience, biotechnology, and biochemistry, *71*(4), 971-976.

[36] Weerawatanakorn, M., Hung, W. L., Pan, M. H., Li, S., Li, D., Wan, X., & Ho, C. T. (2015). Chemistry and health beneficial effects of oolong tea and theasinensins. Food Science and Human Wellness, *4*(4), 133-146.

[37] Wang, Y., Kong, D., Gao, Y., Ying, L., Huang, Q., & Xu, P. (2019). Chemical characterization and bioactivity of phenolics from Tieguanyin oolong tea. Journal of food biochemistry, *43*(7), e12894.

[38] Dou, J., Lee, V. S., Tzen, J. T., & Lee, M. R. (2007). Identification and comparison of phenolic compounds in the preparation of oolong tea manufactured by semifermentation and drying processes. Journal of Agricultural and Food Chemistry, *55*(18), 7462-7468.

[39] Miguel, M. G., Cruz, C., Faleiro, L., Simões, M. T., Figueiredo, A. C., Barroso, J. G., & Pedro, L. G. (2010). Foeniculum vulgare essential oils: chemical composition, antioxidant and antimicrobial activities. *Natural product communications*, 5(2), 1934578X10 00500231.

[40] Wan, X., Li, D., & Zhang, Z. (2008a). Antioxidant properties and mechanism of tea polyphenols. Tea and tea products chemistry and health-promoting properties, 131-155.

[41] Wan, X., Zhang, Z., & Li, D. (2008b). Chemistry and Biological Properties of Theanine. Tea and tea products: chemistry and health-promoting properties, 255.

[42] Sakakibara, H., Honda, Y., Nakagawa, S., Ashida, H., & Kanazawa, K. (2003). Simultaneous determination of all polyphenols in vegetables, fruits, and teas. Journal of agricultural and food chemistry, *51*(3), 571-581.

[43] Verloop, A. J., Gruppen, H., Bisschop, R., & Vincken, J. P. (2016). Altering the phenolics profile of a green tea leaves extract using exogenous oxidases. Food chemistry, *196*, 1197-1206.

[44] Lee, B. H., Nam, T. G., Park, N. Y., Chun, O. K., Koo, S. I., & Kim, D. O. (2016). Estimated daily intake of phenolics and antioxidants from green tea consumption in the Korean diet. International journal of food sciences and nutrition, *67*(3), 344-352.

[45] Liu, Z., Bruins, M. E., Ni, L., & Vincken, J. P. (2018). Green and black tea phenolics: Bioavailability, transformation by colonic microbiota, and modulation of colonic microbiota. Journal of agricultural and food chemistry, *66*(32), 8469-8477.

[46] Sun, A. Y., Wang, Q., Simonyi, A., & Sun, G. Y. (2008). Botanical phenolics and brain health. Neuromolecular medicine, *10*(4), 259-274.

[47] Wan, X. C., Huang, J. Z., and Shen, S. R., Eds. 2003. Tea biochemistry, 3rd ed., 9-20, 180-94. Beijing: China Agriculture Press.

[48] Wang, H., Provan, G. J., & Helliwell, K. (2000). Tea flavonoids: their functions, utilisation and analysis. Trends in Food Science & Technology, *11*(4-5), 152-160.

[49] Arts, I. C., van de Putte, B., & Hollman, P. C. (2000). Catechin contents of foods commonly consumed in The Netherlands. 1. Fruits, vegetables, staple foods, and processed foods. Journal of agricultural and food chemistry, *48*(5), 1746-1751.

[50] Tanaka, T., Matsuo, Y., & Kouno, I. (2005). A novel black tea pigment and two new oxidation products of epigallocatechin-3-O-gallate. Journal of

agricultural and food chemistry, *53*(19), 7571-7578.

[51] Goodman, B. A., Yeretzian, C., Stolze, K., & Wen, D. (2013). Quality aspects of coffees and teas: Application of electron paramagnetic resonance (EPR) spectroscopy to the elucidation of free radical and other processes. Agricultural Sciences, *4*(08), 433.

[52] Luca, S. V., Macovei, I., Bujor, A., Miron, A., Skalicka-Woźniak, K., Aprotosoaie, A. C., & Trifan, A. (2020). Bioactivity of dietary polyphenols: The role of metabolites. Critical reviews in food science and nutrition, 60(4), 626-659.

[53] Rawel, H. M., & Kulling, S. E. (2007). Nutritional contribution of coffee, cacao and tea phenolics to human health. Journal für Verbraucherschutz und Lebensmittelsicherheit, *2*(4), 399-406.

[54] Socha, R., Bączkowicz, M., Fortuna, T., Kura, A., Łabanowska, M., & Kurdziel, M. (2013). Determination of free radicals and flavan-3-ols content in fermented and unfermented teas and properties of their infusions. European Food Research and Technology, *237*(2), 167-177.

[55] Granja A, Frias I, Rute Neves A et al (2017) Therapeutic potential of epigallocatechin gallate nanodelivery systems. *Bio Med Res Int* ID5813793:1-15.

[56] Rahmani AH, Al Shabrmi FM, Allemailem KS et al (2015) Implications of green tea and its constituents in the prevention of cancer via the modulation of cell signalling pathway. *Bio Med Res Int* ID925640:1-12.

[57] Sano, J., Inami, S., Seimiya, K., Ohba, T., Sakai, S., Takano, T., & Mizuno, K. (2004). Effects of green tea intake on the development of coronary artery disease. Circulation Journal, *68*(7), 665-670.

[58] Wang, Q. M., Gong, Q. Y., Yan, J. J., Tang, J. J., Wang, M. W., Yang, Z. J., & Wang, L. S. (2010). Association between green tea intake and coronary artery disease in a Chinese population. Circulation Journal, 0912140554-0912140554.

[59] Fung, S. T., Ho, C. K., Choi, S. W., Chung, W. Y., & Benzie, I. F. (2013). Comparison of catechin profiles in human plasma and urine after single dosing and regular intake of green tea (Camellia sinensis). British journal of nutrition, *109*(12), 2199-2207.

[60] Santhakumar AB, Battino M, Alvarez-Suarez JM (2018) Dietary polyphenols: structures, bioavailability and protective effects against atherosclerosis. Food Chem Toxicol 113:49-65.

[61] Gokulakrisnan A, Jayachandran Dare B, Thirunavukkarasu C (2011) Attenuation of the cardiac inflammatory changes and lipid anomalies by (-)-epigallocatechin-gallate in cigarette smokeexposed rats. Mol Cell Biochem 354(1-2):1-10.

[62] Wang W, Zhang ZZ, Wu Y et al (2018) (-)-epigallocatechin-3-gallateameliorates atherosclerosis and modulates hepatic lipid metabolic gene expression in apolipoprotein E knockout mice: involvement of TTC39B. Front Pharmacol 9(195):1-10.

[63] Chikara S, Nagaprashantha LD, Singhal J et al (2018) Oxidative stress and dietary phytochemicals: role in cancer chemoprevention and treatment. Cancer Lett 413:122-134.

[64] Liang, Y. C., Lin-Shiau, S. Y., Chen, C. F., and Lin, J. K. 1997. Suppression of extracellular signals and cell proliferation through EGF receptor binding by (–)-epigallocatechin-3-gallate in human A431 epidermoid carcinoma cells. J. Cell. Biochem. 67:55-65.

[65] Chen, Y. C., Liang, Y. C., Lin-Shiau, S. Y., Ho, C.-T., and Lin, J. K. 1999.

Inhibition of TPA-induced protein kinase C and transcription activator protein-1 binding activities by theaflavin-3,3'-digallate from black tea in NIH3T3 cells. J. Agric. Food Chem. 47:1415-1421.

[66] Nomura, M., Kaji, A., He, Z., Ma, W. Y., Miyamoto, K., Yang, C. S., and Dong, Z. 2001. Inhibitory mechanisms of tea polyphenols on the UVB-activated phosphatidyl-inositol 3-kinase dependent pathway. J. Biol. Chem. 276:46624-46631.

[67] Kanadaswami, C., Lee, L. T., Lee, P. P. H., Hwang, J. J., Ke, F. C., Huang, Y. T., & Lee, M. T. (2005). The antitumor activities of flavonoids. In vivo, *19*(5), 895-909.

[68] Yamauchi R, Sasaki K, Yoshida K (2009) Identification of epigallocatechin-3-gallate in green tea polyphenols as a potent inducer of p53-dependent apoptosis in the human lung cancer cell line A549. Toxicol In Vitro 23(5):834-839.

[69] Capellino S, Straub RH, Cutolo M (2014) Aromatase and regulation of the estrogen to androgen ratio in synovial tissue inflammation: common pathway in both sexes. Ann N Y Acad Sci 1317(1): 24-23.

[70] Suzuki T, Pervin M, Gotoe S et al (2016) Beneficial effects of tea and the green tea catechin epigallocatechin-3-gallate on obesity. Molecules 21(1305): 1-13.

[71] Rocha A, Bolin AP, Cardoso CAL et al (2016) Green tea extract activates AMPK and ameliorates white adipose tissue metabolic dysfunction induced by obesity. Eur J Nutr 55:2231-2244.

[72] Fei Q, Gao Y, Zhang X et al (2014) Effects of Oolong tea polyphenols, EGCG, and EGCG3" Me on pancreatic α-amylase activity in vitro. J Agric Food Chem 62:9507-9514.

[73] Vernarelli JA, Lambert JD (2011) Tea consumption is inversely associated with weight status and other markers for metabolic syndrome in US adults. Eur J Nutr 52(3):1039-1048.

[74] Dryden, G. W., Song, M., & McClain, C. (2006). Polyphenols and gastrointestinal diseases. Current opinion in gastroenterology, *22*(2), 165.

[75] Cabrera, C., Artacho, R., & Giménez, R. (2006). Beneficial effects of green tea—a review. Journal of the American College of Nutrition, *25*(2), 79-99.

[76] Đudarić L, Fuzinac-Smojver A, Muhvic D et al (2015) The role of polyphenols on bone metabolism in osteoporosis. Food Res Int 77:290-298.

[77] Rigacci S, Stefani M (2015) Nutraceuticals and amyloid neuro-degenerative diseases: a focus on natural phenols. Expert Rev Neurother 15:41-52.

[78] Pogačnik L, Pirc K, Pamela I et al (2016) Potential for brain accessibility and analysis of stability of selected flavonoids in relation to neuroprotection in vitro. Brain Res 1651:17-26.

[79] Alipour M, Malihi R, Hosseini A et al (2018) The effects of catechins on related risk factors with type 2 diabetes: a review. Prog Nutr 20:12-20.

[80] Pastoriza S, Mesias M, Cabrera C et al (2017) Healthy properties of green and white teas: update. Food Funct 8:2650-2662.

[81] Manach, C., Williamson, G., Morand, C., Scalbert, A., & Rémésy, C. (2005). Bioavailability and bioefficacy of polyphenols in humans. I. Review of 97 bioavailability studies. The American journal of clinical nutrition, *81*(1), 230S-242S.

[82] Chan, K. Y., Zhang, L., & Zuo, Z. (2007). Intestinal efflux transport kinetics of green tea catechins in Caco-2 monolayer model. Journal of Pharmacy and Pharmacology, *59*(3), 395-400.

[83] Yong Feng, W. (2006). Metabolism of green tea catechins: an overview. Current drug metabolism, *7*(7), 755-809.

[84] Williamson, G., & Manach, C. (2005). Bioavailability and bioefficacy of polyphenols in humans. II. Review of 93 intervention studies. The American journal of clinical nutrition, *81*(1), 243S-255S.

[85] Brune, M., Rossander, L., & Hallberg, L. (1989). Iron absorption and phenolic compounds: importance of different phenolic structures. European journal of clinical nutrition, *43*(8), 547-557.

[86] Kim, E. Y., Ham, S. K., Shigenaga, M. K., & Han, O. (2008). Bioactive dietary polyphenolic compounds reduce nonheme iron transport across human intestinal cell monolayers. The Journal of nutrition, *138*(9), 1647-1651.

[87] Hurrell, R. F., Reddy, M., & Cook, J. D. (1999). Inhibition of non-haem iron absorption in man by polyphenolic-containing beverages. British Journal of Nutrition, *81*(4), 289-295.

[88] Merhav, H., Amitai, Y., Palti, H., & Godfrey, S. (1985). Tea drinking and microcytic anemia in infants. The American journal of clinical nutrition, *41*(6), 1210-1213.

[89] Razagui, I.B., Barlow, P., Izmeth, M.G., & Taylor, K. (1991). Iron status in a group of long-stay mentally handicapped menstruating women: some dietary considerations. European journal of clinical nutrition, *45* 7, 331-340.

[90] Zijp, I. M., Korver, O., & Tijburg, L. B. (2000). Effect of tea and other dietary factors on iron absorption. Critical reviews in food science and nutrition, *40*(5), 371-398.

[91] Kim, S. (2019). Biopolyphenolics in textile. Advances in Textile Bio-technology, 159-183.

[92] Cheng, T. H., Liu, Z. J., Yang, J. Y., Huang, Y. Z., Tang, R. C., & Qiao, Y. F. (2019). Extraction of functional dyes from tea stem waste in alkaline medium and their application for simultaneous coloration and flame retardant and bioactive functionalization of silk. ACS Sustainable Chemistry & Engineering, *7*(22), 18405-18413.

[93] Bonet-Aracil, M. Á., Díaz-García, P., Bou-Belda, E., Sebastiá, N., Montoro, A., & Rodrigo, R. (2016). UV protection from cotton fabrics dyed with different tea extracts. Dyes and Pigments, *134*, 448-452.

Chapter 21

Bioactive Phenolic Compounds from Indian Medicinal Plants for Pharmaceutical and Medical Aspects

Bhanu Kumar, Ankita Misra and Sharad Srivastava

Abstract

Ayurveda is an age old system of medicine which utilizes thousands of medicinal plants, rich in secondary metabolites for their therapeutic benefits and phenolic compounds are important one. Plant phenolic compounds are one of the major group of phytomolecules having tremendous therapeutic and nutraceutical potential. Indian medicinal plants like *Emblica*, *Terminalia* spp., *Withania*, *Tinospora* etc. are some of the potential source of bioactive phenolics and had been used from ages in various *Ayurvedic* formulations and were scientifically validated too. In this contribution, a brief account of some common Indian medicinal plants rich in bioactive phenolics are summarized along with their therapeutic action on human health and disease. The vast array of phenolics in these plants makes them a suitable candidate for modern medicine, nutraceutical supplements, immuno-modulatory formulations etc. With the advent of modern separation tools and techniques, it is now possible to identify, isolate and purify desired phytoconstituents from plant extracts. This further opens the avenues of utilizing medicinal plants or plant constituents/metabolites as super food for strengthening the body and maintaining the healthy work-life balance. The need of the hour is to identify therapeutically potential phenolics rich plants and development of herbal formulations for human welfare.

Keywords: ayurveda, Indian System of Medicine, herbal drug market, bioactive phenolic compounds, herbal formulations, plant metabolites

1. Introduction

Secondary metabolites are the chemical compounds synthesized in plants having minor role in life processes and do not have direct role in normal metabolism and development of the plant but often have ecological roles e. g., attractants of pollinators and chemical defenses against microbes, insects and higher predators [1]. Several medicinal plants by virtue of their secondary metabolites acquire a number of biological and therapeutic activities. These phytochemicals influence the metabolic activities of human and animals and have been used as drugs since centuries. The secondary metabolites are classified in three chemically distinct groups- terpenes, phenolic compounds, and nitrogen containing compounds.

Phenolic compounds are a vast group of plant secondary metabolites, present in almost every plant in varying quantities. These are chemically heterogeneous and the derivatives include some very important compounds such as flavonoids, tannins, lignins, anthocyanins etc. [2]. They show huge diversity in the structure ranging from simple structures, e.g. phenolic acids, to polyphenol compounds such as flavonoids, which consist of several functional groups (**Figure 1**). Phenolic compounds are very crucial for the diverse therapeutic actions in medicinal plants, color and flavor of fruits etc. During food processing and storage, plant phenolics are converted to a variety of derived compounds. The flavonoids are the largest group of phenolic compounds and play major role in defense and pigmentation. These are reported to be synthesized in the case of occurrence of infection. The activities of different groups of flavonoids are different as per their structural class. These are more commonly known for their antioxidant potential by scavenging of free radicals mediated by the functional hydroxyl groups. Flavonoids are supposed to protect from infections, and also prevent from cardiovascular, cancer and age related problems [3]. This group of molecules are also responsible for the color, taste and fragrance to the flowers and fruits and in this way attract pollinators also.

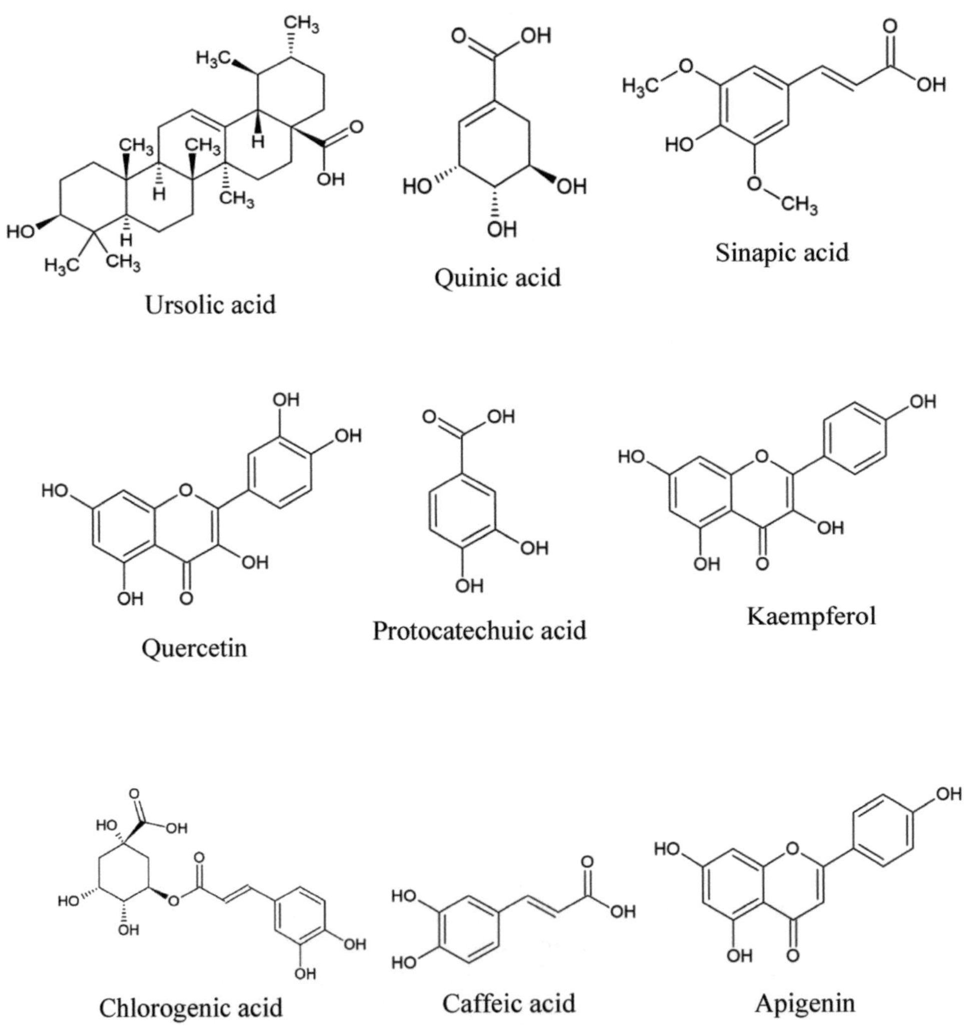

Figure 1.
Some common phenolic acids.

It is a well-established fact that the secondary metabolites in plants are meant to enable the plants to sustain their lives in various kinds of environmental conditions. Phenolic and flavonoid compounds are reported to exhibit strong antioxidant potential by different mechanisms. The hydroxyl group present in the phenolic compounds are excellent hydrogen donors. The reactive oxygen and nitrogen species react rapidly with hydrogen donating moieties which lead to stoppage of generation of new free radicals [4]. Phenolic compounds have also been found to chelate the metal ions involved in the formation of free radicals and hence produce antioxidant effect. Apart from antioxidant potential, dietary supplement of phenolic compounds have also been found to act as anticancer agents [5], provide protection from cardiovascular and several autoimmune diseases [6].

2. Traditional medicine and Indian Systems of Medicine (ISM)

The term "traditional medicine" (TM) refers to the ways of protecting and restoring health that existed before the arrival of modern medicine [7]. It incorporates plant, animal and mineral-based medicines, spiritual therapies and manual techniques designed to treat illness or maintain wellbeing [8, 9]. Many rural and ethnic communities in India use their indigenous knowledge for the treatment of various kinds of ailments [10]. They form the unwritten repository of health practices that have been verbally passed on to the next generations for at least one century and continue even today [11]. The World Health Organization (WHO) has also recognized the important role of traditional medicine in developing countries. WHO accepts that traditional medicine will continue to play an important part in providing services to very large number of people, particularly in rural areas [12].

India is a repository of vast traditional knowledge and a deep rooted system of indigenous medicine. According to a report from Government of India, about 75% of Indian population including majority of tribal and ethnic communities are mostly dependent on the traditional knowledge and practices for primary health care needs [13]. The system of medicines which are considered to be Indian in origin or the systems of medicine, which have come to India from outside and got assimilated in to Indian culture are known as Indian Systems of Medicine [14]. The age old Indian traditional medicine system "*Ayurveda*" is very extensive in terms of the plants used, owing to the great phytodiversity of the country. In addition, it is unique in having a well-defined conceptual framework which identifies it from many other traditional medicine systems [15]. In India, there are five recognized systems of medicine namely *Ayurveda, Yoga and naturopathy, Unani, Siddha and Homeopathy* (AYUSH).

The source of knowledge inculcated in *Ayurveda* finds its roots in one of the four ancient Indian texts (Veda), *Atharvaveda* which includes 114 hymns for prevention, treatment and cure of various diseases [14]. Two major written records of *Ayurveda* are *Charak Samhita* (for medicine) and *Shushrut Samhita* (for surgery). The fundamental principle of Ayurveda is to consider the physical and psychological status of body for maintaining healthy condition. It suggests ways to for proper life style and living in harmony with the nature. In Southern part of India, Siddha system of medicine is practiced. It has a close connection with the Tamil civilization and hence especially popular in Tamil Nadu. The experts of this medicine system are called *Siddhars* that means they have achieved excellence in this system of medicine. The siddha system of medicine uses drugs of metal and mineral origin to a large extent and with time it has included drugs of different other systems. The Siddha system resembles with Ayurveda as far as treatment procedures are concerned. The *Unani* system of medicine originated in Greece, established by Greek philosopher Hippocrates. This system of medicine was introduced in India by the Arabs. Homeopathy was brought to India in the 18th Century, however, it has been very well received by Indians and made a part of Indian System of Medicine.

3. Medicinal plants: india's potential

India is one of the most diverse countries in the world having a rich repository of high value, endemic and rare medicinal plants [16]. In terms of plant diversity, India ranks tenth in the world and fourth in Asia [17]. The reason behind this vast diversity is the presence of different climatic conditions such as alpine in Himalayas to arid zones in Rajasthan. There are tropical forests in the Western Ghats while plateaus, mountains and valleys in North-Eastern states. Apart from varying topography, soil, rainfall, temperature, humidity conditions also differ from place to place which give rise to huge phytodiversity. The microclimatic variations further lead to differences in the phenology, metabolism, physiology, chemical profile and even morphology of plants in addition to growth pattern across the geography [18].

According to an estimate, more than 45,000 plant species are commonly found in India out of which flowering plants constitute around 15,000-18,000; members of bryophytes are around 1800; algal species are 2500; 1600 lichens; 23,000 fungal species exist in India [19, 20]. The surveys conducted by several workers have revealed that approximately 20,000 plant species are having one or the other medicinal properties [13, 21]. From Indian Himalayan Region (IHR) itself, 357 species of medicinal plants belonging to 237 genera and 98 families were recorded. Asteraceae, Lamiaceae, Rosaceae, and Ranunculaceae were the dominant families in the IHR region [22].

4. Market potential of herbal drugs

The use of herbal medicines has risen dramatically all over the world. Global sales of herbal products were anticipated to be US $ 60 billion in 2000 as per Secretariat of the Convention on Biological Diversity. The sale of herbal medicines is predicted to increase at an average annual growth rate of 6.4% [23]. In 2008, the global market for herbal remedies was about US $83 billion with a steady growth rate ranging between 3% and 12% per annum [24]. The market of herbal drugs has seen a good tendency of growth at a fast rate worldwide. There are several factors responsible for this growth like increased general awareness in people to protect from the side effects of synthetic medicine, more inclination of masses towards Ayurveda and herbal treatment; improvement in quality, proof of efficacy and safety of herbal medicines and high cost of synthetic medicines [25].

In India the medicinal plant market is mostly unorganized at present. Most of the herbal drug manufacturers procure the raw material from wild by overexploitation of available natural resources. Due to unavailability of sufficient quantity of raw material, adulteration of inferior quality raw material or similar looking plant species to the genuine drug is common practice in many of the herbal drug industries [26]. The value of medicinal plants related trade in India is US $ 5.5 billion, although its share in the global export market of herbal drugs is less than 0.5 per cent. The export potential of China in medicinal plants is nearly INR 18,000–22,000 Crores. India exports crude drugs mainly to developed countries like USA, Germany, France, Switzerland, UK and Japan. The principal Indian herbal drugs exported to foreign countries include *Aconite, Aloe, Belladonna, Acorus, Cinchona,*

Cassia tora, Dioscorea, Digitalis, Ephedra, Plantago and *Senna* etc. About 165 herbal drugs and there extract are exported from India [27].

5. Some examples of medicinal plants rich in phenolics

There are numerous Indian medicinal plants which are rich in bioactive secondary metabolites along with phenolic compounds. A glimpse of some of them are compiled in **Table 1**. A few plants rich in particular groups of phenolic acid are discussed below:

5.1 *Emblica officinalis* Gaertn. syn. *Phyllanthus emblica*

Emblica officinalis (fam.–Euphorbiaceae), commonly known as Indian gooseberry or *Amla,* is a very famous Ayurvedic medicinal plant highly rich in Vitamin C. It supports healthy metabolism, digestion, nourishes the heart and respiratory system, promotes healthy skin, eyes and hair, and builds immunity. It is a key ingredient of many well-known Ayurvedic formulations such as *Chyawanprah* and *Triphala* [54]. The fruits and leaves of Amla are highly rich in ascorbic acid, phenolic compounds, flavonoids, tannins etc. (**Figure 2**). Even the roots contain ellagic acid and lupeol [55]. The major phenolic compounds include gallic acid, quercetin, apigenin, ellagic acid, chebulinic acid etc. The phenolic content among the three ingredients of *Triphala* is highest in *E. officinalis* [56]. The fruit pulp is also rich in tannin content which gives it astringent properties [57]. The bioactive phenolic compounds impart several therapeutic effects to this plant such as anti-bacterial activity, anti-fungal activity, antioxidant and free radical scavenging activity, insecticidal activity, immunomodulatory activity, anti-inflammatory activity, anti-diabetic and hypoglycemic activity [58].

5.2 *Terminalia chebula* Retz

Terminalia chebula (fam.- Combretaceae), popularly known as *Harad*, is a widely used traditional medicine in Ayurvedic practice and the fruits are one of the ingredients of herbal formulation *Triphala*. It is a very well-known rejuvenating herb. In classical texts, it is reported as natural detoxifier, promotes bowel movement, improves digestion, anti-aging, and good for eyes. It has been suggested to take 1–3 grams of *Harad* fruit powder with a cup of hot water in case of irritable bowel disease associated with low digestion strength. The dried ripe fruit is used as a remedy for heart disorders, urinary disorders and asthma. Major phenolic compound present in *T. chebula* are chebulic acid, chebulagic acid, shikimic acid, ferulic acid, vanillic acid, p-coumaric acid, caffeic acid, gallic acid, ellagic acid, tannic acid (**Figure 3**) [59, 60]. Major flavonoids in *T. chebula* are rutin, quercetin, isoquercetin, luteolin, 3′-methoxyquercetin, pelargonidin [61–63]. It is also known for antioxidant, hepatoprotective, neuroprotective, cytotoxic, antidiabetic, anti-inflammatory activities among others [64]. The flavonoid content and antioxidant properties in *T. chebula* is greater than the rest of the two ingredients of *Triphala* and the order is *T. chebula* > *E. officinalis* > *T. belerica* [56].

5.3 *Terminalia bellirica* Roxb

Terminalia bellirica (fam.- Combretaceae), commonly known as *Beheda,* is another very crucial ingredient of *Triphala*. Its medicinal properties are well recognized across various traditional medicine systems and has been described in

Sl. No.	Plant name	Plant part	Major phenolic compounds reported	Ref.
1.	*Acorus calamus* Linn.	Rhizome	α-asarone and β-asarone	[28]
2.	*Achyranthes aspera* L.	Whole plant	Quinic acid, Shikimic acid, Gallic acid, Chlorogenic acid, Acetylsalicylic acid, quercetin, kaempferol	[29]
3.	*Aloe vera* (L.) Burm.f. (syn. *Aloe barbadensis* Mill.)	Fleshy leaves	Sinapic acid, Quercetin, Kaempferol, Apigenin, Gallic acid, Protocatechuic acid, Catechin, Vanillic acid, Epicatechin, Syringic acid, Chlorogenic acid, Gentisic acid, Caffeic acid, Coumaric acid, Ferulic acid, Rutin, Miricetin	[30]
4.	*Andrographis paniculata* (Burm.f.) Wallich ex Nees	Whole plant	Apigenin, Onisilin, Andrographidine C, Luteolin	[31]
5.	*Bacopa Monnieri* (L.) Pennel	Whole plant	Chlorogenic acid, neochlorogenic acid, caffeic acid, Apigenin, Quercetin, Ursolic acid, Luteolin	[32, 33]
6.	*Berberis aristata* DC.	Root	Anthocyanin, Rutin, Chlorogenic acid	[34, 35]
7.	*Biophytum sensitivum* DC.		Caffeic acid, Ferulic acid, Gallic acid, Chlorogenic acid, Rutin	[36]
8.	*Centella asiatica* (L.) Urban	Whole plant	Rutin, Quercetin, kaempferol, Chlorogenic acid	[37]
9.	*Costus speciosus* J. Koenig (Sm.)	Rhizome	Ferulic acid, Coumarin, Phloroglucinol, Orcinol, Catechin, Quercetin, Rutin, Luteolin, Kaemferol	[38, 39]
10.	*Cuculigo orchiodes* Gaert.	Rhizome	2,6-dimethoxy benzoic acid, curculigoside A, curculigoside B, curculigine A, curculigine D	[40]
11.	*Curcuma longa* Linn.	Rhizome	Curcumin, Demethoxycurcumin, Bisdemethoxycurcumin, Caffeic acid, Ferulic acid, o-Coumaric acid, p-Coumaric acid	[41, 42]
12.	*Datura metel* L.	Fruits	Gallic acid, Vanilic acid, Quercetin and Ferulic acid	[43, 44]
13.	*Diplocyclos palmatus* (L.) Jeffry	Fruits	Chlorogenic acid, gallic acid, caeffic acid and protocateuchic acid.	[45]
14.	*Elephantopus scaber*	Aerial part	Gallic acid, Proto catechuic acid Chlorogenic acid, Ferulic acid	[46]
15.	*Emblica officinalis* Gaertn.	Fruit	Ascorbic acid, Tannic acid, Gallic acid, Geraniin, Quercetin, Isocorilagin, Kaemferol	[47–49]
16.	*Gymnema sylvestre* R. Br.	Leaves	Epigallocatechin, Conduritol, Phloretin, Quercetin, Dihydroquercetin, Gingerol, Hesperetin, Myricetin, Orcinol, Phloretin, Rutin	[50]
17.	*Hedychium spicatum* Buch- Ham. Ex. Smith	Root	Chrysin, Teptochrysin, Ethyl cinnamate, Ethyl-*trans-p*-methoxy cinnamate, *p*-Methoxy cinnamic acid	[51–53]

Table 1.
Some Indian medicinal plants rich in phenolic acids.

Ayurveda, Unani, Siddha, as well as in traditional Chinese medicine. The fruits are useful in the treatment of asthma, bronchitis, hepatitis, diarrhea etc. [65]. Several phytomolecules from the phenolic class such as lignans, ellagic acid, gallic acid,

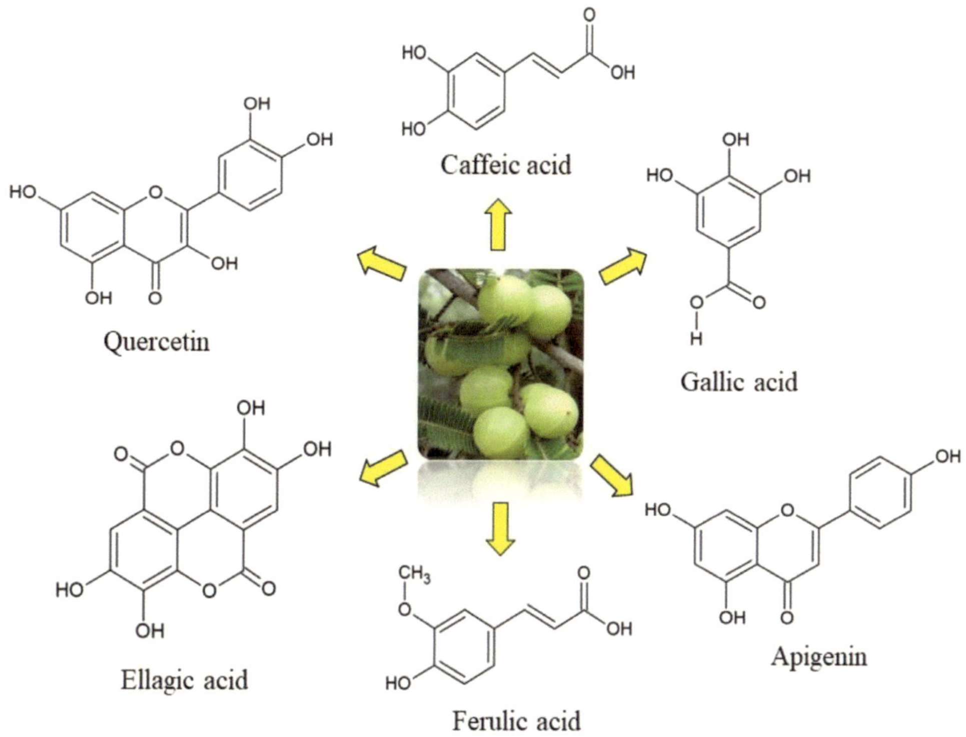

Figure 2.
Some phenolic acids found in Emblica officinalis.

chebulic acid, bellaric acid etc. has been found in the fruits of this plant [66, 67]. Ellagitannins such as corilagin, chebulagic acid, galloylpunicalagin, and digalloyl-hexahydroxydiphenoyl-hexoside were found to be the major components in *T. bellirica* [66].

5.4 *Tinospora cordifolia* (Thunb.) Miers

Tinospora cordifolia (fam.- Menispermaceae), commonly known as *Amrita,* is an age old Ayurvedic remedy for various purposes such as jaundice, diabetes, fever, skin diseases etc. It has been well known for scientifically proven roles as hepatoprotective, antipyretic, anti-oxidant, antimicrobial, anti-diabetic, immunomodulatory, anti-cancer etc. [68]. *T. cordifolia* is rich in many groups of phytomolecules such as alkaloids (berberine, tinosporin, palmetin, jatrorrhizine), terpenes, steroids, glycosides, saponins along with phenol and flavonoids. Major phenolic compounds reported are ellagic acid and kampferol [69].

5.5 *Tribulus terrestris* R. Br

Tribulus terrestris (fam.- Zygophyllaceae), also known as Gokshura is a well-known aphrodisiac in Ayurvedic and traditional Chinese medicine. It is highly rich in flavonoids and different flavonoids has been reported from different parts of the plant such as kaempferol, astragalin, quercetin and rutin from fruits and leaves [70, 71]. In a recent LC–MS study, identification of few phonic and flavonoid compounds has been done such as protocaatechuic acid, scopoletin, caffeic acid, quercetin, ferulic acid, rutin, luteolin, kaempferol, rutinoside etc. [72].

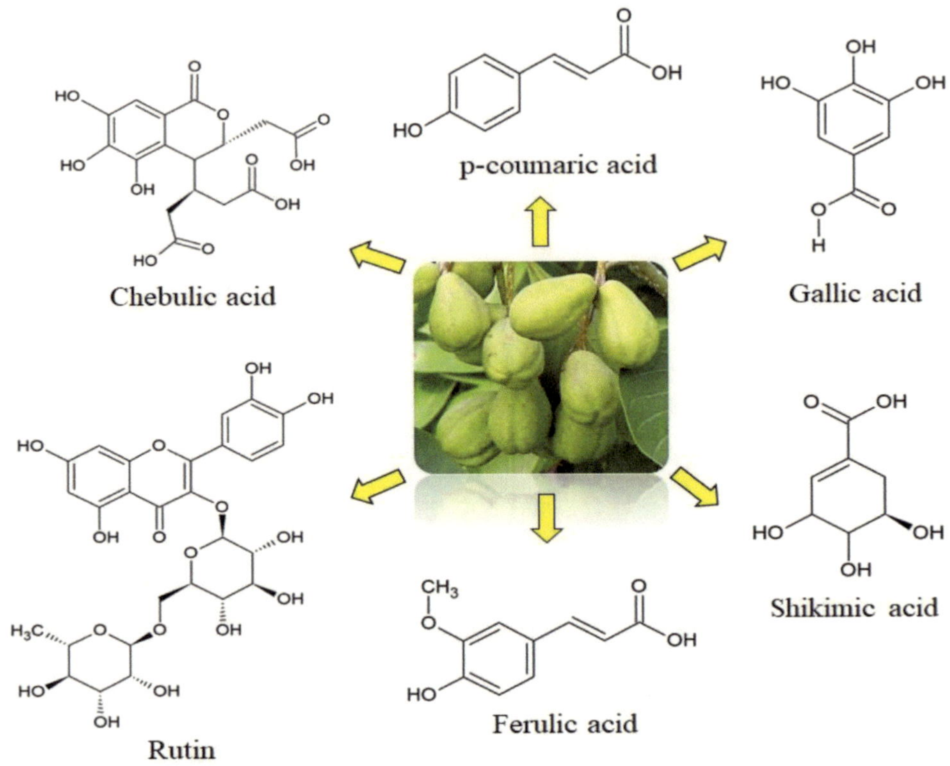

Figure 3.
Some phenolic acids found in Terminalia chebula.

5.6 *Withania somnifera* Dunal

Withania somnifera (fam.- Solanaceae), also known as *Ashwgandha*, is considered as a herbal tonic and health food in Ayurvedic texts and is considered as 'Indian Ginseng'. It is rich in many alkaloids such as withanolides and withaferin. Apart from alkaloids, *W. somnifera* is also rich in several phenolic acids and flavonoids, a few of which are catechin, gallic acid, syringic acid, vanillic acid, p-coumaric acid, benzoic acid, naringenin, and kaempferol [73].

6. Biological activities of plant phenolics and flavonoids

Plant phenolics and flavonoids possess a vast spectrum of biological activities ranging from general body maintenance to specific cure for many ailments such as cancer, diabetes etc. Many reports indicate that consumption of plant derived food material lowers the risk of many kind of diseases including cardiovascular diseases and cancer [74]. Their role as an anti-oxidant has been well established. There are different mechanisms for the anti-oxidant potential, however, radical scavenging via hydrogen atom donation is believed as the main mechanism [75]. The phenolic acids are also recognized as antidiabetic agents as they are able to influence the role of glucose and insulin receptors [76]. Few phenolic acids readily inhibit the activities of α-glucosidase and α-amylase which are responsible to convert dietary carbohydrates into glucose [77–79]. There are several reports showing role of phenolics in treatment of cancer through different mechanisms. Hydroxybenzoic and hydroxycinnamic acids are phenolic acids known for prevention and treatment of cancer [80]. In conclusion, it can be said that these molecules are extremely useful in various beneficial roles for human health.

7. Conclusion

The Indian traditional system of medicine "*Ayurveda*" has rich heritage in mitigating the disease and discomfort of individual with the use of medicinal plants and/plant products. With the advancement in knowledge about plant sciences and sophisticated analytical techniques, the therapeutic potential of medicinal plants can be easily related to their bioactive metabolites. Among the various class of phytomolecules, phenolic compounds are the most abundant, naturally available secondary metabolites having therapeutic potential. Some potential Indian medicinal plants like *Withania somnifera, Tinospora cordifolia,* etc. have multifarious medicinal benefits, are being used from ages and their efficacy is well proven in recent times too. There are reports that suggest a higher intake of vegetables and fruits rich in phenol and flavonoids can lower the risk of diabetes and cardiovascular diseases. In a recent study, the importance of phenolic compounds and their significance in management of type 2 diabetes as well as in human nutrition has been done and found that the role of phenolic compounds are vital for anti-aging, anti-inflammatory and anti-oxidant properties [81].

Up to 19th century, the concept of using medicinal plants in diet is restricted and considered for patients only. However, in recent times the emergence of nutraceutical benefits of plant based products has brought the medicinal plants into our plates. Further, the need of functional foods, dietary supplements, and super foods like Kale, Spirulina, Chia seeds, omega-3 rich foods, *Moringa* leaf powder, has diverted the focus of society in using newer and alternate source of nutrition which can supplement the basic traditional food. Therefore, looking to the relevance, indeed there is need of incorporating medicinal plants rich in phenolics compounds into our daily diet for promoting health and wellbeing.

8. Future perspective

With a growing awareness about the benefits of herbal products across the globe, now a days there is huge demand of food supplements, nutraceuticals, health promoting herbal medicines etc. than ever before. Most of these supplements are rich in phenolic compounds that attribute anti-oxidant potential, health promoting effect, immunomodulatory potential, and maintain general health. More recently, the race of boosting one's immunity enlightens the world about using *Ayurvedic* medicinal plants as dietary supplements, in cohesion with guidelines of regulatory bodies. Keeping this scenario in mind, there is an opportunity for the scientific community to explore the potential of phenolics from medicinal plants with the advent of improved techniques of extraction and purification of phenolics. In a recent study, analysis of phenolic compounds in different parts of *Amaranthus cruentus* was done and found that the type and quantity of phenolic compounds varies across plant parts and also the harvesting time. The quantity of phenolic compounds was found higher in the tender and mature leaves and rutin was found as the most abundant compound in the vegetative part [82]. This kind of study can help us get the better quality and quantity of secondary metabolites from the plants which will be detrimental for developing an efficacious herbal product. The process further needs to be scaled up for their optimum industrial scale prospection for human welfare.

Author details

Bhanu Kumar, Ankita Misra and Sharad Srivastava*
Pharmacognosy Division, CSIR - National Botanical Research Institute, Lucknow, Uttar Pradesh, India

*Address all correspondence to: sharad_ks2003@yahoo.com

References

[1] Taiz L, Zeiger E. Plant physiology. Third Edition. Sinauer Association Inc., California, U.S.A. 2005;690.

[2] Shanker K, Gupta MM, Srivastava SK, Bawankule DU, Pal A, Suman, Khanuja PS. Determination of bioactive nitrile glycoside(s) in drumstick (Moringa oleifera) by reverse phase HPLC. Food Chem 2007;105: 376-382.

[3] Kumar S, Pandey AK. Chemistry and biological activities of flavonoids: An overview. The Sci World J 2013; Article ID 162750, doi.org/10.1155/2013/162750.

[4] Paya M, Halliwell B, Hoult JRS. Interactions of a series of coumarins with reactive oxygen species. Scavenging of superoxide, hypochlorous acid and hydroxyl radicals. Biochem Pharmacol 1992;44:205-214.

[5] Yang CS, Landau JM, Huang MT, Newmark HL. Inhibition of carcinogenesis by dietary polyphenolic compounds. Ann Rev Nutr 2001;21: 381-406.

[6] Han X, Shen T, Lou H. Dietary polyphenols and their biological significance. Int J Mol Sci 2007;8: 950-988.

[7] Pushpangadan P. 'Arogyappacha' (*Trichopus zeylanicus* Gaerin), the "Ginseng" of Kani tribes of Agashyar hills (Kerala) for ever green health and vitality. Anc Sci Life 1988;8:13-26.

[8] Pushpangadan P, Rajasekharan S, Subramaniam A, Latha PG, Evans DA. Further on the pharmacology of *Trichopus zeylanicus*. Anc Sci Life 1995;14:127-135.

[9] Anonymous, WHO guidelines on good agricultural and collection practices (GACP) for medicinal plants. Geneva, Switzerland. 2003; ISBN 92 4 1546271.

[10] Pushpangadan P, George V, Ijinu TP, Chithra MA, Govindarajan R. Resurgence of traditional medicine systems with special reference to Indian systems of medicine and modern scientific developments. J Tradit Med Clin Natur 2018;7:276.

[11] Shankar D, Unnikrishnan PM, Venkatsubramanian P. Need to develop inter-cultural standards for quality, safety and efficacy of traditional systems of medicines. Curr Med 2007;1499-1505.

[12] Payyappallimana U. Role of Traditional medicine in primary health care: An overview of perspectives and challenges, Yokohama J Soc Sci 2010;14:1-22.

[13] Kala CP, Dhyani PP, Sajwan BS. Developing the medicinal plants sector in northern India: challenges and opportunities. J Ethnobiol Ethnomed 2006;2:1-15.

[14] Prasad LV. Indian System of Medicine and Homoeopathy Traditional Medicine in Asia. In: Chaudhury RR, Rafei UM, editors. WHO- Regional Office for South East Asia- New Delhi; 2002. P. 283-286.

[15] Ravishankar B, Shukla VJ. Indian systems of medicine. Afr. J. Trad. CAM 2007;4:319-337.

[16] Kamboj VP. Herbal medicine. Curr Sci 2000; 78:35-39.

[17] Singh JS, Chaturvedi RK. Diversity of Ecosystem Types in India: A Review. Proc Indian Natn Sci Acad. 2017;83: 569-594.

[18] Ncube B, Finnie JF, Van Staden J. Quality from the field: The impact of environmental factors as quality determinants in medicinal plants. S Afr J Bot. 2012;82:11-20.

[19] Bharucha E. Textbook of environmental studies. Hyderabad (India): Universities Press (India) Private Limited; 2006.

[20] Sharma A, Shanker C, Tyagi LK, Singh M, Rao ChV. Herbal medicine for market potential in India: an overview. Acad J Plant Sci 2008;1:26-36.

[21] Mukherjee PK. Quality control of herbal drugs. 1st ed. New Delhi: Business Horizons Pharmaceutical Publications; 2008.

[22] Sharma J, Gairola S, Sharma YP, Gaur RD. Ethnomedicinal plants used to treat skin diseases by Tharu community of district Udham Singh Nagar, Uttarakhand, India. J Ethnopharmacol 2014;158:140-206.

[23] Inamdar N, Edalat S, Kotwal VB, Pawar S, Herbal drugs in milieu of modern drugs. Int J Green Pharm 2008;2:2-8.

[24] Zhang J, Wider B, Shang H, Li X, Ernst E. Quality of herbal medicines: Challenges and Solutions. Complementary Therapies in Medicine. 2012;20:100-106.

[25] Calixto JB. Efficacy, Safety, Quality Control, Marketing and Regulatory Guidelines for Herbal Medicines (Phytotherapeutic Agents). Braz J Med Biol Res 2000;33:179-189.

[26] Dubey NK. Flora of BHU Campus, Banaras Hindu University. BHU Press, Varanasi, India; 2004.

[27] Prajapati ND, Purohit SS, Sharma AK, Kumar TA. A Handbook of Medicinal Plants. Agrobios; 2003.

[28] Rana TS, Mahar KS, Pandey MM, Srivastava SK, Rawat AKS. Molecular and chemical profiling of 'sweet flag' (*Acorus calamus* L.) germplasm from India. Physiol Mol Biol Plants. 2013; 19:231-237.

[29] Narayan C, Kumar A. Identification and characterization of phenolic compounds in hydro methanolic extract of *Achyranthes aspera* (HMEA) by UPLC and MALDI-TOF-MS and in vivo antioxidant activity. Oriental Pharmacy and Experimental Medicine. 2013;13:51-59.

[30] López A, De Tangil MS, Vega-Orellana O, Ramírez AS, Rico M. Phenolic constituents, antioxidant and preliminary antimycoplasmic activities of leaf skin and flowers of *Aloe vera* (L.) Burm. f. (syn. *A. barbadensis* Mill.) from the Canary Islands (Spain). Molecules. 2013;18:4942-4954.

[31] Hossain MD, Urbi Z, Sule A, Rahman KM. *Andrographis paniculata* (Burm. f.) Wall. ex Nees: a review of ethnobotany, phytochemistry, and pharmacology. The Scientific World Journal. 2014;1:2014.

[32] Muszyńska B, Łojewski M, Sułkowska-Ziaja K, Szewczyk A, Gdula-Argasińska J, Hałaszuk P. *In vitro* cultures of *Bacopa monnieri* and an analysis of selected groups of biologically active metabolites in their biomass. Pharmaceutical biology. 2016;54:2443-2453.

[33] Jeyasri R, Muthuramalingam P, Suba V, Ramesh M, Chen JT. *Bacopa monnieri* and their bioactive compounds inferred multi-target treatment strategy for neurological diseases: a cheminformatics and system pharmacology approach. Biomolecules. 2020;10:536.

[34] Zuzanna BZ, Agnieszka SA, David AD, Lukasz OL, Dorota AD. Fundamentals of the use of *Berberis* as a medicinal plant. European Journal of Clinical and Experimental Medicine. 2018;1:41-46.

[35] Mokhber-Dezfuli N, Saeidnia S, Gohari AR, Kurepaz-Mahmoodabadi M. Phytochemistry and pharmacology of

Berberis species. Pharmacognosy reviews. 2014;8-8.

[36] Kumar B, Shukla PK, Niranjan A, Misra A, Rawat AK, Srivastava S. RP-HPLC quantification of five phenolic compounds in *Biophytum sensitivum* (L.) DC. (Oxalidaceae) and their biological evaluation. Ind J Tradit Knowl 2017;16:319-324.

[37] Alqahtani A, Tongkao-on W, Li KM, Razmovski-Naumovski V, Chan K, Li GQ. Seasonal variation of triterpenes and phenolic compounds in Australian *Centella asiatica* (L.) Urb. Phytochemical analysis. 2015;26: 436-443.

[38] Karthikeyan J, Reka V, Giftson RV. Characterization of bioactive compounds in *Costus speciosus* (koen). by reverse phase HPLC. Int J Pharm Sci Res 2012;3:1461-1465.

[39] Thabit ZA. Evaluation of some bioactive effect of phenolic compounds in *Costus speciosus* rhizome extract. Iraqi Journal of Science. 2018 29;59:38-43.

[40] Jiao L, Cao DP, Qin LP, Han T, Zhang QY, Zhu Z, Yan F. Antiosteoporotic activity of phenolic compounds from *Curculigo orchioides*. Phytomedicine. 2009;1:874-881.

[41] Burman V, Kanaujia H. Characterization of phenolic compounds of turmeric using TLC. J Pharmacogn Phytochem 2019;2:994-998.

[42] Chattopadhyay I, Biswas K, Bandyopadhyay U, Banerjee RK. Turmeric and curcumin: Biological actions and medicinal applications. Current science. 2004;10:44-53.

[43] Céspedes-Méndez C, Iturriaga-Vásquez P, Hormazábal E. Secondary metabolites and biological profiles of Datura genus. Journal of the Chilean Chemical Society. 2021;66:5183-5189.

[44] Javid M, Aziz A, Azhar MF, Qayyum A. Antioxidant, antibacterial, phytochemical composition of leaves and roots extracts of *Datura alba*. Zeitschrift Fur Arznei-& Gewurzpflanzen. 2017;22:165-168.

[45] Misra A, Shukla PK, Kumar B, Niranjan A, Rawat AK, Srivastava S. Simultaneous-HPLC quantification of phenolic acids in traditionally used ayurvedic herb *Diplocyclos palmatus* (L.) Jeffry. Pharmacog J. 2017;9(4).

[46] Shukla PK, Misra A, Kumar B, Niranjan A, Srivastava S. Simultaneous RP-HPLC quantification of four phenolics in *Elephantopus scaber* L. and their *in vitro* pharmacological validation. Ind J Pharma Educ Res. 2020;54:368-373.

[47] Muthuraman A, Sood S, Singla SK. The antiinflammatory potential of phenolic compounds from *Emblica officinalis* L. in rat. Inflammo pharmacology. 2011;19:327-334.

[48] Liu X, Cui C, Zhao M, Wang J, Luo W, Yang B, Jiang Y. Identification of phenolics in the fruit of emblica (*Phyllanthus emblica* L.) and their antioxidant activities. Food chemistry. 2008;109:909-915.

[49] Kumar VN, Vibha, Ashwani K. A Comparative study of heavy metals in *Emblica officinalis, Phyllanthus emblica* and *Azadirachta indica.* Int Res J Biol Sci. 2013:2:16-19.

[50] Parveen S, Ansari MH, Parveen R, Khan W, Ahmad S, Husain SA. Chromatography based metabolomics and *in silico* screening of *Gymnema sylvestre* leaf extract for its antidiabetic potential. Evidence-Based Complementary and Alternative Medicine. 2019; Article ID 7523159, DOI https://doi.org/10.1155/2019/7523159.

[51] Reddy PP, Rao RR, Rekha K, Babu KS, Shashidhar J, Shashikiran G,

Lakshmi VV, Rao JM. Two new cytotoxic diterpenes from the rhizomes of *Hedychium spicatum*. Bioorg Med Chem Lett. 2009;19:192-195.

[52] Reddy PP, Tiwari AK, Rao RR, Madhusudhana K, Rao VR, Ali AZ, Babu KS, Rao JM. New labdane diterpenes as intestinal α-glucosidase inhibitor from antihyperglycemic extract of *Hedychium spicatum* (Ham. Ex Smith) rhizomes. Bioorg Med Chem Lett. 2009;19:2562-2565.

[53] Suresh G, Poornima B, Babu KS, Yadav PA, Rao MS, Siva B, Prasad KR, Nayak VL, Ramakrishna S. Cytotoxic sesquiterpenes from *Hedychium spicatum*: Isolation, structure elucidation and structure–activity relationship studies. Fitoterapia. 2013;86:100-107.

[54] Rai N, Tiwari L, Sharma RK and Verma AK. Pharmaco-botanical profile on *Emblica officinalis* Gaertn. – A Pharmacopoeial Herbal Drug. STM Journals. 2012;1:29-41.

[55] Khan KH. Roles of *Emblica officinalis* in Medicine - A Review. Bot Res Int 2009;2:218-228.

[56] Hazra B, Sarkar R, Biswas S, Mandal N. Comparative study of the antioxidant and reactive oxygen species scavenging properties in the extracts of the fruits of *Terminalia chebula*, *Terminalia belerica* and *Emblica officinalis*. BMC Complementary and alternative medicine. 2010;10:1-5.

[57] Kumar GS, Nayaka H, Dharmesh SM, Salimath PV. Free and bound phenolic antioxidants in amla (*Emblica officinalis*) and turmeric (*Curcuma longa*). J Food Compos Anal. 2006;19:446-452.

[58] Hasan MR, Islam MN, Islam MR. Phytochemistry, pharmacological activities and traditional uses of *Emblica officinalis*: A review. Int Curr Pharm J. 2016;5:14-21.

[59] Riaz M, Khan O, Sherkheli MA, Khan MQ, Rashid R. Chemical constituents of *Terminalia chebula*. Nat Prod Ind J. 2017;13:112.

[60] Chang CL, Lin CS. Phytochemical composition, antioxidant activity, and neuroprotective effect of *Terminalia chebula* Retzius extracts. Evidence-Based Complementary and Alternative Medicine. 2012. Article ID 125247, DOI: https://doi.org/10.1155/2012/125247.

[61] Bag A, Bhattacharyya SK, Chattopadhyay RR. The development of *Terminalia chebula* Retz. (Combretaceae) in clinical research. Asian Pac J Trop Biomed. 2013;3: 244-252.

[62] Prakash SD, Satya SN, Vangalapati M. Purification of chebulinic acid from *Terminalia chebula* species by column chromatography. J Chem Biol Phys Sci. 2012;2:1753.

[63] Singh C. 2α-hydroxymicromeric acid, a pentacyclic triterpene from *Terminalia chebula*. Phytochemistry. 1990;29:2348-2350.

[64] Nigam M, Mishra AP, Adhikari-Devkota A, Dirar AI, Hassan MM, Adhikari A, Belwal T, Devkota HP. Fruits of *Terminalia chebula* Retz.: A review on traditional uses, bioactive chemical constituents and pharmacological activities. Phytotherapy Research. 2020;10: 2518-2533.

[65] Singh, M.P., Gupta, A., Sisodia, S.S. Ethno and modern pharmacological profile of Baheda (*Terminalia bellerica*): A review. Pharm Chem J. 2018;5:153-162.

[66] Valsaraj R, Pushpangadan P, Smitt UW, Adsersen A, Christensen SB, Sittie A, Nyman U, Nielsen C, Olsen CE. New anti-HIV-1, antimalarial, and antifungal compounds from *Terminalia bellerica*. Journal of natural products. 1997;60:739-742.

[67] Khatoon S, Singh N, Srivastava N, Rawat A, Mehrotra S. Chemical evaluation of seven Terminalia species and quantification of important polyphenols by TLC. JPC-Journal of Planar Chromatography-Modern TLC. 2008;21:167-171.

[68] Tiwari P, Nayak P, Prusty SK, Sahu PK. Phytochemistry and pharmacology of *Tinospora cordifolia*: A review. Systematic Reviews in Pharmacy. 2018;9:70-78.

[69] Singh B, Sharma P, Kumar A, Chadha P, Kaur R, Kaur A. Antioxidant and *in vivo* genoprotective effects of phenolic compounds identified from an endophytic *Cladosporium velox* and their relationship with its host plant *Tinospora cordifolia*. J Ethnopharmacol. 2016;194:450-456.

[70] Panova D, Tomova M. Screening of *Tribulus terrestris* L. for phenolic compounds. Farmtsiya. 1970;20:29-31.

[71] Bhutani SP, Chibber SS, Seshadri PR. *Tribulus* L. Phytochemistry. 1969;8:299-303.

[72] Kucuk S, Goger F, Iscan G, Ozdemir F, Kirimer N, Incesu Z. Determination of phenolic compounds, cytotoxic, antioxidant and anticandidal effects of *Tribulus Terrestris* L.(Zygophyllaceae). children. 2017;208:382-94.

[73] Alam N, Hossain M, Khalil MI, Moniruzzaman M, Sulaiman SA, Gan SH. High catechin concentrations detected in *Withania somnifera* (ashwagandha) by high performance liquid chromatography analysis. BMC Complementary and Alternative Medicine. 2011;1:1-8.

[74] Wahle KW, Brown I, Rotondo D, Heys SD. Plant phenolics in the prevention and treatment of cancer. Bio-Farms for Nutraceuticals. 2010:36-51.

[75] Stevenson DE, Hurst RD. Polyphenolic phytochemicals–just antioxidants or much more? Cell Mol Life Sci. 2007;64:2900-2916.

[76] Prasad CV, Anjana T, Banerji A, Gopalakrishnapillai A. Gallic acid induces GLUT4 translocation and glucose uptake activity in 3T3-L1 cells. FEBS letters. 2010;584:531-536.

[77] Hanhineva K, Torronen R, Bondia-Pons I, Pekkinen J, Kolehmainen M, Mykkanen H, Poutanen K. Impact of dietary polyphenols on carbohydrate metabolism. Int J Mol Sci. 2010;11: 1365-1402.

[78] Jung UJ, Lee MK, Park YB, Jeon SM, Choi MS. Antihyperglycemic and antioxidant properties of caffeic acid in db/db mice. Journal of pharmacology and experimental therapeutics. 2006;318:476-483.

[79] Jung EH, Ran Kim S, Hwang IK, Youl Ha T. Hypoglycemic effects of a phenolic acid fraction of rice bran and ferulic acid in C57BL/KsJ-db/db mice. J Agr Food Chem. 2007;55:9800-9804.

[80] Huang WY, Cai YZ, Zhang Y. Natural phenolic compounds from medicinal herbs and dietary plants: potential use for cancer prevention. Nutrition and cancer. 2009;62:1-20.

[81] Lin D, Xiao M, Zhao J, Li Z, Xing B, Li X, Kong M, Li L, Zhang Q, Liu Y, Chen H. An overview of plant phenolic compounds and their importance in human nutrition and management of type 2 diabetes. Molecules. 2016;21(10):1374.

[82] Manyelo TG, Sebola NA, Hassan ZM, Mabelebele M. Characterization of the phenolic compounds in different plant parts of *Amaranthus cruentus* grown under cultivated conditions. Molecules. 2020;25(18):4273.